ORGANIZATIONAL BIOSYNTHESIS

Symposium on

ORGANIZATIONAL BIOSYNTHESIS

A SYMPOSIUM

Held at the Institute of Microbiology
of Rutgers · The State University
with support from the
NATIONAL SCIENCE FOUNDATION

Edited by

Henry J. Vogel

J. Oliver Lampen

Vernon Bryson

1967

Academic Press • New York and London

ACADEMIC PRESS INC.
111 Fifth Avenue, New York, New York 10003

United Kingdom Edition published by
ACADEMIC PRESS INC. (LONDON) LTD.
Berkeley Square House, London W.1

Library of Congress Catalog Card Number: 68-16516

PRINTED IN THE UNITED STATES OF AMERICA

List of Participants

ABRAMSON, CARL, The Ogontz Campus, Pennsylvania State University, Abington, Pennsylvania

ADAMSKI, ROBERT, Institute of Microbiology, Rutgers University, New Brunswick, New Jersey

ADELBERG, EDWARD A., Department of Microbiology, Yale University, New Haven, Connecticut

AJL, SAMUEL, Albert Einstein Medical Center, Philadelphia, Pennsylvania

ALBRECHT, ALBERTA M., Sloan-Kettering Institute for Cancer Research, Rye, New York

ALPERS, DAVID H., Department of Medicine, Massachusetts General Hospital, Boston, Massachusetts

ALVARADO, FRANCISCO, Department of Physiology, Rutgers Medical School, New Brunswick, New Jersey

ANFINSEN, CHRISTIAN B., National Institutes of Health, Bethesda, Maryland

APOSHIAN, H. V., Department of Microbiology, Tufts University Medical School, Boston, Massachusetts

ARMSTRONG, R. W., Department of Chemistry, Rutgers University, New Brunswick, New Jersey

ARONOVITCH, JACOB, Department of Bacteriology, Harvard Medical School, Boston, Massachusetts

AUSTRIAN, ROBERT, Department of Research Medicine, University of Pennsylvania School of Medicine, Philadelphia, Pennsylvania

AVERS, CHARLOTTE J., Department of Biological Sciences, Rutgers University, New Brunswick, New Jersey

BACON, DONALD F., Department of Microbiology, Massey University of Manawata, Palmerston North, New Zealand

BALCAVAGE, WALTER, Department of Physiological Chemistry, Johns Hopkins University School of Medicine, Baltimore, Maryland

BARKLIS, S. S., Research Department, St. Barnabas Medical Center, Livingston, New Jersey

BARRATT, RAYMOND W., Department of Biological Sciences, Dartmouth College, Hanover, New Hampshire

BARRY, GUY T., Squibb Institute for Medical Research, New Brunswick, New Jersey

BAUMBERG, S., Department of Genetics, University of Leeds, Leeds, England

BAUTZ, E. K. F., Institute of Microbiology, Rutgers University, New Brunswick, New Jersey

BAYER, MANFRED, Institute for Cancer Research, Philadelphia, Pennsylvania

BECKER, BENJAMIN, Department of Biology, Hamilton College, Clinton, New York

BECKER, MILTON J., Department of Biology, Massachusetts Institute of Technology, Cambridge, Massachusetts

BEISER, SAM M., Department of Microbiology, Columbia University, New York, New York

BERBERICH, MARY ANNE, National Institutes of Health, Bethesda, Maryland

BERBERICH, L. R., National Institutes of Health, Bethesda, Maryland

BERGQUIST, A., Department of Zoology, University of Texas, Austin, Texas

BERNLOHR, R. W., Department of Microbiology, University of Minnesota, Minneapolis, Minnesota

BHATTACHARJEE, JNANENDRA K., Albert Einstein Medical Center, Philadelphia, Pennsylvania

Billen, Daniel, Biology Department, University of Texas, Houston, Texas

Bloch, Konrad, Department of Chemistry, Harvard University, Cambridge, Massachusetts

Bogorad, Lawrence, The Biological Laboratories, Harvard University, Cambridge, Massachusetts

Boyden, Alan A., Department of Zoology, Rutgers University, New Brunswick, New Jersey

Boyden, Mabel G., Department of Zoology, Rutgers University, New Brunswick, New Jersey

Bramwell, Jonathan, Department of Microbiology, College of Physicians and Surgeons, Columbia University, New York, New York

Brawerman, George, Department of Biochemistry, Yale University School of Medicine, New Haven, Connecticut

Brody, S., Rockefeller University, New York, New York

Brotzman, B., Department of Zoology, University of Texas, Austin, Texas

Brown, William E., Squibb Institute for Medical Research, New Brunswick, New Jersey

Brunstetter, Frank H., University of Colorado Medical Center, Denver, Colorado

Bryson, Vernon, Institute of Microbiology, Rutgers University, New Brunswick, New Jersey

Burger, Max, Department of Biology, Princeton University, Princeton, New Jersey

Cairns, John, Cold Spring Harbor Laboratory of Quantitative Biology, Cold Spring Harbor, New York

Cantoni, G. L., National Institutes of Health, Bethesda, Maryland

Cavalieri, L. F., Sloan-Kettering Division, Cornell Medical College, New York, New York

Cerami, Anthony, Rockefeller University, New York, New York

Chai, N., Department of Physics, Kansas State University, Manhattan, Kansas

Chang, H. C., Department of Biochemistry, New York University College of Medicine, New York, New York

Chesbro, William, Department of Biology, University of New Hampshire, Durham, New Hampshire

Chiga, Masahiro, Department of Biochemistry, Rutgers Medical School, New Brunswick, New Jersey

Cirillo, Vincent P., Department of Biological Sciences, State University of New York, Stony Brook, Long Island, New York

Clark, Harold E., Department of Plant Biology, Rutgers University, New Brunswick, New Jersey

Clarke, C. H., Department of Zoology, University of Texas, Austin, Texas

Cohen, Edward P., Department of Microbiology, Rutgers Medical School, New Brunswick, New Jersey

Cohen, Larry W., Department of Zoology, Pomona College, Claremont, California

Cohen, Seymour S., Department of Therapeutic Research, University of Pennsylvania Medical School, Philadelphia, Pennsylvania

Consigli, R. A., Department of Physics, Kansas State University, Manhattan, Kansas

Cook, Elizabeth A., Department of Bacteriology, Douglass College, Rutgers University, New Brunswick, New Jersey

Cora, Elisa, Institute of Microbiology, Rutgers University, New Brunswick, New Jersey

Cordes, Helmuth, Squibb Institute for Medical Research, New Brunswick, New Jersey

Cramer, John W., Department of Pharmacology, Yale University School of Medicine, New Haven, Connecticut

Crane, Robert, Department of Physiology, Rutgers Medical School, New Brunswick, New Jersey

CUMMINS, JOSEPH E., Department of Zoology, University of Washington, Seattle, Washington

CZAKO-HEINCZ, MARIA, Institute of Microbiology, Rutgers University, New Brunswick, New Jersey

DALLNER, GUSTAV, Department of Pathology, Karolinska Institutet, Stockholm, Sweden

DAVERN, CEDRIC I., Cold Spring Harbor Laboratory of Quantitative Biology, Cold Spring Harbor, New York

DAVIS, FRANK F., Department of Physiology and Biochemistry, Rutgers University, New Brunswick, New Jersey

DAVIS, ROWLAND H., Department of Botany, University of Michigan, Ann Arbor, Michigan

DEBUSK, A. GIB, Department of Biological Sciences, Florida State University, Tallahassee, Florida

DE LUCIA, PAULA, Cold Spring Harbor Laboratory of Quantitative Biology, Cold Spring Harbor, New York

DEMAIN, ARNOLD L., Merck, Sharp and Dohme Research Laboratories, Rahway, New Jersey

DONOVAN, JAMES E.,* Institute of Microbiology, Rutgers University, New Brunswick, New Jersey

DREYER, DONALD A., Institute of Microbiology, Rutgers University, New Brunswick, New Jersey

DUDA, GEORGE, Division of Biology and Medicine, U.S. Atomic Energy Commission, Washington, D.C.

DURHAM, NORMAN N., Biology Branch, Division of Biology and Medicine, U.S. Atomic Energy Commission, Washington, D.C.

DUTCHER, JAMES D., Squibb Institute for Medical Research, New Brunswick, New Jersey

* Deceased

EAKIN, E. A., Department of Zoology, University of Texas, Austin, Texas

EBERHART, BRUCE M., Department of Biology, University of North Carolina, Greensboro, North Carolina

EBERLE, HELEN, Department of Physics, Kansas State University, Manhattan, Kansas

EICHHOLZ, ALEXANDER, Department of Physiology, Rutgers Medical School, New Brunswick, New Jersey

EIKENBERRY, E. F., Department of Biology, Massachusetts Institute of Technology, Cambridge, Massachusetts

EISENSTADT, JEROME M., Department of Microbiology, Yale University, New Haven, Connecticut

ELLIS, LILLIAN N., Department of Chemistry, Douglass College, Rutgers University, New Brunswick, New Jersey

ERICKSON, ROBERT J., Institute of Microbiology, Rutgers University, New Brunswick, New Jersey

ETHIRAJ, S., Institute of Microbiology, Rutgers University, New Brunswick, New Jersey

FEDERMAN, MICHELINE, Rutgers Medical School, New Brunswick, New Jersey

FELICETTI, LUCIANO, Rockefeller University, New York, New York

FLAVIN, MARTIN, National Institutes of Health, Bethesda, Maryland

FRESCO, JAQUES, Department of Chemistry, Princeton University, Princeton, New Jersey

FURMANSKI, PHILIP, Albert Einstein Medical Center, Philadelphia, Pennsylvania

FURTH, JOHN J., Pathology Department, University of Pennsylvania Medical School, Philadelphia, Pennsylvania

GALIS, ANNA, Department of Microbiology, College of Physicians and Surgeons, Columbia University, New York, New York

GALSTON, ARTHUR W., Department of

Biology, Yale University, New Haven, Connecticut

GANESAN, A. T., Department of Genetics, Stanford University School of Medicine, Palo Alto, California

GARREN, LEONARD D., National Institutes of Health, Bethesda, Maryland

GARRO, ANTHONY J., Department of Microbiology, Columbia University, New York, New York

GASCÓN, SANTIAGO, Institute of Microbiology, Rutgers University, New Brunswick, New Jersey

GEIDUSCHEK, E. PETER, Department of Biophysics, University of Chicago, Chicago, Illinois

GERBER, NANCY N., Institute of Microbiology, Rutgers University, New Brunswick, New Jersey

GILES, NORMAN H., Department of Biology, Yale University, New Haven, Connecticut

GILVARG, CHARLES, Department of Chemistry, Princeton University, Princeton, New Jersey

GLICK, M. C., Department of Therapeutic Research, University of Pennsylvania Medical School, Philadelphia, Pennsylvania

GOLDBERG, ALAN, Department of Biology, Princeton University, Princeton, New Jersey

GOLDBERGER, ROBERT F., National Institutes of Health, Bethesda, Maryland

GORDON, JULIAN, Rockefeller University, New York, New York

GORDON, RUTH E., Institute of Microbiology, Rutgers University, New Brunswick, New Jersey

GORSKI, JACK, Department of Biology, Princeton University, Princeton, New Jersey

GOTS, JOSEPH S., Department of Microbiology, University of Pennsylvania School of Medicine, Philadelphia, Pennsylvania

GOTTESMAN, M., National Institutes of Health, Bethesda, Maryland

GRANICK, SAM, Rockefeller University, New York, New York

GREGOLIN, CARLO, Department of Biochemistry, New York University College of Medicine, New York, New York

GROBMAN, ARNOLD B., College of Arts and Sciences, Rutgers University, New Brunswick, New Jersey

GROUPÉ, V., Institute of Microbiology, Rutgers University, New Brunswick, New Jersey

GUNSALUS, ANN, Cold Spring Harbor Laboratory of Quantitative Biology, Cold Spring Harbor, New York

GURNEY, CLIFFORD W., Department of Medicine, Rutgers Medical School, New Brunswick, New Jersey

HANDLER, PHILIP, Department of Biochemistry, Duke University Medical Center, Durham, North Carolina

HANSON, KENNETH R., Connecticut Agricultural Experimental Station, New Haven, Connecticut

HARVEY, RICHARD A., Department of Biochemistry, Rutgers Medical School, New Brunswick, New Jersey

HENDLIN, DAVID, Merck, Sharp and Dohme Research Laboratories, Rahway, New Jersey

HEPPEL, LEON A., National Institutes of Health, Bethesda, Maryland

HESS, EUGENE L., Metabolic Biology Program, National Science Foundation, Washington, D.C.

HEYMANN, HANS, CIBA Pharmaceutical Company, Summit, New Jersey

HOFFEE, PATRICIA, Department of Molecular Biology, Albert Einstein College of Medicine, New York, New York

HOWELL, R. RODNEY, The Johns Hopkins Hospital, Baltimore, Maryland

HOYER, BILL H., National Institutes of Health, Bethesda, Maryland

HSU, ROBERT Y., Department of Biochemistry and Microbiology, Rutgers University, New Brunswick, New Jersey

HUTCHISON, DORRIS J., Sloan-Kettering Institute for Cancer Research, Rye, New York

IMPERATO, SAVIERIO, Institute of Microbiology, Rutgers University, New Brunswick, New Jersey

INCEFY, GENEVIEVE, Sloan-Kettering Division, Cornell Medical College, New York, New York

JENSSEN, WARREN D., Institute of Microbiology, Rutgers University, New Brunswick, New Jersey

JIMENEZ-ZAMUDIO, LUIS A., Institute of Microbiology, Rutgers University, New Brunswick, New Jersey

JOHNSTON, JAMES A., Rutgers University, New Brunswick, New Jersey

JOKLIK, WOLFGANG K., Department of Cell Biology, Albert Einstein College of Medicine, New York, New York

JONES, EVAN E., Nutritional Biochemistry Section, Department of Animal Science, North Carolina State University, Raleigh, North Carolina

JONES, MARY ELLEN, Department of Biochemistry, School of Medicine, University of North Carolina, Chapel Hill, North Carolina

JONES, RAYMOND, Department of Biological Sciences, State University of New York, Stony Brook, Long Island, New York

JOSEPHS, MELVIN J., Chemical and Engineering News, Washington, D.C.

KAJI, AKIRA, Department of Microbiology, University of Pennsylvania, Philadelphia, Pennsylvania

KAPLAN, TAMAR, Albert Einstein Medical Center, Philadelphia, Pennsylvania

KASAI, T., Institute of Microbiology, Rutgers University, New Brunswick, New Jersey

KATOH, K., Department of Biology, Massachusetts Institute of Technology, Cambridge, Massachusetts

KATZEN, HOWARD, Merck, Sharp and Dohme Research Laboratories, Rahway, New Jersey

KELLER, FRED A., Institute of Microbiology, Rutgers University, New Brunswick, New Jersey

KENNEY, FRANCIS T., Biology Division, Oak Ridge National Laboratory, Oak Ridge, Tennessee

KONDO, MASATOSHI, Department of Microbiology, University of Illinois, Urbana, Illinois

KRAUSZ, LEON M., Department of Physiology and Biochemistry, Rutgers University, New Brunswick, New Jersey

KREDICH, NICHOLAS M., National Institutes of Health, Bethesda, Maryland

KROGMANN, DAVID W., Molecular Biology Program, National Science Foundation, Washington, D.C.

KUCHLER, ROBERT J., Department of Bacteriology, Rutgers University, New Brunswick, New Jersey

LAMPEN, J. OLIVER, Institute of Microbiology, Rutgers University, New Brunswick, New Jersey

LANDMAN, OTTO E., Department of Biology, Georgetown University, Washington, D.C.

LANE, M. D., Department of Biochemistry, New York University College of Medicine, New York, New York

LARK, C., Department of Physics, Kansas State University, Manhattan, Kansas

LARK, K. G., Department of Physics, Kansas State University, Manhattan, Kansas

LASKIN, ALLEN I., Squibb Institute for Medical Research, New Brunswick, New Jersey

LASKY, LAURENCE J., Institute of Microbiology, Rutgers University, New Brunswick, New Jersey

LEATHEM, JAMES H., Bureau of Biological Research, Rutgers University, New Brunswick, New Jersey

LECHEVALIER, H. A., Institute of Microbiology, Rutgers University, New Brunswick, New Jersey

LECHEVALIER, MARY P., Institute of Microbiology, Rutgers University, New Brunswick, New Jersey

LEE, CHI-HANG, Institute of Microbiology, Rutgers University, New Brunswick, New Jersey

LENGYEL, PETER, Department of Molecular Biophysics, Yale University, New Haven, Connecticut

LE PAGE, R. N., Department of Zoology, University of Texas, Austin, Texas

LEVINTHAL, MARK, Department of Biology, Johns Hopkins University, Baltimore, Maryland

LEWIS, MARY E., Institute of Microbiology, Rutgers University, New Brunswick, New Jersey

LIBERFARB, RUTH M., Institute of Microbiology, Rutgers University, New Brunswick, New Jersey

LIETMAN, PAUL S., The Johns Hopkins Hospital, Baltimore, Maryland

LIPMANN, FRITZ, Rockefeller University, New York, New York

LITWACK, GERALD, Biochemistry Department, Temple University School of Medicine, Philadelphia, Pennsylvania

LUCAS-LENARD, JEAN, Rockefeller University, New York, New York

LUCK, DAVID J. L., Rockefeller University, New York, New York

LYNEN, FEODOR, Max-Planck-Institut für Zellchemie, München, Germany

MACMILLAN, J. D., Department of Biochemistry and Microbiology, Rutgers University, New Brunswick, New Jersey

MAHLER, INGA, Graduate Department of Biochemistry, Brandeis University, Waltham, Massachusetts

MALING, BARBARA, Merck, Sharp and Dohme Research Laboratories, Rahway, New Jersey

MANSON, LIONEL A., The Wistar Institute, Philadelphia, Pennsylvania

MARMUR, JULIUS, Department of Biochemistry, Albert Einstein College of Medicine, New York, New York

MARSHECK, WILLIAM J., JR., Institute of Microbiology, Rutgers University, New Brunswick, New Jersey

MATTOON, JAMES R., Department of Physiological Chemistry, School of Medicine, Johns Hopkins University, Baltimore, Maryland

MAZAITIS, ANTHONY J., Institute of Microbiology, Rutgers University, New Brunswick, New Jersey

MCAUSLAN, BRIAN R., Department of Biology, Princeton University, Princeton, New Jersey

MCDANIEL, LLOYD E., Institute of Microbiology, Rutgers University, New Brunswick, New Jersey

MCFALL, ELIZABETH, Department of Microbiology, New York University School of Medicine, New York, New York

MCGUIRE, JOSEPH M., Department of Experimental Medicine, Yale University School of Medicine, New Haven, Connecticut

MCLELLAN, WILLIAM L., Institute of Microbiology, Rutgers University, New Brunswick, New Jersey

MCNAMARA, PETER, Institute of Microbiology, Rutgers University, New Brunswick, New Jersey

MENON, MIRAN, Institute of Microbiology, Rutgers University, New Brunswick, New Jersey

MESELSON, MATTHEW, The Biological Laboratories, Harvard University, Cambridge, Massachusetts

METZENBERG, ROBERT L., Department of Physiological Chemistry, University of Wisconsin Medical School, Madison, Wisconsin

MILLER, DON, Graduate Department of Biochemistry, Brandeis University, Waltham, Massachusetts

MINOCHA, H. C., Department of Physics, Kansas State University, Manhattan, Kansas

MOLDAVE, KIVIE, Department of Biochemistry, University of Pittsburgh, Pittsburgh, Pennsylvania

MONTENECOURT, BLAND S., Institute of Microbiology, Rutgers University, New Brunswick, New Jersey

MONTIE, THOMAS C., Albert Einstein Medical Center, Philadelphia, Pennsylvania

MORA, JAIME, Department of Botany, University of Michigan, Ann Arbor, Michigan

MUNKRES, KENNETH D., Los Alamos Scientific Laboratory, University of California at Los Alamos, Los Alamos, New Mexico

NAKADA, DAISUKE, Central Research Department, E. I. Du Pont de Nemours, Wilmington, Delaware

NAKANO, MASAYASU, Institute of Microbiology, Rutgers University, New Brunswick, New Jersey

NASON, ALVIN, Department of Biology, Johns Hopkins University, Baltimore, Maryland

NASS, MARGIT M. K., Department of Therapeutic Research, University of Pennsylvania Medical School, Philadelphia, Pennsylvania

NEUMANN, NORBERT P., Rutgers Medical School, New Brunswick, New Jersey

NICKERSON, W. J., Institute of Microbiology, Rutgers University, New Brunswick, New Jersey

NISHIZUKA, YASUTOMI, Department of Medical Chemistry, Kyoto University Faculty of Medicine, Kyoto, Japan

NOMURA, MASAYASU, Laboratory of Genetics, University of Wisconsin, Madison, Wisconsin

NOVICK, RICHARD P., Public Health Research Institute of the City of New York, New York, New York

O'BRIEN, JOAN M., Institute of Microbiology, Rutgers University, New Brunswick, New Jersey

OHAD, ITHAK, Department of Biological Chemistry, The Hebrew University, Jerusalem, Israel

OMURA, TSUNEO, Institute for Protein Research, Osaka University, Osaka, Japan

ORTIGOZA, ETHEL G., Instituto Politecnico Nacional, Escuela National de Ciencias Biologicas, Mexico City, Mexico

ORTIGOZA-FERADO, JORGE, Instituto Politecnico Nacional, Escuela National de Ciencias Biologicas, Mexico City, Mexico

PALADE, GEORGE E., Rockefeller University, New York, New York

PALCZUK, NICHOLAS C., Department of Zoology, Rutgers University, New Brunswick, New Jersey

PARK, RODERIC B., Department of Botany, University of California, Berkeley, California

PEARSON, JOHN W., Institute of Microbiology, Rutgers University, New Brunswick, New Jersey

PEDERSEN, CARL E., Institute of Microbiology, Rutgers University, New Brunswick, New Jersey

PETERMANN, MARY L., Sloan-Kettering Division, Institute for Cancer Research, New York, New York

PIH, KATHERINE D., Institute of Microbiology, Rutgers University, New Brunswick, New Jersey

PLAUT, GERHARD W. E., Department of Biochemistry, Rutgers Medical School, New Brunswick, New Jersey

PLESCIA, O. J., Institute of Microbiology, Rutgers University, New Brunswick, New Jersey

PRAMER, DAVID, Department of Biochemistry and Microbiology, Rutgers

University, New Brunswick, New Jersey

PRICE, CARL A., Department of Plant Biology, Rutgers University, New Brunswick, New Jersey

QUIGLEY, JOHN, Department of Therapeutic Research, University of Pennsylvania Medical School, Philadelphia, Pennsylvania

RACKER, EFRAIM, Section of Biochemistry and Molecular Biology, Division of Biological Sciences, Cornell University, Ithaca, New York

RAINA, AARNE, Department of Therapeutic Research, University of Pennsylvania Medical School, Philadelphia, Pennsylvania

RASKAS, HESCHEL J., The Biological Laboratories, Harvard University, Cambridge, Massachusetts

RASKOVA, JANA, Institute of Microbiology, Rutgers University, New Brunswick, New Jersey

REBOUD, ANNE-MARIE, Sloan-Kettering Division, Institute for Cancer Research, New York, New York

REDDI, K. K., Rockefeller University, New York, New York

REILLY, EUGENE, Institute of Microbiology, Rutgers University, New Brunswick, New Jersey

RICH, ALEXANDER, Department of Biology, Massachusetts Institute of Technology, Cambridge, Massachusetts

RILEY, MONICA, Department of Biological Sciences, State University of New York, Stony Brook, Long Island, New York

ROBERTS, RICHARD B., Department of Terrestrial Magnetism, Carnegie Institution of Washington, Washington, D. C.

ROGERS, PALMER, Department of Microbiology, University of Minnesota, Minneapolis, Minnesota

ROSE, IRWIN A., Institute for Cancer Research, Philadelphia, Pennsylvania

ROSENBERG, BARBARA, Sloan-Kettering Division, Cornell University Medical College, New York, New York

ROSSELET, J. P., Carter-Wallace, Inc., Cranbury, New Jersey

ROTH, JAY S., Department of Biochemistry, University of Connecticut, Storrs, Connecticut

RUTNER, ALAN, Department of Biochemistry, New York University College of Medicine, New York, New York

RYDER, ELENA, Department of Biochemistry, New York University College of Medicine, New York, New York

SADASIV, EILEEN C., Institute of Microbiology, Rutgers University, New Brunswick, New Jersey

SCHACHTELE, CHARLES, Department of Microbiology, University of Minnesota, Minneapolis, Minnesota

SCHAECHTER, MOSELIO, Department of Microbiology, Tufts University Medical School, Boston, Massachusetts

SCHAFFNER, CARL P., Institute of Microbiology, Rutgers University, New Brunswick, New Jersey

SCHLATTER, RICHARD, Rutgers University, New Brunswick, New Jersey

SCHLESINGER, R. W., Department of Microbiology, Rutgers Medical School, New Brunswick, New Jersey

SCHLOM, JEFFREY, Institute of Microbiology, Rutgers University, New Brunswick, New Jersey

SHIFRISS, O., College of Agriculture and Environmental Sciences, Rutgers University, New Brunswick, New Jersey

SHULL, GILBERT M., Squibb Institute for Medical Research, New Brunswick, New Jersey

SIEGEL, ELI, Department of Biochemistry, Albert Einstein College of Medicine, New York, New York

SIEKEVITZ, PHILIP, Rockefeller University, New York, New York

SIMPSON, MELVIN V., Department of

Biological Sciences, State University of New York, Stony Brook, Long Island, New York

SOMBERG, ETHEL W., Newark College of Arts and Sciences, Rutgers University, Newark, New Jersey

SOODAK, MORRIS, Graduate Department of Biochemistry, Brandeis University, Waltham, Massachusetts

SPRINSON, DAVID B., Department of Biochemistry, College of Physicians and Surgeons, Columbia University, New York, New York

SPYRIDES, GEORGE J., Central Research Department, E. I. Du Pont de Nemours, Wilmington, Delaware

STAEHELIN, T., Department of Zoology, University of Michigan, Ann Arbor, Michigan

STARKEY, ROBERT L., Department of Biochemistry and Microbiology, Rutgers University, New Brunswick, New Jersey

STENT, GUNTHER S., Department of Molecular Biology, University of California, Berkeley, California

STETTEN, DEWITT, JR., Rutgers Medical School, New Brunswick, New Jersey

STETTEN, MAJORIE R., Rutgers Medical School, New Brunswick, New Jersey

STEVENS, THOMAS M., Rutgers Medical School, New Brunswick, New Jersey

STILL, CECIL C., Department of Plant Biology, Rutgers University, New Brunswick, New Jersey

STOLLAR, VICTOR, Department of Microbiology, Rutgers Medical School, New Brunswick, New Jersey

STOUDT, THOMAS H., Merck, Sharp and Dohme Research Laboratories, Rahway, New Jersey

STRASSMAN, MURRY,* Albert Einstein Medical Center, Philadelphia, Pennsylvania,

STROHL, WILLIAM A., Department of Microbiology, Rutgers Medical School, New Brunswick, New Jersey

*Deceased.

STRUMEYER, DAVID H., Department of Biochemistry and Microbiology, Rutgers University, New Brunswick, New Jersey

SUEOKA, NOBORU, Department of Biology, Princeton University, Princeton, New Jersey

SUEOKA, TAMIKO, Department of Biology, Princeton University, Princeton, New Jersey

SUHADOLNIK, ROBERT, Albert Einstein Medical Center, Philadelphia, Pennsylvania

SUPUT, JELENA, Institute of Microbiology, Rutgers University, New Brunswick, New Jersey

TAKEDA, MASAO, Rockefeller University, New York, New York

TATA, J. R., National Institute for Medical Research, Mill Hill, London, England

TATUM, E. L. Rockefeller University, New York, New York

TAYLOR, M. WIGHT, Department of Biochemistry and Microbiology, Rutgers University, New Brunswick, New Jersey

TEUBER, M. Department of Molecular Biology, Albert Einstein College of Medicine, New York, New York

THOMAS, CHARLES A., The Biological Laboratories, Harvard University, Cambridge, Massachusetts

TKACZ, JAN S., Institute of Microbiology, Rutgers University, New Brunswick, New Jersey

TOMKINS, GORDON M., National Institutes of Health, Bethesda, Maryland

TOPPER, YALE J., National Institutes of Health, Bethesda, Maryland

TORREY, HENRY C. The Graduate School, Rutgers University, New Brunswick, New Jersey

TRAUB, P., Laboratory of Genetics, University of Wisconsin, Madison, Wisconsin

UMBREIT, WAYNE W., Department of Bacteriology, Rutgers University, New Brunswick, New Jersey

UNGER, LEON, Department of Biochemistry, University of Oklahoma Medical Center, Oklahoma City, Oklahoma

URM, ELO, Institute of Microbiology, Rutgers University, New Brunswick, New Jersey

VOGEL, HENRY J., Institute of Microbiology, Rutgers University, New Brunswick, New Jersey

VOGEL, RUTH H., Institute of Microbiology, Rutgers University, New Brunswick, New Jersey

VON EHRENSTEIN, GÜNTER, Department of Biophysics, Johns Hopkins University School of Medicine, Baltimore, Maryland

VORBECK, MARIE L., Department of Microbiology, Temple University School of Medicine, Philadelphia, Pennsylvania

WAGNER, ROBERT P., Department of Zoology, University of Texas, Austin, Texas

WAINIO, WALTER W., Department of Physiology and Biochemistry, Rutgers University, New Brunswick, New Jersey

WAINWRIGHT, STANLEY D., Department of Biochemistry, Dalhousie University, Halifax, Nova Scotia, Canada

WAKABAYASHI, K., Rockefeller University, New York, New York

WALKER, JAMES R., Department of Biology, Princeton University, Princeton, New Jersey

WARNER, JONATHAN R., Department of Biochemistry, Albert Einstein College of Medicine, New York, New York

WEGENER, WARNER S., Albert Einstein Medical Center, Philadelphia, Pennsylvania

WEINBAUM, G., Albert Einstein Medical Center, Philadelphia, Pennsylvania

WEISS, SAMUEL B., Department of Biochemistry, Argonne Cancer Research Hospital, University of Chicago, Chicago, Illinois

WILLIAMS, CURTIS A., Rockefeller University, New York, New York

WILSON, JAMES F., Department of Biology, University of North Carolina, Greensboro, North Carolina

WOLIN, M. J., Department of Dairy Science, University of Illinois, Urbana, Illinois

WOODWARD, DOW O., Department of Biological Sciences, Stanford University, Stanford, California

WOODWARD, VAL W., Biology Department, Rice University, Houston, Texas

Preface

Until quite recently, there were two domains in biology that seemed separated by an almost impenetrable curtain: the unseen molecules of the biological chemist and the subcellular particles of the microscopist—clearly visible but tantalizingly outside the molecular range. The curtain has now been penetrated from both sides, and a continuity of the particulate with the molecular can be glimpsed. The fundamental particles of biology are being successfully explored in terms of organizational biosynthetic events at the definitive molecular level. Thus, particles of heredity and of protein synthesis as well as of energy metabolism and of photosynthesis are yielding insights into their formation and function, and so are membranes and other aggregate-type structures of the cell.

A symposium on "Organizational Biosynthesis" was held at the Institute of Microbiology of Rutgers, The State University, September 8 to 10, 1966, with support from the National Science Foundation. The meeting took place in conjunction with the Rutgers Bicentennial activities. The proceedings of the symposium are contained in this volume.

We sincerely thank Dr. Fritz Lipmann for his delivery of the Opening Address. The contributions of the session chairmen, Dr. C. A. Thomas, Jr., Dr. S. B. Weiss, Dr. A. Rich, Dr. R. W. Schlesinger, Dr. N. Sueoka, Dr. C. B. Anfinsen, Dr. F. Lynen, Dr. K. Bloch, Dr. R. B. Roberts, and Dr. E. Racker, and of Dr. S. S. Cohen, who made his presentation as an Evening Lecture, are also gratefully acknowledged. We very much appreciate an address, "University Scientists and Public Policy," given by Dr. P. Handler, who was introduced by Dr. DeWitt Stetten, Jr.

For their fine help in connection with many details of the meeting, we are indebted to Mr. E. R. Isaacs and the other staff members of the Institute of Microbiology who were involved.

June, 1967

HENRY J. VOGEL
J. OLIVER LAMPEN
VERNON BRYSON

Contents

PART IV

INTEGRATION OF METABOLISM

PART V

JUXTAPOSITION PHENOMENA I

PART VI

JUXTAPOSITION PHENOMENA II

PART VII

Membrane Synthesis and Associated Phenomena

PART VIII

Chloroplast Synthesis

PART IX

Ribosome Synthesis

PART X

Mitochondrion Synthesis

ORGANIZATIONAL BIOSYNTHESIS

Introductory Remarks

J. Oliver Lampen

Director, Institute of Microbiology

I am delighted to have the opportunity of welcoming you once again to the Institute for this Symposium on Organizational Biosynthesis. I see quite a few faces that have been here before and many new ones. I hope you will all enjoy this occasion.

You will also realize that this symposium is special in that it is a part of the celebration of the Bicentennial of Rutgers University. In addition to the last twenty years as a state university that may be familiar to most of you, Rutgers has spent most of its two hundred years as a colonial college and as a private institution. Science began to be taught here about one hundred and fifty years ago, apparently under rather inauspicious circumstances. I will read you a comment discovered recently by the Bicentennial Office. It was made in 1823 by John H. Livingston who was then President of Rutgers. This was about a year after Chemistry had been added to the curriculum. He said: "They talk of their oxygen, nitrogen, and hydrogen. Fools! It is nothing but matter, and that is all they know." With that emphatic support, progress seems assured! But I trust that you also join with me in good wishes to the University and in hopes that it will continue to provide academic and social leadership in the years ahead.

The present symposium is the third in a series. The first two emphasized the structural and biological role of individual macromolecular components. Now we would like to turn toward the interaction of these components and particularly to their function as integrated structures. In order to provide a sound basis for these discussions, we were delighted to obtain a man who has contributed significantly to both aspects of this reciprocal relationship. Dr. Fritz Lipmann is known to all of you, and certainly any statement of his qualifications for speaking or any attempt to predict what he will say would be redundant. But I should note that he has taken to heart the recent exhortation of one L. B. Johnson that basic scientists turn their thoughts toward the applied aspects of their work, and he will discuss "Molecular Technology."

Opening Address

Molecular Technology

Fritz Lipmann

The Rockefeller University, New York, New York

I should like to begin by reminiscing about the time I started in biochemistry. There was a good deal of talk at that time about structure in a rather vague sense. In his early work on respiration Warburg (27) made the prophetic proposition that it occurred at a boundary between cytoplasm and particles. In the classic paper of Krebs and Henseleit (9) on the urea cycle, there is a paragraph about the relationship of biosynthesis to structure. Finding that liver extract was practically inactive, they concluded that biosynthesis must be linked to structure. The structural elements that were missing in urea synthesis, however, were the mitochondria, and when one began to understand the feed-in of energy and could substitute mitochondrial structure with ATP, this cycle could be split into a sequence catalyzed by soluble enzymes (21) which could be put together to regain the whole.

More recently, after going wholeheartedly into homogenization and extraction of enzymes, and then crystallizing and characterizing them, long reaction sequences have often been solubilized. Historically most important was the extraction of alcoholic fermentation from yeast and of glycolysis from muscle. I think one can say that yeast or muscle extracts are essentially homogeneous systems, and it is almost surprising that, by random collision, enzymes can find their substrates and each other to bring about an exceedingly rapid overall reaction. And yet, in recent years, there has been more and more dissatisfaction with exhaustive homogenization. One has begun to realize that there may be a limit to which this can be driven. Transport across membranes, the synthesis of many macromolecules, and complex biosynthetic paths are nonrandom processes and do depend on structure. Various expressions have recently come into use to indicate structuration; it might be helpful to insert here a short glossary (Table I).

To illustrate the point I want to make: It had been realized for some time that, to understand coupling between oxidation and phosphoryla-

TABLE I
EXPRESSIONS RELATING TO STRUCTURATION

Random	Structural
Homogeneous	Heterogeneous
Isotropic	Anisotropic, polarized
Scalar	Vectorial

tion, a subdivision of mitochondrial structure gave very helpful but limited information about intermediary reactions. To compound these partial reactions into a whole, however, it was essential to recognize that, owing to the anisotropic nature of the system, it tolerates only limited destruction (10, 20); its technical operation depends on specially arranged macromolecular assemblies.

I should like here to make a brief detour. It has impressed me to see the expressions "miniaturization" and "superminiaturization" appear in many engineering advertisements that I see in *Scientific American*. This shows that there is a trend in man-made technology to converge on the molecular technology of the living cell. The engineers realize that they are still far away from building a computer as compact as the human brain, but they are hopeful, and envious; by looking occasionally into their Journals, I see how they strive to condense machines into smaller and smaller spaces. I should like to show you an example of their rather slow start. In electronics, a technique of printing intricate circuits on small areas, say of postage stamp size, has recently been developed in the form shown in Fig. 1. This is what I would think of as a first approximation to molecular technology. But there is still a large gap between man-made and organismic technology. Such miniaturization is very far

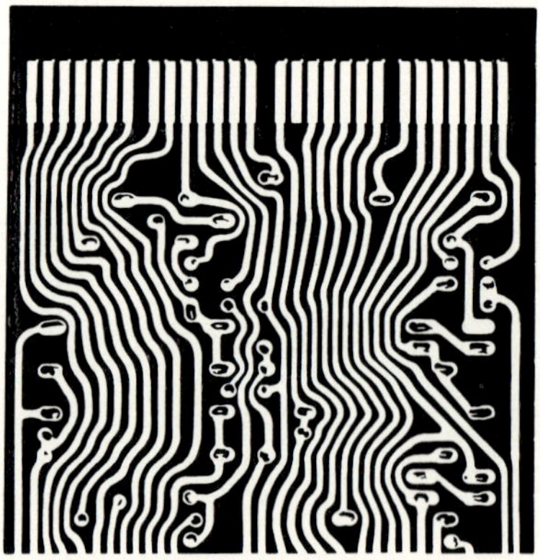

FIG. 1. Pattern of electronic circuit (18).

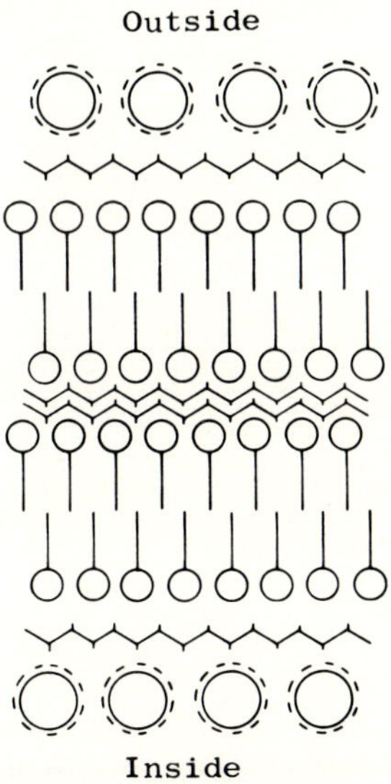

FIG. 2. Alternation of two phospholipid double layers and protein layers in mitochondrial membrane (26).

from the Angstrom scale; it is still on the millimeter level. The engineers are beginning to recognize that one of the reasons for the gap is that they work with hardware and the organism works in wet systems. A difference between a dry and a wet system appears to be that the latter lends itself to almost unlimited miniaturization.

Returning now to our main topic, I will approach it by running through a few significant examples. By way of generalization, I like to say that intracellular structures, the centers of heterogeneous reaction systems, can be divided into two main classes: those that are composed of lipids and proteins melted together in a membrane-like structure, and those that are composed largely of proteins. The latter include the nucleoproteins, such as ribosomes (20) and chromosomes, and protein particles similar in size to ribosomes as, for instance, the multienzyme system for fatty acid synthesis (11), and for α-keto acid oxidation (13). I will concentrate, however, on examples of the membranous type for the reason that we have much better visual information about these structures. Their size is well within the resolution of the electron microscope, and it is electron microscopy that has led us to understand how these structures function because it has made it possible for us to see them. This is not yet easily done with smaller particles such as the ribosomes, and that is why we are not as well ahead in interpreting their activity from the structural point of view.

I will start by showing a schematic representation of a lipoprotein membrane (Fig. 2). What is seen here are two double layers of phospholipid; the hydrophobic fatty acids are oriented inward, and the hydrophilic phosphoric acids point outward. In the middle there is a stratum of protein flanked by the two anisotropic layers which carry a thin coat of

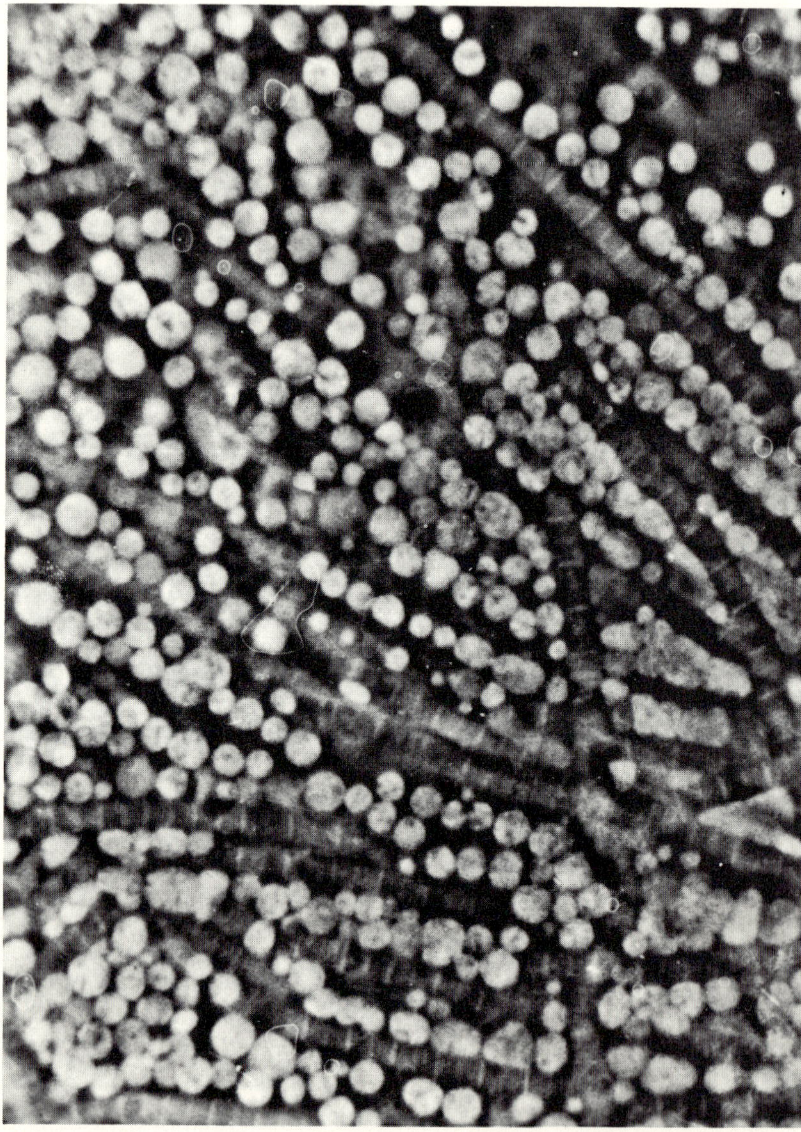

FIG. 3. Electron micrograph of flight muscle of house fly.

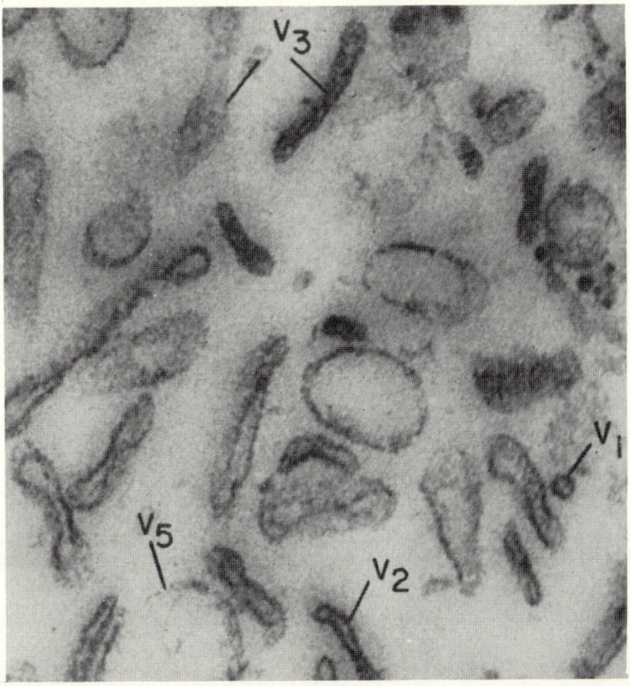

FIG. 4. Electron micrograph of bottom layer of rabbit muscle fraction containing Ca^{++} concentration and relaxation activities (3). The numbers V_1, V_2, V_3, V_5 point out various cuts of vesicles.

protein on each side. Looking at this picture of a membrane, one has to distinguish between down, its inner side, and up, its outer side. In the scheme, all protein appears as indistinct peptide chains that look deceivingly alike. The essence of the membrane, however, is that the inner and outer coatings as well as the middle each have their specific tasks, and the polarization of the membrane results from their differences.

I should like to turn to a concrete example, using an analysis of the relaxing factor of skeletal muscle on which Dr. Ebashi (3) worked in my laboratory. To introduce the muscle, I should first like to show in Fig. 3 a rather crude, low-magnification electron micrograph of a very hard-working machine, the flight muscle of a house fly. One sees the contractile fibers and, tightly hugging them, the sarcosomes, the power supply of the muscle. One can see only a vague structuration inside these mitochondria, which has now been resolved much better. Along these fibers runs Palade's (16) endoplasmic reticulum (ER), a network of tubular channels not visible here. By ultracentrifuge separation one can single out an ER fraction, first isolated by Kielley and Meyerhof (8), who found it to contain a magnesium-linked ATPase recognized as lipoprotein. Ebashi (2) then found that this same fraction carried the relaxing factor, paralleled by an ATP-dependent calcium concentration effect. He found that, if it was suspended in a calcium salt solution, using radioactive calcium, and then spun down after incubation, the calcium was almost completely concentrated in the sediment. When such a sediment is cut, the electron micrograph looks as it appears in Fig. 4 (3).

I have to mention here that a very fortunate feature of such an homogenate of tubules is that it yields self-sealed vesicles (16), which makes it suitable for use in functional studies because its contents are sealed in. What appears in the picture are cut-through vesicles. In some cases one can see double-layered membranes, which reflect the structure shown in Fig. 2. This vesicle suspension functions as a calcium pump. For this, the intact structure is needed; when the lipid membrane is destroyed by deoxycholate, both Ca^{++} fixation and the relaxation effect disappear. Ebashi's primary interest was the factor that causes the muscle

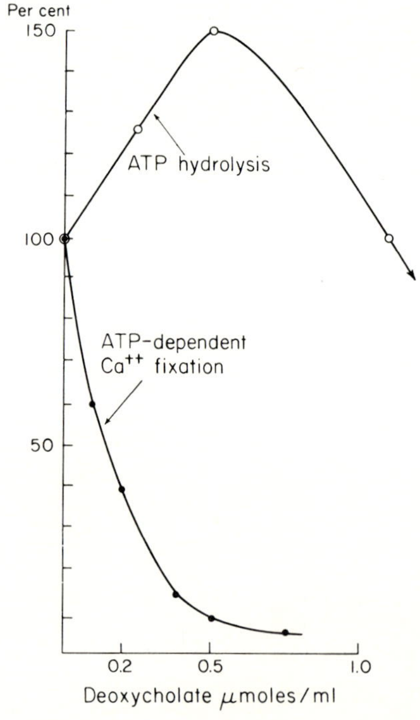

FIG. 5. Effect of deoxycholate on ATP hydrolysis and Ca^{++} concentration of fraction shown in electron micrograph of Fig. 4.

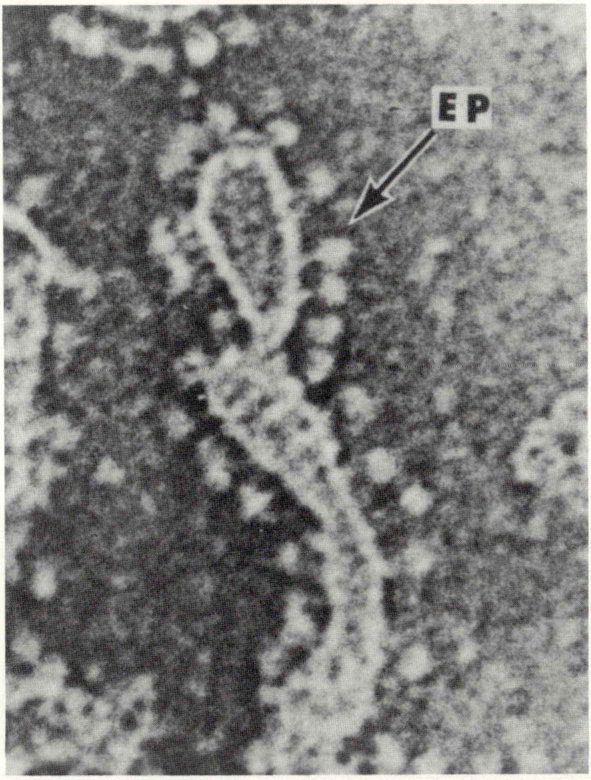

FIG. 6. High-magnification electron micrograph of crista in a beef heart mitochondrion (4). EP stands for elementary particles.

fiber to relax after contraction (2); this presumably is due to the removal of calcium. His purpose in coming to my laboratory was to study the interdependence of Ca^{++} concentration and ATP hydrolysis. What can be seen in Fig. 5 is that the relaxing activity as well as calcium fixation disappear after deoxycholate treatment. When a deoxycholate-treated fraction is looked at in the electron microscope, the vesicular structures of Fig. 4 have disappeared and only a homogeneous fibrous layer remains. On the other hand, it is significant that at the same time as the Ca^{++} fixation disappears an increase in ATPase results. We think this means that ATP is uncoupled from its functional link, and therefore hydrolysis increases. Eventually, with higher deoxycholate concentration, the ATPase also disappears.

I turn now to the mitochondrial membrane, and introduce it with a high-resolution electron micrograph (Fig. 6), from Fernandez-Moran

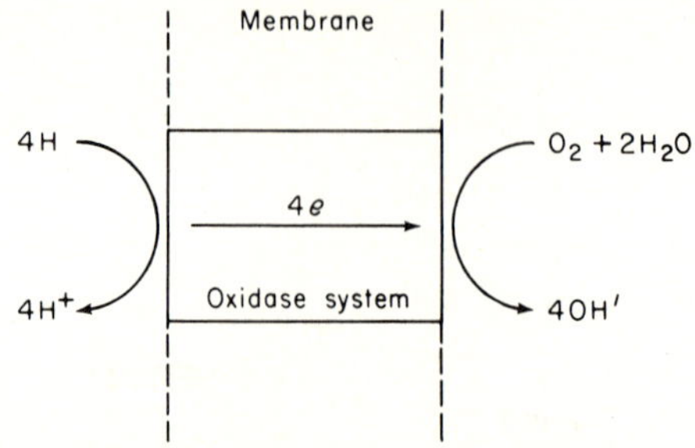

FIG. 7. Model for charge separation by respiration; electron transport across anisotropic membrane (12, cf. 15).

et al. (4), of one of the inside spikes that fold into the mitochondrial cavity. One can see here doorknob-like structures connected to the lipid membrane. This annex turned out to be one of Racker's factors (19, 20), an ATPase isolated from the mitochondrion. More specifically, this represents the anisotropic, reversible ATPase (14) in Mitchell's interpretation of oxidative phosphorylation by chemiosmotic coupling. Here, oxidative phosphorylation is related to a charge separation whereby respiration functions in setting up a fuel cell-like potential gradient. Lundegardh (12) was the first to propose an analogous interpretation of respiration that is schematized in Fig. 7 (cf. 15). The electrons are permitted to pass through the membrane, but hydrogen ions are kept on one side and hydroxyl on the other. As proposed (14), respiratory energy

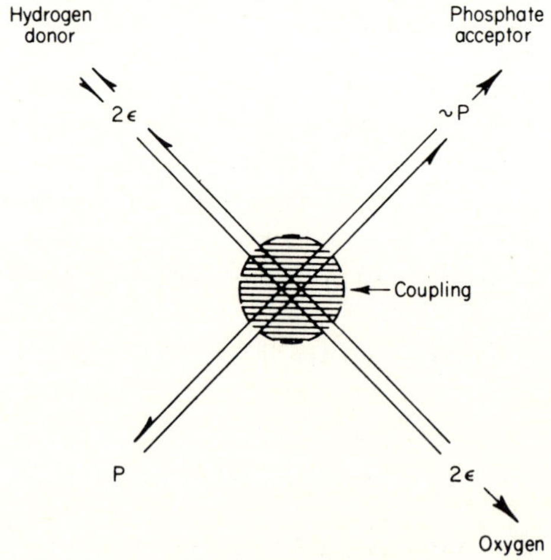

FIG. 8. Coupling scheme (6).

is made usable through this device by transformation of a redox potential into a group potential, eventually a phosphoryl potential.

The next few figures present this interpretation as an example of the role of structure in energy transformation. Figure 8 is an old, simple scheme I once drew (6); it makes coupling a crossing-over of electron flow and group activation, and recognizes that this must be a reversible reaction. Working in reverse, one can visualize ~P driving a Ca^{++} concentration pump such as was discussed above. The essential mitochondrial function, however, is to convert electron flux into group potential; Mitchell's ingenious interpretation of this appears in Fig. 9. It has kindled the imagination of many people. The scheme presented here

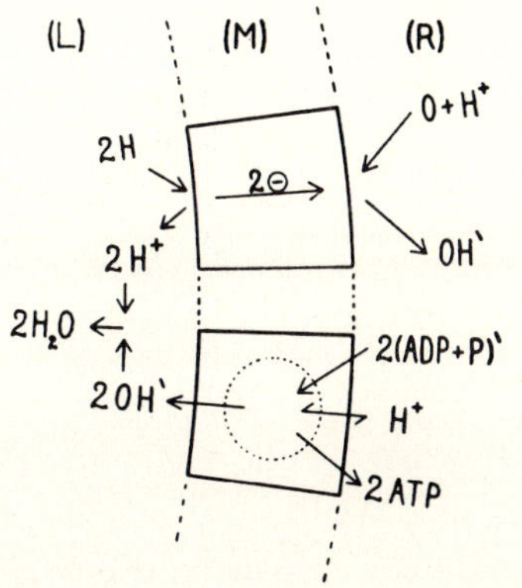

FIG. 9. Mitchell's scheme for combination of respiratory charge separation and phosphorylation (14).

from Mitchell (14) should be taken as a first approximation. It has given a welcome new turn to coupling and is now well supported experimentally (7, 24). In the upper half one recognizes the fuel cell. In the lower part this couples by a "neutralization" device with phosphorylation by pulling, as depicted in Fig. 10, an hydroxyl away from inorganic phosphate, probably with an intermediary phase in between (14).[1]

Finally, I should like to turn to the polarization of a membrane, in this case for transporting proteins. Many organisms produce proteins for "export." The pancreas, for example, produces enzymes that are shed into the intestinal tract. Redman *et al.* (23), studying the excretion of amylase in pancreas, have found it to be synthesized directly into excretory channels. More recently, Sabatini *et al.* (25) have worked on

[1] Recent work, particularly by Watanabe and Brodie (28), makes it attractive to speculate on an initial phenol activation to a phenolium that then interacts with inorganic phosphate.

an analogous phenomenon, namely, the secretion of plasma proteins by liver. Early work by Peters (17) suggested that albumin, which is most abundantly synthesized in the liver and has been studied extensively (1), is channeled into the ER. The mechanics of this transport were clarified by Sabatini *et al.* (25), who looked at microsomal structures in the electron microscope and found the following situation (Fig. 11). One sees the double-layer structure of the tubular membrane with the ribosomes attached to it. The orientation of the ribosome is important; it is always the large subunit that is linked to the membrane, and it is this part on which the growing peptide chain was localized (25).

Probing deeper into the problem of protein excretion, Redman and Sabatini (22) used puromycin for testing whether a nascent polypeptide chain could be detected in the microsomal vesicle: in other words, to see if the protein really was growing into the tubule. They used an homogenized ER fraction containing the self-sealed tubule fragments (16). These vesicles, sealed fragments of the ER similar to those shown in Fig. 4 but charged with ribosomes, are what we generally call the microsomes. After incubation of a microsome suspension with radioactive amino acid to create nascent chains, puromycin was added to release them. It was found that the nascent chains are not released into the supernatant but remain enclosed in the microsomal vesicles. This proves that growth of the peptide chain is polarized and immediately directed to the inside of the tubule. Drs. Ganoza and Williams in our laboratory did some studies on the formation of a number of plasma proteins with liver microsomal fractions (5). By use of immunoelectrophoresis they found that various plasma proteins were retained by the microsomal fraction. In preliminary studies, again by means of immunoelectrophoresis but using antibodies against liver instead of plasma protein, they found these antibodies to react with radioactively marked proteins present in the protoplasmic

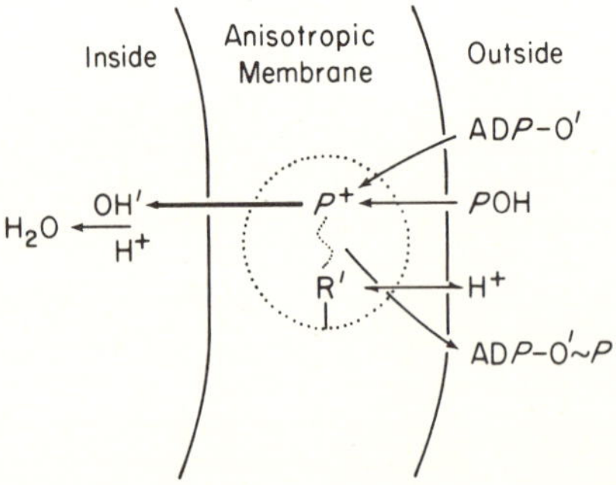

FIG. 10. Amplification of lower part of Fig. 9, after Mitchell (14).

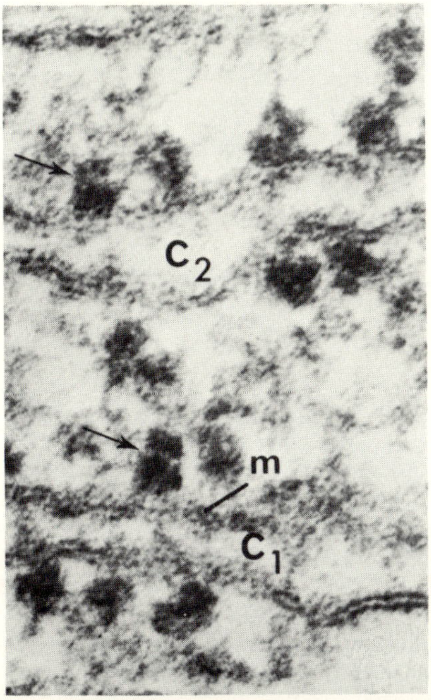

FIG. 11. Electron micrograph of endoplasmic reticulum of liver cell carrying attached ribosomes (25). C_1 and C_2 indicate a narrow and a wide cisterna; m is the membrane. The arrow in the lower part of the picture indicates the junction between the larger part of the ribosome sitting on the membrane and the smaller part on top of it.

supernatant rather than in the microsomal fraction. This indicates that the liver's own proteins are not channeled into the ER.

I have chosen a few examples of simple as well as more complex polarized systems found in cells. I have presented sketchy interpretations by converting electron micrographs into schemes of a functioning machinery; the detailed working mechanisms of such systems remain to be elaborated. With these examples I have aimed at making you realize that to explain fully the workings of an organism we have to stop homogenizing somewhere in order to retain the polarization that means structure.

REFERENCES

1. CAMPBELL, P. N., AND KERNOT, B. A., *Biochem. J.*, **82**, 262 (1962).
2. EBASHI, S., *Arch. Biochem. Biophys.*, **76**, 410 (1958).
3. EBASHI, S., AND LIPMANN, F., *J. Cell Biol.*, **14**, 389 (1962).
4. FERNANDEZ-MORAN, H., ODA, T., BLAIR, P. V., AND GREEN, D. E., *J. Cell Biol.*, **22**, 63 (1964).

5. Ganoza, M. C., Williams, C. A., and Lipmann, F., *Proc. Natl. Acad. Sci. U.S.*, **53**, 619 (1965).
6. Hoch, F. L., and Lipmann, F., *Proc. Natl. Acad. Sci. U.S.*, **40**, 909 (1954).
7. Jagendorf, A. T., and Uribe, E., *Proc. Natl. Acad. Sci. U.S.*, **55**, 170 (1966).
8. Kielley, W. W., and Meyerhof, O., *J. Biol. Chem.*, **183**, 391 (1950).
9. Krebs, H. A., and Henseleit, K., *Z. Physiol. Chem.*, **210**, 33 (1932).
10. Lehninger, "The Mitochondrion," Benjamin, New York, 1964.
11. Lipmann, F., Nishizuka, Y., Gordon, J., Lucas-Lenard, J., and Gottesman, M., this volume.
12. Lundegardh, H., *Arkiv Botan.*, **32A**, 12, 1 (1945).
13. Lynen, F., this volume.
14. Mitchell, P., *Nature*, **191**, 144 (1961); *Biol. Rev. Cambridge Phil. Soc.*, **41**, 445 (1966).
15. Mitchell, P., *J. Gen. Microbiol.*, **29**, 25 (1962).
16. Palade, G. E., and Siekevitz, P., *J. Biophys. Biochem. Cytol.*, **2**, 171 (1956).
17. Peters, T., Jr., *J. Histochem. Cytochem.*, **7**, 224 (1959).
18. Piel, G., *in* "Science and the University," special publication of Brandeis University, 1966.
19. Racker, E., Tyler, D. D., Estabrook, R. W., Conover, T. E., Parsons, D. F., and Chance, B., *in* "Oxidases and Related Redox Systems, Vols. 1 and 2," (T. E. King, H. S. Mason, and M. Morrison, eds.), p. 1077. Wiley, New York, 1965.
20. Racker, E., "Mechanisms in Bioenergetics," Academic Press, New York, 1965.
21. Ratner, S., *Advan. Enzymol.*, **15**, 319 (1954).
22. Redman, C. M., and Sabatini, D. D., *Proc. Natl. Acad. Sci. U.S.*, **56**, 608 (1966).
23. Redman, C. M., Siekevitz, P., and Palade, G. E., *J. Biol. Chem.*, **241**, 1150 (1966).
24. Reid, R. A., Moyle, J., and Mitchell, P., *Nature*, **212**, 257 (1966).
25. Sabatini, D. D., Tashiro, Y., and Palade, G. E., *J. Mol. Biol.*, **19**, 503 (1966).
26. Sjöstrand, F., *Radiation Res.*, *Suppl.*, **2**, 349 (1960).
27. Warburg, O., *Ergeb. Physiol.*, **14**, 253 (1914).
28. Watanabe, T., and Brodie, A. F., *Proc. Natl. Acad. Sci. U.S.*, **56**, 940 (1966).

PART I

ORGANIZATIONAL DNA SYNTHESIS

Chairman's Remarks

C. A. THOMAS, JR.

Department of Biophysics,
Johns Hopkins University,
Baltimore, Maryland

This morning's session will be devoted to the organized biosynthesis of the bacterial chromosome. This topic is an appropriate beginning for this symposium because we now have a reasonable picture for some of the major features of this process. This picture is derived from a number of lines of evidence—the most graphic being the autoradiographs of John Cairns.

The scheme in Fig. 1 presents at least three problems:

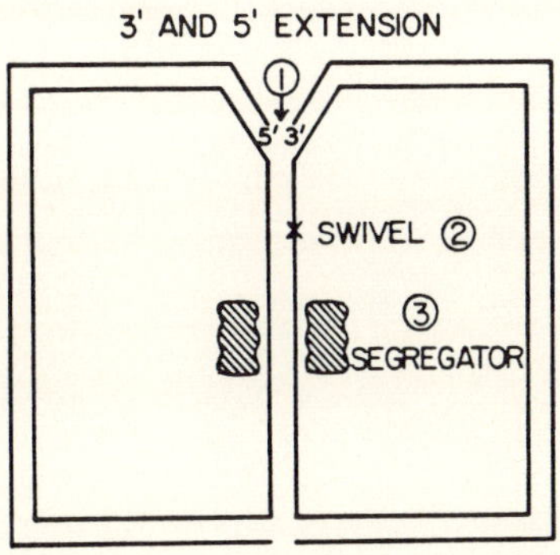

FIG. 1. The bacterial chromosome—three problems.

1. It implies that synthesis can take place by the simultaneous extension of the 3′-ended and 5′-ended polynucleotide chains (no one has been able to circumvent this problem).

2. If the DNA molecule rotates during synthesis (and there is no direct evidence that it does) then there must exist a rather sophisticated swivel, since the DNA molecule must have some torsional rigidity resulting from the two polynucleotide chains.

3. There must be some method to assure the segregation of one (or a given number) of daughter chromosomes into each of the two daughter cells.

These three problems, diagrammatically indicated in Fig. 1, will be of major concern to the speakers this morning.

Particulate Fractions in Macromolecular Synthesis and Genetic Transformation

A. T. GANESAN

Department of Genetics,
Stanford Medical School,
Palo Alto, California

INTRODUCTION

In order to elucidate how cellular components recognize, interact, and coordinate with each other during cellular division and genetic recombination, we have studied, as an example, the transformable bacterium *Bacillus subtilis*. The current evidence suggests that macromolecular biosynthesis involves particulate fractions associated with the cellular membrane (14, 19, 26, 54).

DEOXYRIBONUCLEIC ACID

In the case of mouse fibroblasts and HeLa cells, the DNA polymerase was shown to be of a particulate nature (1, 33). The newly synthesized DNA in these mammalian cells was found to be preferentially lost in the interfacial coagulum formed during deproteinization, suggesting a firm association of the nascent DNA molecules with protein. The nature of this protein is not known (5).

Goldstein and Brown (21) found that, in *Escherichia coli*, synthesis of DNA occurred "predominantly in a particulate fraction." They also observed that the newly synthesized DNA—that is, DNA which is enriched in counts from a recent pulse of labeled precursors—was resistant to sonic degradation. Hanawalt and Ray (22) found that the region of the DNA molecule involved in replication was very fragile. However, the same authors also observed that the newly replicated DNA was firmly associated with proteins, which could be released only after treatment with proteolytic enzymes. Recently Smith and Hanawalt (46) reported the fractionation of *E. coli* DNA into nascent and nonreplicating species in a sucrose gradient. The former was found to be rapidly

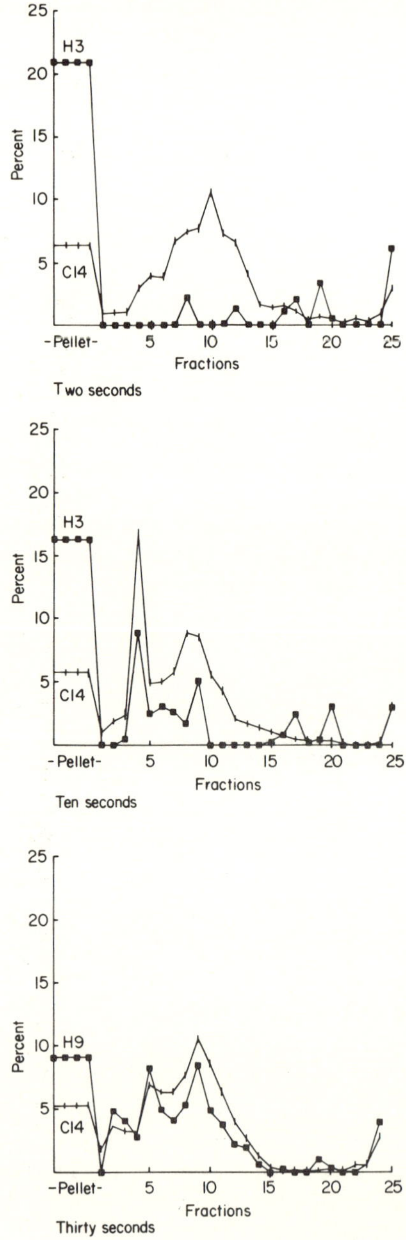

FIG. 1. Sedimentation patterns of lysates of pulse-labeled cells. The plot is computer-generated (on a Calcomp Digital Plotter reading IBM 7090 output). See Table I.

sedimenting. Sodium deoxycholate was found to dissociate the nascent DNA from a protein complex.

Bacillus subtilis DNA (19) showed the same type of fractionation as that of *E. coli.* Thymine- and tryptophan-requiring cells were grown in the presence of C^{14}-thymidine for three generations and then exposed for fractions of a minute to H^3-labeled thymidine. This procedure labels the nascent DNA with H^3 relative to C^{14}. Careful lysis of the azide-killed cells, followed by centrifugation in sucrose gradients, clearly revealed the preferential association of the newly synthesized DNA with a rapidly sedimenting fraction. This fraction was always found in the pellet of the gradients (Fig. 1 and Table I).

TABLE I

SUCROSE GRADIENT SEDIMENTATIONS OF LYSATES OF PULSE-LABELED CELLS[a]

Time of (pulse seconds)	Total H^3 (cpm)	Total C^{14} (cpm)	DNA (mμmoles)	New DNA (mμmoles)
2	223	1.2×10^3	2.14	1.3×10^{-4}
10	426	1.2×10^3	2.14	2.6×10^{-4}
30	876	1.2×10^3	2.14	5.1×10^{-4}
Time of (pulse seconds)	**H^3 in pellet (%)**	**C^{14} in pellet (%)**	**H^3 in sup. (%)**	**C^{14} in sup. (%)**
2	83.68	25.38	16.32	74.62
10	65.30	23.07	34.7	76.93
30	36.37	21.06	63.63	78.94

[a] Thymidine-H^3 was added for 2, 10, or 30 seconds. The pellet is represented as if it were evenly distributed among four fractions. Each fraction consists of 3 drops (0.25 ml). The distributions are normalized as percentages of the total counts.

The same results were obtained by using up to five times the amount of DNA and H^3 pulse counts. In eight experiments, the shortest pulse always set is given because it was done at the same time with the same cells and in the same gradient run, so all the conditions are comparable. In these eight experiments, the amount of C^{14}-labeled non-replicating DNA in the pellet varied from 20 to 60%. It was difficult to predict the amount of nonreplicating DNA in the pellet, while the behavior of nascent DNA was easily predicted. (Sup. refers to supernatant.)

In these experiments we consistently lost 8 to 20% of the nascent DNA, without any detectable loss of nonreplicating (C^{14}) DNA, as judged by acid-precipitable tracer counts. A control experiment, in which cells were pulsed with H^3-thymidine, as described above, during different stages of growth, revealed that, as the culture approached the stationary phase, the isolation of nascent DNA became more difficult. There was a maximum of 70% loss at the last time point. This observation suggested that, in the stationary phase, the nascent DNA might be more susceptible to nuclease or that nuclease production is also localized in the sedimentable fractions, during the stationary phase of growth.

Kinetic data from the pulse experiments are presented in Table I. The amount of acid-insoluble counts incorporated during pulses of different lengths suggests a constant rate of DNA synthesis, after the first 2 seconds. From the counts (H^3) of pulsed DNA that appeared in the supernatant of the gradients during different times of increasing exposure to the tracer, it was possible to calculate the transit time of DNA. The transit time is defined as the time in which a constant proportion of the nascent DNA was bound to the rapidly sedimenting fractions. This was found to be 10.5 seconds, corresponding to 160 to 200 nucleotides at the site of synthesis. This observed rate was found to be slower than the normal replication rate for *B. subtilis* by a factor of 5.7. This may be due to the washing procedures introduced before exposure to H^3-thymidine, which retards the growth initially.

If the replicating point(s) is attached to a specific site on the "membrane-cell wall" or to a unique, rapidly sedimenting fraction, it should be possible to chase the "pulse" label from this fraction by subsequent growth in cold thymidine. Such a procedure should displace the pulse counts (H^3) to the supernatant. If the H^3-thymidine pulse were chased by subsequent growth in C^{14}-thymidine-containing medium, the ratio H^3/C^{14} should decrease markedly in the pellet. The results presented in Table II show the displacement of the pulse label from the rapidly sedimenting fraction to the supernatant in both cases of subsequent growth.

If the association of nascent DNA with a particulate fraction involves a component of protein nature, then treatment of the lysate prior to

TABLE II

H^3 PULSE CHASED IN C^{14}-THYMIDINE MEDIUM[a]

Experiment	Pellet	Supernatant
C^{14}	24%	76%
H^3 pulse	9.06%	90.94%
[H^3 (100%) = 4.6 × 10^3 cpm, while C^{14} = 1 × 10^4 cpm]		
H^3 PULSE CHASED IN COLD THYMIDINE MEDIUM		
C^{14}	43%	57%
H^3	42.4%	57.6%
[H^3 (100%) = 8.7 × 10^3 cpm, while C^{14} = 1.7 × 10^3 cpm]		

[a] These experiments were also tried with different amounts of DNA per gradient. More than 10 mμmoles of DNA in a 5-ml gradient (approximately 2 × 10^9 cell equivalents) leads to aggregation so that most of the DNA sediments out as a gel in the pellet. This experiment refers to a 10-second pulse, as mentioned in Fig. 1 and Table I.

fractionation with a proteolytic enzyme should release the nascent DNA to a slowly sedimenting component. Figure 2 shows the results of such an experiment. A culture of bacteria grown in C^{14}-thymidine-containing medium was pulsed for 5 seconds with H^3-thymidine, then lysed and treated with pronase and sedimented in a sucrose gradient. The DNA molecules sedimented in two components, a fast-moving and a slow-moving species. The fast-sedimenting species included the majority of the nonreplicating DNA molecules, while the slow-moving species was relatively enriched in nascent DNA. This enrichment corresponded to 63% of the total nascent DNA added to the gradient after a 5-second pulse.

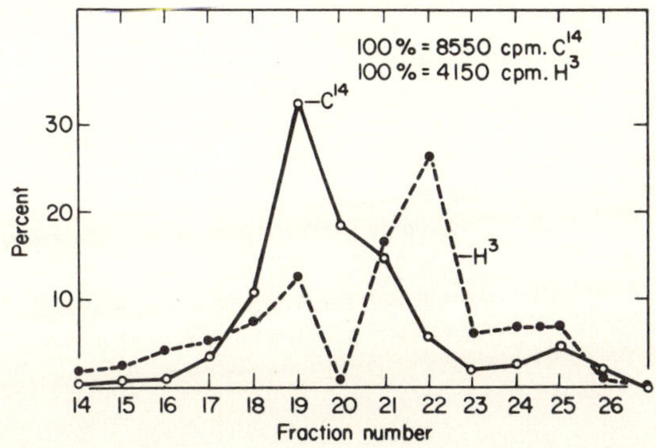

FIG. 2. Dissociation of nascent DNA from the pellet. The gradients were performed as outlined in experiments pertaining to Fig. 1. The lysate was treated with 50 μg of pronase per milliliter for 2 hours. Shorter treatments were found to be inadequate. A total of thirty-one fractions were collected from a column 5-ml gradient, and each fraction was acid-precipitated and counted as in ref. 19. Here we have used eight times the amount of DNA that was used in Fig. 1. The treatment with pronase abolished the aggregation normally encountered in these experiments.

Distribution of DNA Polymerase Activity

It has been shown that DNA polymerase is bound to DNA in *E. coli* and *B. subtilis* (7, 19, 27, 41). In experiments involving pulse labeling of the newly synthesized DNA, we found that 25 to 50% of the total assayable polymerase was associated with the rapidly sedimenting fraction. In lysates of *B. subtilis,* DNA is always found to be associated with proteins. To obtain high-molecular-weight DNA from *B. subtilis,* Massie and Zimm (36) used pronase in high concentrations over a period of several hours. The resulting DNA molecules were free of protein and RNA and had a molecular weight of 250 megadaltons.

In our hands, a carefully lysed preparation, when centrifuged in CsCl solution, still exhibited 8 to 10% of the total assayable polymerase firmly bound to the DNA. We have shown earlier that the binding of DNA polymerase to DNA and membrane fractions depends on the concentrations of Mg^{++} used (19). Free polymerase, found in the crude lysate of *B. subtilis* protoplasts, showed linear kinetics of synthesis with regard to time, when a constant amount of enzyme was used, and with regard to the amount of enzyme over a constant time period. These assays contained an excess of the four deoxytriphosphate substrates; dAT copolymer (copolymer of deoxyadenylate and thymidylate) was used as a primer. As shown in Table III, the reaction was completely dependent on the

TABLE III
EFFICIENCY OF CRUDE LYSATE PREPARATION TO INCORPORATE DEOXYTRIPHOSPHATES[a]

Amount of lysate (0.002 ml = 2.1 μg of protein) (ml)	H^3-dATP incorporated (cpm)	Time (minutes) (amount of lysate 0.001 ml)	Assay 1	Assay 2
0.002	230	0	55	42
0.004	696	2	427	116
0.006	1298	5	567	154
0.008	1769	10	1154	168
0.010	2569	15	1604	184
0.020	4098	30	2838	200
Omit Mg^{++}	31			
Omit TTP	34			
0-minute sample	36			

[a] Reaction time was 30 minutes; primer was dAT copolymer. On addition of exogenous primer (*B. subtilis* DNA), the counts were 216 at 30 minutes. No increase was observed. Assay 1 refers to assay with dAT copolymer; assay 2 refers to assay with the endogenous primer and four deoxytriphosphates.

In assay 1, the incorporation increased steadily up to 180 minutes. At this time the amount of synthesis, as judged by the acid-precipitable counts, was equivalent to 25% of the amount of primer added. The assays were performed according to ref. 42. The only difference was that the assay mixture was reduced to half the amount that was used. The specific activity of H^3-dATP was 1.0×10^5 cpm per 10 mμmoles. In assay 2, increasing the amount of lysate did not increase in the incorporation.

presence of the deoxytriphosphates and Mg^{++}. There is a striking difference between the *E. coli* and *B. subtilis* preparations in the proportionality relation observed in these kinetics, between the amount of synthesis and the amount of enzyme added. During the purification of *E. coli* DNA polymerase, fractions I to IV gave poor proportionality in the reaction when dAT copolymer was used as a primer (42). This was due to the abundance of nucleases in these fractions. Our crude preparations seem to be relatively free of nucleases as judged by this criterion, in agreement with earlier observations (41). When endogenous DNA—

that is, the DNA that is bound to the rapidly sedimenting fraction—was allowed to act as a primer for synthesis, adding the four deoxytriphosphates, the lysate showed less incorporation, by a factor of 10 to 20, than when dAT copolymer was added. Addition of exogenous *B. subtilis* DNA did not increase the incorporation significantly. Sedimentation of the lysate in a sucrose gradient resulted in a pellet fraction at the bottom of the gradient. This fraction, although showing polymerase activity as mentioned earlier, was free of at least two "cytoplasmic" enzymes, tryptophan synthetase and ornithine transcarbamylase. The top fraction of the gradient contained these enzymes and some nucleases, as judged by the loss of transforming activity when this fraction was incubated with a purified *B. subtilis* DNA. The sediment in the bottom of the tube was refractionated on a second sucrose gradient. The resulting pellet material was designated as "purified fraction" and was used in the experiments reported below.

An electron micrograph of such a membrane fraction is shown in Fig. 3. Here, molecules of DNA and fragments of cell membrane are visible. No intact or partially lysed cells were detected in this fraction. The polymerase associated with this fraction showed peculiar properties. It could be released from its association with rapidly sedimentable material by extensive dialysis at 4°C against EDTA solution (0.01 *M*) in the presence of β-mercaptoethanol (0.001 *M*). With no exogenous primer added, DNA synthesis by this fraction was very rapid and reached a plateau in a few minutes (Table IV). The synthesized product was then progressively degraded, as observed by the release of acid-soluble counts, and then there was a second cycle of synthesis. During the synthesis there was no detectable loss of template primer DNA, as measured by acid-precipitable C^{14} counts of the primer DNA. Because of the oscillatory kinetics of synthesis, the specific activity is not rigorously defined for this preparation. However, using the maximum incorporation observed in 1 minute of synthesis, one can calculate a maximum specific activity of 2400 units per milligram of protein. (The units correspond to those defined by Richardson *et al.*, 42.) With the "purified" fraction, dilution of the membrane–DNA–polymerase complex leads to a loss in activity. We cannot, therefore, define any practical units.

The dAT copolymer was a more efficient primer of synthesis by polymerase in the starting lysate than was endogenous DNA. The purified preparation, however, was able to catalyze the incorporation of added deoxytriphosphates equally well in the presence or in the absence of dAT copolymer (Table V); the addition of the copolymer did not increase the rate of synthesis. The inefficiency of the preparation to use the copolymer as a primer for synthesis was not because the primer was

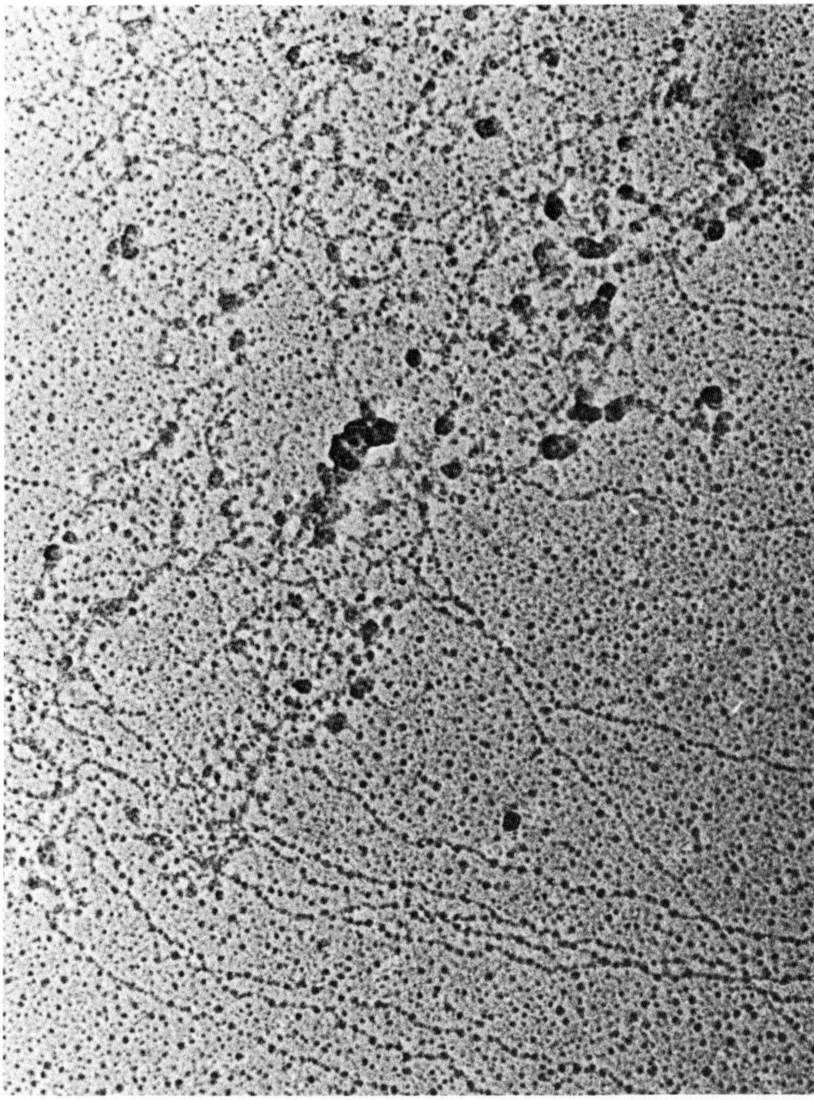

FIG. 3. *Bacillus subtilis* (strain SB 566 thy^-, try_2^-). Electron micrograph of the "purified" fraction from the lysate. This fraction was active in polymerase and possessed the majority of the nascent DNA. It is difficult to trace the specific regions associated with the membrane fragments. However, both are distinguishable. The arrow indicates DNA molecules (magnification 76,800×).

degraded by the nucleases. We have reisolated the primer and tested its priming activity in another reaction.

It was possible that, in the assays in which endogenous DNA was allowed to act as a primer, the observed rapid incorporation reflected a

TABLE IV
PATTERN OF INCORPORATION OF DEOXYTRIPHOSPHATES BY THE "PURIFIED" MEMBRANE FRACTION[a]

Time (minutes)	H^3-dATP incorporated (cpm)
0	43
1	2197
2	742
3	29
4	173
5	2280
10	2158
Omit one triphosphate	50
Omit Mg^{++} (+EDTA)	39
Pretreat with pronase	66

[a] The amount of fraction used was 0.002 ml containing 0.8 to 1.0 μg of protein. The amount of endogenous primer present per assay corresponds to 0.45 mμmole of DNA. The specific activity of H^3-dATP was 1.0×10^5 cpm per 10 mμmoles. The assay contained all the four triphosphates. The conditions were as described in ref. 42. The modifications were that 1 *M* Tris buffer, pH 8.5, was used, and the total amount of substrates was half the amount that was used previously (42).

repair type of synthesis of molecules previously degraded by enzymes such as exonuclease-3 (Hurwitz,[1] personal communication). An assay performed by Dr. Stuart Linn of the Biochemistry Department, Stanford, to detect any exonuclease activity in these fractions, revealed the absence of such an activity. Here we used P^{32}-labeled *E. coli* DNA as a substrate.

We have carried out experiments in which the polymerase preparation was "purified" from lysates of cells grown in N^{15} and D_2O-supple-

TABLE V
EFFICIENCY OF DIFFERENT PRIMERS TO DIRECT DEOXYTRIPHOSPHATE INCORPORATION[a]

	cpm H^3-dATP incorporated (cpm)		
Experiment	Exogenous primer (denatured *B. subtilis* DNA)	dAT copolymer	Endogenous primer
1	760	1480	2158
2	460	1670	1868
3	622	1290	2432

[a] The counts represent incorporation into acid-precipitable form after 10 minutes, as mentioned in Table IV. The proportions of dAT copolymer and denatured DNA were the same as prescribed in ref. 42.

[1] Dr. J. Hurwitz, Albert Einstein College of Medicine, Bronx, New York.

mented medium. The DNA associated with the purified fraction has a buoyant density of 1.753 gm/cc in CsCl solution. We used such a fraction in a reaction with $N^{14}H^{1}$-deoxytriphosphates, and the synthesized product was partially characterized by buoyant density. Preliminary results indicate that DNA molecules of densities less than 1.753 are synthesized. Present studies are directed toward a more detailed understanding of the nature of the synthesized product and also the oscillatory kinetics of synthesis. The observed kinetics may be due to irrelevant nuclease contamination; on the other hand, it is possible that such a nuclease action has some function in the process of DNA replication *in vivo.*

Organization of DNA in the Cell

Bacillus subtilis cells approaching the stationary phase contain three "nuclei" per cell. These were identified histochemically by using acridine orange staining, and basic staining after RNase treatment. The nucleus is about 1 micron in diameter and can be isolated as a compact mass of DNA filaments, after the careful lysis of lysozyme-induced protoplasts, followed by sedimentation at relatively low speeds (17, 48). One can stabilize the nuclear DNA bundle by low concentrations of polyamines such as spermine (17). Methods of careful lysis and isolation of DNA such as this have recently been used in experiments with several bacterial systems designed to study the role of DNA in biosynthesis, and its organization and relation to other cellular components *in vivo,* as well as in physical studies of high-molecular-weight DNA molecules (3, 10, 17, 19, 20, 35, 40, 47).

The chemical composition of the isolated "nuclei" from *B. subtilis* is given in Table VI. There is 2.07×10^{-9} μg of DNA per "nucleus" as determined by chemical methods. This value corresponds to 1.8×10^{6} nucleotide pairs, or a molecular weight of 1.2×10^{9} daltons. The molecular weight of highly purified DNA used in our transformation experiments usually varies between 10 and 20 million daltons. A value of 20 million corresponds to 60 such molecules per genome. Using a value of

TABLE VI

DNA, RNA, and Protein Content of Total Lysate and Nuclear Bodies[a]

Fraction	DNA	RNA	Protein
Lysates	1.0	7.5	19.0
Nuclear fraction	1.0	2.4	9.0

[a] Generally there is a loss of 4 to 8% of the total DNA during the washings and centrifugations. Taking the DNA content as unity, RNA and protein are given to rounded significant figures.

1000 nucleotide pairs per gene, we obtain an estimate of about 1800 genes per chromosome, linearly arranged on the polynucleotide chain.

Electron Microscopy of Protoplasts and DNA

Protoplasts were lysed on a monolayer of denatured protein (28). [Lysozyme was used as the protein in these experiments, instead of cytochrome *c* (20).] Figure 4 shows one such preparation. The heavily shadowed area in one corner presumably represents the DNA-containing region. A few free filaments are seen at the sides. In this stage the DNA is very compact and the cell membrane is not completely disrupted, giving a clear display of ribosomes, probably as one would find them *in vivo.* Figure 5 shows the complete disruption of a protoplast, with the DNA seen to be complexed with the cellular membrane. In these photographs, the DNA exhibits very few free ends. The firm binding of the total genome to the membrane at the initial stages of lysis suggests an association of the two components in the cell.

Figure 6 represents a well-spread preparation, with DNA relatively free of membrane fragments. Several areas show compact configurations of DNA chains. An important feature is the presence of flagella in these areas. In our experience, the presence of flagella indicates the location of a small fragment of cell membrane. These flagella are attached to the membrane and persist even after the removal of the cell wall with lysozyme. The contour length of the DNA is easily measured in these types of photographs to be at least 500 microns. An estimate including all the overlapping regions yields a maximum length of 700 to 800 microns, which corresponds to a molecular weight of 1.4 to 1.6×10^9 daltons. This estimate agrees with the autoradiographic measurement obtained by Dennis and Wake (15), and with the estimate obtained by chemical analysis. This value is very similar to the length of the genome of *Hemophilus influenzae* (6, 35). Sedimentation analysis of the DNA from the protoplasts, using coliphage lambda DNA as a standard, gives a sedimentation coefficient between 80 and 90 Svedberg units, corresponding to a molecular weight of over 500 million. This implies that there is only one break per molecule during isolation. Careful deproteinization and RNase treatment of the above high-molecular-weight DNA results in smaller DNA molecules of 200 million daltons average molecular weight.

The efficiency of high-molecular-weight DNA to transform genetically linked markers has been compared to that of DNA purified according to the usual procedure and to DNA which had been hydrodynamically sheared (39). A striking difference between these preparations is the correlation (Table VII) between high molecular weight and efficiency to transform all four markers in the aromatic linkage group (39). Although

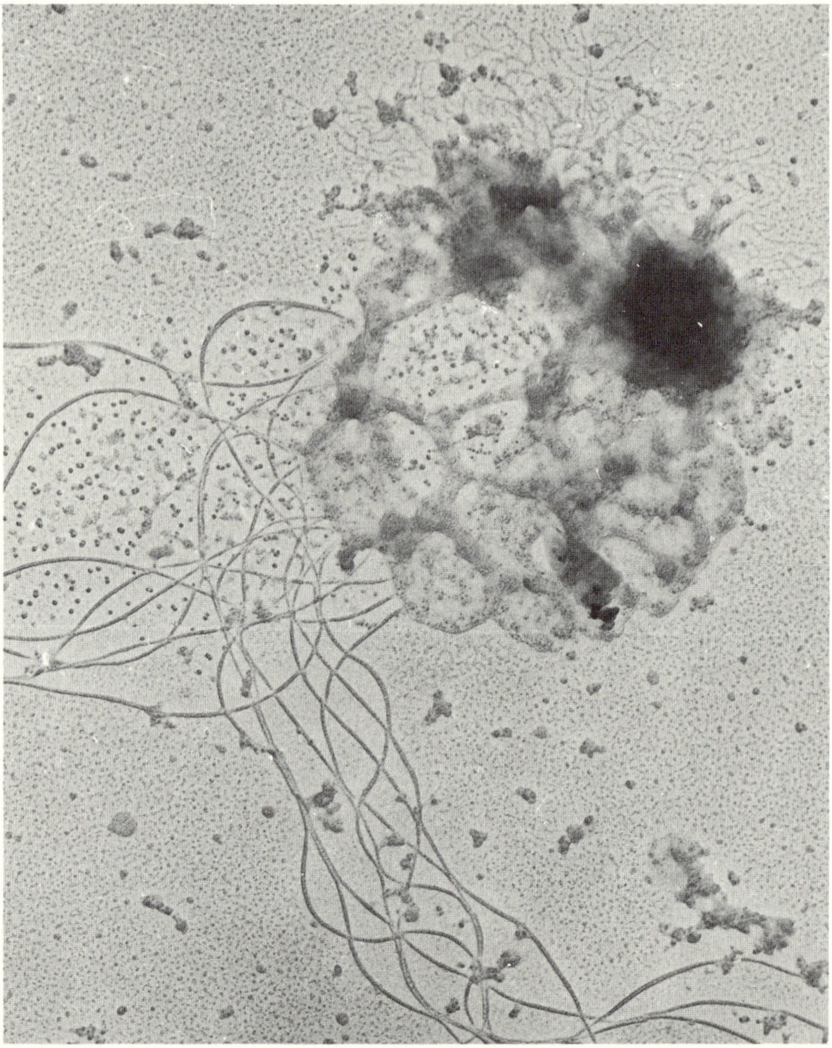

FIG. 4. *Bacillus subtilis* (strain SB 168 try$^-$), a partially bursted protoplast. The DNA filaments are seen next to the heavily shadowed region, which presumably represents the bulk of DNA. The flagella are seen attached to the membrane. The bipartite nature of free 70 S ribosomes is seen, and the aggregation of polysomes close to membranes is easily localized (magnification 38,640×).

reduction of the molecular weight lowers this multiple transfer efficiency, the ability to transform single markers increases. The sheared samples exhibited, when compared to an unsheared DNA, a reduction in total transforming activity to less than 10% (39). But among the transformants

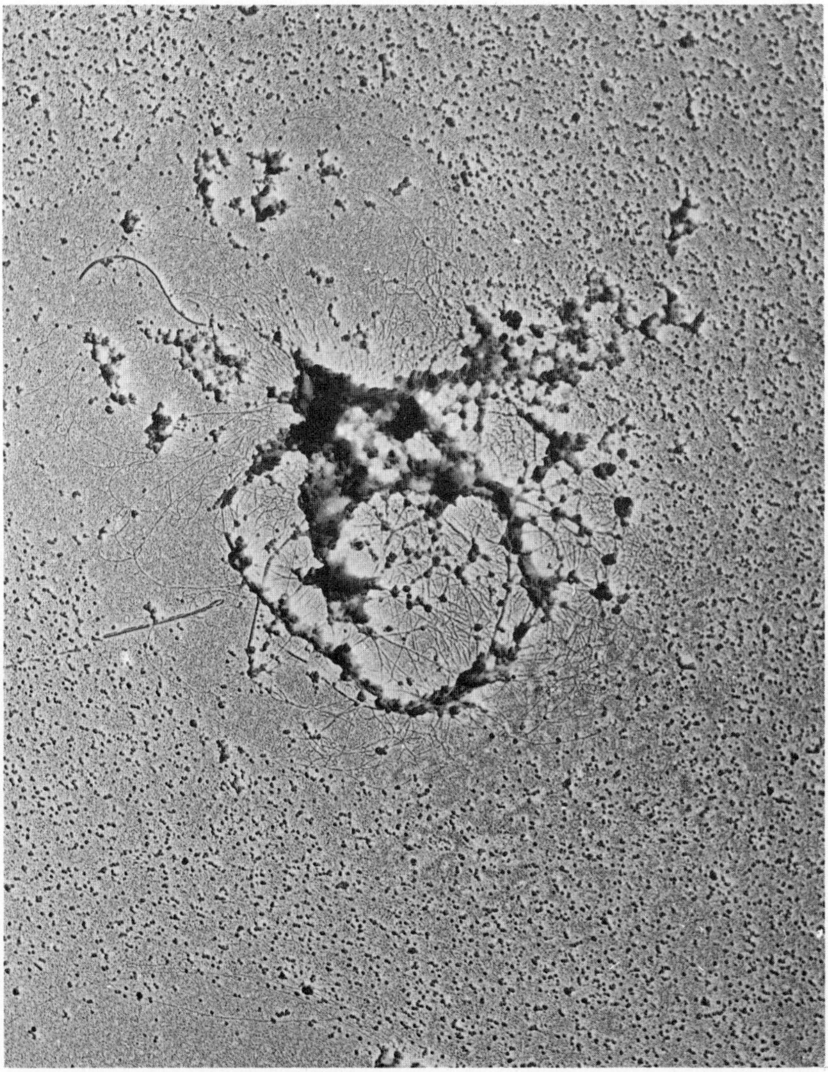

FIG. 5. *Bacillus subtilis* (strain SB 168 try$^-$), a bursted protoplast. The DNA molecule is bound firmly to the membrane and can be easily sedimented at slow speeds. This type of compact nuclei was often observed, probably similar to what was observed by Spiegelmen *et al.* (48) in *B. megaterium* (magnification 15,600×).

for a single marker, only 2 to 3% carry all four linked markers. In *Hemophilus influenzae,* Berns and Thomas (6) observed that high-molecular-weight DNA preparations showed linkage between markers that were previously observed not to be linked. Relevant to the above observation

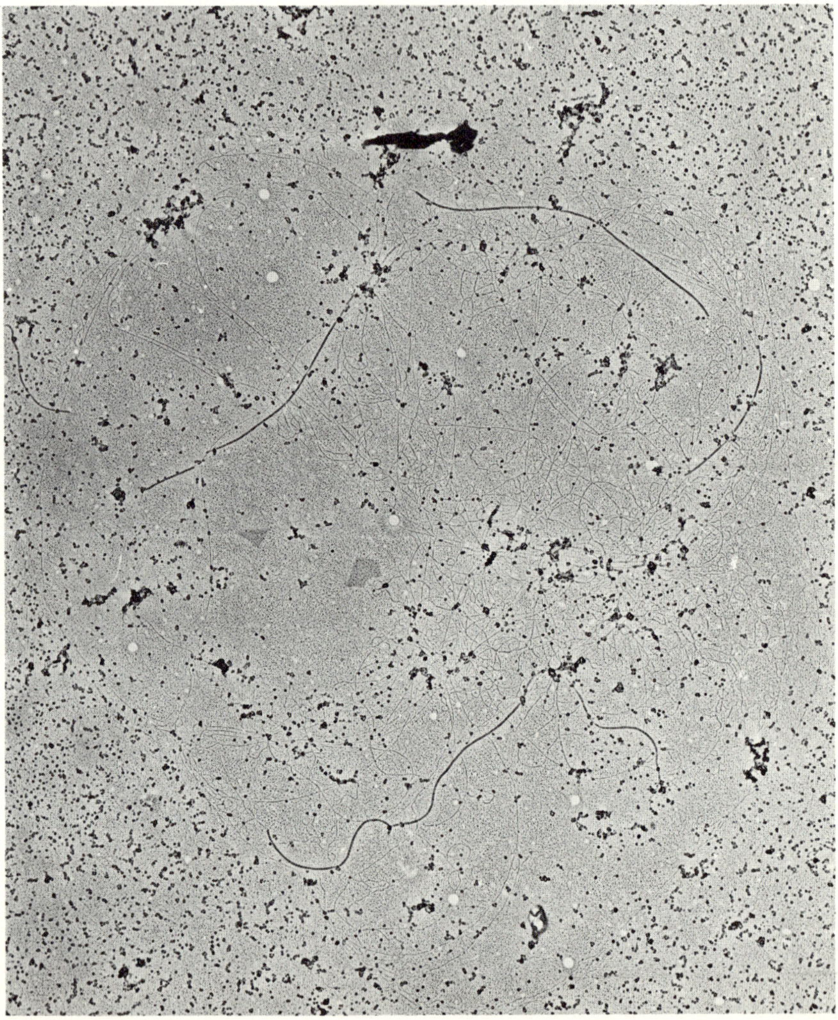

FIG. 6. *Bacillus subtilis* (strain SB 566 thy$^-$, try$^-$). A well-spread DNA preparation of one protoplast, presumably one genome. The length of the DNA molecule can be easily measured (see text). The flagella are found in regions where ribosomes are also localized. Since most of the ribosomes scatter to the periphery, the genome itself occupies the center of each lysis event (magnification 15,600×).

TABLE VII
EFFICIENCY OF THREE DNA PREPARATIONS TO TRANSFORM AROMATIC LINKAGE GROUP IN *Bacillus subtilis*[a]

Sample	Molecular weight	Cotransfer of aro_2–tyr_1 markers by DNA samples Try_2^+ transformant genotypes (% of total transformers)							
		1111	1110	1101	0111	1100	0110	0101	0100
1	200×10^6	67[b]	2.9	2.3	15.7	7.4	1.4	0.3	2.9
2	18×10^6	24	4.7	1.0	14.0	24.0	5.1	1.0	20
3	8×10^6 (sheared)	1.8	2.08	0.33	8.9	22.0	8.0	1.4	50

[a] The aromatic linkage group carries at least nine genes specifying the biosynthesis of aromatic amino acids. This linkage group also includes a histidine marker. The assay strain used here carried four markers, aro_2, try_2, his_2 and tyr_1 (shikimic acid, indole, histidine, tyrosine), all linked but widely distributed on the same group (39). The efficiency of a DNA preparation to transform all four markers was tested by using nonsaturating concentrations of the sample. Transformants for try^+ were selected. Minimum of 300 such try^+ transformants were stroked onto nutrient agar plates. These were replicated to appropriately supplemented media to determine the remainder of the genotype. The donor markers are designated as 1, and the recipient markers as 0. Sample 3 was sheared according to ref. (39). Sucrose sedimentation runs were performed as in ref. (6), using H^3-labeled lambda phage DNA (molecular weight 33 million), kindly provided by Mr. J. Champoux of the Biochemistry Department, Stanford University. DNA preparation 1 was made by lysing the protoplasts and treating the lysates with pronase for several hours (36). Sample 2 DNA was made as described in ref. (39).

[b] Values as high as 76% have been obtained in some experiments.

is the demonstration of a clear correlation between transformation efficiency and single-strand molecular weight by Bodmer (8a).

RIBONUCLEIC ACID

The nuclear fraction mentioned earlier contained both RNA and protein. As in *E. coli* (44), 30 to 40% of the total RNA was found in this nuclear fraction. The bulk of this RNA is possibly contributed by the ribosomes which are bound to the cellular membrane. In the electron microscope pictures (Fig. 4) the polysomes are predominantly associated with the membrane, and the smaller ribosomes are free. When exponentially growing cells were pulsed with H^3-uridine for 30 seconds, 75% of the labeled RNA was found in this nuclear-membrane fraction. In the electron microscope studies of protoplast lysates we observed restricted regions along the DNA molecules in which ribosomes were linearly arranged. In an *in vitro* system, Byrne *et al.* (11) observed a rapidly sedimenting DNA–ribosome–messenger RNA complex. Our electron micrographs indicate that the same association may exist *in vivo*. A brief exposure of *E. coli* cells to P^{32}-phosphate resulted in the labeling of RNA associated with membrane of the cell. After T2 phage infection, the RNA in the membrane fraction was similar in its base composition to the T2 DNA (51, 52). Recently Yudkin and Davis (54) found both messenger and ribosomal RNA in the membrane fractions of *B. megaterium.*

Proteins

The nuclear fraction we have described in *B. subtilis* contains up to 50% of the cellular proteins. Besides membrane, flagellar, and other proteins known to be near the cell wall, this fraction also contains DNA and RNA polymerase. Membrane preparations are known to incorporate amino acids into polypeptides more actively than do free ribosomes (4, 40, 44, 47). In *E. coli,* exposure of the cells to radioactive amino acids for a very short time resulted in the labeling of the membrane-bound ribosomes (23). In higher organisms the ribosomes are associated with the endoplasmic reticulum (45). All these observations suggest that complexes of cellular components associated with membranes are involved in macromolecular biosynthesis. Since *B. subtilis* is transformable, we have studied the process of genetic recombination with the aim of isolating specific protein–DNA complexes which may be involved in the intermediate stages of recombination. Some of our results are described in detail in what follows.

Replication and Recombination

Escherichia coli cells requiring thymine and certain amino acids for growth undergo "thymineless death" in the presence of the required amino acids when thymine is removed from the growth medium during the DNA replication cycle. Cells that have finished DNA replication are immune to this killing. If the same strain is grown in the presence of thymine and in the absence of the required amino acids, no loss in viability occurs; these cells finish the round of DNA replication in progress and do not initiate another (2, 34). Current evidence indicates that protein synthesis is required before each round of replication (31, 32, 53).

We have performed several kinds of experiments to study the nature of genetic transformation of thymine- and amino acid-requiring mutants of *B. subtilis* under conditions in which DNA replication in competent or transformed cells is inhibited by thymine starvation or controlled by starvation for required amino acids. In one set of experiments, we examined the effects of thymine starvation on competent cells before and after treating them with DNA. Competent cells of a thymine- and tryptophan-requiring strain were exposed to wild-type donor DNA for 15 minutes, washed well, and resuspended in a medium containing tryptophan but without thymidine (Fig. 7). Growth of the cells was followed for 300 minutes. We observed that:

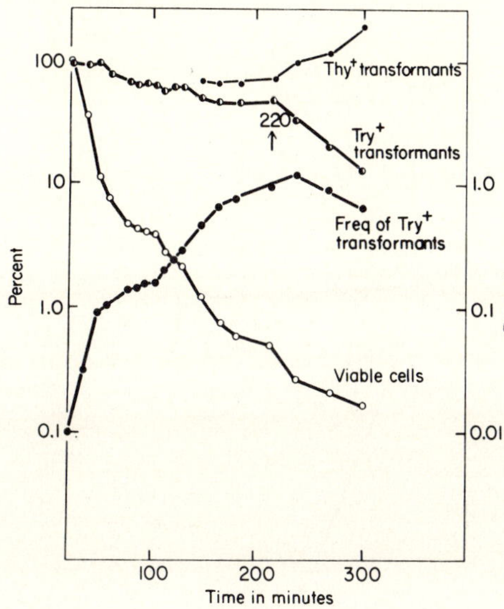

FIG. 7. Fate of transformants during thymineless death. Twenty milliliters of competent cells of SB 566 (try$^-$, thy$^-$) were added to 50 μg of SB 19 wild-type DNA. (After purification, the DNA was centrifuged in CsCl and then dialyzed first against 2 *M* NaCl overnight and then against 0.015 *M* sodium citrate to remove any contaminating thymine.) After 15 minutes the cells were treated with pancreatic DNase (50 μg/ml) for 10 minutes, washed two times with minimal medium + glucose, and then resuspended in 20 ml of minimal medium supplemented with glucose and 2 μg of L-tryptophan per milliliter. Aliquots were plated every 15 minutes for viability and number of try$^+$ and thy$^+$ transformants up to a period of 300 minutes.

1. There was an exponential loss of viable cells, due to thymineless death.

2. Transformants were relatively immune to thymineless death; by 200 minutes, the frequency of transformants, relative to viable cells, had increased by a factor of 100. The rate of decrease in both viability and transformants depends on the aeration and temperature.

3. By 200 minutes, transformants started to divide, shown by the bifunctional curve for tryptophan⁺ and thymine⁺ transformants. The thymine⁺ transformants were able to divide because the medium contained tryptophan. The tryptophan⁺ transformants underwent thymineless death, because the medium lacked thymine. These two markers are unlinked, and hence the double transformants were very rare. In some cases, the thymine⁺ cells also died or exhibited a long lag before replication. The loss might be due to the production of phage.

In another experiment, a competent culture was transferred to a medium with no thymidine, but containing the required amino acid. Samples were taken at short intervals and exposed to H^3-labeled DNA. After DNase (pancreatic deoxyribonuclease-1) treatment, the uptake of donor DNA was measured, and the numbers of viable cells and transformants were determined. The results (Fig. 8) showed that the ability

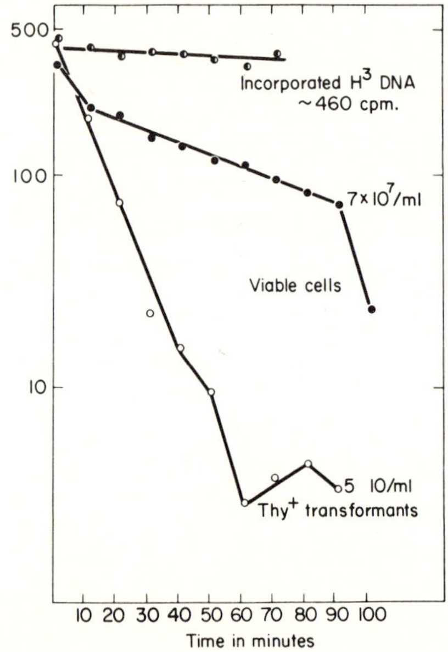

FIG. 8. Efficiency of competent cells to take up DNA and transform during thymineless death. Six milliliters of SB 566 (try_2^-, thy^-) competent culture was washed and resuspended as mentioned in Fig. 7, and allowed to undergo thymineless death. Every 10 minutes 0.5 ml of the culture was withdrawn and added to a tube containing 1 μg of H^3-labeled DNA (2×10^5 cpm/μg). The cells were allowed to react with the DNA for 5 minutes and then treated with DNase; they were washed, and 0.1 ml was used to plate for viability and transformants. Three-tenths milliliter was acid-precipitated and counted for the uptake of donor DNA. The counts indicated in the figure represent that amount present in 0.3 ml.

to integrate donor DNA was lost more rapidly than was viability; the ability to take up DNA remained the same. This experiment shows that the process of integration can be disassociated from the uptake of donor DNA. It also argues against a recently suggested model (8) that DNA synthesis is suspended in competent cells. If this were true, competent cells should be relatively resistant to thymineless death; but they were observed not to be so. The immunity of transformants to thymineless death depends on the addition of donor DNA and is not a general property of the competent cells before they take up DNA.

Other experiments were designed to examine the kinetics of various events in transformation under conditions of amino acid starvation. Competent cells requiring thymine and tryptophan were washed and resuspended in a medium containing thymidine and lacking tryptophan. The resuspension medium also contained H^3-labeled donor DNA from wild-type cells. The main results were that:

1. On resuspension in the above medium, the rate of DNA synthesis increased rapidly up to 40 minutes. From then on, synthesis was observed to continue at a slow rate.

2. A fraction of the donor DNA interacted immediately with the cell, presumably with the cell wall, and became resistant to degradation by added DNase.

3. Transformants could be obtained only after a period of 3 to 4 minutes from the time of DNA addition. This lag might be the time taken for DNA to be absorbed into the cell.

4. The donor DNA inside the recipient cells (8) gradually became acid-soluble by 40 minutes. There was no loss or increase of transformants at this point. There was also no loss in the viable cells.

To trace the fate of donor DNA taken up by competent cells in the absence of amino acids, we used DNA isolated from cells grown in N^{15}, D_2O-, and H^3-thymidine-supplemented medium (DNA density 1.753 in CsCl solution) in a transformation experiment such as that just described. This enabled pycnographic separation of donor and recipient DNA species (light DNA density = 1.703 in CsCl solution). Samples were taken at 0, 8, and 40 minutes after addition of DNA, lysed, and the partially purified lysates centrifuged in CsCl (Table VIII). The results of these experiments were:

1. At 0 minute, in the lysate, the donor DNA banded in a broad region of the gradient, extending all the way from native light DNA to native heavy DNA density positions.

2. By 8 minutes, there was a small amount of donor label in the position of the gradient corresponding to the density of heavy denatured DNA.

TABLE VIII

PYCNOGRAPHIC SEPARATION OF DONOR MOLECULES INSIDE THE CELL DURING AMINO ACID STARVATION (SEE TEXT)[a]

Time (minutes)	*Donor counts distributed in density regions* Denatured heavy	Native heavy	Hybrid	Native light	Total cpm
	Density (gm/cc)				
	1.768	1.753	1.726	1.703	
0	—	2050	2380	400	4380
8	860	1470	1260	630	4220
40	—	480	570	660	1710
Donor and recipient activity in all three points	—	—	—	+	

Donor and recipient biological activity found in the region of native light position

Time (minutes)	try^+ (donor)	tyr^+ (recipient)
0	11	1.5×10^4
8	32	1.4×10^4
40	66	1.8×10^4

[a] Competent cells of SB 566 (thy^-, try^-) cells were exposed to a H^3-labeled heavy DNA (see text) isolated from a shikimic acid- and tyrosine-requiring mutant (SB 532). The added DNA concentration was 2 μg of 1.5×10^8 cells per milliliter. The specific activity of the DNA was 2.0×10^5 cpm/μg. Thirty milliliters of cells was washed and resuspended in a medium containing thymidine (10 μg/ml) but no trytophan. The DNA was added at 0-minute to these washed cells. Ten milliliters of the cells was withdrawn, treated with DNase, and washed well (9). There was an error of 2 minutes at the first datum. Similarly aliquots were taken at 8 and 40 minutes. The washed cells were treated with lysozyme and RNase (9) and then banded in CsCl. Approximately 47 to 50 fractions were collected in each case. Fifty microliters of each fraction was acid-precipitated, and the rest was used for biological assays to determine donor and recipient activity. The SB 566 cells were thy^-, try_2^-, and tyr_1^+, while the donor was try_2^+. The donor and recipient activities were assayed on a mutant strain that contained both tyr_1^- and try_2^- mutations. The activities given in the table (bottom section) corresponds to the total observed in the native light position. The donor activity represents that amount of try_2^+ activity present in 0.05 ml of each fraction assayed in that region, added to include the observed distribution of the donor–recipient complex. The tyr_1^+ activity is expressed in the same way, except that each assay used 0.02 ml.

3. By 40 minutes, most of the donor DNA was degraded, except the fraction that was complexed with the recipient DNA.

4. In all the time points analyzed, donor biological activity was found only in these recipient DNA regions, and nowhere else.

5. When the donor–recipient DNA complex, formed at 40 minutes, was denatured in alkali (9), neutralized to pH 8.0, and banded in CsCl solution, 60% of the donor counts were released as a completely de-

natured heavy species. This observation suggested that the majority of the donor molecules were noncovalently linked to the recipient DNA. We are currently analyzing the genetic properties of the complex before and after covalent linkage establishment (as judged by alkaline denaturation of the complex taken at different times).

When competent cells are transformed in the usual way, without being starved for amino acids that are required, and the donor–recipient DNA complex is isolated, as early as 10 minutes most of the donor DNA is found to be covalently linked to recipient cell DNA (9). In view of this, the above result (point 5) has encouraged us to examine the possibility that amino acid starvation during transformation interferes with the normal formation of physically recombinant molecules. If recombination in *B. subtilis* transformation occurs in several steps, molecules in intermediate stages of recombination are probably very transient; however, if this process of recombination is inhibited by depriving newly transformed cells of amino acids, such intermediate complexes might accumulate. We are testing at present the usefulness of this technique to detect and isolate these complexes, if they occur.

Since these experiments also indicated that stable recombinants could be obtained during the transformation period in an amino acid-starved culture, we used this technique to study the relationship between recombination in transformation and DNA replication. In particular, we wanted to see whether recombination occurs at any particular region of the chromosome, relative to regions of active DNA synthesis. If newly transformed cells are deprived of amino acids, they will complete one round of DNA replication, and not start another. In this respect *B. subtilis* behaves similarly to *E. coli.* Therefore this approach makes it possible to examine the integration of transforming DNA with respect to DNA synthesis in a single replication cycle.

DNA synthesized during and after transformation can be selectively labeled with 5-bromodeoxyuridine (5BdU), and the newly synthesized "hybrid" molecules (density 1.751 in CsCl) identified by preparative pycnography. The density distribution of the integrated donor atoms or donor biological activity will reflect the distribution of sites of integration with respect to the portion of the recipient cell chromosome which has been duplicated at the time of transformation, and that which will be duplicated in the rest of the cycle. If integration is preferentially in the already replicated region, donor material should be found primarily associated with molecules of normal light density (1.703). If it occurs in the nonreplicated or replicating regions, donor material should be found associated with molecules of "hybrid" density or intermediate between light and hybrid density regions (8).

A competent culture of *B. subtilis,* requiring thymine and two amino acids, tryptophan and phenylalanine, was transferred to a medium containing 5BdU, instead of thymine, and lacking both tryptophan and phenylalanine. The medium also contained very low amounts of H^3-labeled DNA isolated from a thymine$^-$, tryptophan$^+$, phenylalanine$^-$ strain. The low amount of DNA might reduce its contribution to the mononucleotide pool of the recipient cells. The system of genetic markers was chosen to minimize reversion during the course of experiment. The transformed cells, after 10 minutes, were treated with DNase and washed and resuspended in the same medium. (The cells were washed with medium containing the base analog in the same concentration as the growth medium to avoid diluting the intracellular pool of 5BdU. The DNase solution also contained the base analog.)

The culture was incubated at 37°C, and samples were taken at various times up to 300 minutes. The cells were lysed, and the lysates were banded in CsCl.

The distribution of recipient cell DNA between light and hybrid densities showed that by 300 minutes most of the cells had finished one cycle of replication and had not started another. At the last time point, donor atoms appeared to be equally distributed between material of light and hybrid density. Donor biological activity (try$^+$) had a similar pycnographic profile. In separate experiments of this kind we were able to measure the density distributions of genetically recombinant molecules, which were distributed in a similar way. The results are consistent with the hypothesis that integration is at random with respect to regions of DNA synthesis. They do not support the model that integration preferentially occurs at replicating points (8). Since very low amounts of DNA were used in these experiments, accordingly the number of recombinants scored was low, compared to the recipient activity. It is possible that the concentration of DNA added might have some influence on the replication of recipient genome.

In the experiments in which cells were transformed with a light (density 1.703) DNA and simultaneously starved for the required amino acids, we observed that recipient and donor DNA, instead of being distributed in a unimodal peak of density 1.703, was often distributed in two peaks. One was of normal density, and the other at a lighter density between 1.695 and 1.698. In one experiment, we treated the material from the lighter density peak with phenol and banded in CsCl. It now appeared to have normal light (1.703) buoyant density. We assume this material to be a DNA protein complex, which survives our partial purification (this includes treatment with pronase at 50 μg/ml for 15 minutes at 37°C) and does not dissociate during 68 hours of centrifugation in CsCl. This complex might be similar to that reported earlier by Rolfe

(43). We also observed that, if amino acids were added to the deprived, transformed culture, the lighter species gradually disappeared to form a unimodal normal density peak.

Samples from the normal and lighter-than-normal density peaks were examined under an electron microscope (20), using the earlier described monolayer method. The two electron micrographs (Fig. 9) are strikingly different. The DNA in the sample taken from the 1.695 density stratum is shown in unusual configuration and is associated with non-DNA structures, which are completely absent in the sample from normal density strata.

We are currently analyzing the structure of the complex and examining its possible relation to replication and recombination.

Discussion

Biochemical knowledge of the chemistry of the bacterial membrane and its association with DNA, polysomes, and various enzymes is rudimentary at present. The mechanism of DNA replication *in vivo* is still unknown, although significant contributions to the problems have been made by several groups of investigators. It has been shown that the bacterial genome duplicates sequentially and semiconservatively as one coordinated unit (12, 37, 50). There appears to be a strict requirement for protein synthesis before each round of replication (32, 53).

It has been suggested that DNA synthesis occurs in a particulate complex at a specific site in the cell (25). Our electron micrographs of fractions rich in newly synthesized DNA showed the association of DNA molecules with membrane components. In thin sections of bacterial cells, the nucleus appears to be attached to the cell wall membrane, in association with large membranous structures called mesosomes (16, 24, 26). These mesosomes have been implicated in cellular division and transport of the mitotic products to the daughter cells, the genome presumably being attached to this organelle (25).

Biochemical experiments with bacteria strongly suggest the participation of a particulate fraction, presumably the cell wall membrane in DNA replication. The association of polymerase with the fast-sedimenting fraction in sucrose gradients may yield information regarding the *in vivo* mechanism of DNA synthesis. We are beginning to isolate the components of these complexes. At present it is not possible to reconstitute the complex. Our lack of understanding of the nature of the membrane complex makes it difficult to obtain reproducibly active preparations from one experiment to another. Since dilution of the preparations results in poor kinetics in some experiments, it is premature to draw any conclusions from the kinetic information obtained.

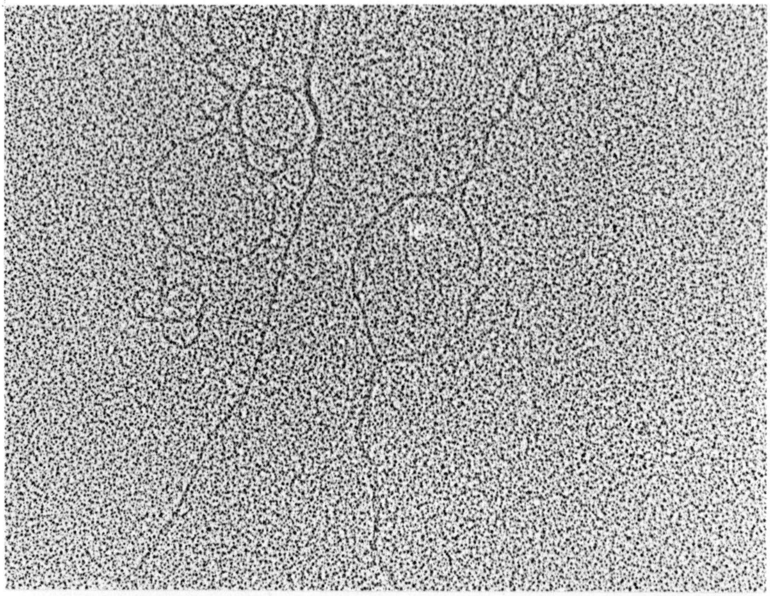

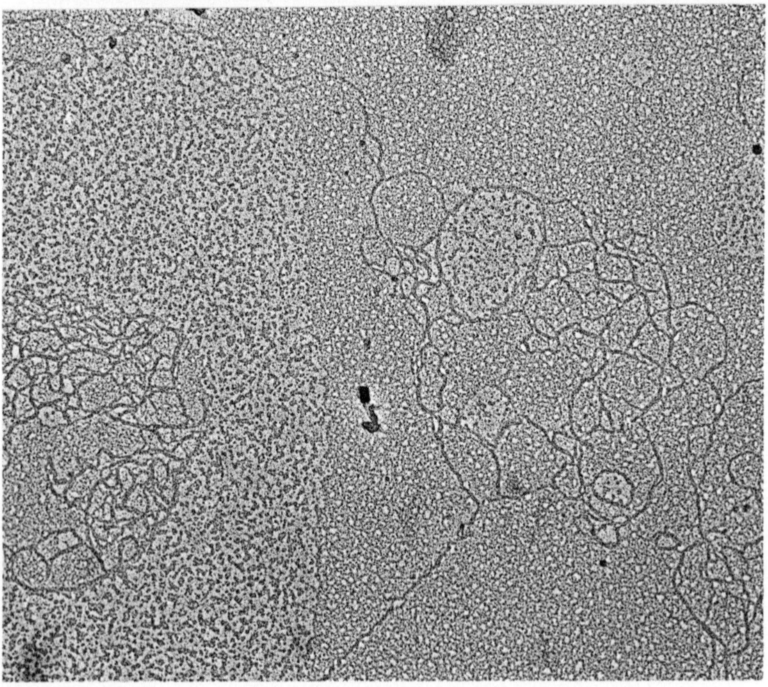

In vivo DNA replicates at a much faster rate—adding 3000 nucleotide units per second (12)—than it is synthesized *in vitro*. The product obtained in the *in vitro* system, using *B. subtilis* DNA as a primer, was biologically inactive (42). Thermal denaturation of these molecules is spontaneously reversible. They appear to have a highly convoluted structure. This might reflect interruptions and a lack of control in coherent linear replication.

The abnormal behavior of the *in vitro* system has raised some question as to whether the DNA polymerase now studied is the crucial enzyme of genetic replication; however, its association with an organized structure may provide the essential basis of continued orderly replication. The characterization of the product made by the membrane-bound polymerase may give a clue to the mechanism of replication. We are presently analyzing the physical and biological properties of this product.

The question recently discussed (26)—whether the DNA molecule moves through a fixed site during replication, or whether the enzyme machinery moves along the DNA molecule—has not been resolved by our experiments. In *E. coli*, the model of conjugational replication (25) proposed that the DNA molecule of the male cell revolves through a complex of enzymes, which is bound to the site of zygotic contact. Conjugation experiments with temperature-sensitive mutants, defective in DNA synthesis (29), show that inhibition of DNA synthesis also results in the inhibition of DNA transfer from male to female cells. Autoradiographic experiments (13) suggest that the male chromosome of *E. coli*, transferred during conjugation, is composed of both conserved and newly synthesized DNA, and that the new DNA is synthesized during the act of conjugation. Thus, there is some evidence for the involvement of replication with conjugation in *E. coli*.

Recently Bodmer (8) proposed a model in which recombination and replication in *B. subtilis* occur in the same region of the DNA molecule. If this is so, it should be possible to isolate genetically recombinant molecules along with nascent DNA, from transformed cells. Experiments discussed earlier here described a zone sedimentation method to isolate the nascent DNA, but this proved to be a technically difficult approach. Instead, the technique of amino acid starvation and density labeling

---

FIG. 9. Two photographs taken from different regions of the CsCl gradient. The top photograph shows the DNA molecules (presumably aggregated) from a region that corresponds to the density strata of normal *Bacillus subtilis* DNA (see the text). The bottom one shows the molecules found in the region of light strata (1.695). The background is different from that of the lysozyme which was used. This is presumably contributed by the protein already bound with the DNA molecules to shift the donor–recipient complex to a lighter position.

following transformation was used to locate the site of integration of transforming DNA with respect to replicating regions of the chromosomes of competent cells. The present evidence suggests that recombination can occur at parts of the genome other than the replicating points (18). Careful density transition experiments in which replicating points were labeled before transformation also strongly suggest no preference for normal replicating regions as the site of recombination (30). Both Bodmer (8) and Laird (30) used density labeling to characterize the fate of donor–recipient complex after transformation with a light DNA. In the first case (8), in which 5BdU was used, donor activity was found at a density intermediate between the light and Bu (5-bromouracil) hybrid positions before transition to a hybrid density. However, in this system mixed utilization of BdU and thymidine from the degraded donor DNA may have occurred. In the second case (30), the recipient cells were grown in light medium (N^{14}, H^1) prior to transformation with a light DNA. The subsequent growth was followed in a heavy medium (N^{15}, H^2) after transformation. The donor–recipient complex, before complete transition into a hybrid strata, occupied an intermediate density position (between light native and hybrid DNA). This intermediate density peak, when sheared, led to the detection of only one kind of molecule, namely hybrid. In our experiments, the DNA samples were always mildly sheared before centrifugation. This was found to give reproducible patterns. This might be a reason why the donor activity was found in the hybrid position, even very early. The other reason may be due to the physiological alteration of the competent cell, which in our case is in a state of amino acid deprivation.

The following is a summary of the results from the experiments in which amino acid starvation or thymine deprivation was used to control DNA synthesis during transformation.

The donor DNA enters the cell, and most of it is degraded within a period of 40 to 60 minutes. The degraded pool of nucleotides may be reutilized by the cells for synthesis of DNA.

When competent thymine-requiring cells were allowed to undergo thymineless death and were tested for the ability to take up DNA and to be transformed, as a function of time, we found that the cells took up DNA, but did not form transformants. The ability to recombine was lost at a much faster rate than was the viability of the population. This implies that deprivation of thymidine results in the immediate loss of the ability to integrate the donor DNA in those cells which have taken it up.

When DNA was added to the competent cells and then thymine was removed, we found that the transformed cells were relatively immune to thymineless death (8). We also found that competent cells are immune to thymineless death only after taking up DNA and not before. We explain this by proposing that cells which have taken up DNA are able to use the thymidine from degraded molecules to sustain a slow rate of replication or switch to new replicating points. Even after the uptake of DNA, a gradual loss of transformants occurred over a period of 2 to 3 hours. We assume this to be because of the eventual exhaustion of the intracellular pool of thymine; in the case of the thymine transformants, if the pool is exhausted before the gene for thymine synthesis is expressed, the cells undergo thymineless death. The loss could be reduced by depriving transformed cells of required amino acids. It is also possible that some of the loss could be due to failure to result in a recombination event. According to the current model of DNA replication this might be because a certain fraction of transformed cells are able to finish a cycle of replication with the thymine available; they cannot begin a new cycle of DNA synthesis and are immune to thymineless death.

If reutilization of the intracellular pool of donor DNA occurs, we have to be cautious in our interpretations of experiments involving the density labeling of the donor DNA molecules. Such experiments require that a very small amount of DNA be added to the cells to minimize the intracellular pool of degraded DNA.

Our experiments indicated that competent cells degrade the donor DNA molecules that are *not* complexed with recipient DNA (Table VIII). We also found that a large proportion of the donor DNA present inside the cell at early times after transformation lacked biological activity. Similar observations were made by Bodmer (personal communication). Since we found the density distribution of heavy donor DNA reisolated from transformed cells to be very broad in all the strata from native light to native heavy positions, we assume that the excess DNA must be somehow modified by the cells before degradation. Such a modification might include single-strand breaks, or stable association with protein (this would explain the shift to lighter densities).

Nester and Stocker (38) and Stocker (49) observed that the biochemical expression of genetic characters takes place 2 to 3 hours after the addition of DNA to the competent cells. We observed (9) that covalent and genetic linkage was established between donor and recipient DNA within a few minutes. Why is there a long lag between the time of integration and expression of a gene? This may have to do with the transcription process. That is, the transformed cells are unable to transcribe the introduced character in the same replication cycle in which they have undergone recombination. It is possible that the recombination occurs first by complementary pairing with a region that is structurally similar to the regions that are presumed to be involved in transcription—that is, transient and reversibly denatured sections. Consistent with this suggestion is the inhibition of transformation by actinomycin D added after the entry of donor DNA into the cell. During this time there was no loss in viability (18). Analysis of the pycnographic profiles of the irreversibly fixed donor DNA during different stages of inhibition of transformation by the antibiotic, or after otherwise altering the metabolic pattern of competent cells during transformation (for example, amino acid deprivation), may help in the study of intermediate stages in recombination. These studies are currently in progress.

Acknowledgments

I am grateful to Professor Joshua Lederberg for his suggestions and encouragement. I am also indebted to Mr. Willard Spiegelman for his comments and critical reading of the manuscript. The discussions and comments of Dr. W. F. Bodmer and the expert assistance of Mrs. Susan Grether and Miss Nancy Buckman are gratefully acknowledged.

Part of the electron microscope studies described here were previously reported by Dr. A. Kleinschmidt.

This work was supported by National Institutes of Health Training Grant 5TI GM-295-08 and Grant G. M.-14108-01.

References

1. Bach, M. K., *Federation Proc.*, **22**, 645 (1963).
2. Barner, H. D., and Cohen, S. S., *J. Bacteriol.*, **72**, 115 (1956).
3. Barr, G. C., and Butler, J. A. V., *Biochem. J.*, **88**, 252 (1963).
4. Beljanski, M., and Ochoa, S., *Proc. Natl. Acad. Sci. U.S.*, **44**, 494 (1958).
5. Ben-Porat, T., Steere, A., and Kaplan, A., *Biochim. Biophys. Acta*, **61**, 150 (1962).
6. Berns, K. I., and Thomas, C. A., *J. Mol. Biol.*, **11**, 476 (1965).
7. Billen, D., *Biochem. Biophys. Res. Commun.*, 3, 179 (1962).
8. Bodmer, W. F., *J. Mol. Biol.*, **14**, 534 (1965).
8a. Bodmer, W. F., *J. Gen. Physiol.*, **49**, 233 (1966).
9. Bodmer, W. F., and Ganesan, A. T., *Genetics*, **50**, 717 (1964).
10. Butler, J. A. V., and Godson, G. N., *Biochem. J.*, **88**, 176 (1963).
11. Byrne, R., Levin, J. G., Bladen, H. A., and Nirenberg, M. W., *Proc. Natl. Acad. Sci., U.S.*, **52**, 140 (1964).
12. Cairns, J., *Cold Spring Harbor Symp. Quant. Biol.*, **28**, 43 (1963).
13. Caro, L. G., and Gross, J. D., cited in ref. 25.
14. Cronenwett, C. S., and Wagner, R. P., *Proc. Natl. Acad. Sci. U.S.*, **54**, 1643 (1965).
15. Dennis, E. S., and Wake, R. G., *J. Mol. Biol.*, **15**, 435 (1966).
16. Fuhs, G. W., *Bacteriol. Rev.*, **29**, 277 (1965) (see also the discussion by Fitz-James).
17. Ganesan, A. T., Ph.D. Thesis, Stanford University, 1963.
18. Ganesan, A. T., in preparation.
19. Ganesan, A. T., and Lederberg, J., *Biochem. Biophys. Res. Commun.*, **18**, 824 (1965).

20. GANESAN, A. T., KLEINSCHMIDT, A. K., AND HELLMAN, W., in preparation.
21. GOLDSTEIN, A., AND BROWN, B. J., *Biochim. Biophys. Acta,* **53,** 19 (1961).
22. HANAWALT, P. C., AND RAY, D. S., *Proc. Natl. Acad. Sci. U.S.,* **52,** 125 (1964).
23. HENDLER, R. W., AND TANI, J., *Biochim. Biophys. Acta,* **80,** 294 (1964).
24. VAN ITERSON, W., *Bacteriol. Rev.,* **29,** 299 (1965).
25. JACOB, F., BRENNER, S., AND CUZIN, F., *Cold Spring Harbor Symp. Quant. Biol.,* **28,** 329 (1963).
26. JACOB, F., RYTER, A., AND CUZIN, F., *Proc. Roy. Soc.,* **B164,** 267 (1965).
27. KADOYA, M., MITSUI, H., TAKAGI, Y., OTAKA, E., SUZUKI, H., AND OSAWA, S., *Biochim. Biophys. Acta,* **91,** 36 (1964).
28. KLEINSCHMIDT, A. K., LANG, D., AND ZAHN, R. K., *Z. Naturforsch.,* **16b,** 730 (1961).
29. KOHIYAMA, M., LAMFROM, H., BRENNER, S., AND JACOB, F., *Compt. Rend.,* **257,** 1979 (1963).
30. LAIRD, C., Ph.D. Thesis, Stanford University, 1966.
31. LARK, K. G., REPKO, T., AND HOFFMAN, E. J., *Biochim. Biophys. Acta,* **76,** 9 (1963).
32. LARK, C., AND LARK, K. G., *J. Mol. Biol.,* **10,** 120 (1964).
33. LITTLEFIELD, J. W., McGOVERN, A. P., AND MORGENSON, K. B., *Proc. Natl. Acad. Sci. U.S.,* **49,** 102 (1963).
34. MAALØE, O., *Cold Spring Harbor Symp. Quant. Biol.,* **26,** 45 (1961).
35. MACHATTIE, L. A., BERNS, K. I., AND THOMAS, C. A., *J. Mol. Biol.,* **11,** 648 (1965).
36. MASSIE, H. R., AND ZIMM, B. H., *Proc. Natl. Acad. Sci. U.S.,* **54,** 1636 (1965).
37. MESELSON, M., AND STAHL, F. W., *Proc. Natl. Acad. Sci. U.S.,* **44,** 671 (1958).
38. NESTER, E. W., AND STOCKER, B. A. D., *J. Bacteriol.,* **86,** 785 (1963).
39. NESTER, E. W., GANESAN, A. T., AND LEDERBERG, J., *Proc. Natl. Acad. Sci. U.S.,* **49,** 61 (1963).
40. NISMAN, B., AND FUKUHARA, H., *Compt. Rend.,* **249,** 2240 (1959).
41. OKAZAKI, T., AND KORNBERG, A., *J. Biol. Chem.,* **239,** 259 (1964).
42. RICHARDSON, C. C., SCHILDKRAUT, C. L., APOSHIAN, H. V., AND KORNBERG, A., *J. Biol. Chem.,* **239,** 222 (1964).
43. ROLFE, R., *Proc. Natl. Acad. Sci. U.S.,* **49,** 386 (1963).
44. SCHLESSINGER, D., *J. Mol. Biol.,* **7,** 569 (1963).
45. SIEKEVITZ, P., AND PALADE, G. E., *J. Biophys. Biochem. Cytol.,* **7,** 619 (1960).
46. SMITH, D., AND HANAWALT, P. C., *Biophys. Soc. Abstr.* (1965).
47. SPIEGELMAN, S., *in* "Recent Progress in Microbiology" (G. Tunevall, ed.), p. 81, Almqvist & Wiksell, Uppsala, 1959.
48. SPIEGELMAN, S., ARONSON, A. I., AND FITZ-JAMES, P. C., *J. Bacteriol.,* **75,** 102 (1958).
49. STOCKER, B. A. D., *J. Bacteriol.,* **86,** 797 (1963).
50. SUEOKA, N., *in* "Cell Synchrony," p. 38. Academic Press, New York, 1966.
51. SUIT, J. C., *Biochim. Biophys. Acta,* **72,** 488 (1963).
52. SUIT, J. C., *J. Bacteriol.,* **84,** 1061 (1962).
53. YOSHIKAWA, H., *Proc. Natl. Acad. Sci. U.S.,* **53,** 1476 (1965).
54. YUDKIN, M. D., AND DAVIS, B., *J. Mol. Biol.,* **12,** 193 (1965).

The Bacterial Chromosome as a Unit of Structure and Replication

CEDRIC DAVERN, JOHN CAIRNS, PAULA DE LUCIA, AND ANN GUNSALUS

Cold Spring Harbor Laboratory of Quantitative Biology, Cold Spring Harbor, New York

GENOME, CHROMOSOME AND NUCLEUS

A physical parallel of the circular map (33) for the linkage group comprising the *Escherichia coli* genome has been revealed in autoradiographs of H^3-labeled DNA released from lysozyme-lysed cells (7, 9, 10). These autoradiographs show that the unreplicated *E. coli* chromosome contains about 2.8×10^9 daltons of DNA. This amount of DNA, when corrected where necessary for continuous duplication of the chromosome, agrees with estimates of the DNA content of the bacterial nucleus (19, 26). Since, under certain conditions of culture, viable cells have only one nucleus (38), there must be only one chromosome per genome, which agrees with the genetic evidence of one linkage group per genome.

It is conceivable that the unreplicated chromosome may contain more than one copy of the genome. Indeed, an observation compatible with such a notion has been made by Fulton (20), who found that a second round of marker transfer begins, under conditions of prolonged conjugation, with the conservation of genetic linkage across the Hfr site. While there have been a number of direct attempts to estimate the DNA content of a single copy of the genome, and other data are available from which such an estimation might be made, the uncertainties and errors attending these estimates are such that a definitive answer to this question cannot yet be given. Nevertheless, in the following review, the nature of some of these difficulties is discussed.

The first estimate of the DNA equivalent of the genome was derived from measuring the efficiency with which P^{32} decay in DNA interrupts marker transfer in bacterial mating (31). The direct correlation of this sensitivity of a marker to P^{32} decay and its distance (as measured by its time of transfer) from the origin of transfer suggested that these data may be used for estimating the number of phosphorus atoms between various markers. Assuming an inactivation efficiency equivalent to that observed for DNA P^{32} decay killing of the bacteria, this calculation led to a value about three times as high as that which we now know to be the maximum mass of the chromosome. The uncertainty in the calculation lies in the assumption that P^{32} decay has equal efficiency for killing and for interruption of marker transfer. Two estimates of the proportion of label transferred to the female from males whose DNA was labeled before mating (22) were made by Silver (56), one for F′lac transfer and the other for normal conjugation with an Hfr. In each case this proportion agreed with that expected from the proportion of the genome calculated to have been transferred, assuming one genome per chromosome.

Because of his uncertainty about some of the factors entering into these calculations, Silver felt the agreement to be largely fortuitous. Indeed, if corrected for the recent findings of Gross and Caro, that only one of the pre-existing strands of the mating chromosome is transferred during conjugation (24; see also 13), the amount of label transferred in Silver's experiment would correspond to twice that calculated from the proportion of the genome transferred.

Gross and Caro (24) computed the rate of DNA transfer during conjugation by a quantitative autoradiographic study of UV-killed female cells which had been mated with H^3-thymidine-labeled males in the presence of H^3-thymidine. The rates observed varied, with one exception, from 5 to 8 microns/min. If we assume 89 minutes to be the minimum time for total genome transfer, these transfer rates correspond to a length of DNA ranging from 450 to 700 microns for the total genome. Although this amounts to only about half the length of the chromosome, these data are not precise enough for drawing such a specific quantitative conclusion, and as such the result is not sufficient support for the unorthodox view that there are two genomes per chromosome.

Another approach to this question is provided by quantitative annealing experiments which measure the proportion of the organism's total DNA that is homologous to a specific RNA or DNA molecule of known molecular weight.

Yankofsky and Spiegelman (59) showed that 23 S *E. coli* ribosomal RNA (MW 1.2×10^6) was homologous to about 0.2% of the *E. coli* DNA, indicating one 23 S homology site per 6×10^8 daltons of DNA. Similar results were obtained by Attardi *et al.* (2). Without knowledge of the number of such sites per genome, this information cannot be used to assess the DNA equivalent of the genome.

The quantitative annealing experiments made with lambda DNA against DNA derived from lysogenic and nonlysogenic strains of *E. coli* (12) gave a value of 5×10^9 daltons for the amount of *E. coli* DNA per molecule of lambda integrated into the chromosome. This value would be somewhat too high even if the integrated lambda were always the last region of the chromosome to be duplicated. From this result and that of Gross and Caro, it is not clear whether the chromosome contains one or two genomes.

Molecular Basis for the Physical Continuity of the Chromosome

Electron micrographs of lysed cells of *Micrococcus lysodeikticus* (34) and *Hemophilus influenzae* (44), whose contents have been spread out in a protein monolayer, show continuous filaments of uniform diameter. For *Hemophilus*, the contour length of such filaments corresponds to the DNA content of the cell. These observations, together with those made on the autoradiographs of *E. coli* DNA (9, 10), while suggesting that the bacterial chromosome may contain a single DNA molecule running its entire length, do not rule out the possibility of a sequence of shorter DNA molecules joined end to end by protein linkers (17). However, when Berns and Thomas (4) isolated DNA from *Hemophilus influenzae* in the presence of the proteolytic enzyme pronase, using methods specifically designed to minimize shear, they found the sedimentation and buoyant density properties of this DNA to be incompatible with the presence of protein linkers because these properties correspond to those expected for pure DNA with a molecular weight of 4×10^8 daltons, or half the estimated DNA content of the nucleus. In contrast, the DNA they isolated from *E. coli* by the same method had a molecular weight of no more than 1.3×10^8 daltons. Massie and Zimm (45) obtained qualitatively similar results with minimal shear isolation in the presence of pronase for the DNA of *Bacillus subtilis* and *E. coli*. They concluded that the bacterial chromosome consists of some eight to ten subunits joined together by pronase-susceptible protein linkers. Although minimizing shear, the methods used above were not free of it. Thus, the size of the isolated DNA molecule may have been more a function of shear fragmentation than the consequence of dissolution of a larger structure by pronase attack.

To avoid shear, a method for both extracting and measuring the size of the isolated DNA in the same system was devised (14). The method involved release of the DNA from *E. coli* spheroplasts by lysis directly on top of a CsCl solution in a centrifuge tube. Lysis was achieved by the combined action of pronase and sodium dodecyl sarcosinate (Sarkosyl N.L.-97, Geigy Chemicals). The size of the isolated DNA was then determined indirectly by density gradient analysis (47) of DNA released from cells that had incorporated the density label, 5-bromouracil (BU), into their DNA for about one-third of a replication cycle. If the chromosome could be isolated in one piece, then the buoyant density of such a molecule should be intermediate between that of normal and BU hybrid DNA. In the experiment, apparently only a small proportion of the DNA was released from the lysed spheroplasts. Nevertheless, this DNA banded in a unimodal distribution at the intermediate density, indicating the isolation of the entire DNA content of the replicating entity in one piece. Deliberate shearing of the lysate prior to centrifugation led to quantitative recovery of the DNA in the classical density distribution expected after a density transfer experiment (46)—namely, into bands corresponding to normal and BU hybrid DNA. Since gentle shear alone will free the spheroplast-associated DNA from its complex, this DNA, to be accessible to shear, must have been extended well beyond the confines of whatever remained of the spheroplast after lysis. This observation, together with the one on the unfragmented nature of the released DNA, shows that the spheroplast-associated DNA derives from incompletely released chromosomes, whose DNA can exist in no more pieces than there are points of association with the membrane. If, as is likely, there is only one point of association per chromosome (52–54), then this observation confirms the previous one and so leads to the conclusion that the *E. coli* chromosome does not consist of a sequence of subunits joined by pronase-susceptible linkers.

Since the replicating form of the circular chromosome (Fig. 1) could sustain as many as two complete breaks without falling apart into two fragments, the above results cannot altogether rule out the existence of a protein discontinuity in the DNA molecule.

Replication of the Bacterial Chromosome

Under normal conditions of culture, DNA replication in the *E. coli* chromosome proceeds throughout most of the cell cycle (38, 40, 55) in a semiconservative and sequential fashion (46), usually by means of a single growing point (8, 10, 43, 49). Furthermore, each replication cycle seems to be initiated from the same site on the chromosome (41). The selective association of nascent DNA with the membrane fraction after incomplete extraction of the DNA (21, 23, 25) led to the suggestion that the growing point may maintain some fixed association with the region of the cell membrane. Indeed, a single point of connection has been observed in electron micrographs of serial sections of *B. subtilis* (52, 53) and *E. coli* (54), wherein each nucleus is seen to be connected to the cell membrane through an invagination of membrane called a mesosome. Whether this association is at the growing point is yet to be determined.

The manifestations of regulation and order for *in vivo* DNA replication are absent from *in vitro* DNA synthesis systems. It is not known whether the degeneracy of *in vitro* systems is due to damage of the DNA polymerase (6) during isolation, or to some abnormal state of the

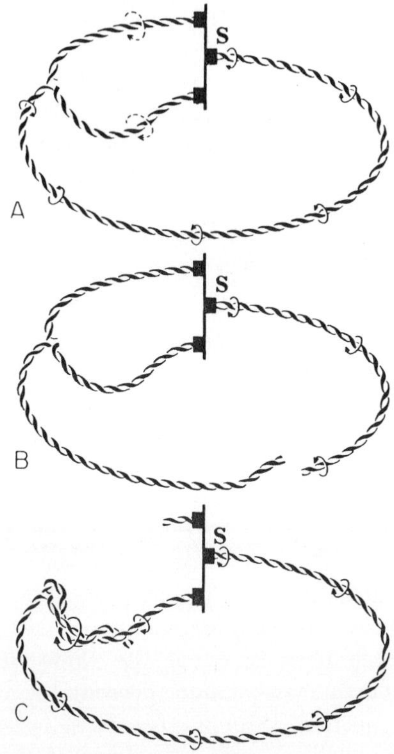

FIG. 1. (*A*) Our working hypothesis supposes that at the end of each bacterial chromosome there is a device, here marked S, that spins the parental duplex about its axis so that the two parental polynucleotide chains are continually being unwound at the replicating fork before acting as templates. Such an unwinding action should cease following a break either (*B*) in front of or (*C*) behind the replicating fork, and DNA synthesis should stop forthwith. From Cairns and Davern (11), with permission from Academic Press.

DNA template *in vitro,* or to lack of some ancillary mechanism. Little more can be said at this stage other than to catalog the deficiencies of the *in vitro* system. The specific activity of a single polymerase molecule, when compared with that yielded by a single cell, indicates that each cell must contain at least 300 such molecules (3). Since there are so few sites of replication *in vivo,* it is therefore possible that the *in vitro* polymerase is a subunit of an unstable polymerous complex existing *in vivo.* Indeed, the *in vitro* polymerase has a turnover number of 50 (42), which is lower than the *in vivo* rate of DNA replication by a factor of about 100 (10). However, if each cell contains about 300 such molecules, these would be sufficient in number for the replication of three chromosomes per cell. The polymerase is limited in its biochemical versatility *in vitro,* in that only one of the expected polymerization reactions has been observed, namely the addition of DNA monomers to the free 3′-OH of the deoxyribose (1, 48, 51). Despite the precision with which the *in vitro* polymerase copies a template (5, 58), the product is branched and displays abnormal denaturation properties (48, 51). Even *in vivo,* fragments of DNA gaining entry into cells by means of transformation or conjugation are unable to replicate in the recipient cells (30). These observations led Jacob and Brenner (29) to postulate that orderly replication is an attribute of a complex particle (the replicon) which behaves as an integral unit in the control of its replication. A scheme for the replicon's control system (30) has been derived from studies with temperature-conditional mutants of DNA replication and conjugation transfer for the bacterial chromosome (35, 36) and episome (30) replicons and for DNA replication mutants in temperate phage replicons (31). The model for this control system so far involves a modulatable cytoplasmic element (initiator), specified by the chromosome, which initiates replication by combination with a recognition site (the replicator) located at the origin of replication of the chromosome. Replicons seem not to be completely autonomous in some situations, because DNA strands synthesized at the same time but in separate replicons can maintain an association, presumably by means of a common membrane connection, through successive cycles (32).

The "Spinner" Hypothesis for Chromosome Replication

It has long been thought that frictional drag may be a formidable problem for the rotation which is believed to accompany the replication of a DNA duplex. Various swivel devices have been proposed (16, 17) by which rotation of the entire molecule could be avoided.

Now that it seems that the *E. coli* chromosome consists of a single DNA molecule in the form of a circle, the rotation accompanying replication presents topological difficulties in addition to the frictional ones which may again be resolved by swivel devices (10). This possibility led to the suggestion that such a swivel could be the seat of an active mechanism that drives the rotation and hence the duplication of the chromosome (11). The simplest version of this "spinner" hypothesis (Fig. 1) supposes the existence of such a machine at the end of the parent limb of the chromosome. Rotation of this limb relative to the two daughter limbs causes a torque to be propagated to the replicating fork where it is converted to unwinding of the parent helix, thus presenting a template suitable for replication. Breakage of the chromosome ahead of the fork will cut the fork off from the source of torque, while breakage of either daughter limb could arrest the conversion of torque to unwinding at the fork. If normal DNA replication is dependent upon some distant unwinding mechanism, then such breakage should stop DNA synthesis. The possibility that an unwinding machine may be one of the elements of the replicon was tested indirectly by using P^{32} decay in the DNA as the agent for effecting two-strand breaks (15, 57) in the molecule. Since DNA P^{32} decay in strain 15 thymine⁻ arginine⁻ uracil⁻ (15 TAU) did not seem to reduce its capacity for DNA synthesis despite the fact that it is killed with an efficiency of 1/50 (19), we surmised that *E. coli* may possess mechanisms that can repair two-strand breaks. So we proceeded to examine the effects of DNA P^{32} decay in the recombinationless (rec⁻) strain 28-152 given to us by Meselson. Using a thymine auxotroph of this rec⁻ strain isolated by Ptashne, we found that DNA P^{32} decay arrested the DNA synthesis capacity of this strain immediately with an efficiency of 1/30. A similar sensitivity of DNA synthesis capacity to P^{32} decay was shown by Hill's radiation-sensitive strains B_{s-1} and B_{s-2} (27), but not by the uvr⁻ strains A, B, and C (28) of Howard-Flanders (Table I).

TABLE I
SURVIVAL AND DNA SYNTHESIS CAPACITY AFTER THE DECAY OF ~90 P^{32} ATOMS PER AVERAGE REPLICATING CHROMOSOME

Strain	Survival (%)	DNA Synthesis Capacity (%)
B	4.8	77
B_{s-1}	0.6	12
B_{s-2}	0.7	2
rec⁻	2.3	12
rec⁺	27.0	100
uvr A	24.0	100
uvr B	27.0	100
uvr C	24.0	100

Pulse-labeling experiments, where P^{32} was incorporated into different parts of the chromosome relative to the replicating fork, showed that P^{32} decay occurring anywhere in the chromosome was uniformly effective in arresting DNA synthesis, thus fulfilling the prediction of the hypothesis.

There is no direct evidence that the effective P^{32} decay lesion here is a two-strand break. Even to entertain this possibility, we must reconcile our observations with those of others. First, Fuerst and Stent (19) attributed the killing of radiation-resistant strains of *E. coli* to two-strand breaks arising from DNA P^{32} decay with an efficiency of 1/50. If the primary lesions (two-strand breaks?) responsible for this killing are the same as those that both kill and arrest DNA synthesis with the higher efficiency of $\frac{1}{30}$ in the rec⁻ strain, then we must postulate that, whereas the physical integrity of the broken chromosome is restored by repair mechanisms in radiation-resistant strains, their genetic integrity is not always also restored. Second, if the interruption of marker transfer in mating following DNA P^{32} decay in radiation-resistant males (18) is to be attributed to two-strand breaks, then we must suppose that a mating male chromosome is not accessible to repair mechanisms, because the efficiency of P^{32} decay in interrupting marker transfer is three times that of its efficiency of killing in the same strain (37).

Whatever the nature of the DNA P^{32} decay lesion that arrests DNA synthesis in a certain class of radiation-sensitive *E. coli* strains, we can at least conclude that DNA replication can be arrested immediately by some decay event in the chromosome distant from the replicating fork. Although this sensitivity of DNA replication to distant events is compatible with the spinner hypothesis, it by no means excludes other models which could account for such sensitivity.

DNA Breakdown as a Consequence of DNA P^{32} Decay

Not only is DNA synthesis arrested in P^{32}-decay-killed rec⁻ cells, but their DNA is broken down during incubation in culture medium (11). Such P^{32}-decay-induced breakdown has now been observed in the radiation-sensitive strains B_{s-1} and B_{s-2}, but not in the radiation-resistant strain 15 TAU.

The extent of DNA breakdown in a rec⁻ culture is correlated with the proportion of P^{32}-killed cells it contains (Fig. 2), suggesting that the DNA P^{32} lesion responsible for inactivating replication, and hence lethality, was also responsible for rendering the chromosome vulnerable to nuclease attack.

The kinetics of this breakdown was approximately exponential with a half-time of 30 to 40 minutes, irrespective of whether the breakdown was followed in P^{32} pulse-labeled or uniformly labeled chromosomes. These kinetics could either be a reflection of the kinetics of breakdown for individual P^{32}-damaged chromosomes or the consequence of some kind of asynchronous initiation of a rapid breakdown process among the damaged chromosomes.

If the damaged chromosomes break down synchronously and slowly, then it would be possible to inquire into the pattern of breakdown. Since the breakdown kinetics of pulse and uniformly labeled DNA are similar, and independent of the number of hits sustained by the chromosomes (Fig. 3) it is unlikely that the site of the effective P^{32} decay is a preferred site from which a slow synchronous breakdown proceeds. Nor

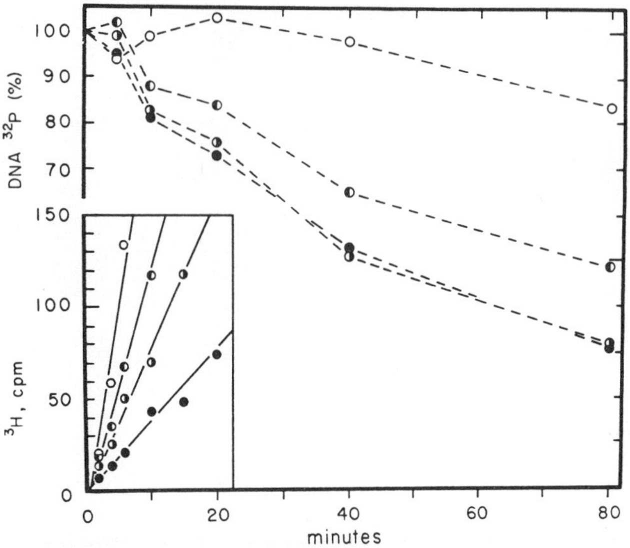

FIG. 2. The loss of TCA-insoluble DNA P^{32}, and, inset, the loss of ability to incorporate H^3-thymidine by *E. coli* K12 rec⁻ that had been labeled with P^{32} (16 mC/mg) for 4 hours and stored until 0% (○), 4% (◐), 9% (◑), and 18% (●) of the P^{32} atoms had decayed. From Cairns and Davern (11), with permission from Academic Press.

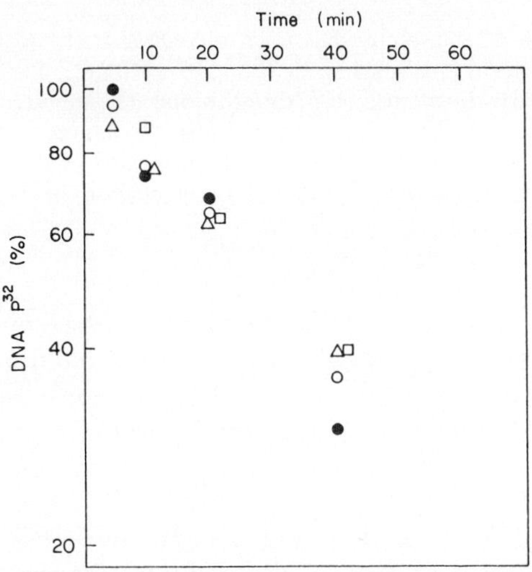

FIG. 3. DNA breakdown after various amounts of P^{32} decay in *E. coli* K12 rec⁻ thy⁻ cells uniformly labeled with P^{32} (16 mC/mg). The surviving fraction of DNA was calculated after subtraction of that portion of the DNA attributable to surviving cells. Breakdown measured after periods of storage allowing for an average of 1 (●), 2 (○), 3 (△) and 7 (□) lethal P^{32} decays per cell.

does it seem likely that such breakdown could proceed from the replicating fork, because similar kinetics were observed for the breakdown of H^3-DNA and P^{32}-DNA for chromosomes whose DNA was labeled with P^{32} and pulse-labeled with H^3-thymidine in different parts of the chromosome relative to the fork (Fig. 4).

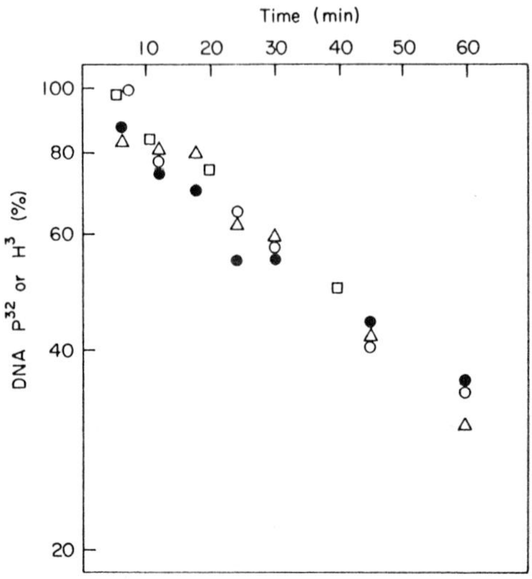

FIG. 4. Loss of TCA-insoluble DNA P^{32} and DNA H^3 by *E. coli* K12 rec⁻ thy⁻ that had been uniformly labeled with P^{32} (16 mC/mg) for 3.75 hours, then pulse-labeled with H^3-thymidine (6.7 C/mM) for 11 minutes. Washed samples were stored away at —78°C immediately, and after 30-minute and 50-minute chases in the presence of unlabeled thymidine and P^{32}. After 18% of the P^{32} atoms decayed, DNA breakdown was followed during 37°C incubation of these samples after thawing and dilution at least six fold into fresh culture medium. DNA P^{32} (△), DNA H^3: no chase (○), 30-minute chase (●), and 50-minute chase (□).

Similarly, no difference in the kinetics of breakdown of H^3-DNA could be observed among samples of cells that differed in the location of the label with reference to the origin of replication (Table II). In this experiment the H^3 pulse served the dual purpose of marking a region of the chromosome and acting as a nonlocalized source of radiation damage (50).

Thus, if breakdown is slow and synchronous, it must be at least random in its origin (or initiate at many different sites) or random in its progress.

Certainly, the overall rate of breakdown is not determined simply by the rate of action of responsible nucleases, because the progress of breakdown can be arrested at any time by 0.005 *M* dinitrophenol or by

TABLE II

H^3-DECAY-INDUCED BREAKDOWN OF H^3-DNA FOR *Escherichia coli* K12 REC⁻ THY⁻ CELLS PULSE-LABELED WITH H^3-THYMIDINE (5.9 C/mM) AT VARIOUS TIMES AFTER INITIATION OF DNA REPLICATION FOLLOWING TERMINALIZATION OF THE DNA REPLICATION CYCLE BY INCUBATION IN THE PRESENCE OF 50 μG OF CHLORAMPHENICOL PER MILLILITER (39) FOR 2 HOURS FOLLOWED BY 1 HOUR OF INCUBATION IN THE ABSENCE OF THYMIDINE

Poststorage incubation time (min)	DNA surviving (%) as TCA-insoluble H^3[a]		
	0–10-minute pulse	15–25-minute pulse	30–40-minute pulse
0	100	100	100
6	90	92	92
20	86	97	97
30	73	82	84
40	67	76	75
60	68	60	66

[a] Initiation of DNA synthesis after the synchronization treatment was asynchronous—the relative H^3-thymidine incorporation being 1–8–16 for the 0-minute, 15-minute, and 30-minute 10-minute pulses, respectively. The cells were stored for 17 days to allow H^3 decay inactivation.

a combination of 0.01 *M* iodoacetate and 0.01 *M* cyanide. This observation and that of the exponential character of the breakdown kinetics are compatible with a process that, though very rapid, is unleashed at different times in different bacteria—perhaps according to the dictates of some energy-consuming clock.

Whatever the mechanism of breakdown, the integrity of the chromosome seems to be as essential for its stability as it is for its ability to replicate.

ACKNOWLEDGMENT

This work was supported by the National Science Foundation.

REFERENCES

1. ADLER, J., LEHMAN, I. R., BESSMAN, M. J., SIMMS, E. S., AND KORNBERG, A., *Proc. Natl. Acad. Sci. U.S.*, **44**, 641 (1958).
2. ATTARDI, G., HUANG, P., AND KABAT, S., *Proc. Natl. Acad. Sci. U.S.*, **53**, 1490 (1965).
3. BALDWIN, R., *in* "The Bacteria" (I. C. Gunsalus and R. Stannier, eds.), Vol. V, p. 327, Academic Press, New York, 1964.
4. BERNS, K. I., AND THOMAS, C. A., *J. Mol. Biol.*, **11**, 476 (1965).
5. BESSMAN, M. J., LEHMAN, I. R., ADLER, J., ZIMMERMAN, S. B., SIMMS, E. S., AND KORNBERG, A., *Proc. Natl. Acad. Sci. U.S.*, **44**, 633 (1958).
6. BESSMAN, M. J., LEHMAN, I. R., SIMMS, E. S., AND KORNBERG, A., *J. Biol. Chem.*, **233**, 171 (1958).

7. Bleecken, S., Strohbach, G., and Sarfert, E., *Z. Allgem. Mikrobiol.*, **6**, 121 (1966).
8. Bonhoeffer, F., and Gierer, A., *J. Mol. Biol.*, **7**, 534 (1963).
9. Cairns, J., *J. Mol. Biol.*, **6**, 208 (1963).
10. Cairns, J., *Cold Spring Harbor Symp. Quant. Biol.*, **28**, 43 (1963).
11. Cairns, J., and Davern, C. I., *J. Mol. Biol.*, **17**, 418 (1966).
12. Cowie, D. B., and McCarthy, B. J., *Proc. Natl. Acad. Sci. U.S.*, **50**, 537 (1963).
13. Curtiss, R., III, and Charamella, L. J., *Genetics*, **54**, 329 (1966).
14. Davern, C. I., *Proc. Natl. Acad. Sci. U.S.*, **55**, 792 (1966).
15. Davison, P. F., Freifelder, D., Hede, R., and Levinthal, C., *Proc. Natl. Acad. Sci. U.S.*, **47**, 1123 (1961).
16. Delbrück, M., and Stent, G., *in* "The Chemical Basis of Heredity" (W. D. McElroy and B. Glass, eds.), p. 699, Johns Hopkins Press, Baltimore, 1957.
17. Freese, E., *Cold Spring Harbor Symp. Quant. Biol.*, **23**, 13 (1958).
18. Fuerst, C. R., Jacob, F., and Wollman, E. L., *Compt. Rend.*, **243**, 2162 (1956); Jacob, F., and Wollman, E. L., "Sexuality and the Genetics of Bacteria," p. 221, Academic Press, New York, 1961.
19. Fuerst, C. R., and Stent, G., *J. Gen. Physiol.*, **40**, 73 (1956).
20. Fulton, C., *Genetics*, **52**, 55 (1965).
21. Ganesan, A. T., and Lederberg, J., *Biochem. Biophys. Research Commun.*, **18**, 824 (1965).
22. Garen, A., and Skaar, P. D., *Biochim. Biophys. Acta*, **27**, 457 (1958).
23. Goldstein, A., and Brown, B. J., *Biochim. Biophys. Acta*, **53**, 438 (1961).
24. Gross, J. D., and Caro, L., *J. Mol. Biol.*, **16**, 269 (1966).
25. Hanawalt, P. C., and Ray, D. S., *Proc. Natl. Acad. Sci. U.S.*, **52**, 125 (1964).
26. Hershey, A. D., and Melechen, N. E., *Virology*, **3**, 207 (1957).
27. Hill, R., and Simson, E., *J. Gen. Microbiol.*, **24**, 1 (1961).
28. Howard-Flanders, P., Boyce, R. P., and Theriott, L., *Genetics*, **53**, 1137 (1966).
29. Jacob, F., and Brenner, S., *Compt. Rend.*, **256**, 298 (1963).
30. Jacob, F., Brenner, S., and Cuzin, F., *Cold Spring Harbor Symp. Quant. Biol.*, **28**, 329 (1963).
31. Jacob, F., Fuerst, C. R., and Wollman, E. L., *Ann. Inst. Pasteur*, **93**, 724 (1957).
32. Jacob, F., Ryter, A., and Cuzin, F., *Proc. Roy. Soc.*, **B164**, 267 (1966).
33. Jacob, F., and Wollman, E. L., *Symp. Soc. Exptl. Biol.*, **12**, 75 (1958).
34. Kleinschmidt, A., Lang, D., and Zahn, R. K., *Z. Naturforsch.*, **16b**, 730 (1961).
35. Kohiyama, M., Cousin, D., Ryter, A., and Jacob, F., *Ann. Inst. Pasteur*, **110**, 465 (1966).
36. Kohiyama, M., Lamfrom, H., Brenner, S., and Jacob, F., *Compt. Rend.*, **257**, 1979 (1963).
37. Krisch, R., *Genet. Res.*, **6**, 454 (1965).
38. Lark, C., *Biochim. Biophys. Acta*, **119**, 517 (1966).
39. Lark, C., and Lark, K. G., *J. Mol. Biol.*, **10**, 120 (1964).
40. Lark, K. G., and Lark, C., *J. Mol. Biol.*, **13**, 105 (1965).
41. Lark, K. G., Repko, T., and Hoffman, E. J., *Biochim. Biophys. Acta*, **76**, 9 (1963).

42. Lehman, I. R., Bessman, M. J., Simms, E. S., and Kornberg, A., *J. Biol. Chem.*, **233**, 163 (1958).
43. Maaloe, O., *in* "Phage and the Origins of Molecular Biology" (J. Cairns, G. S. Stent, and J. D. Watson, eds.), p. 265, Cold Spring Harbor Laboratory of Quantitative Biology, Cold Spring Harbor, 1966.
44. MacHattie, L., Berns, K. I., and Thomas, C. A., *J. Mol. Biol.*, **11**, 648 (1965).
45. Massie, H. R., and Zimm, B. H., *Proc. Natl. Acad. Sci. U.S.*, **54**, 1636 (1965).
46. Meselson, M., and Stahl, F. W., *Proc. Natl. Acad. Sci. U.S.*, **44**, 671 (1958).
47. Meselson, M., Stahl, F. W., and Vinograd, J., *Proc. Natl. Acad. Sci. U.S.*, **43**, 581 (1957).
48. Mitra, S., and Kornberg, A., *in* Macromolecular Metabolism, *J. Gen. Physiol.*, **49**, No. 6, part 2, p. 59 (1966).
49. Pritchard, R. H., and Lark, K. G., *J. Mol. Biol.*, **9**, 288 (1964).
50. Rachmeler, M., and Pardee, A. B., *Biochim. Biophys. Acta*, **68**, 62 (1963).
51. Richardson, C. C., Schildkraut, C. L., and Kornberg, A., *Cold Spring Harbor Symp. Quant. Biol.*, **28**, 9 (1963).
52. Ryter, A., and Jacob, F., *Compt. Rend.*, **257**, 3060 (1963).
53. Ryter, A., and Jacob, F., *Ann. Inst. Pasteur*, **107**, 384 (1964).
54. Ryter, A., and Jacob, F., *Ann. Inst. Pasteur*, **110**, 801 (1966).
55. Schaechter, M., Bentzon, M. W., and Maaløe, O., *Nature*, **183**, 1207 (1959).
56. Silver, S. D., *J. Mol. Biol.*, **6**, 349 (1963).
57. Stent, G. S., and Fuerst, C. R., *J. Gen. Physiol.*, **38**, 441 (1954).
58. Wake, R. G., and Baldwin, R. L., *J. Mol. Biol.*, **5**, 201 (1962).
59. Yankofsky, S. A., and Spiegelman, S., *Proc. Natl. Acad. Sci. U.S.*, **48**, 1466 (1962).

Chromosome Segregation and the Regulation of DNA Replication[1]

K. G. Lark, Helen Eberle, R. A. Consigli, H. C. Minocha, N. Chai, and C. Lark

Kansas State University, Department of Physics, Manhattan, Kansas

Chromosome replication in plants, animals, and microorganisms proceeds via a semiconservative mechanism of replication (21, 7, 29, 27, 24). In those systems tested this replication occurs sequentially in that all portions of the genetic material of the cell are replicated once, before any portion is replicated twice (21, 27).

In higher organisms, DNA synthesis occupies a restricted portion, S, of the cell division cycle, whereas in bacteria DNA synthesis may occur throughout the division cycle.

However, as we shall see, even in bacteria the initiation of DNA synthesis requires certain synthetic events, which presumably occur during the period between mitosis and "S" in the division cycle of higher organisms. In the experiments to be described below, we have investigated the fixation of the chromosome to the bacterial cell and the possible role of this attachment in the initiation of DNA replication. We have begun to extend these experiments to studies on animal cells.

The Initiation of Chromosome Replication in Bacteria

When auxotrophic bacteria are starved of required amino acids they cease to synthesize protein and RNA and, eventually, DNA (22, 20). The amount of DNA synthesized suggested that such cells were completing a cycle of chromosome replication, but not initiating a new cycle (20). Subsequent experiments using isotopic and density labels confirmed this hypothesis (19, 25). They also demonstrated that in bacteria which contain one replicating chromosome and one about to begin replication the one which is replicating will finish its replication cycle; but the other cannot initiate a new replication cycle (17). These results strongly implicated protein or possibly RNA in the initiation of the

[1] This paper is dedicated to Max Delbrück on his 60th birthday.

replication of individual chromosomes. They are also consistent with the finding that animal cells, similarly starved for amino acids, are all blocked in the G_1 phase of their cell division cycle (23).

Auxotrophic bacteria, which are starved for thymine, will not synthesize DNA, although both RNA and protein synthesis continue. When thymine is restored to these cells DNA synthesis resumes, but at an accelerated pace. It has been demonstrated (25) that this is the result of the premature initiation of a new cycle of chromosome replication (see Fig. 1). A peculiarity of this phenomenon is that the new replication cycle is restricted to only one of the two available sister chromatids (15). Inducing this premature cycle of replication also may be blocked by starvation of essential amino acids, indicating that the induction process requires protein or RNA synthesis (25).

Although amino acid starvation blocks RNA as well as protein synthesis, the block to RNA synthesis may be reversed by simultaneous treatment with high concentrations of chloramphenicol. We have found that this does not permit the initiation of new cycles of DNA replication. Therefore, it may be assumed that it is protein synthesis which is essential to the initiation process.

Finally, treatment of cells with phenethyl alcohol (PEA) also prevents the initiation of a new cycle of chromosome replication, although this treatment allows the completion of replication already underway (30, 18). The effect of PEA is of especial interest, since it does not inhibit net RNA or protein synthesis completely, but allows them to increase for 5 to 6 hours (equivalent to seven or eight generation periods) at a synthetic rate equal to approximately 40% of that observed under normal growth conditions.

TABLE I

PHYSIOLOGICAL EFFECTS OF VARIOUS TREATMENTS AFTER WHICH DNA REPLICATION PROCEEDS FROM A COMMON REGION OF THE CHROMOSOME

1. Amino acid starvation:	Blocks protein synthesis, RNA synthesis, and membrane and wall synthesis.
2. Thymine starvation:	Blocks DNA synthesis, but allows protein, RNA, cell membrane, and cell wall synthesis.
3. Phenethyl alcohol treatment:	Allows protein and RNA synthesis, damages cell membrane (increased fragility of cells).

The effects of these treatments are summarized in Table I and Fig. 1. All these experiments implicate a need for protein synthesis in the initiation of chromosome replication at a unique region of the chromosome.

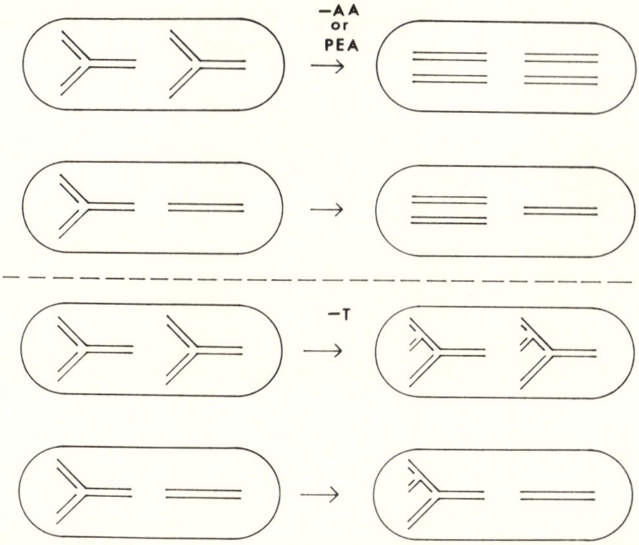

FIG. 1. Schematic representation of the effects on chromosome replication in *Escherichia coli* of amino acid starvation (−AA), phenethyl alcohol treatment (PEA), or thymine starvation (−T). Cells growing in glucose medium—two replicating chromosomes—or in succinate medium—one replicating chromosome—are shown.

The Need for More Than One Type of Protein in the Initiation Process

Jacob *et al.* (9) proposed a model for the initiation of DNA replication whereby two types of protein play a role. The one, a "structural" protein, involves the attachment of the chromosome to the cell membrane; the other, the initiator, acts upon the attached chromosome to initiate replication.

Studies on the initiation process indicate that two such types of protein may be distinguished experimentally on the basis of the susceptibility of initiation to inhibitors.

5-Fluorouracil or low concentrations of chloramphenicol (25 μg/ml) will block the premature initiation of replication if these agents are present during thymine starvation. On the other hand, they permit the reinitiation of replication after amino acid starvation if they are present when amino acids are restored (13).

Various amino acid analogs produce similar effects. Thus, when ethionine, or either 5-methyltryptophan or 7-azatryptophan, is added 80 minutes after the onset of starvation of an *Escherichia coli* auxotroph for methionine or for tryptophan, respectively, chromosome replication is readily reinitiated (C. Lark, personal communication). On the other hand, these analogs will block premature chromosome replication if substituted for methionine or tryptophan during starvation for thymine. These effects (summarized in Table II) imply that initiation of chromosome replication requires the synthesis of two amino acid-containing compounds. The synthesis of one of these is not inhibited by amino acid analogs, low concentrations of chloramphenicol, or 5 fluorouracil. This compound could be involved in the attachment of the chromosome to the cell surface.

TABLE II

EFFECT OF VARIOUS INHIBITORS ON THE REINITIATION OR PREMATURE INITIATION OF CHROMOSOME REPLICATION IN *Escherichia coli* 15T⁻ METH⁻ ARG⁻ TRYPT⁻

Inhibitor	Physiological action	Effect on initiation of chromosome replication
Ethionine 7-Azatryptophan 5-Methyltryptophan	Each inhibits protein and RNA synthesis[a]	Each inhibits the induction of a premature replication cycle if present during thymine starvation. They allow reinitiation of one replication cycle after amino acid starvation and several replication cycles after PEA treatment.[a]
Chloramphenicol (25 μg/ml) (13, 18)	Inhibits protein but not RNA synthesis.	Effects same in all respects to amino acid analogs.
5-Fluorouracil (13, 18)	Interferes with RNA and protein synthesis, but allows some net synthesis. Causes lysis after PEA treatment.	Inhibits the induction of a premature replication cycle if present during thymine starvation. Allows the reinitiation of one replication cycle after amino acid starvation.

[a] These results were obtained in experiments in which each amino acid analog was used *in place* of a required amino acid. Other required amino acids were all present.

Chromosome replication after PEA treatment is also reinitiated in the presence of chloramphenicol or of amino acid analogs (18; C. Lark, personal communication). However, the extent of replication which follows initiation is much greater than after amino acid starvation and is proportional to the period of PEA treatment. It can be reduced by the presence of either chloramphenicol or amino acid analogs during PEA treatment (see Table III). Thus, PEA may block the synthesis or activity of an attachment structure needed for initiation of chromosome replica-

TABLE III

EFFECT OF INHIBITORS ADDED DURING PEA TREATMENT ON THE AMOUNT OF DNA SYNTHESIZED IN CHLORAMPHENICOL AFTER THE REMOVAL OF PEA[a]

Inhibitor present during PEA treatment	Increase in DNA during 200 minutes of incubation after removal of PEA[b]
None	187%
DL-Ethionine (40 μg/ml)	17%
DL-7-Azatryptophan (40 μg/ml)	37%
Chloramphenicol (25 μg/ml) (18)	47%

[a] Cultures of *E. coli* $15T^-$ $meth^-$ arg^- $trypt^-$ were grown in ^3H-thymine (5 μC./2 μg/ml), and 0.25% PEA was added. Incubation was continued for 120 minutes in PEA alone or in PEA plus the inhibitors indicated.

The amino acid analogs were added in place of either methionine or tryptophan. Each of the cultures was then collected, washed on membrane filters, and transferred into medium containing chloramphenicol (25 μg/ml). Radioactivity was measured at the time of transfer out of PEA and after 200 minutes of incubation.

[b] Chloramphenicol (25 μg/ml) present during the post-PEA incubation.

tion but not the synthesis of the second component (initiator). This latter component, the synthesis of which is sensitive to chloramphenicol, can accumulate during PEA treatment and then may be used in subsequent cycles of chromosome replication.

Although the data are insufficient to build a detailed model of the interaction of the two components in initiating chromosome replication, it seems clear that there are at least two such entities, one of which could be involved in attaching the chromosome to a specific site in the bacterium.

THE DEPENDENCE OF CHROMOSOME CONTENT ON BACTERIAL GROWTH RATE

If chromosome replication is limited by attachment to the bacterial cell surface, we can expect that a stoichiometric relationship may exist between the ability of a cell to support chromosome replication and its growth rate, or ability to synthesize protein. That this is the case was previously indicated by the data on *Salmonella typhimurium* of Schaechter *et al.* (26). More recent data on *E. coli* and *Bacillus subtilis* (12, 6) have confirmed and extended this. Measurements of the DNA content at different growth rates have been correlated with autoradiographic measurements of the number of conserved units of DNA (deoxyribonucleotide strands) present per cell (Table IV and Fig. 2). It can be seen that, when the growth rate decreases beyond a certain point, the cell loses the ability to maintain two chromosomes, and the chromosome content drops to one. This is accompanied by a change in the nuclear

TABLE IV

DNA CONTENT OF *Escherichia coli* AND *Bacillus subtilis* AS A FUNCTION OF THE GROWTH RATE[a]

Medium	Generation time (minutes)	DNA content (10^{-14} gm/cell)	Average number of conserved units of DNA per cell
	Escherichia coli		
L-broth	22	2.6	16
Glucose-M9 (12)	45	1.4	8
Succinate-M9 (12, 16, 17)	70	1.1	6–8
Aspartate-M9 (12)	120	0.7	—
Proline-M9 (12)	180	0.5	—
Acetate-M9 (12)	270	0.55	3[b]
	Bacillus subtilis (6)		
Penassay	27	3.3	—
Glucose	80	2.0	7
Aspartate	160	1.8	—
Succinate	300	0.84	—
Acetate	550	0.92	4.5

[a] The DNA content per cell was measured by the diphenylamine reaction. The number of conserved units per cell was determined by autoradiography. A replicating chromosome has four such units.

[b] The number of conserved units of DNA per cell and the DNA content per cell indicate that these cells are making DNA only during the latter part of the division cycle. The DNA content of a single replicating *E. coli* chromosome has been calculated as 0.65×10^{-14} gm (1).

pattern of the cell (Fig. 3), which corresponds to the change in chromosome content.

A series of subtle changes are observed in *E. coli* when the growth rate is reduced. At rapid growth rates—40 minutes' generation time—the cell contains two chromosomes, both of which are replicating at the same time. At somewhat slower rates—70 minutes' generation time—the cell contains two chromosomes, but only one is replicating at any time. At still slower growth rates the cells contain only one chromosome which is replicated throughout the cell cycle, whereas at the slowest growth rates they contain one which does not appear to be replicated until the later half of the cell cycle (Fig. 2). These slowly growing cells are at least superficially similar to cells of higher organisms in restricting chromosome replication to a portion of the cell division cycle.

It is important that in the *E. coli* system it is not the chromosome content as such which is regulated by the growth rate, but rather the

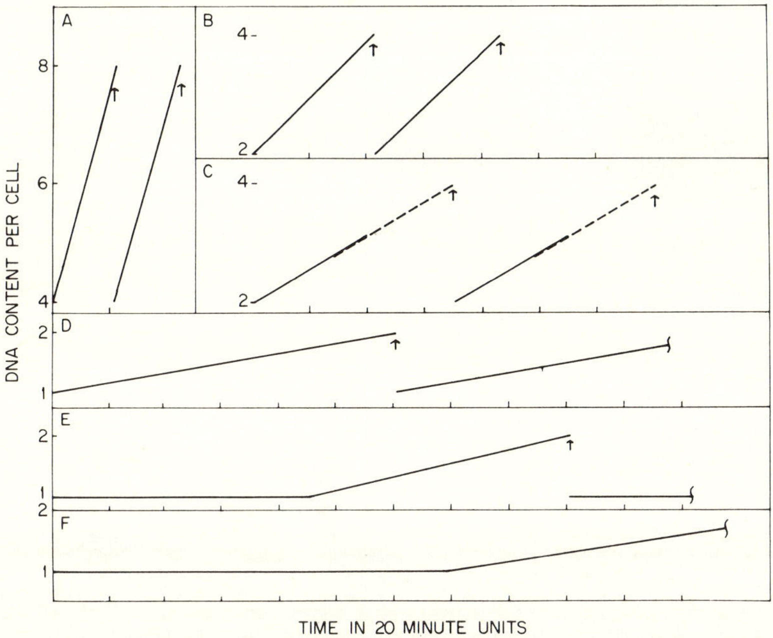

FIG. 2. A schematic representation of the dependence of chromosome replication in *Escherichia coli* on growth rate (12). Growth in (*A*) L-broth, (*B*) glucose-M9, (*C*) succinate-M9: The alternate replication of the two chromosomes is shown by use of dashed and solid lines. (*D*) aspartate-M9, (*E*) proline-M9, (*F*) acetate-M9: The time of cell division as inferred from a study of autoradiographs of pulse-labeled cells is indicated by an arrow.

ability of cells to initiate new cycles of chromosome replication relative to cell division. The data suggest that the growth rate imposes a stoichiometric restriction on chromosome replication.

Thus, the chromosome content of cells at a given growth rate is the result of a competition between the rate of initiation of DNA replication and cell division. It is not remarkable that cytological examination (Fig. 3) reveals a few cells in rapidly growing cultures with the nuclear pattern of slow growth and a few in slowly growing cultures with the nuclear pattern of fast growth.

SEGREGATION OF CHROMOSOMES IN BACTERIA

Genetic experiments have shown that chromosome segregation occurs in bacteria. Thus, the products of new genetic events, mutational or recombinational, are eventually sorted into separate progeny cells. Early autoradiographic experiments (7, 32) indicated that this was not a

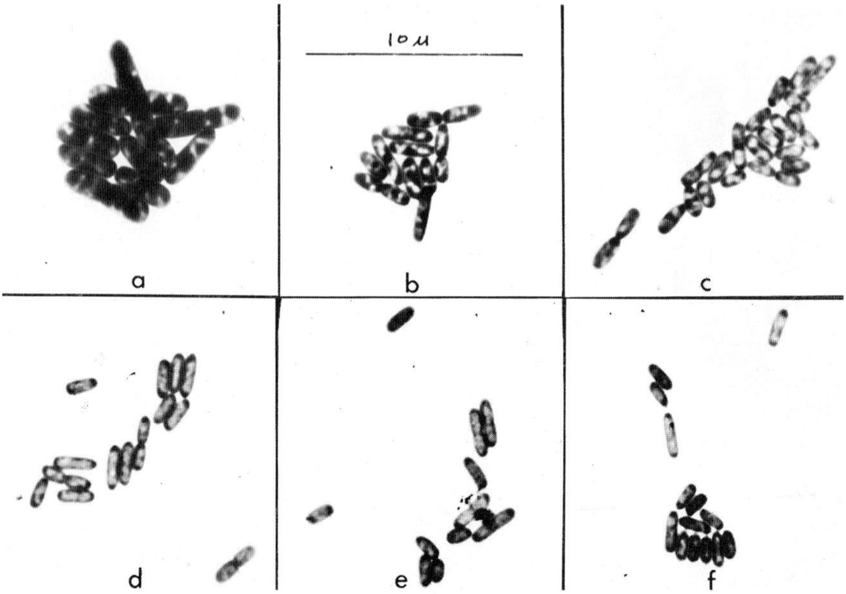

FIG. 3. Electron micrographs of *Escherichia coli* cells from cultures growing in: (*a*) L-broth, (*b*) glucose-M9, (*c*) succinate-M9, (*d*) aspartate-M9, (*e*) proline-M9, or (*f*) acetate-M9 medium. Cells were fixed with osmium tetroxide and collected by filtration on a collodion membrane (11). The nuclear areas appear as areas of low electron density. It should be noted that cells containing either four, two, or one chromosome (see Fig. 2) fall into characteristic classes.

haphazard process. Two experiments in our laboratory illustrate the precision of this segregation.

Escherichia coli growing in L-broth contains an amount of DNA equivalent to four replicating chromosomes. Autoradiographic analysis supports this (Table IV). When such cells incorporate a pulse of ^{3}H-thymine, eight conserved units of DNA are labeled which are eventually separated into daughter cells. The distribution of units is not random, since after one, two, or three generations of growth in nonradioactive medium (chase) all the cells (>98%) are radioactive (Table V). After three generations of growth, the cell population has increased eightfold since the incorporation of radioactivity. A detailed study of radioautographs of this third generation showed that all the cells were labeled to an equal extent—that is, the distribution of the number of grains per cell fitted a Poisson distribution. During the next (fourth) generation of growth, nonradioactive cells appeared. From this, it can be concluded that the eight DNA units, labeled with radioactive thymine, are precisely distributed, one each, into the eight progeny cells

TABLE V
PERCENT OF LABELED CELLS OBSERVED ONE, TWO, THREE, FOUR, OR FIVE GENERATIONS AFTER AN L-BROTH CULTURE WAS LABELED WITH A PULSE OF ^{3}H-THYMINE (5)[a]

Generation of growth in non-radioactive medium (chase)	Unlabeled cells
1	1%
2	1%
3	2%
4	22%
5	48%

[a] A culture of *E. coli* 15T$^-$ in L-broth was labeled for 2 minutes with ^{3}H-thymine and then grown in nonradioactive medium. Samples were taken after one, two, etc., generations of growth, and radioautographs were prepared. These were developed after different exposures. The data shown are for an exposure to yield on the average 10 to 20 grains per labeled cell. Many of the labeled cells at the fourth generation contained only one or two grains. When a lower exposure was used and the grains per cell were counted, a Poisson distribution was obtained for the third-generation sample, but not for the fourth. About 46% of the cells at the fourth generation deviated from a Poisson distribution having less than the expected number of grains per cell (either 0 or 1 per cell). Presumably these were cells that did not receive a radioactive strand of DNA or received only a portion of a radioactive strand as a result of chromosome fragmentation.

which have been produced by three generations of growth. This cannot occur if at each generation radioactive and nonradioactive DNA units are randomly mixed and then distributed into progeny cells.

Another experiment, with *B. subtilis,* dramatically illustrates this same point. When a selected strain of this organism is grown on nutrient agar, it forms a chain of cells (Fig. 4). If cells are prelabeled with ^{3}H-thymine for several generations and then allowed to grow into a chain, the position of radioactive units of DNA along the chain can be located by autoradiography. Examples of this are shown in Fig. 4. When we examined chains of 16 to 32 cells for the location of radioactive DNA, a high proportion of cells at the chain ends were found to contain a radioactive unit. The frequencies of labeled cells found at the chain ends after different conditions of radioactive labeling are tabulated in Table VI. It is evident that the frequency of radioactive cells found at the ends of chains excludes any possible model of segregation in which chromosomes are randomly distributed into daughter cells. Indeed, as we shall see, the data constitute strong evidence for the permanent attachment of a part of the chromosome to a structure present in the cell at the time of radioactive labeling.

As a result of experiments with *E. coli* growing at intermediate

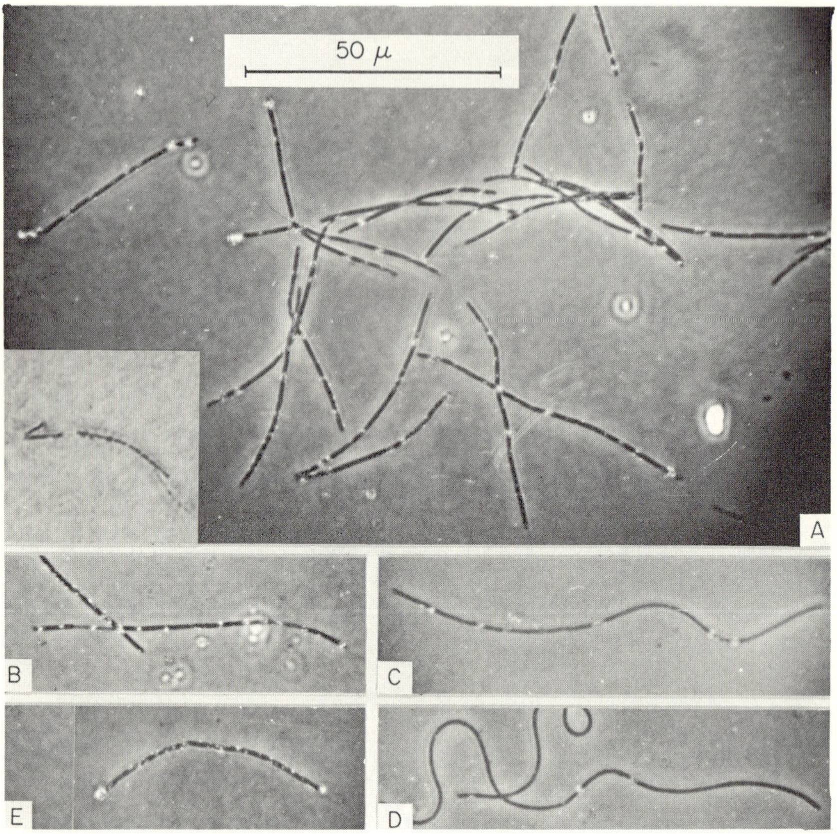

FIG. 4. Photomicrographs of chains of *Bacillus subtilis*. Cultures of *B. subtilis* were grown in different media in which individual cells contained one or two replicating chromosomes (6). The cultures were labeled with ^{3}H-thymine for three generations (continuous label; 200 μC per 40 μg/ml) or for a tenth of a generation period (pulse label; 200 μC per 6 μg/ml). The labeled cells were washed, placed on nutrient agar slides, and allowed to develop into chains of 16 to 32 cells. Autoradiographs were made and exposed until grain clusters appeared over cells containing radioactive DNA.

A, a group of cells from a glucose culture which had been pulse-labeled with ^{3}H-thymine. The inset is a picture of a preparation not dipped in film which shows the septal separation between cells. *B*, an end-labeled chain arising from a continuously labeled cell grown on glucose (two replicating chromosomes). *C*, a chain with unlabeled ends arising from a pulse-labeled cell grown on glucose. *D*, an end-labeled chain arising from a continuously labeled cell grown on acetate (one replicating chromosome). *E*, a chain with unlabeled ends arising from a pulse-labeled cell grown on acetate.

TABLE VI

DISTRIBUTION OF RADIOACTIVE DNA IN CHAINS OF *Bacillus subtilis* ARISING FROM CELLS CONTAINING ONE OR TWO REPLICATING CHROMOSOMES[a]

	Number counted	Average number of grain clusters per microcolony	Fraction with clusters in: Two ends	One end	No end
Cultures of cells with one chromosome (acetate or succinate carbon source)					
Continuous label					
Observed	375	4.6	.49	.40	.11
Calculated			.51	.49	0
Pulse label					
Observed	375	2.2	.07	.38	.55
Calculated			0	.49	.51
Cultures of cells with two chromosomes (glucose, aspartate, or proline carbon source)					
Continuous label					
Observed	750	7.4	.26	.46	.28
Calculated			.20	.50	.30
Pulse label					
Observed	400	4.8	.26	.43	.31
			.30	.50	.20

[a] Radioautographs of the chains shown in Fig. 4 were classified for the number of chains in which both ends were labeled, one end was labeled, or no ends were labeled. The calculated frequencies were obtained from the model in Fig. 9 by assuming that it is three times as probable for a template to attach distal to an already formed septum as next to one. In the case of acetate and succinate cells a correction was also made for the fact that 50% of the chains arise from pairs of cells (see legend to Fig. 9). This is not true for cells with two replicating chromosomes, more than 90% of which occur as single cells. The calculated values for chains from such cells were made by using the probability of obtaining end labeling from cells with a single replicating chromosome (44) and then applying a binomial distribution. These calculated values differ from those proposed previously (5).

growth rates, we have formulated a more precise model of this segregation (14). When this organism is grown in synthetic medium with succinate as an energy source, it contains two chromosomes which are replicated, one after the other, in sequence (17) (Table IV and Fig. 2). A pulse of radioactive thymine, therefore, labels the progeny of one of the chromosomes but not of the other. Subsequent growth in nonradioactive medium results in the separation of the progeny chromosomes into the first generation of daughter cells. We have found that all these cells are radioactive (Table VII) (16). We are forced to conclude, therefore, that in each daughter cell one of the progeny from each chromosome of the parent is present—that is, a radioactive as well as a nonradioactive one (Fig. 5A).

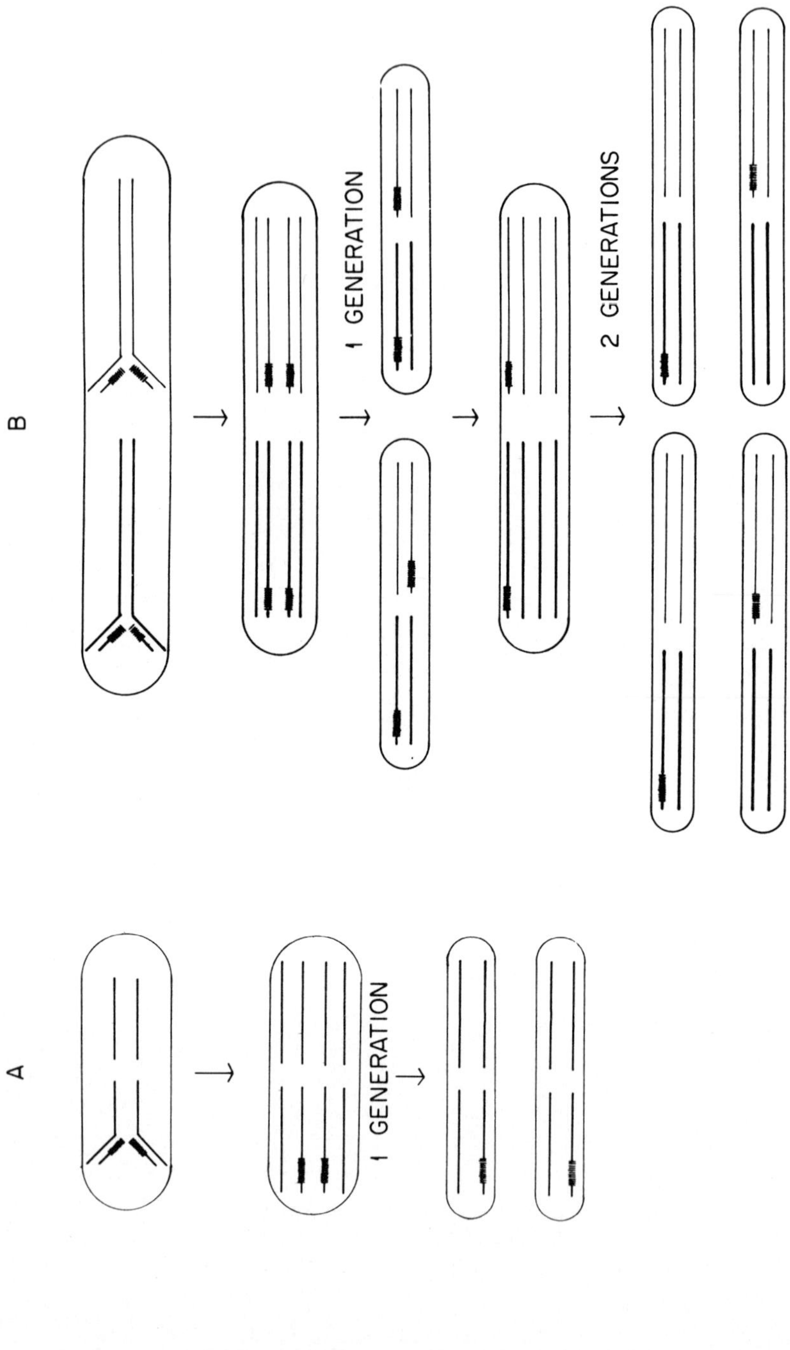
A
1 GENERATION
B
1 GENERATION
2 GENERATIONS

TABLE VII

DISTRIBUTION OF PULSE-LABELED RADIOACTIVE DNA INTO DAUGHTER CELLS DURING SUCCESSIVE GENERATIONS OF GROWTH AFTER LABELING

Succinate growth[a]

Generations of growth after pulse labeling	Percent of cells without a radioactive chromosome
0	5
1	4.3
2	49
3	72
4	87

Shift from glucose into succinate medium[b]

Time after the medium shift at which the cells were pulse-labeled (minutes)	Generations of growth after labeling	Percent of cells without a radioactive chromosome
0	2	1
	3	51
40	1	1
	2	43

[a] A culture growing in succinate medium was pulse-labeled with ^{3}H-thymine and then grown in nonradioactive medium. Samples were taken for autoradiography at zero, one, two, three, or four generations. The autoradiographs were evaluated in terms of a Poisson distribution, and the number of cells with too few grains, lying outside the distribution, was measured. These were assumed to be those cells which did not get a radioactive chromosome or which received only a fraction of a radioactive chromosome. The details of this experiment have been presented elsewhere (16).

[b] A culture growing in glucose medium was transferred into succinate medium and pulse-labeled immediately or after 40 minutes. Each pulse-labeled sample was then grown in nonradioactive medium, and samples were taken for autoradiography at one, two, or three generations. The autoradiographs were evaluated for the frequency of cells without a radioactive chromosome as in *a* above.

FIG. 5. A representation of chromosome segregation in *Escherichia coli* 15T$^-$ based on the data in Table VII (14).

A, growth in succinate medium. A pulse of radioactive thymine labels the two progeny resulting from the replication of one of the two chromosomes. Subsequent segregation places one of these radioactive chromosomes in each daughter cell, together with an unlabeled chromosome.

B, the distribution into daughter cells of the four radioactive progeny resulting from the replication of two chromosomes in a cell. The culture was shifted from glucose into succinate medium and pulse-labeled with ^{3}H-thymine. As the data in Table VII demonstrate, replication within 40 minutes after the shift down proceeds as in model *A* of this figure. Therefore, one of the radioactive progeny of each parent chromosome must eventually be paired with a nonradioactive progeny of the other parent by the second generation. This will give rise to every cell's being radioactive after two generations of growth. In the diagram, one parent is denoted by a heavy pair of lines, the other by a light pair.

At faster growth rates, as in glucose minimal medium, each cell contains two chromosomes, both of which are replicated at the same time (Table IV and Fig. 2). A pulse of radioactive thymine therefore labels the progeny of *both* of the chromosomes. When such cells are transferred to succinate medium, they continue to replicate both chromosomes simultaneously for a short period and then begin to replicate them in alternation (as discussed above). By pulse labeling cells immediately after a shift from glucose into succinate medium we can label all four of the progeny chromosomes in the cell and then follow their segregation into daughter cells during subsequent growth in nonradioactive succinate medium—that is, by the mechanism illustrated in Fig. 5*A*. When this is

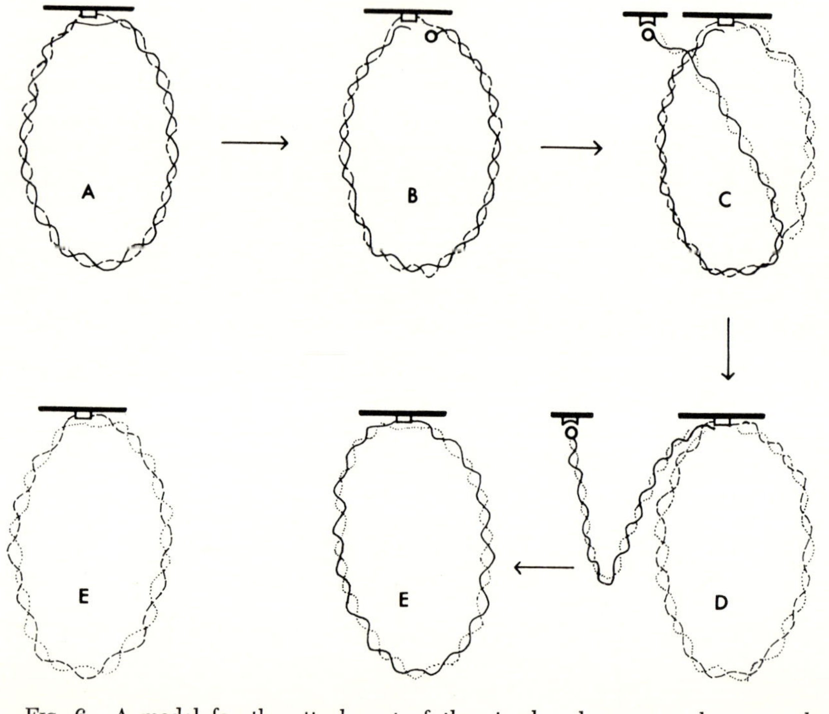

FIG. 6. A model for the attachment of the circular chromosome by one polynucleotide strand. Replication via the attachment of the other strand is shown. Details of this model have been discussed elsewhere (14).

done (Table VII), all the cells are still labeled after two generations of growth. Prior to the second division the progeny of each chromosome will consist of one radioactive and one nonradioactive daughter chromatid (see Fig. 5*B*). Thus, in order to have every cell labeled after two generations of growth in succinate medium, each cell must have received one nonradioactive and one radioactive chromosome. Each of these is the progeny of a different parent chromosome, since the cells are growing in succinate medium (14).

It follows that chromosomes must be paired in the daughter cells in such a way that a chromosome containing a radioactive polynucleotide template made during the previous replication cycle (parent template) is always paired with a chromosome containing a nonradioactive template strand [synthesized two replication cycles previously (grandparent template)].

To explain this, we have proposed the following model: The conserved strands of a DNA molecule are not attached to the cell membrane

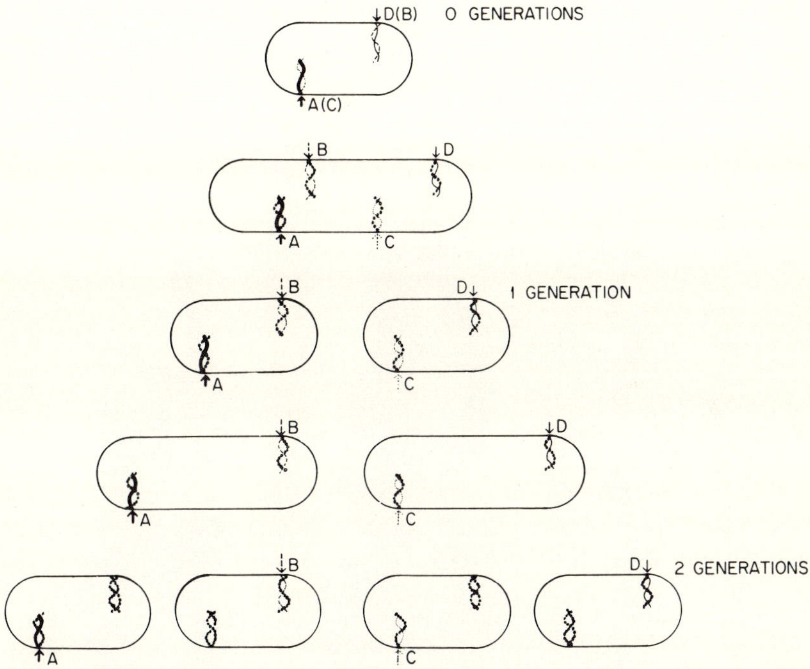

FIG. 7. Segregation and separation of conserved units of DNA in cells from a succinate culture (14). Two chromosomes, containing four conserved units of DNA (A, B, C, and D), are present at the beginning of replication. One conserved unit of each chromosome is permanently attached to the cell surface (that is, A and D). The other two are unattached, represented by parentheses. Upon replication, the conserved units separate as shown, and B and C also attach permanently to the cell surface. Repetition of this process separates the four units into the four cells which have formed by the second generation. This process distinguishes conserved units used as templates for the first time, (C) and (B), from conserved units being used as templates for the second time, A and D. At each division, a daughter cell receives two chromosomes, one formed from each type of template.

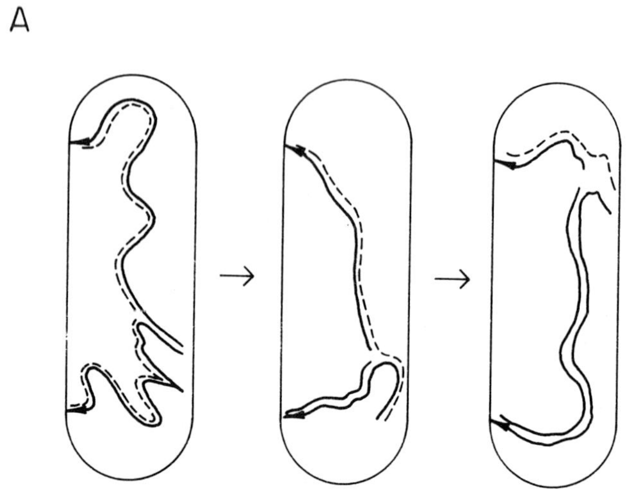

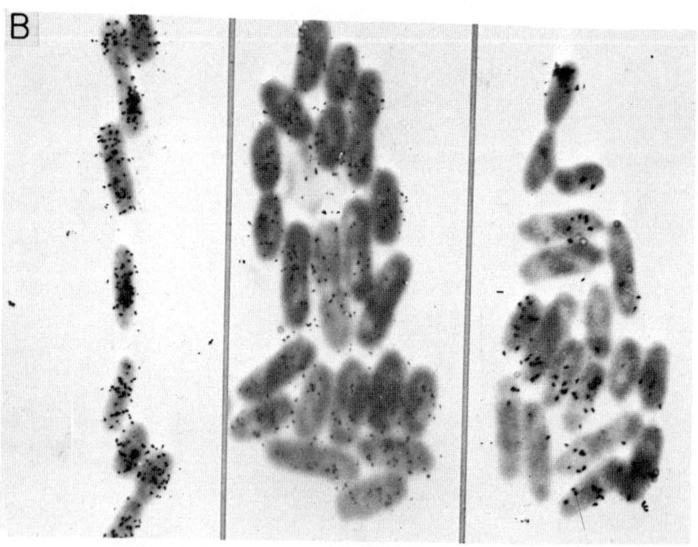

FIG. 8. *A*, The diagrammatic representation shows the expected position of the labeled portion of the chromosome (dashed line) as growth and segregation proceed according to the model in Fig. 7.

B, Electron microautoradiographs of succinate-grown *Escherichia coli* cells labeled with a 4-minute pulse of ^{3}H-thymine (100 μC per 2 μg/ml) and then grown in nonradioactive succinate medium. Samples of cells were taken immediately and after one or two generations of growth. These were fixed with osmium tetroxide, until they are used as a template in replication. To initiate replication, that strand of the double helix which is used as a template for the *first* time is attached permanently to the cell surface. Thus, at the close of a replication cycle each complete chromosome is attached to the cell by one strand of the DNA double helix, but not by the other. The two chromosomes formed are separated as the membrane grows. The "grandparent" template remains with the old portion of the cell surface, while the "parent" template is attached to the newly formed surface (Fig. 6). An appropriate choice for the separation of these templates in cells of a succinate culture will give rise to the segregation pattern observed. This is shown in Fig. 7. As may be seen, the data require that at the onset of replication the most recently synthesized strand of each chromosome becomes attached to a point in the growing cell which is located in the opposite half of the cell from the attachment site of the template from which it was replicated.

This leads to the following experimental prediction. If a pulse of ^{3}H-thymine is given to such cells, the radioactive label should be found in the two daughter chromosomes—that is, throughout the cell. After one generation of growth in nonradioactive medium, it should be found in only one of the progeny chromosomes, but the radioactive strand should be stretched between the old site and its new attachment site—that is, radioactivity should be found throughout the cell. After two generations of growth in nonradioactive medium, the radioactive strand is attached to the cell, and radioactivity should be found in only one end of the cell. This is shown schematically in Fig. 8*A*, and the result of the experiment utilizing electron microautoradiography is shown in Fig. 8*B*. It can be seen that the prediction is confirmed by experiment.

It is obvious that different rates of synthesis of the cell membrane could give rise to different patterns of segregation. Therefore, the important and, we would hope, general aspect of this model is the permanent attachment of the polynucleotide strand when it is first used as a template in replication. By so doing, new chromosomes could always be indexed as they are produced, indicating their destination for a new cell.

This model is strongly supported by the data in Table VI on the distribution of radioactive DNA in chains of *B. subtilis*. Confining our attention to those cells which contain only one replicating chromosome, we find that the majority of cells which have been prelabeled for several

---

and specimens were prepared by collection on collodion membranes. Radioautographs were prepared according to the technique of Caro (2) using Ilford L4 film.

From left to right: zero generation, one generation, and two generations of growth after labeling.

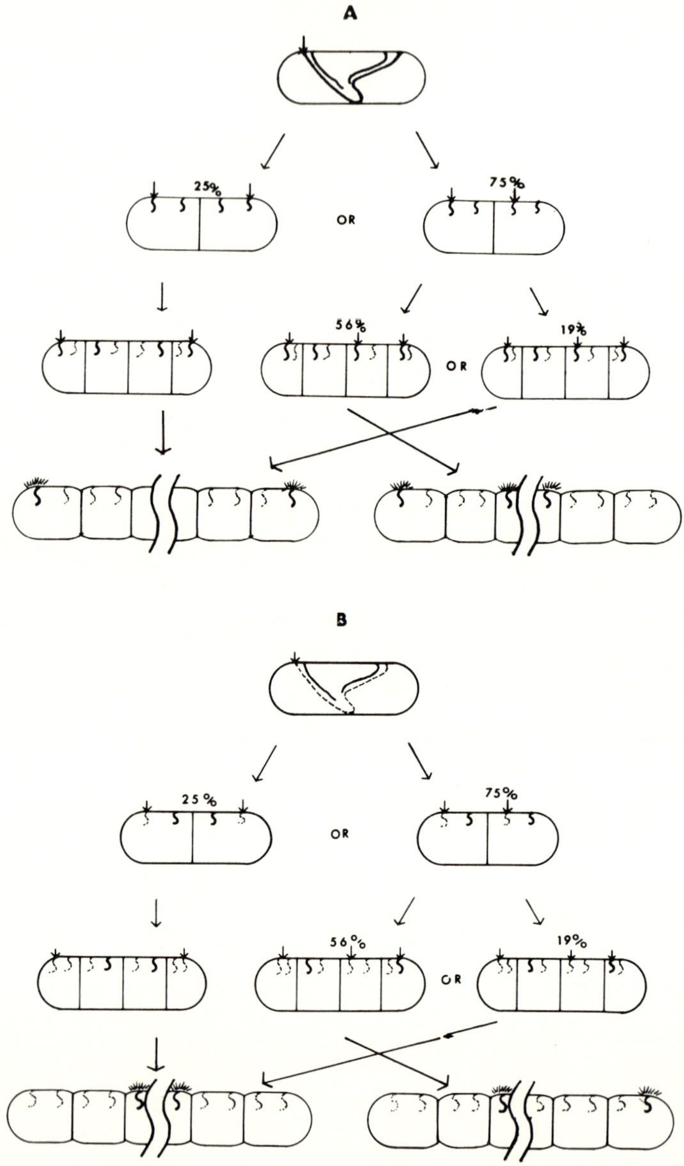

FIG. 9. A model to explain the segregation of radioactive DNA observed in chains of *Bacillus subtilis* (Fig. 4). Chains arising from cells with a single replicating chromosome are shown. The radioactive portion is denoted by the solid line. Chromosome attachment is indicated by an arrow: *A*, continuously prelabeled cells; *B*, pulse-prelabeled cells. The frequencies were chosen to fit the data in Table VI. In calculating the values which appear in that table, we have corrected for the fact that

generations give rise to chains of cells with radioactive DNA at the ends. Cells which were pulse-labeled just before plating on agar give rise to few chains which are labeled on the ends. Studies with fluorescent antibody on division in bacteria in *B. megatherium* and *Streptococcus pyogenes* have shown that cell growth occurs at the center, and sometimes at one end of the cell (4, 3). Similarly, there are indications that the production of new cell membrane material is restricted to the center portion of the cell (10). It is reasonable to assume, therefore, that cells on the ends of the chains contain that material of the original cell surface which is conserved during growth, whereas cells in the middle contain only newly synthesized material. The results in Table VI may thus be explained (see Fig. 9) by assuming that the template strand of DNA is attached to the original cell surface, whereas the newly synthesized strand is not. Upon further replication this newly synthesized strand is attached (when it is itself used as a template) either to the right or to the left of its complementary strand. After cell division, the two chromosomes replicate again. The one containing the "grandparent" template replicates by attaching the new strand to new cell material on the *same* side as in the previous cycle. On the other hand, the new, or "parent," template attaches its other strand either to the right or to the left (thus the septum will form to the right or left of it). After this, the septum again forms on the same side of this chromosome in all subsequent cycles of replication.

According to this model, the rules of chromosome attachment in *B. subtilis* are:

1. When a strand is used as a template for the first time, it becomes permanently attached to the cell surface.

2. At the time of this attachment its relationship to the site of septum formation also is established permanently. Henceforth, all chromosomes containing this polynucleotide strand will appear consistently to the "right" or to the "left" of the division septum.

This will lead to a pattern in which continuously prelabeled cells give

---

50% of the chains arising from cells grown in acetate or succinate medium arose from cell pairs. This was done by extending the freedom of choice for one additional cycle.

To obtain the frequencies calculated for cells with two replicating chromosomes we used a binomial distribution. Thus, the probability of a cell with two replicating chromosomes being labeled at both ends, one end, or neither end can be calculated as the terms obtained in the expansion of $(0.56 + 0.44)^2$. This model is based on the assumption that both of the chromosome templates located at the poles of the cell will have a choice as to which orientation toward the septum they will assume in subsequent replication cycles. The frequencies used are different from those proposed in an earlier paper (5).

rise to cell chains which are labeled on both ends or on one end. A culture prelabeled with a pulse of ^{3}H-thymine immediately before being placed on agar should give rise to cell chains with one end radioactive or with no end radioactive.

The data in Table VI are in good agreement with this prediction. Using this data, we can establish the probability of the two classes of labeled cell chains arising from cells with *one* replicating chromosome. [The calculated values in Table VI and Fig. 9 are somewhat different from those calculated in an earlier paper (5). This change is based on a consideration of the origin of pairs of cells which on agar give rise to single chains of cells (50% of the chains arise from such pairs). The labeling in such chains may be calculated from the model in Fig. 9 by extending it through a third cycle. Thus the probability of obtaining an end-middle labeled chain from a continuously prelabeled pair of cells is $(0.75)^3$.] When this model (Fig. 9) is extended to cells with two chromosomes, it yields a prediction in agreement with the results observed, provided that one assumes that both chromosomes are attached by their "grandparent" template at the center of the cell. Thus, each newly attached chromosome will be located at opposite poles of the cell. As a result, during the subsequent development of the chain of cells the probability that either end will be labeled will be one-half. We have made no attempt to justify this assumption, although models to explain the propagation of such cells can be constructed. This extension is discussed in detail in the legends to Fig. 9 and Table VI.

The attachment of the template strand of the chromosome as a prerequisite to replication explains several aspects of replication:

1. The requirement for a protein whose synthesis is not blocked by amino acid analogs, chloramphenicol, or 5-fluorouracil presumably represents the synthesis of the site to which the template attaches.

2. The restriction of chromosome initiation at slow growth rates would reflect the reduced rate of synthesis of attachment sites.

3. The premature initiation of chromosome replication induced by thymine starvation on one but not both of the partially completed chromatids (see Fig. 1) (15) may be explained as a difference in structure between a permanently attached "grandparent" template and a recently attached "parent" template (14).

4. Recently, it has been found that the parent templates of episomes and chromosomes segregate in such a manner that the parent templates of both are found in one cell, and the grandparent templates in the other (10; Rownd, personal communication). This is most easily explained by assuming that templates of a given generation attach to a common segregation structure.

This last result could also explain the dichotomous chromosome replication observed in *B. subtilis* (31). Thus, if this organism contained two genetic linkage units, they could replicate independently, but nevertheless segregate together by virtue of their attachment to a common segregation structure. This could give the appearance of two replication points on a single segregation unit.

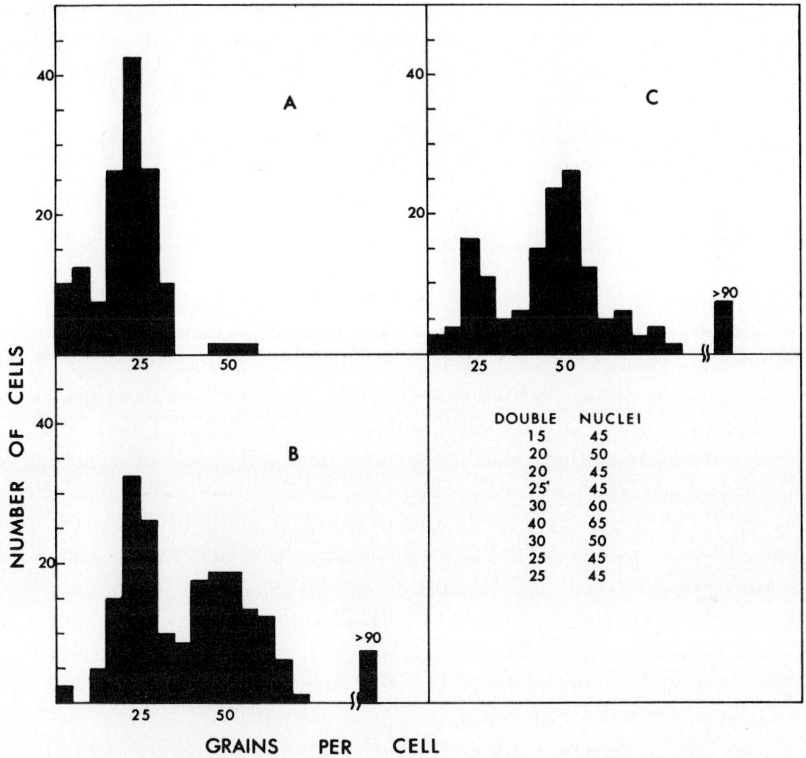

FIG. 10. Distribution of silver grains in autoradiographs of mouse embryo fibroblasts grown as a primary tissue culture in Eagle's medium one, two, or three generations in ^{3}H-thymidine (0.025 μC/ml).

A, 2.5 $\times$ 10^6 cells were used as inoculum and grown for 24 hours in a 100-mm petri dish containing 10 ml of medium overlaying four coverslips. *B*, 1.25 $\times$ 10^6 cells were used as inoculum and grown for 48 hours as in *A*. Radioactive medium was replaced after 24 hours. *C*, 0.63 $\times$ 10^6 cells were used as inoculum and grown for 72 hours as in *A*. Radioactive medium was replaced after 24 hours and 48 hours.

All coverslips were washed and fixed with glutaraldehyde. Autoradiographs were prepared and, after 4 days of exposure, developed. The distributions of grains per cell were counted on populations of: (*A*) 150 cells, (*B*) 200 cells, and (*C*) 150 cells. The numbers in the lower right-hand corner represent the number of grains in pairs of nuclei found in dividing cells in sample *B*. These numbers have been approximated to the nearest multiple of 5.

Segregation of Sister Chromatids in Animal Cells

The observations on the segregation of the progeny of chromosomes and episomes led us to believe that this segregation mechanism might apply to the sister chromatids of higher organisms.

To test this, we used mouse embryo cells grown as a primary tissue culture. We felt that this system would avoid many of the problems of polyploid or aneuploid cells, and would be closest to the segregation system of the animal itself. Moreover, two technical advantages were derived from the use of this system:

1. Such cells spend 60 to 70% of their division cycle in C_1, a period in which there is only one set of $2n$ chromatids per cell.
2. DNA synthesis in such cells is synchronized during the first 24 hours of growth (presumably as a result of the trauma of cell preparation).

Cells were incubated with ^{3}H-thymidine for one, two, or three generations, and the radioactivity per cell was measured by autoradiography. As may be seen in Fig. 10, after one generation most of the cells contain the same amount of radioactive material. After two generations, however, there are two cell populations, one with twice as much radioactivity as the other. Further growth does not result in a change in the amount of label per cell observed in the two cell populations, although there is an increase in the proportion of highly labeled cells. When cells are found in the act of division, the two nuclei show the asymmetry characteristic of the entire population.

In another experiment (Fig. 11) cells were labeled for one generation and then placed in nonradioactive medium for two more generations of growth. Again, two populations of cells were eventually observed, one labeled and the other unlabeled.

These results can be explained only if, at division, the forty pairs of sister chromatids of the mouse are segregated such that those which contain a radioactive template are separated from those which contain a nonradioactive template.

When similar experiments were tried with established tissue culture cell lines, the results were not so definitive. Figure 12 shows the results of experiments with Chinese hamster and HeLa cells. In both of these systems, $S + G_2$ occupied more than 50% of the division cycle. Moreover, the cells are not synchronized with respect to DNA synthesis. As a result, one generation of growth in radioactive medium gives rise to a population of cells which contain either one or two chromatids per chromosome, or, if they are in S, a number in between. If such cells are grown for one generation in radioactive medium, and then transferred

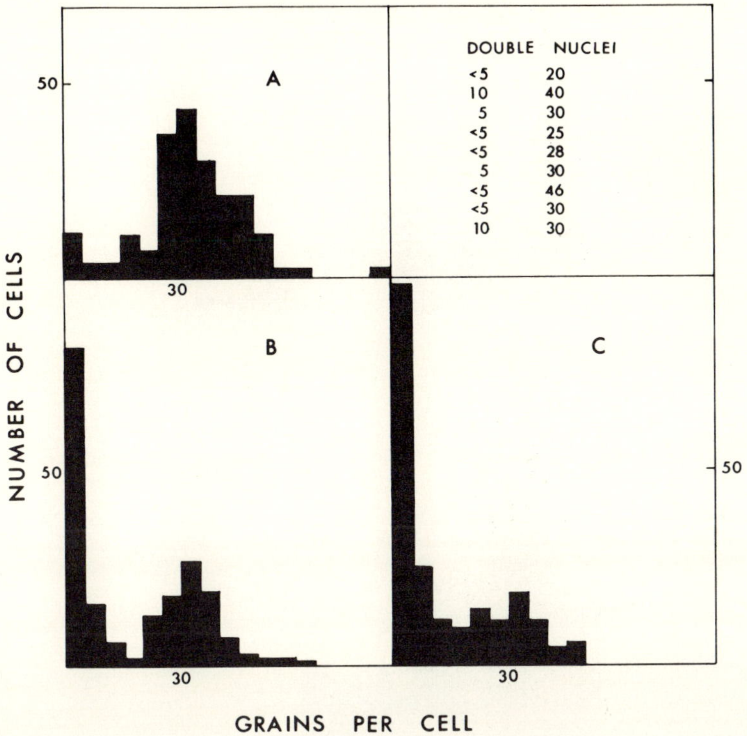

Fig. 11. Distribution of silver grains in autoradiographs of mouse embryo fibroblasts grown as a primary tissue culture in Eagle's medium for one generation in ^{3}H-thymidine (0.025 μC/ml) and for two subsequent generations in nonradioactive medium.

A, 0.3 $\times$ 10^6 cells were used as inoculum and grown for 24 hours in a 35-mm petri dish containing a single coverslip overlaid with 1.5 ml of medium containing ^{3}H-thymidine. *B*, 0.15 $\times$ 10^6 cells were used as inoculum and grown for 24 hours as in *A*, and then the medium was replaced with nonradioactive medium. *C*, 0.008 $\times$ 10^6 cells were used as inoculum and grown for 24 hours as in *A*, and then for 48 hours in nonradioactive medium. The medium was replaced each 24 hours with fresh medium.

Coverslips were washed and fixed, and autoradiographs were prepared and developed after 6 days of exposure. Two hundred cells were counted for each population. The average number of grains per cell were (*A*) 37, (*B*) 18, and (*C*) 12. The numbers in the upper right-hand corner represent the number of grains in paired nuclei found in dividing cells in sample *B*. They have been approximated to the nearest multiple of 5.

to nonradioactive medium, we expect most of them still to be radioactive after one generation of growth in nonradioactive medium. This is because the two sister chromatids are both labeled and will be distributed into daughter cells. At the next division, however, we should produce a class

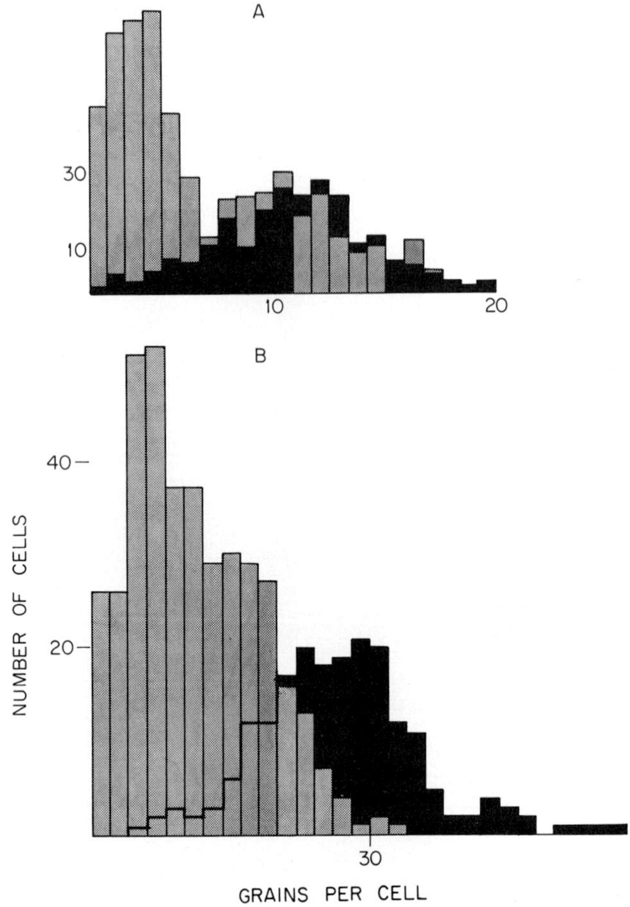

FIG. 12. Distribution of silver grains in autoradiographs of Chinese hamster and of HeLa cells grown for 24 hours in radioactive thymine and then for either 24 or 48 hours in nonradioactive medium.

A, Chinese hamster: ^{3}H-thymine (0.1 μC/ml) was used to label the cells which were grown in Ham's medium. Distribution: After 24 hours chase ■, 260 cells, average number of grains per cell 11.4; after 48 hours chase ▩ 600 cells, average number of grains per cell 4.6.

B, Hela: ^{3}H thymine (0.05 μC/ml) was used to label the cells which were grown in Eagle's medium. Distribution: after 24 hours chase ■ 200 cells, average number of grains per cell 27; after 48 hours chase ▩ 400 cells, average number of grains per cell 12.

of unlabeled cells. A tendency in this direction is obvious in the diploid (8) Chinese hamster cells. However, most of the "unlabeled" cells contain a small amount of radioactive material. In this case, however, the labeled population retains most of its original radioactivity, and there are clearly "hot" and "cold" populations. In the case of the heteroploid HeLa cells this distinction is not clear, although it is obvious that the distribution is very asymmetrical and that sister chromatids are not segregating at random.

Whether the blurring of the distributions in Fig. 12 is due to sister chromatid exchange or to a breakdown of segregation rules is not yet known. Nor is it known whether polyploidy or heteroploidy effects the segregation.

Whatever the explanation, it is clear that, even in established tissue culture lines, there is some tendency to segregate those chromatids together which contain templates synthesized during the same division cycle.

Summary and Discussion

The experiments described for bacteria have led to the conclusion that DNA replication can proceed only after both template strands of the replicating molecule attach to the cell. Of these two, one is already permanently attached from a previous replication cycle. The attachment of the other is a necessary condition for the initiation of replication. Eventually this attachment of the new template becomes permanent. The location of the attachment site may also be permanently oriented with respect to the location of future sites of septum formation and, therefore, of new cell formation.

Such a model, applied to the larger chromosomes of animal cells, could explain their semiconservative replication despite a large number of replicating subunits in each.

If our initial experiments with primary tissue cultures prove to be generally valid, we would conclude that a similar rule governs the attachment of entire sets of chromatids to the mitotic apparatus during chromosome replication in higher organisms. Such an attachment would serve as a practical mechanism for insuring that at mitosis each daughter cell receives one of each chromatid pair. A prediction from this type of attachment would be that mitotic recombinational events on separate chromosomes should appear to be linked if they are located near the centromere.

It is obvious that the mechanism we have described cannot operate during meiosis. Thus, the cell must have an efficient method for detaching and later reattaching chromosomes which is at least specific enough to distinguish one diploid pair from another. If chromosomes remain attached to a protein structure, this structure could break down into subunits at meiosis, only to reassemble in the heterozygote.

The structure to which attachment could occur is unknown in either bacteria or higher organisms. It has been assumed that in bacteria the chromosome attaches to the cell membrane (9). Arguments have also been presented for the attachment of chromosomes of higher organisms to the nuclear membrane (28). It seems probable that the period of the cell division cycle between mitosis and the onset of DNA synthesis (G_1) could be occupied by the synthesis of a new nuclear membrane or some other structure to which the chromosomes could attach. Variability in G_1 may reflect different efficiencies in the synthesis of this structure.

The final definition of the attachment site in bacteria and in higher organisms must await its isolation. This may prove to be a formidable task, since the amount of such material, per cell, will be stoichiometric with the chromosome content—that is, in bacteria only one or two. We hope that this difficulty may be offset by the use of selective inhibitors such as amino acid analogs or chloramphenicol.

Acknowledgments

We wish to thank Dr. E. Kellenberger and Dr. L. Caro for their valuable advice on the preparation of electron micrographs. This research was supported by grants from the National Science Foundation (GB 4833) and from the National Institutes of Health (AI 05711). K.G.L. is the recipient of a Career Development Award (5-K3-GM-2519). H.E. holds a postdoctoral fellowship from the National Institutes of Health (1-F2-GM-31,275).

References

1. Cairns, J., *Cold Spring Harbor Symp. Quant. Biol.*, **7**, 534 (1963).
2. Caro, L., *in* "Methods in Cell Physiology" (D. M. Prescott, ed.), Vol. 1, p. 327, Academic Press, New York, 1964.
3. Chung, K. L., Hawerko, R. Z., and Isaac, P. K., *Can. J. Microbiol.*, **10**, 43 (1964).
4. Cole, R. M., and Hahn, J. J., *Science*, **135**, 722 (1962).
5. Eberle, H., and Lark, K. G., *J. Mol. Biol.*, in press.
6. Eberle, H., and Lark, K. G., *Proc. Natl. Acad. Sci. U.S.*, in press.
7. Forro, F., Jr., and Wertheimer, S. A., *Biochim. Biophys. Acta*, **40**, 9 (1960).
8. Hsu, T. C., and Somers, C. E., *Proc. Natl. Acad. Sci. U.S.*, **47**, 396 (1961).
9. Jacob, F., Brenner, S., and Cuzin, F., *Cold Spring Harbor Symp. Quant. Biol.*, **28**, 329 (1963).
10. Jacob, F., Ryter, A., and Cuzin, F., *Proc. Royal Soc.*, **164**, 267 (1966).
11. Kellenberger, E., *Symp. Gen. Microbiol.*, **10**, 39 (1960).
12. Lark, C., *Biochim. Biophys. Acta*, **119**, 517 (1966).
13. Lark, C., and Lark, K. G., *J. Mol. Biol.*, **10**, 120 (1964).
14. Lark, K. G., *Bacteriol. Rev.*, **30**, 3 (1966).
15. Lark, K. G., and Bird, R., *J. Mol. Biol.*, **13**, 607 (1965).
16. Lark, K. G., and Bird, R., *Proc. Natl. Acad. Sci. U.S.*, **54**, 1444 (1965).
17. Lark, K. G., and Lark, C., *J. Mol. Biol.*, **13**, 105 (1965).
18. Lark, K. G., and Lark, C., *J. Mol. Biol.*, **20**, 9 (1966).

19. Lark, K. G., Repko, T., and Hoffman, E. J., *Biochim. Biophys. Acta,* **76,** 9 (1963).
20. Maaløe, O., and Hanawalt, P. C., *J. Mol. Biol.,* **3,** 144 (1961).
21. Meselson, M., and Stahl, F. W., *Proc. Natl. Acad. Sci. U.S.,* **44,** 671 (1958).
22. Pardee, A. B., and Prestidge, L. S., *J. Bacteriol.,* **71,** 677 (1956).
23. Paul, J., quoted in "Molecular Genetics" J. H. Taylor (ed.), Part 1, p. 177, Academic Press, New York, 1963.
24. Prescott, D. M., and Bender, M. A., *Exptl. Cell Res.,* **29,** 430 (1963).
25. Pritchard, R. H., and Lark, K. G., *J. Mol. Biol.,* **9,** 288 (1964).
26. Schaechter, M., Maaløe, O., and Kjeldgaard, N. O., *J. Gen. Microbiol.,* **19,** 592 (1958).
27. Simon, E. H., *J. Mol. Biol.,* **3,** 101 (1961).
28. Sved, J. A., *Genetics,* **53,** 747 (1966).
29. Taylor, J. H., Woods, P. S., and Hughes, W. L., *Proc. Natl. Acad. Sci. U.S.,* **43,** 122 (1957).
30. Treick, R. W., and Konetzka, W. A., *J. Bacteriol.,* **88,** 1580 (1964).
31. Yoshikawa, H., O'Sullivan, A., and Sueoka, N., *Proc. Natl. Acad. Sci. U.S.,* **52,** 973 (1964).
32. Van Tubergen, R. P., and Setlow, R. B., *Biophys. J.,* **1,** 589 (1961).

Discussion of Part I

C. A. Thomas, O. E. Landman, A. T. Ganesan, E. Racker, J. K. Bhattacharjee, D. Billen, C. I. Davern, H. V. Aposhian, F. Lipmann, C. B. Anfinsen, P. Siekevitz, M. Riley, K. G. Lark, R. P. Novick, N. Sueoka

Chairman Thomas: Do we have any questions at this point?

Dr. Landman: Dr. Ganesan, have you verified that the very short pulse counts that were acid-precipitable and in the pellet fraction, are incorporated in the DNA strands?

Dr. Ganesan: Yes, we have taken the pellet fraction and, as I mentioned, they can be dissociated by treatment with pronase. Such a preparation was banded in a CsCl gradient. The density, transforming activity, and susceptibility to exonuclease treatment after denaturation of the DNA with the pulse counts suggest that they are incorporated into the growing strands of DNA.

Dr. Racker: The high activity for DNA polymerase in the membrane fraction may not be due to a large amount of enzyme present in the fraction, but due to activation by the membrane of a small fraction of the enzyme.

Dr. Ganesan: Yes, we are only observing the incorporation of the label in the deoxytriphosphates into an acid-insoluble form.

Dr. Bhattacharjee: It may interest you to know, Dr. Ganesan, that, in yeast, tetrad analysis of hybrids heterozygous for fifteen markers in chromosome V shows two of the four chromatids often involved (chromatid interference) in meiotic recombination.

Dr. Billen: I would just like to ask one question and make a comment. Dr. Davern, did you try the chloramphenicol experiment with the 15 T^- or the *rec*$^-$? That is, after the P^{32}-induced lesion, did you put in chloramphenicol?

Dr. Davern: We checked chloramphenicol to see if that would inhibit repair mechanisms in 15 T^-, but I did not mention that in the talk. The chloramphenicol experiment there was done with *rec*$^-$.

Dr. Billen: You are aware that this is surprisingly similar to the ionizing radiation study. The results almost mimic that and I would predict, if you have not already looked, that if you add chloramphenicol after the P^{32}-ing to the 15 T^- that you will see degradation which would probably mimic that which would be seen in the *rec*$^-$, as well as probably a result in terms of DNA synthesis which is a lot like that of the *rec*$^-$.

DR. DAVERN: Yes, we set out to try and do that, but in our hands we were not able to make a phenotypic copy of *rec*⁻ with chloramphenicol.

DR. APOSHIAN: In the *rec*⁻, where you had the P^{32} DNA being broken down, did you find out whether the RNA was also being broken down, and is the *rec*⁻ strain of Meselson deficient in endonuclease I or not?

DR. DAVERN: I don't know what the picture on the endonuclease is on this. We did not check RNA breakdown. There wasn't any significant decline in counts which would be largely a reflection of RNA. We know that in these dead cells RNA synthesis can go on for some time but not very long (40 to 50 minutes).

CHAIRMAN THOMAS: I think the answer to your second question about endonuclease I is known. Can anyone give us the answer? I think that Meselson's strain *rec*⁻ does have endonuclease I.

DR. LIPMANN: Dr. Davern, since you speak of the chromosome as a particle, what else is in this particle besides DNA?

DR. DAVERN: There can be many more things in this particle apart from DNA because we are not in a position to rule them out with the observations that have been made so far, nor is anybody else in a position to rule them out. There may be things attached to the chromosome as well as in the linear sequence of the chromosome, and these things could be discovered by, say, getting mutants which affect DNA replication through screening programs which would serve to implicitly define functions and perhaps their location. It seems that direct observation has gotten us about as far are we can go at the moment, and it is not in a position to yield a definitive answer.

DR. LIPMANN: What do you think in regard to proteins?

DR. DAVERN: We have found no evidence for protein but this is because we have not used the appropriate technique to look for it.

DR. ANFINSEN: I just want to ask a question for information. I never really thought about it before, but as you talked, I suddenly realized that I had no understanding of what the physical or metabolic basis of radiosensitivity or radioresistance is. It is hard for me to understand why a strand of P^{32}-labeled DNA in one case breaks down and in another it doesn't. Is there some ready explanation for this?

DR. DAVERN: Repair mechanisms provide the ready out in all these things. It is such an easy hypothesis to fix up all these problems.

DR. SIEKEVITZ: One question in regard to Dr. Anfinsen's comment. You have assumed that the P^{32} causes breaks in DNA. Is it also possible that you have an inactivation of the enzyme so that you would get the same picture, so it isn't caused by breaks in the DNA?

DR. DAVERN: Yes, your suggestion is quite valid and not ruled out. We could have an event, a decay event, anywhere in these chromosomes, which would have a by-product which would then act as a feedback inhibitor of the enzyme. This would be action at a distance of a trivial kind as far as we are concerned, but it could be so.

Dr. Riley: Dr. Lark, would you go through—I tried to visualize it as you went along—the segregation of a recessive mutation in a bacterial cell. How would that work with the segregation of the DNA?

Dr. Lark: Recessive mutations in a succinate cell would be expected to be carried for a very long period of time, theoretically. We have looked at these cells cytologically. The result was very clear cut. L-broth cells had a different nuclear pattern from glucose and succinate, both of which have the same pattern, whereas the proline and slow-growing acetate cells had another different pattern. The point is that the succinate cells and the glucose cells have a few (about 10%) cells with a type of nuclear pattern characteristic of a single chromosome which would lead to a segregation of the recessive character. These are present in the population at all times. About 10% of the acetate and proline cell population have double nuclear areas. So I think that to do the experiment carefully, we have to measure the rate at which a recessive segregates. It is not enough to look for recessives; you must measure their rate of segregation. It is clear that the rate of growth of the glucose cell might be such—that is, the rate of division—that the two progeny of each chromosome are never separated into different cells. We are trying now, autoradiographically, to test glucose cells to see if we can determine if one of the progeny chromosomes gets transferred to the other side of the cell. I mentioned earlier that succinate growth is an abnormal situation. This statement should not be used as a basis for saying that glucose-grown cells have two chromosomes and the two progeny of each segregate into separate cells whereas in succinate they may mix. The types are not that clear and as far as we know, the degree of mixing may depend on growth rate, (not on energy source). We have a similar system in *Salmonella* where the growth rate is a function of the presence or absence of amino acids.

Dr. Novick: Dr. Lark, it seems that the postulated attachment of the newly formed strand must not be nucleotide-sequence-specific, since the strand of opposite polarity is now attached whereas the strand of the other polarity is not attached.

Dr. Lark: It is not polarity-specific, if that is what you mean.

Dr. Novick: Well, it must not be sequence-specific either then, since the nucleotide sequences of the two attachment regions would presumably be different for the two strands if these regions are complementary.

Dr. Lark: It depends on how it is threaded through, doesn't it? If you are threading it through as you replicate, then it is not. If the one strand is threaded up, the other is threaded down, it could be sequence specific.

DR. NOVICK: Alternatively, you can work out a sequence, such that its complement has the same sequence but in the opposite direction. This will be a very special type of genetic structure. It would be interesting to look at genetically.

DR. LARK: Now I understand. This would presumably be at the initiation point after amino acid starvation. We are in the process of doing this with Dr. Reiji Okazaki. We shall make a small bit of BU-DNA after a minus amino acid treatment and then do a nearest-neighbor sequence after separating this DNA by density gradient, but we haven't done this yet. Perhaps such a sequence is unique. I believe Dr. Sueoka is also doing this type of experiment with *B. subtilis*. Perhaps the same sequence will be found in the two systems.

DR. NOVICK: One more point, that is, you would expect to find asymmetric mutations in a sense. If the hypothetical attachment region were to consist of a stretch of DNA whose nucleotide sequence on one strand were the same as that on the other strand read in the opposite direction $\left(\text{e.g.: } \begin{matrix}\overrightarrow{\text{TACGCGTA}}\\ \underleftarrow{\text{ATGCGCAT}}\end{matrix}\right)$, then a mutation would destroy the symmetry and would alter one of the complementary attachment regions in a different way than the other.

CHAIRMAN THOMAS: Dr. Sueoka, do you want to ask your question now?

DR. SUEOKA: Dr. Davern, could you comment on the possible involvement of a single-strand break in your experiment or give your idea about the effect of a single-strand break on your model?

DR. DAVERN: Single-strand breaks of themselves may not provide points of free rotation in that the base pair interactions at this location of the break may act as some sort of fluid clutch and prevent free rotation. This is a possibility, but if excision followed such a single-strand break, then it must become a free region of rotation.

CHAIRMAN THOMAS: But we know that a single-strand nick can convert a superhelix into an ordinary circular helix which must involve a certain number of free rotations.

PART II

ORGANIZATIONAL RNA SYNTHESIS

Chairman's Remarks

SAMUEL B. WEISS

Department of Biochemistry,
Argonne Cancer Research Hospital,
University of Chicago,
Chicago, Illinois

This morning, certain emphasis was placed on the importance of structure and biological function, setting an appropriate tone for the symposium on Organizational Biosynthesis. We heard several papers on DNA synthesis, especially those problems concerning chromosomal replication, and this afternoon we turn our attention toward the second, but no less important nucleic acid, the ribopolynucleotides, and its relationship to protein synthesis. Although the precise mechanism for the assembly of the two essential macromolecules is not clearly understood, one feels that the general pattern by which these macromolecules are assembled is, in a sense, defined. The intimate relationship between protein and RNA biosynthesis has been felt for some time, especially its importance for cell growth. However, it has not received, possibly, the attention which it merits. This afternoon we shall hear two papers which will present some experimental probes into this problem.

Coupled Regulation of Bacterial RNA and Protein Synthesis

GUNTHER S. STENT

Department of Molecular Biology and Virus Laboratory, University of California, Berkeley, California

Translational Control of Transcription

Beckwith's discovery (4) three years ago that o^0 mutants in the *Lac* operon of *Escherichia coli* are not, in fact, the "operator-negative" mutants which Jacob and Monod (22) had supposed them to be, but are, instead, highly polar, suppressor-sensitive nonsense mutants in the operator-proximal sector of the galactosidase cistron, brought about a new way of thinking about the control of bacterial RNA synthesis. For since o^0 mutants do not form detectable amounts of *Lac* operon messenger RNA in the absence of their suppressor mutation, but obviously do so in its presence (3), it became apparent that there exists a feedback connection between the translation of the messenger RNA and its synthesis: The translational block represented by a nonsense codon can block transcription of the DNA sector bearing that codon. The generality of this feedback connection was recently demonstrated by Lodish and Zinder (27), who found that a nonsense mutation in the capsid cistron of an RNA bacteriophage blocks synthesis of the viral, or translated "plus" strand, RNA, but not of the complementary, or nontranslated "minus" strand in the formation of the double-stranded replicative RNA. Thus, whereas previous attempts to understand the regulation of the rate of transcription of DNA mainly focused either on DNA-affined "repressors" that block the template or on RNA-polymerase-affined "inhibitors" that block function of the transcription enzyme, some latter-day theories envisage that the primary site of transcriptional regulation is the ribosome. The purpose of this essay is to examine how the idea that the rate of RNA translation controls the rate of its synthesis can account for the *general* control of cellular RNA synthesis. I shall not dwell here on the possible mechanism of *specific* translational control of induction and repression of individual operons, except to point out that the recent discovery that aminoacyl-transfer RNA is more immediately involved in the repression of enzymes of amino acid biosynthesis than the free amino acid (15, 40) lends support to the view that the expression of some operons is controlled by the translatability of their messenger RNA (2, 43).

The Feedback Model

The point of departure of this discussion is the model shown in Fig. 1, which envisages that ribosomes engage the nascent RNA as soon as it emerges from the RNA polymerase and begin forming a polyribosome aggregate before formation of the RNA molecule is complete. Translation commences at once upon the nascent RNA, and the relative motion of RNA and ribosomes in protein synthesis pulls away the nascent ribopolynucleotide from the DNA–polymerase complex that has spawned it. The following considerations were previously adduced (44) in support of this model:

1. An active *in vivo* removal of the nascent RNA from the DNA–polymerase complex would explain three deficiencies of the usual *in vitro* reaction mixtures of DNA-directed enzymatic RNA synthesis, which make no provision for removal of the reaction product: (*a*) each polymerase molecule synthesizes only one RNA molecule; (*b*) the finished RNA molecule remains attached to the DNA template; and (*c*) the rate of chain growth of the RNA molecules is too low, probably by one, or even two, orders of magnitude (6). [More recently, it was shown that continuous removal of *in vitro*-synthesized nascent RNA by addition of ribonuclease to the reaction mixture endows the polymerase enzyme with a longer active life (25).]

2. Ribosomes appear to be involved not only in protein synthesis but also in RNA synthesis, since in bacteria recovering from severe depletion of their ribosomes, the rate of synthesis of RNA is proportional to the degree to which the intracellular ribosome content has been restored (23, 29, 33).

3. Complexes containing template DNA, nascent RNA, and ribosomes can be isolated from bacteriophage-infected cells (44) and from *in vitro* enzymatic reaction mixtures (11).

4. An important necessary condition of the model is met, in that the chemical direction in which RNA is synthesized is the same as the chemical direction in which the RNA is translated in protein synthesis—namely, from the 5′ terminus of the polynucleotide (7, 47, 41).

I think that by now most men of good will might agree that under conditions of normal bacterial growth the model of Fig. 1 does approximate the real situation for synthesis of *messenger* RNA—that is, that the nascent message is removed from its DNA template by ribosomes engaged in its incipient translation. The homeostatic character of this process is obvious, since the intracellular level of messenger RNA will

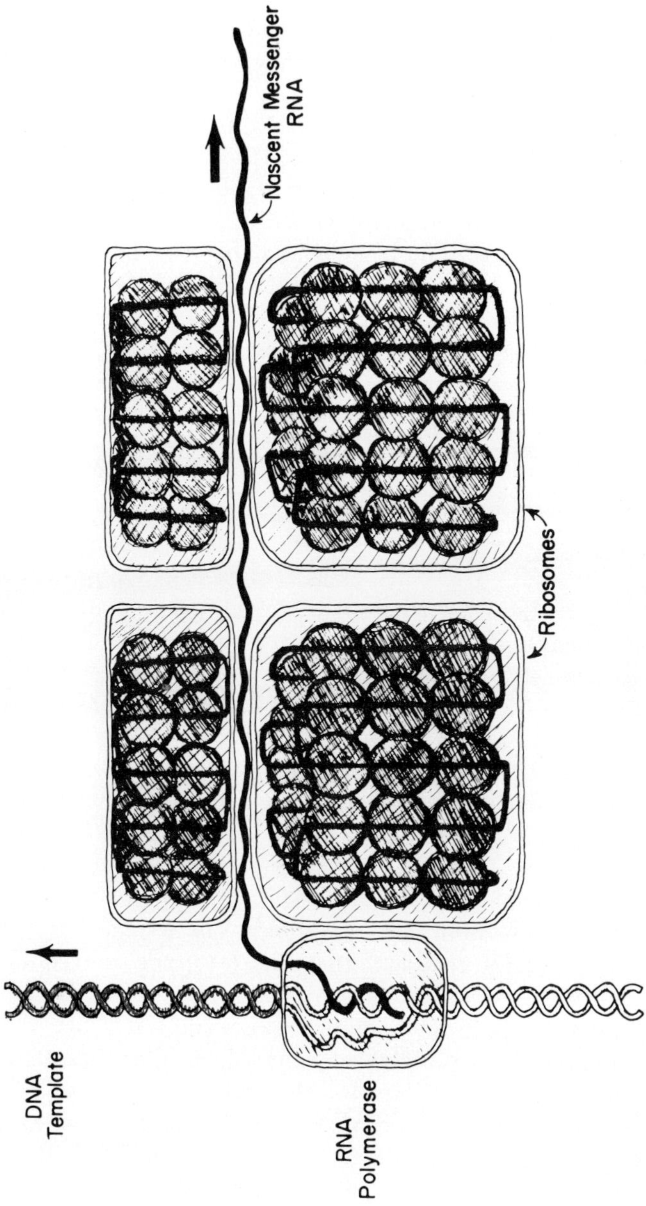

FIG. 1. Schematic representation of the formation of polyribosomes on nascent messenger RNA. From Stent (44).

not exceed the number of ribosomes that can handle its translation. There is rather less general agreement, however, on whether this process obtains also in bacteria in which protein synthesis has been arrested, either by deprivation of a required amino acid or by addition of an antibiotic such as chloramphenicol, and on whether it applies also to the synthesis of ribosomal and transfer RNA, the two classes that actually represent the bulk of cellular RNA. I shall first deal with the second of these doubtful points and focus my discussion on the synthesis of ribosomal RNA, meanwhile assuming, for want of any better idea, that transfer RNA molecules originate as short, nontranslated tail ends of long messenger RNA chains, from which they are later cut away and modified structurally by enzyme action. By this convenient assumption, the model of Fig. 1 is automatically extended to transfer RNA, the only, admittedly weak, justification for this assumption being that the physiological regulation of transfer RNA synthesis appears to follow that of general messenger RNA, and not that of ribosomal RNA (24).

Ribosomal RNA as a Messenger

In order to extend the model of Fig. 1 to the synthesis of ribosomal RNA, it is most convenient to follow Roberts' suggestion (38) that nascent ribosomal RNA also serves as a messenger for protein synthesis —namely, for synthesis of ribosomal protein, prior to its encapsulation into mature ribosomes. The virtues of this concept are obvious: homeostatic adjustment of the intracellular level of ribosomal protein and ribosomal RNA, since the average messenger lifetime of a nascent ribosomal RNA prior to maturation would vary inversely with the size of the intracellular pool of nascent ribosomal protein, in a manner similar to the dependence of the prematuration messenger lifetime of viral RNA on the intracellular phage capsid pool (27). Indeed, the preferential synthesis of ribosomal protein in bacteria which had been allowed to accumulate ribosomal RNA in the absence of protein synthesis (17, 32) seems difficult to explain in any other way. The weaknesses of this concept are, alas, equally obvious: First, the 5000 odd nucleotides of the 70 S ribosomal RNA complement could be expected to code for no more than five to ten different polypeptide chains, each 170 to 340 amino acids in length, a diversity of proteins considerably smaller than that inferred for ribosomal protein on the basis of electrophoretic analyses (50). Since the meaning of this abundance of electrophoretic bands does not seem clear, however, in particular since it is not yet known whether some of these bands may not represent a set of "isopolypeptides" coded for by the several iterated ribosomal RNA cistrons known to be present in the *E. coli* genome (53) and hence subject to genetic drift, it is still permissible to think that the bulk of the ribosomal protein is represented by no more primary polypeptide chains than could be coded by the ribosomal RNA. Second, it was shown more than five years ago that RNA extracted from mature ribosomes does not function as a messenger for *in vitro* amino acid-polymerizing reaction mixtures (34, 48). To save the messenger role of ribosomal RNA one is thus obliged to suppose either that the messenger property of nascent ribosomal RNA is lost in the maturation process, or that it has additional requirements that have not so far been met in the *in vitro* experiments. The report that ribosomal RNA extracted from immature ribosomes *does* function as an *in vitro* messenger (35) seemed to support the former of these suppositions, but more careful examination of this phenomenon, while not disproving the basic idea, has cast doubt on the validity of the interpretation of the earlier experiments (46, 30). The possibility that the latter of these suppositions obtains will be considered presently. In any case, the following discussion assumes that nascent ribosomal RNA *is* the messenger of ribosomal protein, and many of the inferences to be drawn here will be pointless if this assumption should prove to be invalid.

Stringent Control

It has been known for ten years now that normal, or stringent control (RC^{st}), bacteria greatly reduce their net rate of RNA synthesis as soon as they are deprived of a required amino acid (21, 36). It is not easy to ascertain the exact extent to which *de novo* RNA synthesis is actually reduced by amino acid starvation, since the severe reduction in net RNA synthesis renders the intracellular pools of RNA precursors difficult to penetrate by externally added labeled compounds (14). Nevertheless, it has been shown that whatever residual RNA synthesis does proceed under amino acid starvation is distributed in nearly normal proportion among the three major classes of messenger, ribosomal, and transfer RNA (18). The intracellular pools of ribonucleoside triphosphates, presumed to be direct substrates of RNA polymerization, remain at a high level throughout amino acid starvation (19, 14), making it most unlikely that amino acid control is exerted at any level other than polynucleotide synthesis. Ten years ago, Pardee (36) proposed that amino acids control RNA synthesis because RNA and protein are cosynthesized obligatorily, so that stopping one process stops the other. But the idea of an obligatory cosynthesis of protein and RNA was abandoned as soon as the role of messenger RNA in protein synthesis came to be understood. To bring the regulatory effect of amino acids on RNA synthesis in line with the then current ideas of primary regulation of RNA synthesis at the transcription level, a new theory was proposed in 1961—namely, that amino-acid-free transfer RNA is an inhibitor of RNA synthesis (45, 26). One plausible way in which this inhibition was thought to occur was that amino-acid-free transfer RNA, but not aminoacyl transfer RNA, can combine with, and thus neutralize, the RNA polymerase enzyme (49). However, further detailed studies of the interaction of RNA polymerase with transfer RNA showed that, whereas this interaction possesses many interesting and subtle features, it does not really commend itself as the primary reaction by which amino acids produce such drastic changes in the rate of cellular RNA synthesis (8, 9). Nevertheless, the original notion that transfer RNA *does* intervene in the regulation of RNA synthesis was later found to be correct, when Eidlic and Neidhardt (16) showed that an amino acid must be activated and transferred to its cognate transfer RNA before it can exert its stimulation of RNA synthesis. The model of Fig. 1, which is, of course, nothing but a detailed restatement of the old notion of obligatory cosynthesis of protein and RNA, now explains these effects very simply: As soon as the absence of an amino acid, or failure of its transfer to transfer RNA, arrests protein synthesis, progress of the ribosomes over the nascent RNA is halted, and RNA synthesis is arrested in turn.

Chloramphenicol and Relaxed Control

An immediate objection can be raised against the idea that protein synthesis is necessary for RNA synthesis—namely, that bacteria continue their RNA synthesis in the presence of chloramphenicol, while all polypeptide formation is held in abeyance. This objection is more apparent than real, however, since for continuous RNA synthesis the model of Fig. 1 demands only movement of ribosomes over the nascent RNA, presumably effected by the sequential arrival and departure of aminoacyl-transfer RNA species and by the codon-anticodon match at their ribosomal sites for RNA synthesis, but not necessarily formation of peptide bonds. The precise mechanism of action of chloramphenicol still remains unknown, though it seems likely that the inhibition is directly concerned with the formation of the peptide bond. In any case, it appears that, as demanded by the model of Fig. 1, ribosome movement over nascent RNA *does* continue in the presence of chloramphenicol, since under these conditions polyribosomes are both formed and maintained (51, 31) and both entered and left by nascent RNA (12, 13).

A case related to that of chloramphenicol is presented by the relaxed control, or RC^{rel}, mutants of *E. coli* which do continue RNA synthesis even though their protein synthesis has been arrested by starvation for a required amino acid (5, 45). A hypothesis advanced to explain the aberrant behavior of such RC^{rel} mutants was that they possess a "catholic inducer" which can replace the missing amino acid on its homologous transfer RNA species (1). Subsequent studies of the intracellular state of transfer RNA molecules in leucine-starved RC^{rel} bacteria eliminated this hypothesis, however, since they failed to reveal attachment of any unusual substance to the mainly uncharged leucine transfer RNA (54). Thus we are now led to the proposition that during amino acid starvation an intracellular condition obtains in RC^{rel} bacteria, analogous to that created by chloramphenicol, that permits continued movement of ribosomes over the nascent RNA in the absence of appreciable protein synthesis. The findings that upon amino acid starvation RC^{str} bacteria rapidly lose and RC^{rel} bacteria maintain their complement of polyribosomes (31) seems in good accord with these notions.

Nutritional Upshift

Chemical analyses of bacteria growing with very different generation times under different nutritional conditions have led Maaløe and Kjeldgaard (28) to state their fundamental theory of bacterial growth: ". . . the number of ribosomes [per DNA genome] is proportional to the rate of growth and of protein synthesis." They inferred from this finding that the average rate of polypeptide chain growth is constant under a broad spectrum of conditions that engender a more than tenfold variation in the rate of protein synthesis per DNA genome. How then do the bacteria adjust their relative content of ribosomes over the wide range which Maaløe and Kjeldgaard have observed? The model of Fig. 1 (under the proviso that nascent ribosomal RNA is a messenger for ribosomal protein) offers a ready explanation of this phenomenon, provided that it is assumed further that Maaløe and Kjeldgaard's fundamental theorem of constant polypeptide chain growth rate holds for bulk protein *but not for ribosomal protein.* Thus if the average chain growth rate of ribosomal proteins were to rise with the growth rate of the cell, it would follow automatically that the cell would be the richer in both ribosomal protein and ribosomal RNA, the higher its growth rate. Indeed, the transitional behavior of bacteria during their nutritional upshift from slow growth in a poor medium to rapid growth in an enriched medium may be thought to support this view. As the studies of Maaløe and his collaborators (28) had shown, bacteria respond to such an upshift by an immediate acceleration of their RNA synthesis, while, in accord with their fundamental theorem, total protein synthesis accelerates more slowly, as the ratio of ribosomes to DNA genome gradually waxes. Now, as recent experiments by Schleif (42) have shown, the rate of synthesis of ribosomal protein shows the same immediate acceleration following upshift as does the synthesis of RNA, in accord with the view that average rates of chain growth of ribosomal protein and ribosomal RNA are directly coupled.

"Special" Classes of Transfer RNA for Ribosomes

How could variations in nutritional conditions find reflection in a variable average peptide chain growth rate of ribosomal protein (and hence ribosomal RNA) in the face of a constant chain growth rate of the bulk protein of the cell? The most plausible explanation would be that "special" classes of transfer RNA are involved in the translation of ribosomal protein, whose extent of aminoacylation has a dependence on the size of the intracellular amino acid pool different from that of the "ordinary" classes of transfer RNA involved in the translation of the bulk protein. Thus the small amino acid pool obtaining in bacteria growing slowly on a poor substrate would be thought to suffice for an extent of aminoacylation of the "ordinary" transfer RNA class that permits the constant peptide chain growth rate of bulk protein of the fundamental theorem, whereas this pool size would not suffice for adequate aminoacylation of the "special" class. The larger amino acid pool (10) in rich media would, however, allow adequate aminoacylation of the "special" class, and thus lead to a higher chain growth rate of ribosomal protein, and hence of ribosomal RNA. The possibility of the use of "special" classes of transfer RNA in the translation of ribosomal protein might explain, furthermore, why experiments designed to show the messenger property of ribosomal RNA have so far failed. For, as recent studies (52) have shown, some transfer RNA species appear to require special factors for their enzymatic aminoacylation *in vitro*. It is possible, therefore, that the *in vitro* reaction mixtures to which ribosomal RNA was added as a potential messenger did not, in fact, achieve aminoacylation of the "special" transfer RNA classes required for its translation.

Indeed, a simple calculation shows that if nascent ribosomal RNA *is* the messenger for ribosomal protein, its use of synonymous codons would have to be different from that of the messengers of nonribosomal, or bulk, protein. Table I presents the amino acid composition of *E. coli* bulk and ribosomal protein and the codon assignments of each amino acid. From these data, the relative base frequencies of RNA coding for proteins of these compositions have been calculated, *by giving equal weight to all synonymous codon assignments*. This calculation shows that *E. coli* ribosomal and messenger RNA differ in their divergence from the base compositions that would be expected if each coded for its proteins by random choice among synonymous codons: Ribosomal RNA would code for ribosomal protein by favoring G-rich codons, whereas messenger RNA apparently codes for bulk protein by discriminating against G-rich codons.

It thus appears to me that, once the old idea of pre-messenger RNA

TABLE I

COMPOSITIONAL COMPARISON OF *E. coli* PROTEIN AND RNA

A. Mole percent of amino acids in E. coli ribosomal and bulk protein

	Ala	Glu	Val	Lys	Asp	Gly	Leu	Arg	Ile	Thr	Ser	Prol	Phe	Met	His	Tyr	Try	Cys
Codon	GC.	GA^{A}_{G}	GU.	AA^{A}_{G}	GA^{U}_{C}	GG.	${}^{U}_{C}U.$	CG.	AU^{U}_{C}	AC.	UC.	CC.	UU^{U}_{C}	AUG	CA^{U}_{C}	UA^{U}_{C}	UGG	UG^{U}_{C}
Ribosomal (37)	11	10	9.6	9.0	8.3	8.2	7.4	7.3	5.5	5.2	4.4	3.7	3.0	2.4	1.9	1.2	0.7	0.5
Bulk (39)	13	10	5.5	7.0	9.9	7.8	7.9	5.3	4.6	4.7	6.1	4.6	3.3	3.4	1.0	2.1	1.0	1.7

B. Mole percent of nucleotides in E. coli RNA

Calculated[a] composition	A	G	C	U
Ribosomal protein messenger	26	30	22	22
Bulk protein messenger	24	30	23	23

Actual composition	A	G	C	U
Ribosomal RNA (37)	25	32	22	21
Messenger RNA (20)	25	27	24	24

[a] Giving equal weight to all synonymous codon assignments. In these calculations the treatment of an unknown amount of GluN and AspN, whose codons are CA^{A}_{G} and AA^{U}_{C}, respectively, as Glu and Asp leads to an over-estimate of the predicted G content of the messenger. This over-estimate could not affect the main conclusion drawn from this table, however, unless the ratios of Glu/GluN and Asp/AspN in ribosomal and bulk protein were drastically different.

times of obligatory cosynthesis of protein and RNA is refurbished in latter-day terms, a new framework is provided for understanding the regulation of bacterial RNA synthesis.

Acknowledgments

This research was supported by U.S. Public Health Service Research Grants CA 02129 from the National Cancer Institute and GM 11957 from the National Institute of General Medical Sciences.

References

1. Alföldi, L., Stent, G. S., Hoogs, M., and Hill, R., Z. *Vererbungslehre,* **94,** 285 (1963).
2. Ames, B. N., and Hartman, P., *Cold Spring Harbor Symp. Quant. Biol.,* **28,** 349 (1963).
3. Attardi, G., Naono, S., Rouviere, J., Jacob, F., and Gros, F., *Cold Spring Harbor Symp. Quant. Biol.,* **28,** 363 (1963).
4. Beckwith, J. R., *in* Structure and Function of the Genetic Material (H. Stubbe, ed.), Abhandl. *Deut. Akad. Wiss., Berlin, Kl. Med.,* No. 1, 119 (1964).
5. Borek, E., Ryan, A., and Rockenbach, J., *J. Bacteriol.,* **69,** 480 (1955).
6. Bremer, H., and Konrad, M. W., *Proc. Natl. Acad. Sci. U.S.,* **51,** 801 (1964).
7. Bremer, H., Konrad, M. W., Gaines, K., and Stent, G. S., *J. Mol. Biol.,* **13,** 540 (1965).
8. Bremer, H., and Stent, G. S., *Acta Biochim. Polon.,* **13,** 367 (1967).
9. Bremer, H., Yegian, C., and Konrad, M., *J. Mol. Biol.,* **16,** 94 (1966).
10. Britten, R. J., and McClure, F. T., *Bacteriol. Rev.,* **26,** 292 (1962).
11. Byrne, R., Levin, J. G., Bladen, H. A., and Nirenberg, M. W., *Proc. Natl. Acad. Sci. U.S.,* **52,** 140 (1964).
12. Das, H. K., Goldstein, A., and Kanner, L. C., *Mol. Pharmacol.,* **2,** 158 (1966).
13. Donini, P. L., personal communication.
14. Edlin, G., and Newhard, J., *J. Mol. Biol.,* **24,** 225 (1967).
15. Eidlic, L., and Neidhardt, F. C., *Proc. Natl. Acad. Sci. U.S.,* **53,** 539 (1965).
16. Eidlic, L., and Neidhardt, F. C., *J. Bacteriol.,* **89,** 706 (1965).
17. Ennis, H. L., and Lubin, A., *Biochim. Biophys. Acta,* **95,** 624 (1965).
18. Friesen, J. D., *J. Mol. Biol.,* **20,** 559 (1966).
19. Goldstein, D. B., Brown, B. J., and Goldstein, A., *Biochim. Biophys. Acta,* **43,** 55 (1960).
20. Gros, F., Gilbert, W., Hiatt, H. H., Attardi, G., Spahr, P. F., and Watson, J. D., *Cold Spring Harbor Symp. Quant. Biol.,* **26,** 111 (1961).
21. Gros, F., and Gros, F., *Exptl. Cell Res.,* **14,** 104 (1958).
22. Jacob, F., and Monod, J., *J. Mol. Biol.,* **3,** 318 (1961).
23. Kennel, D. E., and Magasanik, B., *Biochim. Biophys. Acta,* **55,** 139 (1962).
24. Kjeldgaard, N. O., and Kurland, C. G., *J. Mol. Biol.,* **6,** 341 (1963).
25. Krakow, J. S., *J. Biol. Chem.,* **241,** 1830 (1966).
26. Kurland, C. G., and Maaløe, O., *J. Mol. Biol.,* **4,** 193 (1962).
27. Lodish, H. F., and Zinder, N. D., *J. Mol. Biol.,* **19,** 333 (1966).
28. Maaløe, O., and Kjeldgaard, N. O., *in* "Control of Macromolecular Synthesis," W. A. Benjamin, New York, 1966.
29. McCarthy, B. J., *Biochim. Biophys. Acta,* **55,** 880 (1962).

30. MANOR, H., AND HASELKORN, R., *J. Mol. Biol.*, **24**, 269 (1967).
31. MORRIS, D. W., AND DEMOSS, J. A., *Proc. Natl. Acad. Sci. U.S.*, **56**, 262 (1966).
32. NAKADA, D., *J. Mol. Biol.*, **12**, 695 (1965).
33. NAONO, S., ROUVIERE, J., AND GROS, F., in press.
34. NIRENBERG, M., AND MATTHASI, J., *Proc. Natl. Acad. Sci. U.S.*, **47**, 1358 (1961).
35. OTAKA, E., OSAWA, S., AND SIBATANI, A., *Biochem. Biophys. Res. Commun.*, **15**, 568 (1964).
36. PARDEE, A. B., AND PRESTIDGE, L., *J. Bacteriol.*, **71**, 677 (1956).
37. PETERMANN, M. L., "The Physical and Chemical Properties of Ribosomes," Elsevier, Amsterdam, 1964.
38. ROBERTS, R. B., *J. Theoret. Biol.*, **8**, 49 (1965).
39. ROBERTS, R. B., ABELSON, P. H., COWIE, D. B., BOLTON, E. T., AND BRITTEN, R. J., *in* Studies of Biosynthesis in *Escherichia coli, Carnegie Inst. Wash. Publ.* No. 607 (1955).
40. ROTH, J. R., SILBERT, D. F., FINK, G. R., VOLL, M. J., ANTON, D., HARTMAN, P. E., AND AMES, B. N., *Cold Spring Harbor Symp. Quant. Biol.*, **31**, 383 (1966).
41. SALAS, M., SMITH, M., STANLEY, W. M. JR., WAHBA, A., AND OCHOA, S., *J. Biol. Chem.*, **240**, 3988 (1965).
42. SCHLEIF, R. F., Ph.D. Thesis, University of California, Berkeley, 1967.
43. STENT, G. S., *Science*, **144**, 816 (1964).
44. STENT, G. S., *Proc. Roy. Soc.*, **B164**, 181 (1966).
45. STENT, G. S., AND BREMER, S., *Proc. Natl. Acad. Sci. U.S.*, **47**, 2005 (1961).
46. SYPHERD, P. S., *J. Mol. Biol.*, **24**, 323 (1967).
47. TERZAGHI, E., OKADA, Y., STREISINGER, G., EMRICH, J., MOUYE, M., AND TSUGITA, A., *Proc. Acad. Sci. U.S.*, **56**, 500 (1966).
48. TISSIERÈS, A., AND HOPKINS, J. W., *Proc. Natl. Acad. Sci. U.S.*, **47**, 2015 (1961).
49. TISSIERÈS, A., BOURGEOIS, S., AND GROS, F., *J. Mol. Biol.*, **7**, 100 (1963).
50. WALLER, J. P., AND HARRIS, J., *Proc. Natl. Acad. Sci. U.S.*, **47**, 18 (1961).
51. WEBER, M., AND DEMOSS, J. A., *Proc. Natl. Acad. Sci. U.S.*, **55**, 1224 (1966).
52. WETTSTEIN, F. O., *Cold Spring Harbor Symp. Quant. Biol.*, **31**, 595 (1966).
53. YANKOFSKY, S. A., AND SPIEGELMAN, S., *Proc. Natl. Acad. Sci. U.S.*, **48**, 1466 (1962).
54. YEGIAN, C. D., STENT, G. S., AND MARTIN, E. M., *Proc. Natl. Acad. Sci. U.S.*, **55**, 839 (1966).

Interdependence of Translation and Transcription in T4-Infected *Escherichia coli*

T. Kasai and E. K. F. Bautz

Institute of Microbiology, Rutgers, The State University, New Brunswick, New Jersey

The regulation of RNA and protein synthesis can be studied at different levels of complexities from the smallest viruses to cells of higher organisms with genomes a thousand times as big.

While a very small virus may possess only a crude and inefficient mechanism for regulating transcription and translation, or even none at all, but may have evolved to rely on the regulatory mechanisms of its host, a highly sophisticated cell type may be too complex to allow an adequate analysis of the most basic forms of regulation.

The bacteriophage T4 represents a compromise between these extremes. While still a virus, its genome is big enough to code for a hundred or more proteins, some being synthesized at different times as well as with varying rates. The production of phage-specific proteins may be regulated at two levels: at the level of transcription or at the level of translation. If regulation were operating at the level of transcription, appearance of messenger RNA and its corresponding polypeptide would be congruent; in other words, successful transcription of a particular gene would automatically insure its translation, and, as long as we grant the message only a short half-life, cessation of transcription would insure cessation of translation. Regulation at the translation level, on the other hand, could allow the production of messengers without their utilization as templates unless the two processes are so intimately connected that changes in the rate of translation will also affect the rate of transcription. Thus, translation could conceivably exert some sort of feedback control over transcription. Accepting a regulatory mechanism involving feedback control of mRNA synthesis, one can reconcile the findings that a *lac* o^0 mutant produces little, if any, *lac*-specific messenger RNA (1), and that the mutant used in these studies later turned out to respond to *ochre* suppressors (4). Several studies on *in vitro* and *in vivo* RNA synthesis have since contributed to the notion of feedback regulation at the translational level (see Stent, this volume). In this paper, we wish to report experimental results which we have interpreted to show such an interdependence between translation and transcription.

The development of phage T4 in its host bacterium proceeds according to a rigid time schedule. During the first 8 minutes after infection, enzymes concerned with the replication of phage DNA are manufactured predominantly, with DNA synthesis starting around 7 minutes after infection. After this time interval, production of early enzymes ceases and is followed by the appearance of structural proteins and late function enzymes such as phage lysozyme.

The deletion method for the isolation of gene-specific mRNA, developed in our laboratory, provides a tool to study transcription at the level of a single gene. This technique depends on the availability of a mutant which deleted either the entire gene in question, or at least the better part of it. The principle of the method is illustrated in Fig. 1. RNA marked with a radioisotope for a given period of time after phage

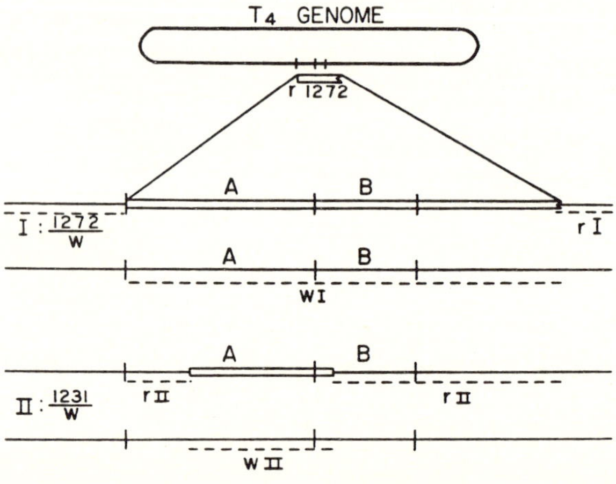

FIG. 1. Schematic illustration of the *deletion method.* Pulse-labeled RNA, isolated from T4 wild-type infected *E. coli* cells, is passed through a two-layer nitrocellulose column containing DNA of a deletion mutant (*r*1272) in the upper layer and DNA of wild-type phage in the lower layer (I:1272/w). After extensive annealing of the RNA sample first with the upper and then with the lower column layer in high salt at 60°C, the two layers are eluted separately with low salt at the same temperature. The lower layer yields RNA homologous to the DNA region missing in the deletion mutant. This RNA can be fractionated further, if necessary, by DNA of a mutant phage stock missing a shorter segment (II:1231/w). Further details are given elsewhere (2).

infection is forced to anneal with single-stranded DNA of a mutant phage stock missing a genetically defined DNA segment. RNA species homologous to the DNA of all those genes which are present in the mutant DNA are adsorbed, whereupon the remaining RNA species, which are homologous to the missing DNA segment, are trapped on wild-type DNA. Experimentally, this can be accomplished by a single passage through a double-layer DNA column (2).

There are two DNA regions in the T4 genome for which deletion mutants are available: the *r*II genes analyzed in great detail by Benzer (5), and the endolysin (*e*) gene, studied by Streisinger and co-workers. The *r*II function, although its mechanism is still obscure, is required early after phage infection (16, 10), whereas lysozyme activity does not appear during the first 6 to 8 minutes after infection. Thus, with mutants available to identify either *r*II- or *e*-specific RNA, one can follow the rate of transcription of these genes at different times during the latent period. It was found that, in wild-type infected *E. coli* cells, transcription of the early function *r*II genes proceeds throughout the latent period at similar rates (Fig. 2*a*)—a result in agreement with the conclusions derived from DNA–RNA hybridization studies by Hall *et al.* (13) involving competition between RNA species synthesized early and late after infection. These authors concluded further that late function genes are not transcribed during the first several minutes after infection. However, contrary to our expectation, transcription of the supposedly late function endolysin gene was found to proceed at a substantial rate within the first 3 minutes after infection. Moreover, it was observed that production of *e*-specific RNA decreased around minute 4. During the second half of the latent period the level of *e*-specific RNA increases steadily (Fig. 2*b*); this increase is paralleled by the appearance of lysozyme activity. Within the limits of the assay, no lysozyme activity could be detected during the first 6 minutes after infection.

The conditional lethal mutant *am*82, which lacks an enzyme essential for DNA replication, also produces the initial burst of *e*-specific RNA in the nonpermissive host *Escherichia coli* B. But this mutant appears to be unable to synthesize lysozyme messenger RNA at later times, in accord with the fact that this mutant never produces functional lysozyme in the restrictive host. In contrast, the *r*II region is transcribed continuously in this mutant, in agreement with the observations (21) that the level of early enzymes continues to increase linearly for more than 1 hour after infection (Fig. 2*a*, *b*).

Table I shows the result of an experiment where the incorporation of H^3-uracil into the precursor pool was terminated 3 minutes after infection by adding an excess of cold uracil. The culture was divided into equal

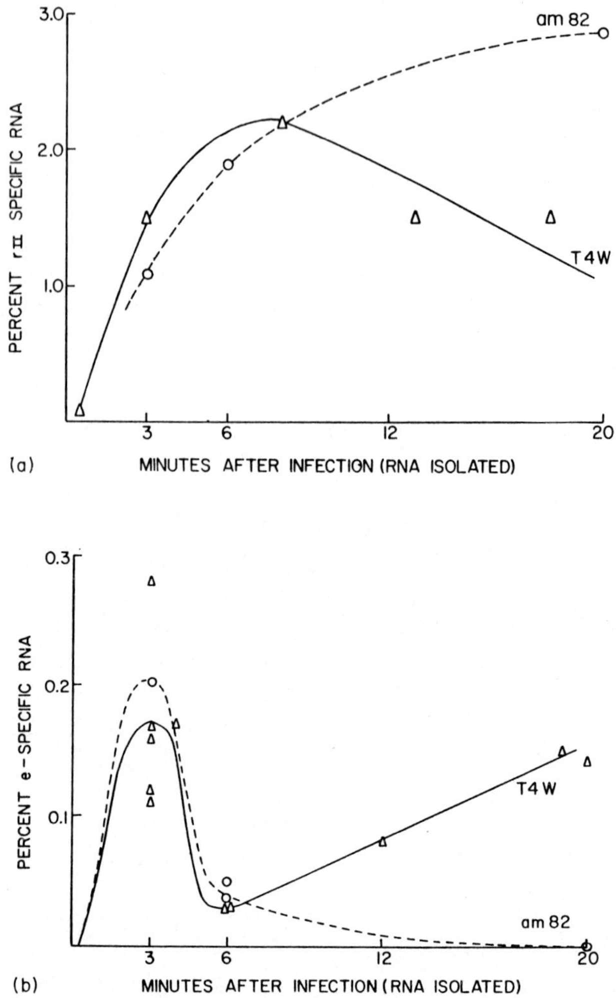

FIG. 2. Percentage of *r*II-specific (*a, top*) and *e*-specific (*b, bottom*) RNA found at different times after infection. Experimental details and some of the points shown are those of Bautz *et al.* (3). See *Note added in proof,* page 123.

portions, and one portion was harvested after 3 minutes (zero time with respect to the addition of the cold precursor), the other 8 minutes after infection. The RNA was extracted from the two portions, and the relative amounts of both, H^3-labeled *e*- and *r*II-specific RNA, were determined. The reduced level of *e*G19-specific RNA at minute 8 indicates that most of the lysozyme messenger RNA is degraded in the absence of protein synthesis, and also that, of the label remaining in the pool, little is

TABLE I
TURNOVER OF EARLY LYSOZYME MESSENGER RNA

Time of labeling (minutes after infection)	*e*G19-specific RNA (% of total hybrid)	*r*1272-specific RNA (% of total hybrid)
A. 1–3[a]	0.204	2.58
B.[b] 1–3 + cold uracil 3–8[a]	0.048	2.49

[a] Cells were harvested at times indicated.
[b] The specific activity of B was 45% of that of A.

reutilized toward the synthesis of new lysozyme messenger RNA. The nucleotide precursor pool in phage-infected cells seems to be extremely tight (14); that is, once incorporated into RNA, the same precursors are recycled preferentially even in the presence of a large excess of purines or pyrimidines in the medium. In view of this fact, we conclude that the constant relative specific activities observed for *r*II-specific RNA are due to continued synthesis rather than to stability of *r*II-mRNA produced before minute 3.

Before drawing any conclusions about possible mechanisms involved in the production and cessation of early messengers, we had to ascertain first that the lysozyme-specific RNA found at early times after infection is indeed the coding strand for the lysozyme protein. The deletion mutant *e*G19 used in our studies spans the entire *e* gene and extends with both ends into neighboring genes. Therefore, it is necessary to exclude the possibility that a neighboring gene, also delected by *e*G19, may be transcribed early, while the *e* gene may be transcribed only late. To test this possibility, we have isolated early and late *e*G19-deletion-specific RNA and incubated it with DNA filters of two other deletion mutants covering a large and a small part of the *e* gene and both extending beyond the right-hand side of the *e* cistron. The two sets of data obtained with early and late *e*G19-specific RNA roughly coincide, indicating that both RNA samples are mainly specific for the *e* gene (Fig. 3).

While the experiment just described tends to exclude the possibility that we are dealing with messengers transcribed from two different genes at early and late times after infection, it could still be possible that early and late *e*-specific RNA might be transcribed from opposite strands. Thus, at minute 3 we might be looking at an RNA species complementary in sequence to the true lysozyme messenger. This possibility is rendered unlikely by the following two experiments:

Experiment I: If the early lysozyme RNA is complementary in sequence to the late lysozyme RNA, it should be possible to anneal the

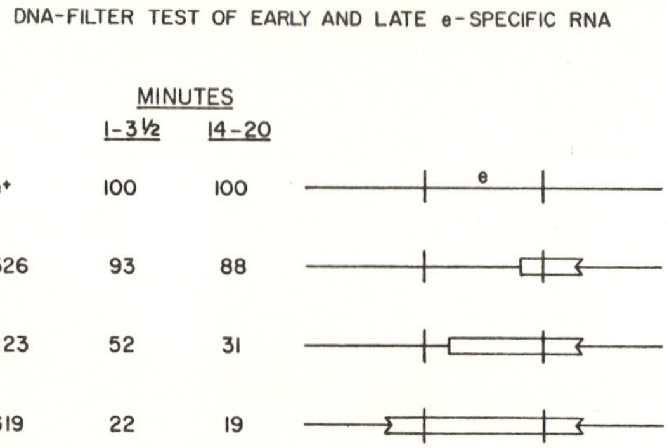

FIG. 3. Annealing test of early and late *e*-specific RNA with DNA of different deletion mutants. The RNA samples, pulse-labeled for the times indicated, were purified by two passages through *e*G19/*e*$^+$ DNA columns. Samples of *e*-specific RNA were incubated with filters containing DNA of *e*$^+$ phage or of the mutants shown.

two RNA species to form double-stranded complexes resistant to ribonuclease. The result from such an annealing experiment is negative—no nuclease-resistant complexes have been found (Table II).

Experiment II: The preliminary amino acid sequence of phage lysozyme which has been established recently by Tsugita *et al.* in Japan and which has been communicated to us by Emrich (9) enables us to write out a tentative nucleotide sequence for lysozyme mRNA. However, because of the degeneracy of the nucleic acid code, almost every third nucleotide has one or more alternatives. Nevertheless, due to the exceptionally high ratio of AT over GC in T4 DNA, one can expect that, whenever there is a choice between A and G or U and C at position three in a codon, A and U will be preferred. While this assumption may not hold for all codons, it should certainly apply for the great majority of them. If we assume that A and U are *always* preferred, it is possible to calculate the distribution of oligonucleotides to be expected from a complete digest of *e*-specific RNA with ribonuclease T1. Besides finding a high incidence of very large fragments of more than ten nucleotides, one

TABLE II
SELF-ANNEALING TEST OF 0- TO 4-MINUTE P^{32}-LABELED *e*-SPECIFIC RNA WITH EXCESS COLD 15-MINUTE RNA[a]

TCA-precipitable	cpm
−RNase	256
+RNase	12

[a] Experimental details as described by Geiduschek *et al.* (11).

should obtain as many as twenty-five trinucleotides per lysozyme messenger molecule, representing roughly 200% of that expected from a random distribution of nucleotides. If, at early times after infection, we were dealing with the anticodon strand (that is, the complement of *e*-RNA), the number of trinucleotides expected from a T1 digest would become reduced to 5, or less than one-half of randomness. When a sample of P^{32}-labeled 0- to 4-minute RNA, which was approximately 50% specific for the lysozyme gene, was digested with ribonuclease T1 and fractionated by DEAE cellulose chromatography in 8 *M* urea of increasing salt concentration, the trinucleotide peak found turned out to be significantly higher than the one obtained from an identical digest of nonspecific T4 RNA (Fig. 4). The distributions of nucleotides for the two

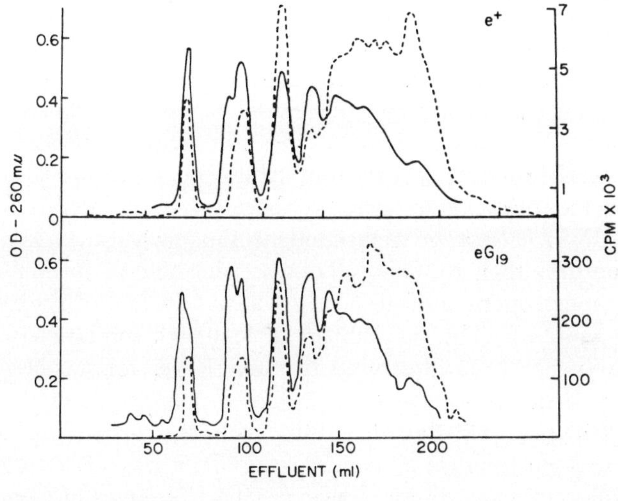

FIG. 4. Elution profile of oligonucleotides resulting from a complete digest with ribonuclease T1 of 0- to 4-minute P^{32}-labeled RNA obtained after passage through an *e*G19/e^+ double-layer DNA column. Upper profile: RNA eluted from e^+ DNA column specific to approximately 50% for region deleted from the genome of mutant *e*G19. Lower profile: Nonspecific T4 RNA, eluted from *e*G19 DNA column. The samples were digested in the presence of 1 mg of yeast RNA for 40 minutes at 37°C at an enzyme-to-substrate ratio of 1:20. The digests were fractionated through DEAE Sephadex columns of the dimensions 0.4 × 90 cm. Elution gradient: 0 to 0.6 *M* NaCl in 8 *M* urea.

RNA samples are shown in Table III. Theoretically, 15% of the lysozyme message should be found in the trinucleotide fraction. The nonspecific RNA yields 9% of the total counts recovered as trinucleotides. With our sample of *e*-specific RNA being roughly 50% pure, we would expect to find 12% of the total radioactivity in the trinucleotide peak. This is the

TABLE III
DISTRIBUTION OF NUCLEOTIDES IN A COMPLETE DIGEST WITH RNASE T1

Peak number (chain length of product)	Percent of total cpm (RNA sample eluted from column)		
	*e*G19	e^+	
1	3.7	4.6	
2	6.0	5.7	
3	9.5	*12.3*	15.7[a]
4	7.6	6.2	
5 and higher	73.2	70.7	

[a] Expected if 100% *e*-specific.

percentage actually observed. The results of these experiments provide indirect but nevertheless strong evidence that the early *e*-specific RNA is indeed the messenger for the structural protein of phage lysozyme.

How then can we explain this initial burst of synthesis of lysozyme messengers? We propose that the lysozyme gene is transcribed together with the early function genes, but, while the latter are also translated, the lysozyme messengers, possibly because of a missing or modified transfer RNA, cannot be translated at this early time. This failure of translation may then result in feedback inhibition of further synthesis of lysozyme messengers, indicated in the sharp decline of lysozyme-specific RNA at minute 6. The mechanism of feedback inhibition for *e*-specific RNA may proceed as illustrated by Stent (this volume, Fig. 1). While protein synthesis proceeds, the ribosomes moving along the messenger RNA will help to pull the RNA away from the DNA template, allowing further RNA-polymerase molecules to produce more RNA copies. In the absence of protein synthesis, however, the ribosomes are unable to pull the messenger off its template. This may cause the halt of further transcription, due to the unavailability of template DNA. The burst of lysozyme messenger synthesis appears to be quite analogous to the overshoot mechanism discussed by Pardee (18)—in fact, it may actually represent an example of the repression-leakage phenomenon observed in bacteria, which, when in synchrony, show an oscillatory behavior of repressed levels of enzymes due to periodic changes in repressor concentrations brought about mainly by a steady increase in cell volume plus a rhythmic change in gene dosage.

The question to be asked now is whether the overshoot observed for lysozyme transcription at early times is representative for all late function genes, or whether there are late function genes which will not be transcribed at all until after the onset of DNA replication. Experiments involving competition for DNA–RNA hybrid formation between pulse-labeled late RNA and excess cold early RNA favor the second alternative. To ascertain the validity of these observations we performed competition experiments identical in design with those reported (13) except that, for the annealing process, the DNA was immobilized on nitrocellulose filters (12). This procedure, we believe, offers the advantage that the T4 DNA filters can be charged with an excess of RNA species produced during a certain time interval after infection, followed by transfer of the DNA filters into a vial containing RNA species isolated after a different time interval. Since the DNA–RNA complexes are stable and cannot be detectably displaced by further incubation with the same RNA species, this procedure should serve to eliminate the possibility that the presence of low levels of some RNA species at one time after infection might be masked by a manifold higher level of the same species produced at a different time.

The results obtained from such a competition experiment are summarized in Fig. 5. Three sets of DNA filters were preincubated with 500 μg each of 3-, 9-, and 15-minute cold RNA, washed, and incubated with tritiated RNA isolated from cells harvested at the same times after infection.

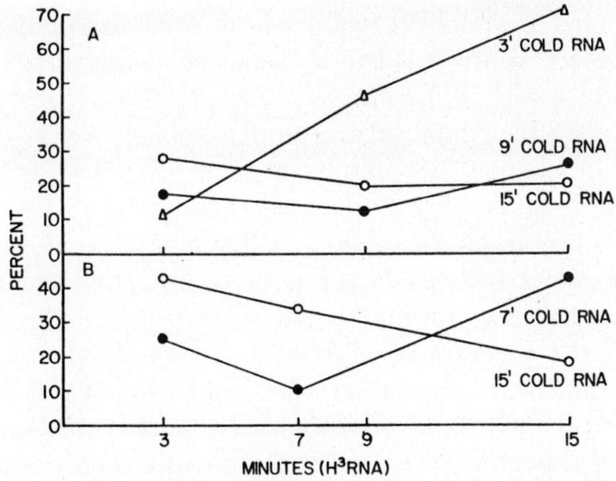

FIG. 5. *A*, nitrocellulose filters charged with 1.0 μg of T4 DNA were incubated in 2xSSC at 65°C for 6 hours with 500 μg of cold RNA extracted at 3 (△———△), 9 (●———●), or 15 (○———○) minutes after infection. The filters were then washed and incubated for another 6 hours at the same temperature with 250 μg of RNA, pulse-labeled for 2 minutes with uracil-6-H^3, and harvested at the times after infection shown on the abscissa. *B*, filters charged with 5 μg of T4 DNA were incubated as above with 115 μg of RNA pulse-labeled with uracil-2-C^{14} for a period of 2 minutes ending at minute 7 (●———●) or 15 (○———○) after infection. Second incubation as in *A*.

After preincubation with 3-minute cold RNA, the 3-minute H^3-RNA could occupy only about 10% of the sites normally available, whereas the 9-minute and 15-minute H^3-RNA could hybridize with the same DNA to increasing percentages. This result shows that, after preincubation with excess early RNA, there are more vacant sites on the T4 DNA for late than for early RNA species.

Therefore, in agreement with earlier competition studies, we must conclude that only part of the T4 genome is being transcribed at the time the lysozyme gene is transcribed for the first time.

Excess amounts of 9-minute and 15-minute cold RNA compete almost as effectively with labeled 3-minute RNA as with labeled late RNA, indicating that early RNA species are present late also. However, the slopes of these two curves between 3 minutes and 9 minutes appeared slightly negative, and we therefore repeated the experiment with subsaturating levels of cold RNA (experiment B). Here, the negative slope is steeper, indicating that those RNA species that predominate early are not the same as the ones occurring most frequently late. Therefore we conclude that there is not only a qualitative but also a quantitative shift of messenger RNA species as phage development proceeds.

When we compare the data on the early transcription of the *e* gene with those on the competition between early and late T4 RNA, we are forced to conclude that not all genes whose functional polypeptides are found only late are controlled by the same mechanism. It seems rather that some, if not most, of the late function genes are not transcribed at all during the first few minutes and are thus likely to be controlled at the DNA level, whereas the endolysin gene, together with the early function genes, appears subject to translational control. Considering the possibility that we might be able to explain almost all relevant data gathered in bacterial systems in terms of translational control affecting the rate of transcription through feedback regulation, it seems almost incredible that the much smaller genome of T4 should command two independent mechanisms to regulate the expression of a relatively small number of genes. However, if one compares the biosynthetic steps during the life cycle of a bacteriophage with those of its host, one recognizes a fundamental difference between the two: bacteria duplicate, whereas the phages multiply. Under constant environmental conditions, and while the cells are in logarithmic growth, a bacterial genome is expressing itself in exactly the same pattern as will its two daughters a generation later—that is, after one round of replication. In contrast, once the phage DNA starts multiplying, there are no biochemically defined "replicative cycles," but rather a cascade of uninterrupted rounds of replication. Prior to the onset of DNA synthesis, the enzymes necessary to initiate and to sustain this cascade are produced. These early enzymes must be specified by the parental genome. Since production of early enzymes is discontinued after the first few rounds of replication, it is obvious that the parental genome is expressed differently than are its daughters, in which different portions of the genome are expressed predominantly. Thus one could imagine that, before replication, the T4 genome is in a differentiated state allowing only part of it to be transcribed. Replication could result in *de*differentiation or *re*differentiation leading to the transcription of all or a different portion (the one coding for late functions) of the genome.

This property of a differentiated genome may be unique for phage and missing in its host. However, it may not be entirely absent from all bacterial species, since one might be able to draw a parallel between the germination of spores and the early stages in phage development as two instances of limited expression of a genome.

Speculating about how the phage genome could accomplish this dedifferentiation, one could think of the *E. coli* RNA-polymerase as having a greater affinity for the initiation sites of early function genes than for those of late function genes, with one of the early genes coding for a phage-specific RNA-polymerase reading late genes exclusively. Alternatively, the structural conformation of DNA itself may play the most important role in regulating the expression of late function genes. This idea is supported by the following two observations: Looking at the circular linkage map of T4, both the early function *r*II genes (7) and the late function lysozyme gene (20) are transcribed counterclockwise (Fig. 6). Gene 23 (specifying the phage coat protein, 6) and the two pairs of genes, 51/27 and 34/35 (concerned with the synthesis and assembly of tail fibers), are read clockwise—in the direction opposite to *e* and *r*II (19). RNA is synthesized only antiparallel to the template DNA strand, implying that, of genes transcribed in opposite directions, opposite DNA strands are serving as templates. Second, Kozinski *et al.* (15) have reported that within the first 6 minutes after infection the parental genome is present as a duplex DNA molecule consisting of two uninterrupted chains. Soon after the onset of DNA replication, however, the daughter DNA molecules develop "nicks"—single-strand breaks which contribute to a change in tertiary structure of the double helix.

Generalizing from the position of the signs in Fig. 6, we may speculate that genes located in the early cluster are read from one DNA strand, while those located in the late cluster are read from the other strand. (According to this generalization, we have to classify the function of the *e* gene as a "delayed early" one.) If we now make the reasonable assumption that during transcription the DNA template is spinning, rather than that the RNA polymerase is moving around it (17), the DNA can be

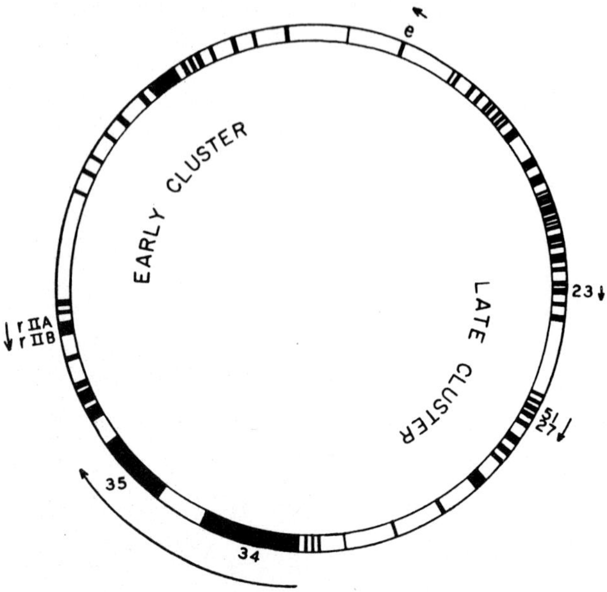

FIG. 6. Circular linkage map of phage T4 (abstracted version of map published by Edgar and Wood, 8).

spun only in one direction in order to become transcribed successfully (if the DNA template were to spin in the wrong direction, it would continuously move away from the enzyme). If we then assume that the host polymerase has a stronger affinity for early function genes, it will start spinning the injected phage genome automatically into that direction, allowing transcription of the early, but not of the late, cluster. Single-strand breaks could act as "swivel points" allowing the two portions of the genome to be read in two directions. If correct, one could expect the late function proteins always to be preceded by the occurrence of single-strand breaks in the template DNA; the validity of this correlation can be checked by experiment.

In summary, it appears as if phage T4 commands two modes of regulating transcription: one, as in the case of the endolysin gene, in the form of a feedback control governed by the rate of translation, the other, as for the genes in the "late cluster," in the form of a transcriptional control possibly exerted through the integrity of the genome.

ACKNOWLEDGMENTS

This work was supported by grants from the U. S. Public Health Service (GM 10395) and NSF (GB 4971). One of us (E. B.) holds a U. S. Public Health Service Research Career Development Award.

Note added in proof:

We have since found that, at late times after infection, the rate of mRNA synthesis has increased severalfold over that found at minute 3; therefore, proportionally more *e*G19-specific RNA is made late.

References

1. Attardi, G., Naono, S., Rouviere, J., Jacob, F., and Gros, F., *Cold Spring Harbor Symp. Quant. Biol.,* **28,** 363 (1963).
2. Bautz, E. K. F., and Reilly, E., *Science,* **151,** 328 (1966).
3. Bautz, E. K. F., Kasai, T., Reilly, E., and Bautz, F. A., *Proc. Natl. Acad. Sci. U.S.,* **55,** 1081 (1966).
4. Beckwith, J. R., *in* "Structure and Function of the Genetic Material" (H. Stubbe, ed.), *Abhandl. Deut. Akad. Wiss. Berlin Kl. Med.,* No. 1, 119 (1964).
5. Benzer, S., *in* "The Chemical Basis of Heredity" (W. D. McElroy and B. Glass, eds.), p. 70, Johns Hopkins Press, Baltimore, 1957.
6. Brenner, S., personal communication, 1964.
7. Crick, F. H. C., Barnett, L., Brenner, S., and Watts-Tobin, R. J., *Nature,* **192,** 1227 (1961).
8. Edgar, R. S., and Wood, W. B., *Proc. Natl. Acad. Sci. U.S.,* **55,** 498 (1966).
9. Emrich, J., personal communication, 1966.
10. Garen, A., *Virology,* **14,** 151 (1961).
11. Geiduschek, E. P., Moohr, J. W., and Weiss, S. B., *Proc. Natl. Acad. Sci. U.S.,* **48,** 1078 (1962).
12. Gillespie, D., and Spiegelman, S., *J. Mol. Biol.,* **12,** 829 (1965).
13. Hall, B. D., Nygaard, A. P., and Green, M. H., *J. Mol. Biol.,* **9,** 143 (1964).
14. Haselkorn, R., personal communication, 1966.
15. Kozinski, A., Kozinski, P., and James, R., *J. Virol.,* **1,** 758 (1967).
16. Krieg, D., *Virology,* **8,** 80 (1959).
17. Maaløe, O., and Kjeldgaard, N. O., "Control of Macromolecular Synthesis," Benjamin, New York, 1966.
18. Pardee, A. B., *in* "Metabolic Control Colloquium," Johnson Research Foundation, 1965.
19. Stahl, F. W., Murray, N. E., Nakata, A., and Crasemann, J. M., *Genetics,* **54,** 223 (1966).
20. Streisinger, personal communication, 1966.
21. Wiberg, J. S., Dirksen, M., Epstein, R. H., Luria, S. E., and Buchanan, J. M., *Proc. Natl. Acad. Sci. U.S.,* **48,** 293 (1962).

Discussion of Part II

S. B. Weiss, J. K. Bhattacharjee, G. S. Stent, C. A. Thomas, T. Kasai, R. P. Novick

Chairman Weiss: Are there any questions?

Dr. Bhattacharjee: I find it difficult for the eucaryotic system where we have a distinct nucleus and a cytoplasm separated by a nuclear membrane; how do you postulate, Dr. Stent, that the synthesis of protein is initiated on the messenger RNA long before messenger RNA separates from the DNA which is inside the nucleus? Now the second question, again has to do wtih the eucaryotic cells where we have the origin of the ribosomal RNA perhaps primarily in the nucleolar RNA, if I am not mistaken, and when you postulate that the ribosomal RNA serves as the messenger for the ribosomal protein, do you postulate that the nucleolar RNA in the case of eucaryotic cells has some kind of genetic message which it can transcribe to the ribosomal RNA?

Dr. Stent: As far as the first question is concerned, my answer will depend on whether protein synthesis occurs in the nucleus, a matter which, I think, is still the subject of controversy. If there *is* protein synthesis in the nucleus, then of course my scheme is readily applicable because the ribosomes would originate in the nucleus, pick up the nascent messenger RNA from the chromosomes, and migrate as polysomal aggregates to the cytoplasm, where they then continue the protein synthesis they began in the nucleus. Indeed, there are indications that the messenger does not migrate to the cytoplasm as free RNA, but already in the nucleus forms part of some kind of ribonucleoprotein aggregate. If, however, it should turn out that there is no protein synthesis in the nucleus, then I would say that, since most of the messenger synthesis appears to proceed in the vicinity of the nuclear membrane, perhaps the ribosomes are on the outside of the membrane and are pulling the messenger through the nuclear membrane from chromosomes into the cytoplasm. Thus, I believe, I have covered myself for both possibilities. And as for the second question, I would, of course, believe that the same messenger function obtains also for the ribosomal RNA of eucaryotes; here the nucleolus would be the place where the cistrons for the ribosomal RNA are located, which is translated into protein just as any other messenger for the generation of ribosomal protein.

Chairman Weiss: I would like to ask one question. I am just a little bit confused with the scheme. In one case, you invoke chloramphenicol when one does not need protein synthesis, just a flow of ribosomes pulling the messenger off; and in the other case, you invoke transfer RNA, whether it is charged or uncharged, which would imply that protein synthesis is needed. Can you clarify this for me?

Dr. Stent: I believe that for motion of ribosomes over nascent RNA, the sequential arrival of aminoacyl transfer RNA is essential. Thus, aminoacyl transfer RNA arrives at the ribosomal "A" site, and then under ordinary conditions it moves to the "P" site as soon as the peptide is formed causing advance of the messenger and display of the next codon at the "A" site. I would imagine that in the presence of chloramphenicol, advance of aminoacyl-tRNA from "A" to "P" site and advance of messenger RNA occurs without formation of the peptide bond, whose very formation the antibiotic appears to inhibit. This view can draw support from the finding that for RNA synthesis to proceed in chloramphenicol, all tRNA's must be aminoacylated. Thus, I believe that also in the presence of chloramphenicol the driving mechanism of nascent RNA through ribosomes is nothing but the successive arrival of aminoacylated transfer RNA's.

Chairman Weiss: Are there any other comments?

Dr. Thomas: Dr. Kasai, I don't know whether you can provide the answer to this or not, but have you got any idea of whether a single infected cell can produce both early and late lysozyme messenger? I was thinking for example, that since the molecules are permuted, the lysozyme gene is near the beginning of some molecules and near the end of others, and I wonder whether there is a position effect that could be playing a role here.

Dr. Kasai: If the T4 genome is randomly permuted, there should not be a relation between permutation and the order of transcription, because we did not detect any *e* gene transcription before one minute after infection, although there is a substantial transcription during this period.

Dr. Novick: I have a question for Dr. Stent. Do the relaxed mutants that you studied make polysomes when starved of their amino acids?

Dr. Stent: Yes, a paper came out about just a month ago by DeMoss and his collaborators and a very similar paper is now in press by Davis and his collaborators, which show that during amino acid starvation the relaxed strain not only forms but also maintains a normal polysome complement. This finding seems perfectly natural to me because obviously if messenger RNA is made polysomes are made. It is a fact, furthermore, that during amino acid starvation of stringent-control strains, while RNA synthesis stops or is greatly reduced, the polysomes fall apart.

Chairman Weiss: If there are no more questions, the session is concluded.

PART III

PROTEIN SYNTHESIS AS HETEROGENEOUS PROCESS

Chairman's Remarks

Alexander Rich

Department of Biology,
Massachusetts Institute of Technology,
Cambridge, Massachusetts

I would like to open the next session of our symposium. Let me remark in passing that the rate of progress in this field of protein synthesis is nothing short of phenomenal. As recently as five years ago, we had only a meager understanding of this process. And today we have a reasonably cogent general scheme whereby we believe we understand most of protein synthesis. We might be wrong, but the evidence available today appears quite sound. The title of this session is Protein Synthesis as a Heterogeneous Process. This is a very apt title. There is, as you know, a branch of chemistry called heterogeneous catalysis which deals with phenomena that occur at interfaces. Protein synthesis is of that type, except that it is not a process involving a solid state, but rather something intermediate between solids and solutions. The ribosome, as you know, is a very large particle, heavily hydrated. It has a great deal of organizational structure and the assembly process which takes place on or in the ribosome is one that we understand in general outline. However, the process has a great many unknowns connected to it. There are many questions that one can pose at this stage. We talk about the *universality* of the genetic code and that appears to be a fact. However, we might also talk about the *universality* of the ribosome which also appears everywhere that protein synthesis occurs. Why is it used universally in this way? The answer is unknown today.

The ribosome has been described as the "black box" of molecular biology. The "black box," to the electronics engineer, is a symbol that describes a certain function; the contents of the black box are not defined. This is very much the state of our understanding of the ribosome today. We understand something about its function. We do not understand how it does it. Over 80% of the RNA of the cell is bound up in the ribosomal RNA. Our ignorance is complete concerning the reason for this large accumulation of RNA an "informational macromolecule" stored within the ribosome.

There are, of course, a great variety of molecular interactions which occur on the surface of ribosomes attendent to protein synthesis. In part, this is related to the great diversity of proteins which are being assembled. Some of them are fibrous proteins in which there is a fair amount of association between the growing polypeptide chains as they are being formed. Others are globular proteins which fall off the ribosome rather quickly after they are completed. It is quite likely that a great deal of effort will be needed to understand fully the molecular details and the durability of this system. Finally, we will eventually have to face the problem of the evolutionary origin of the system and ask ourselves how proteins were made before the ribosome existed. This gets us into a quite different kind of chemistry that presumably will occupy our attention once we understand more fully the details of protein synthesis in contemporary biological systems.

Bacterial Amino Acid Polymerization[1]

Fritz Lipmann, Yasutomi Nishizuka,[2] Julian Gordon,
Jean Lucas-Lenard, and Max Gottesman[3]

The Rockefeller University, New York, New York

When we first became interested in protein synthesis, the ribosome impressed us as an inert stage on which the process of polypeptide synthesis occurred (15). In a way, of course, the ribosome can be described as a chemical reactor on which components are brought together to interact with each other. However, the more we go into the problem, the more we realize that it is an "active" stage, or that the participation of the ribosomal components, and particularly the proteins, is considerably greater than we had anticipated. An increasing number of different "supernatant" factors and ribosomal proteins are now in the process of being identified. In mammalian systems, Arlinghaus *et al.* (3) and Fessenden and Moldave (8) have isolated two supernatant factors. In bacterial systems, with which we are going to deal here, our group first isolated two factors that we called G and T (2, 19*a*); more recently, Dr. Lucas-Lenard has split T into two fractions, unstable T_u and stable T_s (16), so that now we have three altogether. In addition, factors dealing with chain initiation are reported by a number of groups (7, 20, 23), and recently essential proteins have been reversibly extracted from the ribosome (12, 14, 22).

The systems we chose, the poly U-directed polymerization of phenylalanine and poly A-dependent polylysine synthesis, are simplified. Under our conditions, polypeptide chains are started in the absence of initiation factors, and termination is circumvented, since there is no chain release because the polypeptide remains linked to sRNA. What we hope to do with these systems, however, is to concentrate on the mechanism of polypeptide chain elongation.

[1] This work was supported in part by a grant from the National Science Foundation.

[2] Present address: Department of Medical Chemistry, Kyoto University Faculty of Medicine, Kyoto, Japan.

[3] Fellow of the Jane Coffin Childs Memorial Fund for Medical Research. Present address: National Institute of Arthritis and Metabolic Diseases, Bethesda, Maryland.

Instead of thinking of an inert ribosome, we now feel that we may be dealing with a complementarity between ribosomal and supernatant factors, and should consider the multiplicity of ribosomal proteins as part of the polymerization system. One of the reasons for thinking of a complementarity is the specificity of microbial supernatant for microbial ribosomes and animal supernatant for animal ribosomes (Table I) (17).

TABLE I
RIBOSOME SPECIFICITY OF TRANSFER FACTOR (17)

Ribosomes	Supernatant factor	C^{14}-Leucine transferred (cpm)
E. coli	None	90
E. coli	Liver	80
E. coli	*E. coli*	665
Liver	None	21
Liver	*E. coli*	19
Liver	Liver	244

The break appears to be between nucleated and nonnucleated cells, since there is an overlap between plant and animal systems. In spite of the relative specificity of these factors, we do not expect the sequence of reactions to be essentially different in the microbial and nucleated cells; rather, it appears to be a variant of the same scheme.

CHARACTERISTICS OF THE G FACTOR

Complementarity between supernatant and ribosomal fractions is exemplified by the G factor. In the studies of Allende *et al.* (2) and of Conway (6), it was noticed that a specific GTPase appeared to be coincident with a polymerization factor. By further purification of this factor, Nishizuka and Lipmann (19*a*) showed that it could be assayed for either by polymerization or by GTP hydrolysis; it was, therefore, called G. Figure 1 shows the complementarity between G and the ribosome in the hydrolysis of GTP. GTPase activity of G depends on the ribosome, and neither ribosomes alone nor G alone is active; the amounts of G that saturate the ribosome, it may be noted, are rather small. The ribosomes retain slight activity for GTP hydrolysis, but it is of too low an order to appear in the experiment of Fig. 1. Table II shows that GTPase activity depends on the whole ribosome. The 50 S unit has some activity, due possibly to incomplete separation, and the 30 S unit has no activity. The recombined 50 S +30 S ribosome regains full activity.

Figure 2 shows that polymerization can be uncoupled from GTPase by heating the ribosome. Whereas polymerization is rapidly inhibited by heating at 55°C, GTP hydrolysis first increases and then declines with

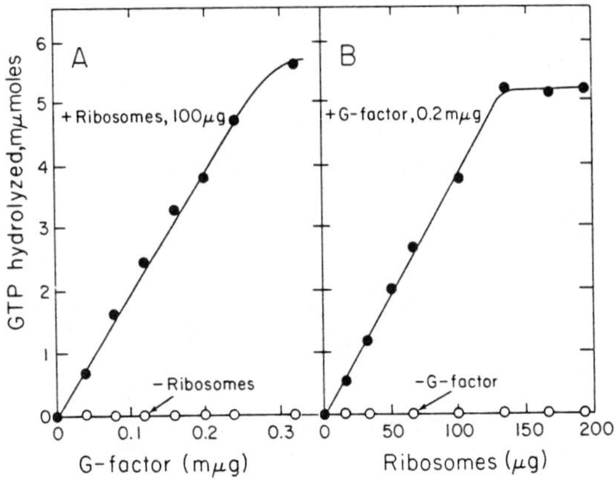

FIG. 1. GTPase catalyzed by the combination of G and ribosomes. A, the reaction was assayed under conditions previously described (19*b*), except that increasing amounts of G were added to the standard amount of ribosomes. *B*, assayed with a constant amount of G (0.2 mμg) and increasing amounts of ribosomes.

further heating. When we assume a functional coupling between phosphoryl transfer and polymerization and the latter is damaged by denaturation, a functional uncoupling of phosphate transfer appears to release GTP hydrolysis. An uncoupling in the crude system used in the experiment of Table II is also indicated by the fact that about forty times as much phosphate is released as amino acid is polymerized. However, when highly *purified* G in limiting concentration and well-washed ribosomes are used, the addition of purified T causes the phosphate released from GTP above the blank to become almost equal to amino acid polymerization (19*a*).

Difference between Mg^{++}- and Ca^{++}-Supplemented Ribosomal Systems

We turn now to experiments by Dr. Gordon in which he used calcium instead of magnesium for polymerization (9). It had been shown (5, 26)

TABLE II
Activities of Ribosomal Subunits (19*b*)

Ribosomes	Phenylalanine incorporated (μμmoles)	GTP hydrolyzed (μμmoles)
70 S ribosomes (150 μg)	27.7	1280
50 S subunits (150 μg)	2.3	504
30 S subunits (150 μg)	6.0	75
50 S (100 μg) + 30 S (50 μg)	29.3	1290

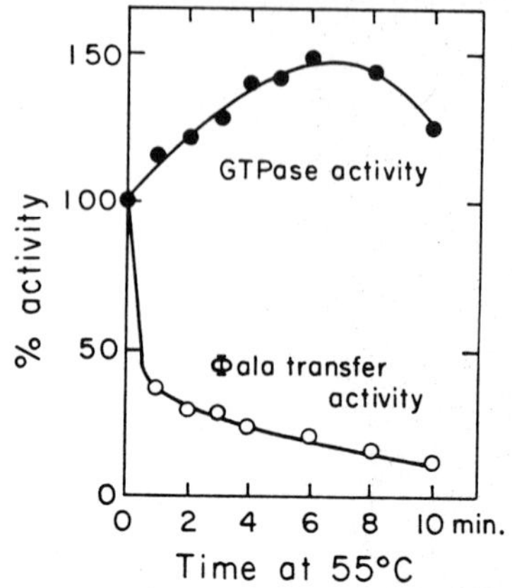

FIG. 2. Uncoupling of GTP split and phenylalanine polymerization by heating of ribosomes. The washed ribosomes were suspended in 0.01 *M* Tris-HCl buffer, pH 7.4, containing 0.1 *M* $MgCl_2$ (10 mg/ml), and heated at 55°C for various periods of time as indicated. These ribosomes were assayed for GTPase activity and polyphenylalanine synthesis with constant amounts of purified factors. The assays were carried out under conditions described in Nishizuka and Lipmann (19*b*).

that calcium can replace magnesium to stabilize the ribosome, but this observation received little attention. The calcium-supplemented system proved to be rather interesting. In Fig. 3 one sees the effects of the replacement of magnesium by calcium; polymerization is better, but, on

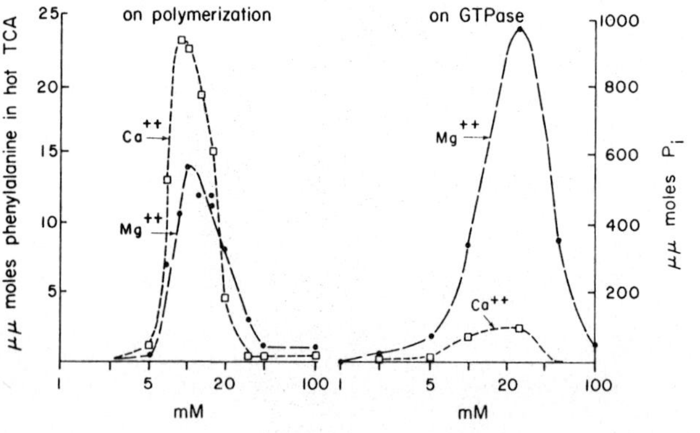

FIG. 3. Effect of nature of divalent cation. Reaction rates were measured after 10 minutes of incubation in 0.25-ml volume incubation mixtures containing 12.5 μmoles of Tris-HCl, pH 7.6, 3 μmoles of 2-mercaptoethanol, 30 μmoles of NH_4Cl, 10 μg of poly U, 100 μg of five-times-washed ribosomes, 100 μg of C^{14}-phenylalanyl-sRNA, 50 μg of 100-fold-purified combined G and T (9), and divalent ions as indicated. Polymerization was measured after addition of 0.25 μmole of GTP, and GTPase after addition of 5 mμmoles of GTP-γ-P^{32}.

the other hand, there is a lower rate of GTP hydrolysis in calcium. This seems to indicate that GTP cleavage and polymerization are more closely coupled in calcium than in magnesium. Figure 4 shows that G is rather closely associated with the ribosomes; in sucrose gradient centrifugation, the combined T factors are washed off, whereas G sticks tenaciously to the ribosome. After repeated washing with 0.5 *M* NH_4Cl, however, most of G is eventually removed. When well-washed ribosomes are assayed for polymerization in Mg^{++} and with added T but in the absence of added G, GTP hydrolysis and polymerization are low, but may be underestimated

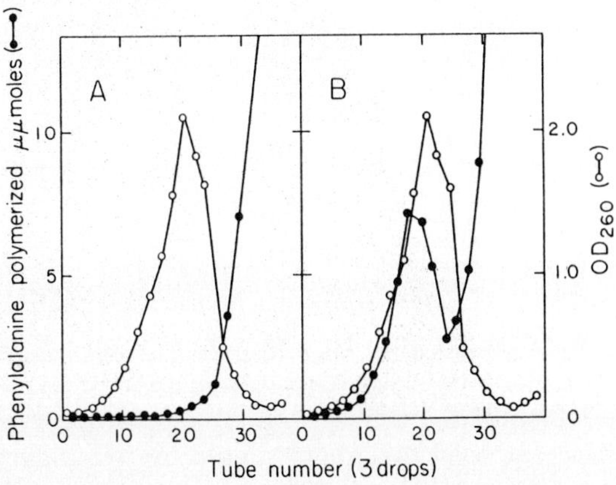

FIG. 4. Retention of T and G on ribosomes. Crude ribosomes were washed once with 0.01 *M* Tris-HCl buffer, pH 7.4, containing 0.01 *M* $MgCl_2$. Five milligrams of the preparation was subjected to a sucrose density gradient centrifugation (5 to 20% exponential, 5 ml) for 90 minutes at 39,000 rpm (Spinco, SW 39 rotor). After collecting 3-drop fractions, the activity of T (*A*) or G (*B*) was assayed in the presence of a saturating amount of the other factor, using 100 μg of collected ribosomes. The assay was carried out under conditions previously described (19*b*).

because we find that the low residual G prompts synthesis of short peptides that easily escape detection because of hot TCA-solubility. Calcium, however, amplifies the activity of residual G, since in its presence there is remarkable polymerization without addition of G (Fig. 5).

PEPTIDE SYNTHESIS AND CHAIN ELONGATION

To many who have been involved in the problem of peptide synthesis, the part played by GTP has been puzzling. Obviously it delivers energy-rich phosphate into the system. The energy needed for peptide synthesis, however, is relatively low (13) and energetically seems to be covered amply by the group potential of the ester link between the carboxyl group and the sRNA (19*b*). Against this background, let us proceed to

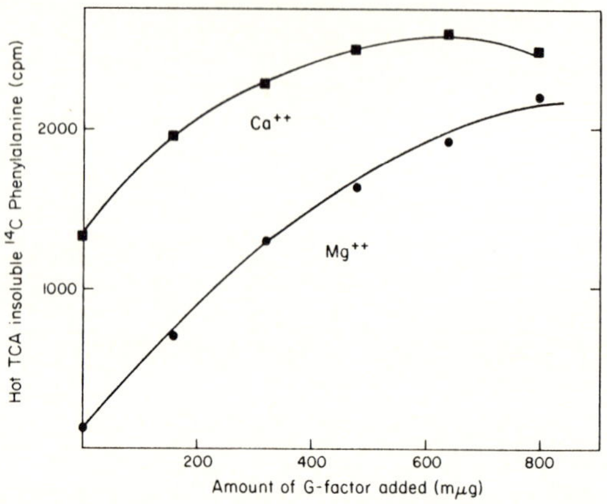

FIG. 5. Effect of nature of divalent cation on transfer activity of G. Reaction mixtures were the same as in Fig. 3, except for the inclusion of 5 μg of purified T and G as indicated.

map an action scheme of the ribosome in Fig. 6, which pictures phenylalanine polymerization on poly U. The poly U is on the 30 S part; on the 50 S portion one finds the two reactants in peptide synthesis, peptidyl-sRNA on the left and phenylalanyl-sRNA on the right. As indicated in the scheme, it is proposed that there should be two different sites on the ribosome, one for peptidyl-sRNA and one for aminoacyl-sRNA (19*b*, 25). For many reasons, one has to expect to find two sRNA's next to each

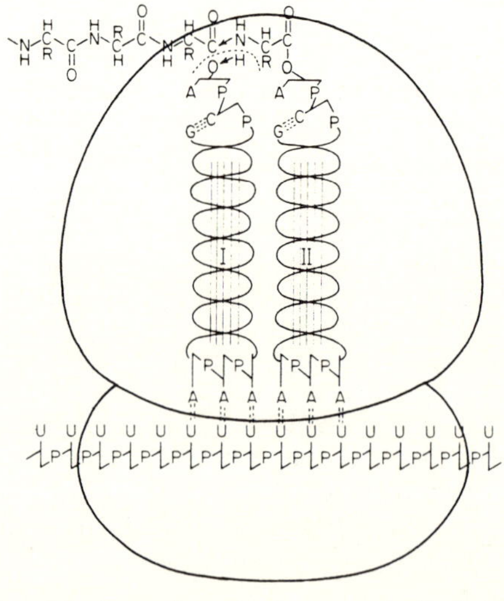

FIG. 6. Phenylalanine polymerization on poly U-charged ribosome.

other on a ribosome to form a new peptide bond. The peptidyl-sRNA linked by its triplet to the messenger determines the next triplet, which attracts the sRNA carrying the amino acid that is to follow; they should settle one beside the other to yield the amino acid sequence prescribed by the triplet sequence in the messenger. In the upper part of Fig. 6 one sees the newly arrived aminoacyl-sRNA, with its amino group, approaching the activated carboxyl group on the neighboring peptidyl-sRNA and displacing sRNA, thereby forming a new peptide bond.

In chain elongation the energy is taken off the ester link in peptidyl-sRNA. The group potential of peptidyl-sRNA is somewhat below that of aminoacyl-sRNA, which is comparable to a ~P in ATP. The difference in ΔF has been assessed to +2 kcal, which at pH 7 still leaves −7 kcal available to form a peptide bond with a group potential of −2–3 kcal (19*b*). If group activation through the ester link to sRNA is sufficient, one has to ask what else the ~P that is released from GTP could do.

Before facing this problem, we should like to discuss studies by Dr. Gottesman of a number of reactions that occur on well-washed ribosomes without added supernatant factors or GTP; Bretscher and Marcker have reported analogous results (4). Among the first to work in this area were Traut and Monro (25), who studied the reaction of puromycin with peptidyl-sRNA. Puromycin adds terminally to the peptidyl-carboxyl group by displacing sRNA. In Fig. 7 it can be seen that puromycin is a mimicry of the aminoacyl-sRNA terminal; its amino group on the O-methyltyrosine fools the peptide-linking enzyme into reacting with the carboxyl of peptidyl-sRNA. This reaction, therefore, can be considered a model for peptide synthesis, with the exception, however, that puromycin

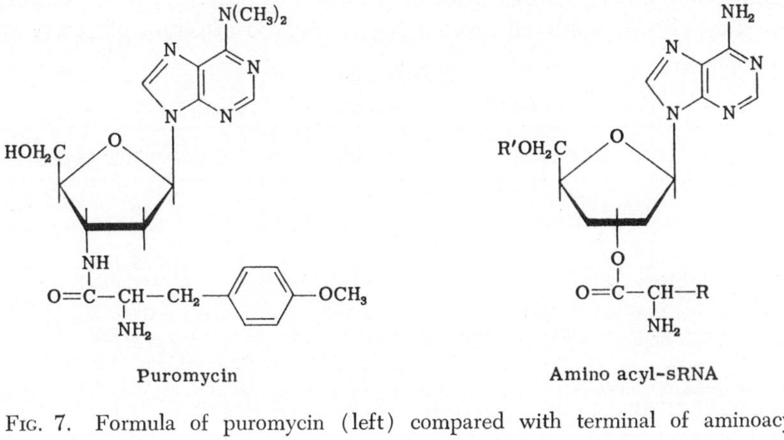

FIG. 7. Formula of puromycin (left) compared with terminal of aminoacyl-sRNA.

does not attach to the messenger RNA. Allen and Zamecnik (1) found that mistreated ribosomes can still carry out puromycin addition, and Rychlik (21) observed that polylysyl-sRNA can recombine with ribosomes charged with poly A and can be released by puromycin. Gottesman (10) found that well-washed ribosomes can be used to study this polylysine release as a model for peptide synthesis. It is rather indicative that triple-A (Table III) supports the binding of polylysyl-sRNA, which then

TABLE III
THE PUROMYCIN-INDUCED RELEASE OF H^3-POLYLYSINE FROM RIBOSOMES CHARGED WITH POLYLYSYL-sRNA[a]

Template	Bound H^3-polylysyl-sRNA (cpm) puromycin		Release
	+	−	(%)
Poly A	534	342	36
ApApA	368	86	77

[a] The system contained: 50 mM Tris-HCl buffer, pH 7.4; 12 mM 2-mercaptoethanol; 160 mM NH_4Cl; 0.4 mM C^{12}-lysine; 20 mM $MgCl_2$; 485 μg of washed ribosomes per milliliter; and H^3-polylysyl-sRNA (570 cpm, specific activity 5.8×10^9 cpm/μmole). Where indicated, 120 μg of poly A per milliliter, ApApA 1.81 OD_{260}/ml, and 4.6×10^{-5} M puromycin were added. Samples of 0.05 ml were incubated for 20 minutes at 24°C. The Nirenberg and Leder (18) Millipore binding assay was used.

reacts with puromycin. Table IV shows that addition of supernatant factors or GTP has little, if any, effect on the puromycin reaction. The synthesis of a peptide bond occurs here without added factors. Since the system is able to form a peptide bond with the amino group of puromycin, it was attractive to try for an addition to polylysyl-sRNA of lysine from lysyl-sRNA. With H^3-lysyl-sRNA, a limited addition of lysine was

TABLE IV
CLEAVAGE OF H^3-POLYLYSYL-sRNA BY PUROMYCIN[a]

Additions	TCA-soluble (%)
None	10
ApApA	36
A_6	54
Poly A	68
Poly A + GDP	64
Poly A + G, T, GTP	73

[a] The system was the same as in Table III, except that 10 mM $MgCl_2$, 4.6×10^{-5} M puromycin, and H^3-polylysyl-sRNA (750 cpm) were used. Where indicated, 0.94 OD_{260}/ml of A_6, 0.25 mM GDP, 0.25 mM GTP, and combined G and T, 4 μl of the ammonium sulfate fractions of supernatant per milliliter (19a), were added. Samples of 0.05 ml were incubated for 10 minutes at 37°C. Assay: precipitation by cold 5% TCA.

TABLE V
THE ELONGATION OF H^3-POLYLYSYL-sRNA BY C^{12}-LYSYL-sRNA ON VARIOUS TEMPLATES[a]

Template	Increase in hot tungstate-precipitable counts (– blank)
None	69
ApApA	72
A_6	210
Poly A	190

[a] The system was as used in Table III except that H^3-polylysyl-sRNA (3750 cpm TCA-insoluble, 300 cpm hot TCA-tungstate-insoluble) was used. After 5 minutes at 37°C, C^{12}-lysyl-sRNA was added in excess, and the samples (0.10 ml) were incubated for an additional 10 minutes at 37°C. Assay: precipitation after heating for 10 minutes at 90°C in TCA-tungstate. Blank values were obtained by omitting C^{12}-lysyl-sRNA from the incubation mixture.

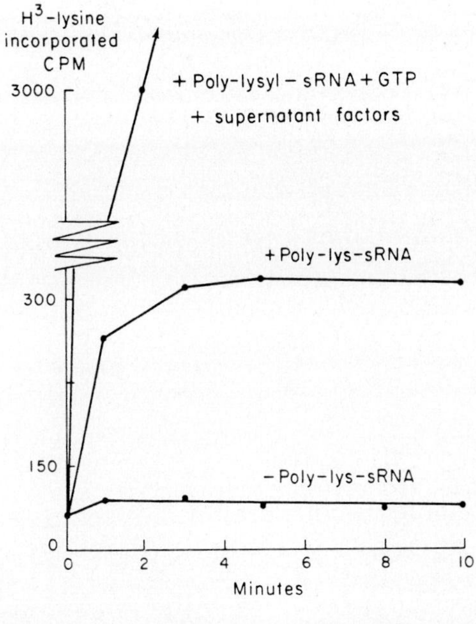

FIG. 8. Limited amino acid addition on well-washed ribosomes in absence of supernatant factors. Conditions were as described for Table III, except that H^3-polylysyl-sRNA was omitted, and 2.5 $\mu\mu$moles of C^{12}-polylysyl-sRNA were added; 10 mM $MgCl_2$ and 120 μg of poly A per milliliter were used. After 5 minutes at 37°C, H^3-lysyl-sRNA (20,000 cpm, specific activity 1.4×10^9 cpm/μmole) and, where indicated, 1.0 mM GTP were added; as supernatant factors DEAE-Sephadex eluate (0.16 mg/ml) as described by Nishizuka and Lipmann (19*b*) was added. Samples of 0.10 ml were further incubated at 37°C. Assay: precipitation by TCA-tungstate after limited alkaline hydrolysis.

found. Table V shows that, in contrast to puromycin addition, A_6 or poly A works but A_3 does not, according to the expectation that one needs at least two triplets to add a new lysine to polylysyl-sRNA. It is of interest that this addition is completely inhibited by chlortetracycline (11, 24). Figure 8 shows that monolysyl-sRNA alone does not polymerize (lower curve). With polylysyl-sRNA, one obtains a limited addition of lysine (middle curve). If, however, supernatant and GTP are added, chain elongation continues with an extensive incorporation of lysine (upper curve). By using double markers in a limited addition experiment, the chromatographic analysis of the different peptides formed is consistent with an addition of one lysine per reacting peptide chain.

Now we will return to the GTP problem and will discuss it with the assistance of Fig. 9. To explain why polypeptidyl-sRNA stops after adding a single lysine, we might argue it does so because the peptidyl-sRNA, after it has added a new amino acid, is on the wrong site, namely

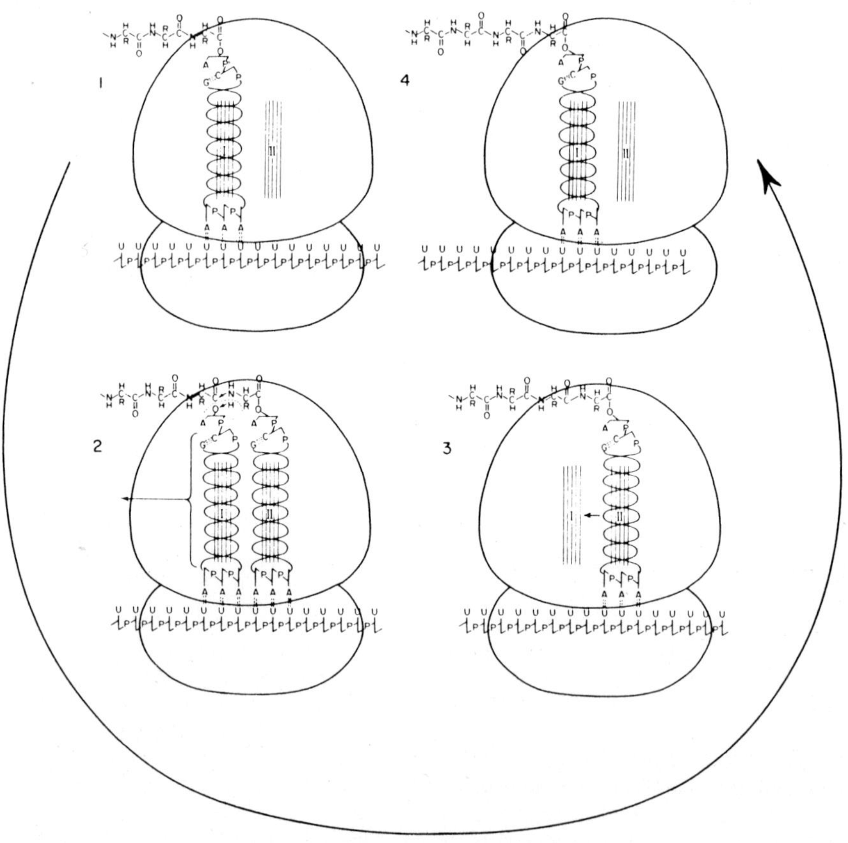

FIG. 9. Peptide condensation cycle.

II instead of I. A translocation has to follow (25)—that is, a transfer of the newly formed peptidyl-sRNA to its right site, site I, and it is proposed that in translocation, the phosphoryl released from GTP may be used. Therefore, supernatant factors and GTP are needed to continue amino acid addition. As we have seen, it is necessary to have the whole ribosome for GTPase activity, and we cautiously propose (cf. also ref. 25) that G + GTP release energy and promote the moving forward to site I of the elongated peptidyl terminal after it has formed through the condensation of the peptidyl residue to aminoacyl-sRNA. Figure 9 indicates, further, that the messenger RNA is pulled along in this movement; that is, the messenger moves one notch ahead and exposes a new triplet to site II, to which a new aminoacyl-sRNA is to be attracted.

FIG. 10. Electron micrograph of 50 S subunits by H. Ris, University of Wisconsin, Madison (personal communication).

The scheme in Fig. 9 was presented at a recent meeting in Holland which Dr. H. Ris attended. He was prompted by this proposition of two sites on the ribosome to mention his observation of two grooves in electron micrographs of 50 S subunits of *Escherichia coli.* He was kind enough to give permission to present his micrograph (Fig. 10). Although the grooves appear to be rather far apart, it is attractive to presume they correspond to site I and II.

References

1. Allen, D. W., and Zamecnik, P. C., *Biochim. Biophys. Acta,* **55**, 865 (1962).
2. Allende, J. E., Monro, R., and Lipmann, F., *Proc. Natl. Acad. Sci. U.S.*, **51**, 1211 (1964).
3. Arlinghaus, R., Shaeffer, J., and Schweet, R., *Proc. Natl. Acad. Sci. U.S.*, **51**, 1291 (1964).
4. Bretscher, M. S., and Marcker, K. A., *Nature,* **211**, 380 (1966).
5. Chao, F., *Arch. Biochem. Biophys.*, **70**, 426 (1957).
6. Conway, T. W., and Lipmann, F., *Proc. Natl. Acad. Sci. U.S.*, **52**, 1462 (1964).
7. Eisenstadt, J. M., and Brawerman, G., *Biochemistry,* **5**, 2777, 2784 (1966).
8. Fessenden, J. M., and Moldave, K., *J. Biol. Chem.*, **238**, 1479 (1963).
9. Gordon, J., and Lipmann, F., *J. Mol. Biol.*, **23**, 23 (1967).
10. Gottesman, M., *J. Biol. Chem.*, in press.
11. Hierowski, M., *Proc. Natl. Acad. Sci. U.S.*, **53**, 594 (1965).
12. Hosokawa, K., Fujimura, R. K., and Nomura, M., *Proc. Natl. Acad. Sci. U.S.*, **55**, 198 (1966).
13. Jencks, W. P., Cordes, S., and Carriuolo, J., *J. Biol. Chem.*, **235**, 3608 (1960).
14. Lerman, M. I., Spirin, A. S., Gavrilova, L. P., and Golov, V. F., *J. Mol. Biol.*, **15**, 268 (1966).
15. Lipmann, F., Bennett, T. P., Conway, T. W., Goldstein, J., Nakamoto, T., and Spyrides, G. J., *in* "New Perspectives in Biology" (M. Sela, ed.), p. 69, Elsevier, Amsterdam, 1964.
16. Lucas-Lenard, J., and Lipmann, F., *Proc. Natl. Acad. Sci. U.S.*, **55**, 1562 (1966).
17. Nathans, D., and Lipmann, F., *Proc. Natl. Acad. Sci. U.S.*, **47**, 497 (1961).
18. Nirenberg, M. W., and Leder, P., *Science,* **145**, 1399 (1964).
19. Nishizuka, Y., and Lipmann, F., (*a*) *Proc. Natl. Acad. Sci. U.S.*, **55**, 212 (1966); (*b*) *Arch. Biochem. Biophys.*, **116**, 344 (1966).
20. Revel, M., and Gros, F., *Biochem. Biophys. Res. Commun.*, **25**, 124 (1966).
21. Rychlik, I., *Biochim. Biophys. Acta,* **114**, 425 (1966).
22. Staehelin, T., and Meselson, M., *J. Mol. Biol.*, **15**, 245 (1966).
23. Stanley, W., Salas, M., Wahba, A., and Ochoa, S., *Proc. Natl. Acad. Sci. U.S.*, **56**, 290 (1966).
24. Suarez, G., and Nathans, D., *Biochem. Biophys. Res. Commun.*, **18**, 743 (1965).
25. Traut, R. R., and Monro, R. E., *J. Mol. Biol.*, **10**, 63 (1964).
26. Ts'o, P. O. P., *in* "Microsomal Particles and Protein Synthesis" (R. B. Roberts, ed.), p. 156, Pergamon, New York, 1958.

On the Problem of Autocomplementarity in Messenger RNA

ALEXANDER RICH

Department of Biology,
Massachusetts Institute of Technology,
Cambridge, Massachusetts

The genetic code is the name applied to the relationship between the sequence of nucleotides in messenger RNA and the amino acids which are polymerized during protein synthesis. One of the great accomplishments of the past five years has been the elucidation of this relationship. The information has been obtained from the laboratories of Nirenberg, Khorana, and others, largely through the use of synthetic polyribonucleotide molecules which can act as artificial messenger RNA in cell-free protein synthetic systems (1). Work by Crick and his collaborators on the mutations produced by acridine orange dyes pointed toward a reading of triplets of nucleotides during the translational process (3). This triplet reading system has been confirmed through the use of synthetic polyribonucleotide templates containing an ordered nucleotide sequence as well as by the study of acridine orange mutations in the production of the bacteriophage lysozyme molecule (6). Through these studies, as well as the studies of the binding of aminoacyl transfer RNA to ribosomes stimulated by nucleotide triplets (1), it has become possible to assign amino acids to most nucleotide triplets (codons), as shown in Table I. The purpose of this paper is to present some speculative considerations regarding the nature of the genetic code and to point out an effect which may have been important in the evolution of codons and the reading mechanism.

There are several generalizations related to the genetic code. First of all, most of the nucleotide triplets or codons appear to be active in the selection of a particular aminoacyl transfer RNA in binding experiments, and they act to incorporate that amino acid into a polypeptide chain. At the present time there is some uncertainty about a few of the triplets, but we can regard most of the code as established (1). Three of the nucleotide triplets appear to be related to polypeptide chain termination in mutants of *Escherichia coli,* and possibly some of these are active in the normal organism. Two codons appear to be

TABLE I
THE GENETIC CODE

Second base → First base (5′-OH end) ↓	U	C	A	G	Third base (3′-OH end) ↓
	PHE	SER	TYR	CYS	U
	PHE	SER	TYR	CYS	C
U	LEU	SER	Term.[a] (Ochre)	Term.[a]	A
	LEU	SER	Term.[a] (Amber)	TRYP	G
	LEU	PRO	HIS	ARG	U
C	LEU	PRO	HIS	ARG	C
	LEU	PRO	GLUN	ARG	A
	LEU	PRO	GLUN	ARG	G
	ILEU	THR	ASPN	SER	U
A	ILEU	THR	ASPN	SER	C
	ILEU	THR	LYS	ARG	A
	MET[b]	THR	LYS	ARG	G
	VAL	ALA	ASP	GLY	U
G	VAL	ALA	ASP	GLY	C
	VAL	ALA	GLU	GLY	A
	VAL[b]	ALA	GLU	GLY	G

[a] Terminates polypeptide chain synthesis in some mutants.
[b] F-met initiating amino acid in *E. coli*.

active in initiating polypeptide chain synthesis in *E. coli* through the incorporation of N-formyl methionine. However, most of the remaining codons are structural in that they specify a particular amino acid in a polypeptide chain. Thus, we may regard a messenger RNA coding for a particular polypeptide chain as having an initiator segment at the beginning 5′-OH end, followed by a group of structural codons, and finally a terminator sequence (1). This is shown diagrammatically in Fig. 1*a*, which also shows the general flow of ribosomes which pass over the messenger RNA during protein synthesis.

One of the remarkable properties of the genetic code is the fact that the amino acid specified by a codon is largely independent of the particular purine or pyrimidine found in the third position (Table I). Thus, in most cases either guanine or adenine is sufficient, or uracil or cytosine. Instead of being a triplet code, we can describe it as a doublet code which is further specified by an additional purine or pyrimidine. This has been interpreted by Crick as associated with an altered type of

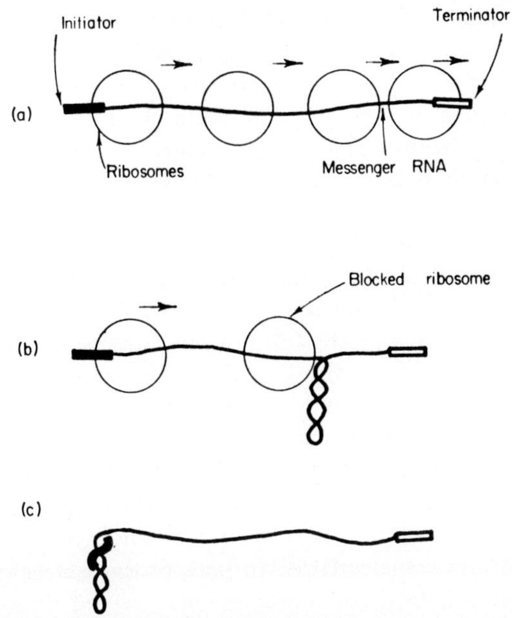

FIG. 1. Schematic diagram of messenger RNA and ribosomes: (*a*) normal protein synthesis, with the small arrows indicating the direction of ribosomal movement over the messenger; (*b*) the autocomplementary messenger sequence blocks the movement of the ribosomes; (*c*) the autocomplementary messenger RNA segment includes the initiator and blocks ribosomal attachment.

hydrogen bonding due to a "wobble" or change in the position of the hydrogen-bonding base in the third position (2). However, this interpretation does not explain what particular utility is derived from this type of ambiguity in the third position. It is possible that the value of this phenomenon can be understood from physical properties of the nucleic acids.

SECONDARY STRUCTURE IN THE NUCLEIC ACIDS

One of the characteristic features of polynucleotides is their ability to form double-stranded helical structures with the polymer backbones organized in antiparallel fashion. This arises directly as a consequence of the physical properties of the nucleotides. The planar purine and pyrimidine residues are stabilized to a considerable degree when they are stacked above each other, held together by van der Waals interactions, a type of hydrophobic binding in which water is excluded from between the bases. In this stacked configuration, the polynucleotide chain assumes a helical configuration as determined by the geometry of the sugar-phosphate backbone. The hydrogen-bonding interaction between the purines and pyrimidines contributes less stabilization energy than the stacking energy, but nonetheless it adds a large enough contribution to stabilize the formation of double helical molecules, especially if they have any appreciable length. Thus, for example, it has been shown that double helical nucleotides have substantial stability even though the chain length is not very appreciable (5). The exact stability is a function of the type of base. However, as the chain length increases above ten or fifteen residues, then substantial amounts of energy are required to break up the structure, as shown by an elevated melting temperature. This tendency to form stable double helical structures is found widely in DNA as well as in RNA and in synthetic polynucleotides. In addition to the double helical form of DNA, considerable secondary or helical structure is found in ribosomal RNA as well as in transfer RNA. In these latter molecules, the secondary structure undoubtedly plays a significant role in bringing about the type of molecular structural specificity required to carry out the biological function of ribosomal RNA and transfer RNA. In both of these structures it is likely that stable double helical configurations are created through the formation of stable loops of RNA in which the double helical segments are formed between the two branches of the loop.

Let us consider, however, the consequences of forming such secondary structures in messenger RNA. The role of messenger RNA is to act as a template for ribosomes which pass over it, adding an amino acid with each reading of a codon triplet. It is believed that all types of polypeptide sequences can be made by ribosomes acting on messenger RNA. A survey of all the known polypeptide structures shows that all dipeptide sequences are possible, and no restraint on sequences has been uncovered in the work done up to the present time (4). Suppose a messenger strand was coding for a sequence of amino acids such as is illustrated in Fig. 2*a*. The messenger RNA coding for this sequence of amino acids might be the sequence of nucleotides shown in Fig. 2*b*. These were obtained by using the appropriate codons from Table I. However, as shown in Fig. 2*b*, this sequence of nucleotides has the capacity to form a double helical loop stabilized by fifteen hydrogen-bonded pairs. Given the magnesium ion and cationic levels found *in vivo,* the double helical segment shown in Fig. 2*b* would form quite a stable structure. If it were stable enough, it might block the passage of ribosomes over the messenger, as shown schematically in Fig. 1*b*. In experiments with synthetic polynucleotides acting as messenger RNA, it is well known that protein synthesis is impaired if the complementary polynucleotide is added to the solution. Thus the addition of poly-

(a) . . . – Asp – Lys – Thr – Asn – Val – Ser – Asp – Val – Gly – Leu – Ile – . . .

Polypeptide chain

(b) . . . – GAU – AAG – ACC – AAC – GUC – UCU – GAC – GUU – GGU – CUU – AUC – . . .

```
... -G-A-U-A-A-G-A-C-C-A-A-C-G-U-C\
     | | | | | | | | | | | | | | |  U\
     | | | | | | | | | | | | | | |    C
     | | | | | | | | | | | | | | |  U/
... -C-U-A-U-U-C-U-G-G-U-U-G-C-A-G/
```

Messenger RNA (autocomplementary)

(c) . . . – GAU – AAA – ACC – AAU – GUC – UCU – GAU – GUU – GGC – CUU – AUU – . . .

Messenger RNA (noncomplementary)

FIG. 2.

adenylic acid decreases the effectiveness of polyuridylic acid as a synthetic messenger RNA. It would be of great interest in this regard to perform experiments in which the degree of protein synthetic impairment is determined as a function of the length of added polynucleotide complementary to the messenger RNA. Such experiments might give us some information concerning the amount of double helical material which the ribosome can "open up" in its transit over the messenger strand. We can call this property of forming double helical segments in messenger RNA "autocomplementarity," implying the property of forming substantial stretches of double helical structure. Of course, we can only appreciate the meaning of "substantial" once the relevant experiments have been carried out to determine the size and composition of the double helical segments which would have the effect of blocking the ribosome. It might be possible that there is a mechanism associated with the ribosome which has the property of opening short segments of double helical pairs on the messenger RNA, but no mechanism of this type is known.

However, such polypeptide sequences need not be an impairment to protein synthesis. Thus, for example, the nucleotide sequence specifying the polypeptide chain of Fig. 2*a* could be that given in Fig. 2*c*. Although this messenger RNA codes for the same polypeptide sequence as the messenger in Fig. 2*b*, it does not form significant autocomplementary stretches. Essentially, this is brought about by an alternative selection of codons from Table I; changing the purine or pyrimidine in the third position does not alter the amino acid which is specified but does have the effect of introducing noncomplementary base pairs which would thereby substantially reduce the stability of the looped structure.

(a) F – Met – Ala – Ser – Val – Gly – Asp – Arg – Ser – His – . . .

Initiating sequence for *E. coli* protein

(b) AUG – GCU – UCG – GUC – GGC – GAC – CGA – AGC – CAU – . . .

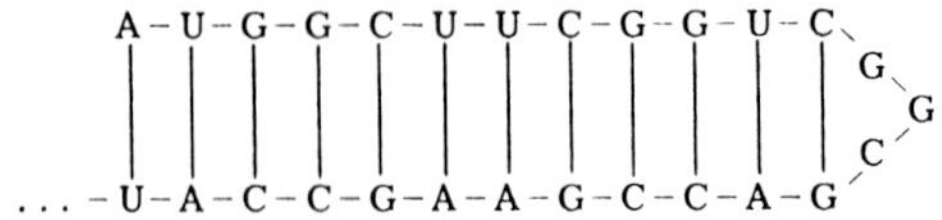

Messenger RNA (autocomplementary)

(c) AUG – GCC – UCG – GUU – GGC – GAC – CGG – AGC – CAC – . . .

Messenger RNA (noncomplementary)

FIG. 3.

Similar drastic results might occur if the initiator portion of the messenger RNA were blocked. This is illustrated in Fig. 3 for a hypothetical sequence of an *E. coli* protein which initiates with N-formyl methionine (Fig. 3*a*). Figure 3*b* shows an autocomplementary sequence which is also illustrated diagrammatically in Fig. 1*c*. Figure 3*c* shows an alternative nucleotide sequence which will not form a large autocomplementary segment blocking the initiator.

We thus postulate that an important component of the selection pressure found in codon assignment for messenger RNA must be the prevention of large segments of autocomplementarity, since, once these are formed, translation of this protein is substantially decreased. One could also imagine an intermediate form of selection pressure whereby smaller segments of autocomplementary messenger RNA are formed which have the effect of slowing down the rate of messenger translation rather than prohibiting it entirely. For example, if one were forming a smaller length of autocomplementary messenger RNA, it could act in a regulatory role during translation. When protein synthesis is carried out actively over this stretch, there would be relatively little impairment of protein synthesis because the ribosomal loading on the messenger RNA would, in effect, prevent the messenger RNA from folding up. However, as soon as the ribosomal density on that messenger is decreased, the messenger strand could then form a small autocomplementary segment which would in turn slow down the rate at which this particular polynucleotide once again becomes available for translation. It might thus give rise to some discontinuity in protein production which the cell might utilize in a control mechanism. For example, local changes in ionic strength could alter the effectiveness of the ribosomal block. Further investigation will be necessary before we can assess the likelihood of such regulatory interactions.

The evaluation of this hypothesis regarding the importance of messenger RNA autocomplementary in effecting protein synthesis will have to await the determination of nucleotide sequences. However, it is relatively easy to devise computer programs which will search out regions of possible messenger autocomplementarity which may be associated with known polypeptide sequences. We would predict that these particular sets of codon assignments would not be chosen. However, a more general prediction is that messenger RNA molecules should have less secondary structure than RNA from other sources. This type of prediction is capable of direct experimental demonstration.

References

1. A reference source for information concerning the genetic code is *Cold Spring Harbor Symp. Quant. Biol.*, **31** (1966).
2. Crick, F. H. C., *J. Mol. Biol.* **19**, 5481 (1966).
3. Crick, F. H. C., Barnett, L., Brenner, S., and Watts-Tobin, R. J., *Nature*, **192**, 1227 (1961).
4. Eck, R. V., and Dayhoff, M. O., *Atlas of Protein Sequence and Structure, National Biomedical Research Foundation,* Silver Springs, Maryland, 1966.
5. Lipsett, M., Heppel, L., and Bradley, D. F., *J. Biol. Chem.*, **236**, 857 (1961).
6. Streisinger, G., Okada, Y., Emrich, J., Newton, J., Tsugita, A., Terzaghi, E., and Inouye, M., *Cold Spring Harbor Symp. Quant. Biol.*, **31**, 77 (1966).

Discussion of Part III

H. J. VOGEL, A. RICH

DR. VOGEL: Is it possible that the ribosome has a mechanism for opening up hydrogen-bonded pairs in messenger RNA, one at a time? If this were the case, it would not matter how long the autocomplementary segment is, since the ribosome might be able to open it up along its entire length.

CHAIRMAN RICH: That is entirely possible; however, I think it unlikely that such a mechanism could open up an autocomplementary segment which includes the initiator; in short, the ribosome would not be attached to the messenger and therefore, would be unable to cleave the hydrogen bonding.

DR. VOGEL: What do you think would happen, in the course of evolution, to a system with a messenger RNA that has a significant autocomplementary segment?

CHAIRMAN RICH: I think that the organism would have difficulty in translating that particular protein. Another mutant which did not have this autocomplementary segment but could still synthesize the same protein would then have a better chance for survival. Thus, I think there would be a significant selection pressure for the adoption of those codons which prevented this type of translational difficulty.

DR. VOGEL: It has been shown that the synthesis of hemoglobin polypeptide chains is discontinuous, that is, there is a break in the rate at which different parts of the α chain become labeled. Could this be attributed to an autocomplementary segment of messenger RNA for the α chain?

CHAIRMAN RICH: I have been looking into this and am attempting to find an autocomplementary sequence for the messenger in that region.

PART IV

INTEGRATION OF METABOLISM

Chairman's Remarks

R. Walter Schlesinger

Department of Microbiology,
Rutgers Medical School,
New Brunswick, New Jersey

I think we had better come to order if this is to end today. It has been a heavy day and for many of us a heavy meal, and it might be appropriate to ease us back into the serious business of organizational biosynthesis by way of a little triviality. I understand that a well-known television personality, in commenting on the explosive rise in the New York birth rate nine months to the day after the big northeastern blackout, observed: "I know there was a blackout, but you certainly can't say that there was a power failure."

By a bit of free association, this pun made me think of metabolic integration and Seymour Cohen. After all, the replication of T-even phages and other viruses depends on the maintenance and operation of intracellular power lines, even though all or most measurable macromolecular synthetic capacities of the cell may be blacked out. So it is the utilization of the cellular metabolic machinery, diverted to fit the restricted coding requirements of the viral genome, which makes the virus-infected cell such an ideal object for the study of biosynthetic processes which have as end products highly organized structures. The conceptual and experimental extrapolation from the lytic phage-bacterium complex and the lysogenic system to animal viruses and malignant transformation clearly holds one important key to the ultimate understanding of cytodifferentiation and specialized cellular function. Seymour Cohen laid the biochemical groundwork for all of this with his elegant early dissection of phage-induced or phage-stimulated enzymes and shut-off of cell-specific synthesis. In keeping with Dr. Lipmann's platform, Dr. Cohen has long since stopped homogenizing and has taken notice of structure and integrated metabolic interaction.

Some Interrelations of Natural Polyamines and Nucleic Acids in Growing and Virus-Infected Bacteria

SEYMOUR S. COHEN AND AARNE RAINA

Department of Therapeutic Research,
University of Pennsylvania School of Medicine,
Philadelphia, Pennsylvania

For many years we have felt that the problems relating to the nature, distribution, and role of the cations participating in cellular structure and function have been neglected. It is instructive to ask, for example, how many nucleates used as substrates or templates have ever been adequately characterized with respect to their cations. Or, in a less serious vein, what has been the ratio of published papers on polyanions such as the nucleates to papers appearing with some data, however inadequate, on the cations associated with these anions?

Our interests in these substances have been heightened by a variety of questions on bacterial metabolism with which we have been concerned; these have included the nature of the cations neutralizing DNA in the bacterial chromosome, the role of the polyamines in phage multiplication, and the mode of action of the basic streptomycinoid drugs. However, except for some brief forays in the past into the biochemical problems concerned with the polyamines, we have not previously had an opportunity to explore this area in a systematic way. We have spent this past year on such an activity—*i.e.*, determining the changes in amounts of natural polyamines in a variety of physiological conditions characterized by one or another pattern of nucleic acid synthesis. These results have suggested to us that the polyamines not only are components of cellular structures but also may play important roles in the control of synthesis of the nucleic acids. In the test of such conclusions it has been possible to show that natural polyamines supplied exogenously to bacteria have profound effects on these syntheses.

POLYAMINES IN HIGHER CELLS, BACTERIA, AND VIRUSES

We do not wish to imply that the subject begins in our laboratory. After all, Leeuwenhoek described the crystals of spermine phosphate in human seminal fluid in 1677. The Tabors have written a most useful review of this subject (34), and they have indicated the kinds of data that have become available since the early 1920's when Leeuwenhoek's crystals were first identified. This literature is best described as fragmentary, indicating that polyamines are present in all tissues examined and in many microorganisms, but little is clearly evident concerning their role. The problem of their role in the life of one organism, *Escherichia coli,* is the subject of our paper. Although it is not our intention to present another review of the status of the field, it is relevant to make some remarks concerning polyamines in other organisms at this time. According to Herbst, one of the lonely pioneers in this field in the very recent past, gram-positive organisms contain little or none of the polyamines (12), while these substances are generally present in fairly high concentrations in gram-negative organisms. It is evident then that, if a role in biosynthesis is ascribed to polyamines in *E. coli,* we shall immediately be confronted with the problem of how this function is effected in *Staphylococcus aureus.* More recently, some pseudomonads and some other bacteria have been found to contain putrescine and to lack spermidine (17).

Figure 1 presents several naturally occurring polyamines. The compounds with which we shall be most concerned are putrescine and

$H_2N—(CH_2)_4—NH_2$
1,4-Diaminobutane
(putrescine)

$H_2N—(CH_2)_5—NH_2$
1,5-Diaminopentane
(cadaverine)

$H_2N—(CH_2)_4—NH—(CH_2)_3—NH_2$
Spermidine

$H_2N—(CH_2)_3—NH—(CH_2)_4—NH—(CH_2)_3—NH_2$
Spermine

FIG. 1. Some naturally occurring polyamines.

spermidine, and their monoacetyl derivatives. Although *E. coli* can generate significant amounts of cadaverine, for example, when grown in relative anaerobiosis on media rich in lysine, this compound has not been produced in detectable amounts under our conditions of study. Spermine is not produced in *E. coli,* although this relatively strong base can be taken up by bacteria. It will apparently displace the weaker bases, putrescine and spermidine, to some extent in cellular structures, and it can also be acetylated (7). This compound is found in animal tissues in amounts approaching or even exceeding that of spermidine. It can be anticipated, therefore, that to the extent that the polyamines are cellular components of significance to metabolism, cells and structures containing spermine—*e.g.,* liver and yeast—might show reactions of a different quality from those of many bacteria.

In 1957 Hershey detected basic substances derived from arginine in the T-even phages (13). These substances were identified as putrescine and spermidine by Ames *et al.* in 1958 (2). Although these and subsequent studies, which showed that the polyamines can account for about 40% of the phage cations, awakened some interest in these substances, the work of Hershey had indicated that polyamines did not possess a genetic role. Furthermore, a study by Ames and Dubin (1) had suggested that in a permeable T-even phage mutant the polyamine could be replaced by Mg^{++} without affecting the activity of the phage particle. Also, many other permeable phages are low in polyamine, suggesting that bound polyamines are not essential for phage activity. Such studies have tended to shift interest to other substances and phenomena hoped to be of a more central significance in virus multiplication. However, since phage nucleic acid is inserted into the cell containing an internal milieu rich in polyamine and is packaged from such a medium, it is not really possible to say from these data that the polyamines do not play crucial physiological roles in virus multiplication.

In certain plant viruses containing RNA, such as turnip yellow mosaic virus, Johnson and Markham (15) discovered a new triamine, bis-(3-aminopropyl)amine, which they were unable to detect in the plant host. The host, Chinese cabbage, contains putrescine and spermidine. It appears possible that an infected plant cell may make a new polyamine, perhaps with a virus-induced enzyme. That a virus might have to go to such trouble to make this compound available could be interpreted to suggest that the new polyamine does play some crucial and specific role in the life of this virus. Unfortunately, it is not yet known whether a new enzyme is actually made, why this triamine rather than spermidine has been selected by virus RNA, or even if this triamine does play any special role in the life of the virus.

Biosynthetic Routes to Putrescine and Spermidine

In this discussion, we should bear in mind the known steps of polyamine biosynthesis in *E. coli*. The route for the biosynthesis of spermidine from putrescine was described by the Tabors, as presented in Fig. 2. The enzymes for these reactions have been shown to be present in *E. coli* (34). It should be noted that putrescine can arise not only from the decarboxylation of ornithine via two different enzymes in *E. coli* (21), but also in a second path involving the hydrolysis of the guanido derivative, agmatine, derived from the decarboxylation of arginine (34). It is known that the synthesis of ornithine is under a variety of metabolic controls, and it may be anticipated that numerous controls will operate on the overall production of putrescine. As an example of

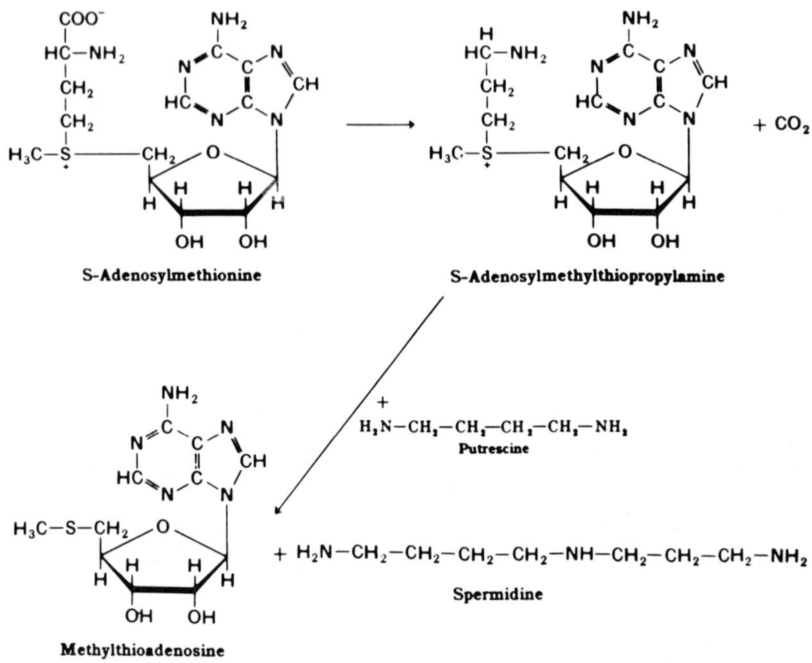

FIG. 2. Biosynthesis of spermidine.

the operation of such a control, it will be seen later that a methionine deficiency results in a sharp reduction of the rate of putrescine formation.

Methionine and S-adenosyl methionine (SAM) are essential precursors in spermidine biosynthesis. It is conceivable, therefore, that the parceling out of the amino acid to protein synthesis, methylation, and biosynthesis of spermidine may also be under a variety of controls. How is methionine distributed among these functions at low methionine concentrations, and what controls on this distribution exist other than the affinities for substrates of the various competing enzymes? As an organism approaches the stationary phase as a result of depleted methionine, in what order are protein synthesis, methylation, and polyamine synthesis slowed and eliminated? It is known that a K^+ deficiency in plants causes the accumulation of putrescine (30). Is this related to the requirement for K^+ in the enzymatic synthesis of S-adenosyl methionine? What controls operate on the apparently crucial SAM decarboxylase, as enzyme that unequivocally begins the shunt of SAM away from methylation and toward spermidine biosynthesis? The mere outline of the reactions in Fig. 2 calls these questions to mind, once you begin to focus on the schema.

It is of some interest to note how the numerous recent papers on methylation have treated this subject without any significant consideration of the metabolically connected polyamines; indeed, many workers with methionine seem not to have been aware of the existence of spermidine or of its derivation from this amino acid. Such phenomena as host-induced modification of some viruses (lambda, T1, etc.), in which methionine has been implicated, and the role of methionine in facilitating the replication of DNA synthesized in the absence of amino acids have been discussed as possible errors in methylation (18), but have not been considered in the literature as possibly relating to the availability of spermidine.

Some Structural Relations of the Polyamines

A few organisms are known which have absolute requirements for polyamines. *Hemophilus parainfluenzae* (11) will grow only if given a polyamine, such as putrescine or spermidine among others; a mutant of *Aspergillus nidulans* will grow very little with spermidine and shows a significant growth response only with putrescine (31). An *E. coli* mutant with an absolute requirement for putrescine has recently been isolated (22). The rarity of such organisms suggests either that polyamines may be readily replaced by other cations, such as Mg^{++}, or that multiple alternative pathways for the polyamines exist, making the isolation of mutants rather difficult, or that a lesion making polyamines unavailable is lethal.

In numerous other organisms, the polyamines stimulate growth. These compounds are known to prevent lysis of fragile organisms and spheroplasts and swelling of mitochondria, and it has been suggested that stimulation of growth by these amines reflects this stabilizing activity (34). In any case, it is at least conceivable that the polyamines play a role in the neutralization of phospholipids in various membranes. A number of devices appear to have been used to modify these membrane anions—*e.g.*, the interaction with basic cytochrome c in mitochondria, the introduction of the strong base choline uniquely into the phospholipids of bacteria with extensively proliferated internal membranes, the use of phosphatidyl ornithine and lysine in some organisms, etc.—and it may be asked if the polyamines do not serve as another approach to this exigency.

Our own interest in the polyamines arose some years ago during an effort to develop a fractionation procedure for a bacterial extract at low ionic strength (5). It was not known what cations neutralize the bacterial chromosome or whether histones or other basic proteins are associated with DNA in the cell. Although work at low ionic strength is essential to such a study, the answers to these questions are still not known. We were soon sidetracked to the problem of determining the distribution of some cations we could measure—*i.e.*, the polyamines—as well as the relation of these compounds to the dissociation of ribosomal components at low ionic strength. We showed the presence of the polyamines on the ribosomes (5), with a somewhat higher spermidine-to-polyamine ratio than is found in the cell as a whole, suggesting a selection of the stronger base, spermidine, for these RNA-rich structures. The polyamines on the ribosomes were shown not to have exchanged readily in the course of isolation with the polyamines of a ribosome-free supernatant fluid derived from a comparable extract. Furthermore, we found that significant concentrations of spermidine (5×10^{-3} *M*) in an extracting fluid could hold the 30 S and 50 S components together to form 70 S monomers in the absence of Mg^{++}. Actually a combination of 5×10^{-3} *M* spermidine and 5×10^{-3} *M* Mg^{++} is better in this activity than either cation alone at this concentration, whereas putrescine could not replace spermidine in promoting this association. We suggested at that time (5) that "Mg^{++} and spermidine are the naturally occurring clasps on the complex ribosomal components existing in the bacteria."

The observation that 30 S and 50 S ribosomes can be held together without exogenous Mg^{++} has been found useful in studies of amino acid incorporation. Nathans and Lipmann (23), for example, observed that the transfer of C^{14}-aminoacyl-sRNA to ribosomes, which is usually studied at a high Mg^{++}, can be effected to a considerable extent by spermidine in the absence of Mg^{++}, suggesting that the limitation in the transfer is that of the state of association of the ribosomes. More recently, Hurwitz has reported that *E. coli* does not concentrate Mg^{++}, although it grows at a reasonable rate at 10^{-5} *M* Mg^{++}. Hurwitz has observed that the spermidine concentration of *E. coli* grown in this way is twenty times that of cells grown at 10^{-3} *M* Mg^{++} (14). Ribosomal-bound spermidine varied inversely with ribosomal-bound Mg^{++}, but ribosomal putrescine did not change in this way. The obvious implication may be drawn that the presence of spermidine has now compensated for Mg^{++} in important respects in assuring normal ribosomal activity in protein synthesis in this organism. Hurwitz also reports (personal communication) that, in the strain of *Pseudomonas* lacking spermidine, ribosomal-bound putrescine now varies inversely with bound Mg^{++}, suggesting that the role of ribosomal clasp in this organism has now been taken over, at least in part, by putrescine.

It is not yet entirely clear whether the formation of polyribosomes occurs with intact 70 S units or with dissociated 30 S and 50 S subunits. If the internal concentrations of Mg^{++} and polyamine determine the state of association of ribosomal subunits, it may be asked if these concentrations do not exert an effect on the rate of protein synthesis by determining the rate of polyribosome formation. These and other effects of polyamines on ribosomes and protein synthesis have been observed by numerous workers (34).

In the course of extracting *E. coli* in the presence of 5×10^{-3} *M* spermidine, we had also observed that the extracts were normal with respect to ribosome content but had only 15 to 25% of their normal DNA content (4, 5). The DNA appears to be selectively retained by the cell debris, a point which seems relevant to the question of the association of the bacterial chromosome and membrane. If the extracts are made from phage-infected cells, a disproportionately large percentage of the phage-induced RNA is also found with the debris, from which it can be extracted with low salt. We have found extraction in 5×10^{-3} *M* spermidine to be useful in preparing certain kinds of soluble extracts—*i.e.*, Mg^{++}-free extracts lacking in certain types of nucleotide donors. In such extractions we have been sedimenting and discarding walls, membranes, DNA, and much of mRNA. Once again it may be asked whether spermidine does not play a structural role in the normal association of the bacterial chromosome and membrane. In this connection we note also that some polyamines markedly stabilize double-stranded DNA, significantly increasing melting temperatures, particularly at low ionic strength. Again it seems pertinent to raise the question of whether this phenomenon relates in any way to the physiology of replication within the organism. However, the general problem of defining the cellular relations of the cations requires a general method of extracting and fractionating under conditions which will not redistribute these constituents.

Some Metabolic Relations of the Polyamines

It has been observed that in regenerating rat liver there is a close correlation between the concentration of RNA and polyamines, as well as in the stimulation of synthesis of these compounds (25). A correlation in the accumulation of polyamines and nucleic acid has also been found in the developing chick embryo (3). At the other end of the scale, in *in vitro* systems, 5×10^{-3} *M* spermidine will stimulate DNA-dependent RNA polymerase, both by maintaining and slightly stimulating the initial rate of RNA formation when polyamine is added at the beginning of the reaction, and also by causing a renewal of activity when the enzyme has apparently ceased polymerization (8, 9, 10). In the latter instance it may be supposed that polyamine either combines with an inhibitory RNA product in solution or helps to release the product from the enzyme or template. Krakow (20) leans to the former interpretation, having shown that these effects are also produced by RNases. We have undertaken to fill in some of the gap between these orders of complexity by studying the relations of the polyamines to synthesis of nucleic acids in bacteria whose patterns of polymer synthesis can be readily manipulated.

For the past ten years we have been exploiting several polyauxotrophic strains of *E. coli*, among which has been strain 15 $T^-A^-U^-$, an organism requiring thymine, arginine, and uracil for growth (16). Omission of one or another requirement from the medium has striking effects on the metabolic pattern of the organism in a mineral medium supplemented with glucose. This organism is unable to make DNA in the absence of thymine, and irreversibly loses the ability to multiply when permitted to metabolize in the thymineless state—*i.e.*, suffers thymineless death. In the absence of uracil, the organism synthesizes and grows at a markedly reduced rate for more than an hour, turning over messenger RNA and some other labile RNA components. In the absence of arginine, the rate of S^{35} incorporation falls to less than 0.5% of normal, and the rate of accumulation of RNA is rapidly reduced about 90% (33). The organism is therefore defined as "stringent." During the accumulation of RNA in the absence of arginine in a period lasting about 90 minutes, a cycle of DNA synthesis is also completed, and the RNA that was made in this period is turned over rapidly. A small proportion of this RNA appears in the 30 S and 50 S ribosomes, at about 1% of the normal rate of formation of ribosomal RNA, and the remainder is found in a broad band of RNA sedimenting at $<$30 S (10). These patterns of stringent RNA synthesis can be relaxed by the antibiotics chloramphenicol and streptomycin (32). What happens to the polyamines in these and other alterations of the biosynthetic pattern of this organism?

In beginning the analyses, we soon discovered that strain TAU produced mainly four components—putrescine, spermidine, and their monoacetyl derivatives—and that these substances were also excreted into the medium (24). Putrescine and monoacetyl putrescine appeared in the medium continually, monoacetyl spermidine began to be found in the medium when accumulation of RNA in the cells tended to come to a maximum, and small amounts of free spermidine appeared in the medium somewhat later. Therefore, each sample required estimates of polyamine in both cells and medium. Strain TAU was also observed to differ from both strain B and strain K12, each of which excreted far less polyamine into the medium, and differed from TAU and each other with respect to the relative ratios of total putrescine to total spermidine (total = free + conjugated).

Various methods for the analysis of these components have been summarized elsewhere (24, 34). Our methods for the four components have recently been published (24); they involve extractions from the cells with acid, followed by extraction with butanol from an alkaline solution. Separation by paper electrophoresis is used before and after acid hydrolysis, and the separated polyamines are finally stained by ninhydrin, extracted, and estimated spectrophotometrically. Corrections are applied for minor losses in extraction. The problem of estimating the smaller polyamines in bacteria, and bacterial cultures relatively rich in putrescine, differs in some important respects from that of estimating the polyamines in animal cells. In such cells, in contrast to bacteria, spermine and spermidine are the major components for which putrescine is a precursor present in smaller amount.

During exponential growth of TAU in complete mineral medium supplemented by glucose, thymine, arginine, and uracil, the ratio of total polyamine nitrogen to RNA phosphorus varies within the narrow limits of 0.21 to 0.23. The average intracellular concentration of total putrescine was 10.8 μmoles, and that of total spermidine 4.7 μmoles per gram wet weight. About 20% of the total putrescine and 60% of the total spermidine within the cells occurred as monoacetyl derivatives under these conditions.

Although methionine is not essential to the growth of TAU, the presence of methionine in the growth medium results in a markedly higher content of free and total spermidine and of RNA per turbidity unit. Ron and Davis (26) have also recently observed a specific stimulation by methionine of RNA synthesis in a variety of strains. However, these workers failed to estimate polyamines. In our experiments the addition of methionine to an already complete medium stimulates the synthesis of both spermidine and RNA.

Polyamines in Thymine Deficiency

In a culture (about 2×10^8 viable bacteria per milliliter) deprived of thymine, synthesis of RNA and protein continued for an hour. During this period free and total putrescine also continued to accumulate and ceased abruptly within the cells at this time, as presented in Table I. Approximately twice as much putrescine, of which 90% was in the free state, appeared in the medium in this time, and continued to spill out into the medium at essentially the same rate for several additional hours.

TABLE I
POLYAMINES OF A THYMINELESS CULTURE

Time (hours)	Turbidity (Klett 420)	Polyamines (μmoles/150 ml)							
		Cells				Medium			
		Putrescine		Spermidine		Putrescine		Spermidine	
		Free	Total	Free	Total	Free	Total	Free	Total
0	90	0.96	1.30	0.25	0.50	—	—	—	—
1	147	2.25	3.05	0.60	1.31	5.48	6.10	0	0.31
2	159	2.31	3.01	0.93	1.91	15.50	17.60	0.13	0.61
3	159	2.55	3.33	0.95	2.18	21.80	23.90	0.16	0.88

On the other hand, free and total spermidine accumulated in the cells for 2 hours, reaching a level of twice the initial internal concentration. Indeed, total but not free spermidine also increased in the third hour. The ratio of free and total putrescine to spermidine thus changed from 3.8 and 2.6 initially to 2.5 and 1.6, respectively, after 2 hours. About a quarter of the spermidine was found in the medium, in the ratio of 1 part free spermidine to 5 parts monoacetyl derivative.

Thus, synthesis of neither polyamine is stopped by cessation of DNA synthesis initially and later of RNA synthesis. For the first hour cellular polyamine parallels the increase in RNA. An excess of polyamine, produced after RNA synthesis stops, is not retained by the cells, putrescine being eliminated mainly in the free state and spermidine mainly as the monoacetyl derivative. Eventually excess spermidine tends to displace putrescine somewhat from the cell and accumulates to a greater extent than does putrescine during thymineless death, resulting in a significant decrease in the ratio of putrescine to spermidine. We call to your attention the facts that cells killed by thymine starvation are able to make DNA and RNA on readdition of thymine and that these syntheses occur more readily than does that of protein (6). This is at least consistent with our suggestion that an increased intracellular spermidine level might have effects on ribosome function.

POLYAMINES IN ARGININE DEFICIENCY

As presented in Fig. 3, elimination of arginine from the growth medium distinctly changed the polyamine pattern of the cells, under conditions in which the total RNA content was not extensively altered. The putrescine content increased by about 60% within 1 hour and remained elevated during further incubation, whereas the concentration

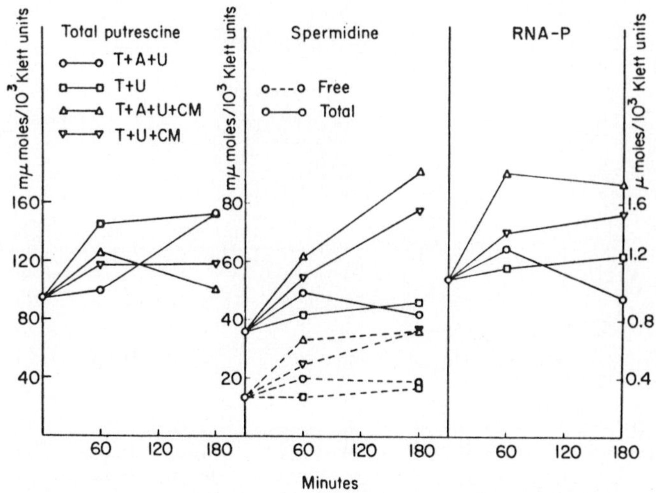

FIG. 3. The effect of arginine deprivation and chloramphenicol on the polyamine and RNA content of *E. coli* strain TAU. The cells were harvested in exponential growth and incubated for 3 hours in media with various additions as indicated in the graph. At zero time the concentration of cells was 2.9×10^8 cells/ml. Aliquots were taken at 60 and 180 minutes and analyzed as described in this paper. The polyamine contents are expressed as millimicromoles and RNA-P as micromoles per 10^3 Klett units (approximately 9 mg wet weight). T = thymine, A = arginine, U = uracil, CM = chloramphenicol.

of spermidine remained at the initial level. This resulted in a significant increase in the ratio of putrescine to spermidine. A similar accumulation of putrescine, without any change in spermidine, was also found when growth was stopped by exhaustion of glucose in a complete medium. In both cases, in which internal putrescine was sharply increased, there was also an extrusion of large amounts of putrescine into the medium, mainly in the unconjugated state. As in thymine deprivation, much less spermidine, almost entirely as the monoacetyl derivative, was found in the medium. It is clear that in this organism, blocked between ornithine and citrulline, the route to putrescine need not pass through arginine.

The elimination of arginine from the medium inhibited spermidine biosynthesis somewhat without significantly decreasing the continuing biosynthesis of putrescine. These data are presented in Table II. The culture containing arginine had exhausted its glucose at the 2-hour point.

EFFECT OF CHLORAMPHENICOL

As shown in Fig. 3, this antibiotic produced a considerable increase in free and total spermidine, but only moderately increased the putres-

TABLE II
POLYAMINES IN AN ARGININE-DEFICIENT CULTURE

Culture	Time (hours)	Turbidity (Klett 420)	Putrescine (μmoles/150 ml) Cells Free	Putrescine Cells Total	Putrescine Medium total	Spermidine (μmoles/150 ml) Cells Free	Spermidine Cells Total	Spermidine Medium total
T + U + A	0	85	0.79	1.02	—	0.17	0.43	—
	1	159	1.79	2.15	3.77	0.40	0.83	0.29
	2	258	2.55	3.02	11.35	0.64	1.13	0.43
	3	261	2.56	3.12	17.10	0.63	1.13	0.45
T + U	0	85	0.79	1.02	—	0.17	0.43	—
	1	94	0.96	1.31	5.15	0.13	0.45	0.38
	2	101	0.89	1.20	14.25	0.14	0.47	0.56
	3	107	1.16	1.50	26.00	0.18	0.52	0.81

cine content of the cells. At the end of the 3-hour incubation, chloramphenicol-treated cells contained about twice as much spermidine as those without chloramphenicol, approaching a concentration of total spermidine close to 10 m*M*. This increased accumulation of spermidine was seen both in the presence and in the absence of arginine. In the latter case, the antibiotic stimulated net production of spermidine, and this continued even after accumulation of RNA had stopped.

EFFECT OF STREPTOMYCIN

We have shown that streptomycin is lethal to strain TAU in the absence of arginine and of protein synthesis (32), indicating that the errors in translation produced by this antibiotic are irrelevant to the lethal effects of the antibiotic. When TAU is exposed to streptomycin in the absence of arginine, the development of lethality closely parallels the stimulation or relaxation of RNA synthesis (Fig. 4). We have recently shown that the RNA whose production is stimulated by the antibiotic is mainly 16 S ribosomal RNA, similar to that stimulated by chloramphenicol (10), and have observed also that the antibiotic has relatively small effects in a system unable to make ribosomal RNA—*i.e.*, T-even phage infection (10).

In cultures of strain TAU containing arginine, streptomycin kills rapidly. It can be seen in Fig. 5 that in such a system putrescine is lost rapidly and can be recovered in the medium. Free and total spermidine are not so lost but in fact accumulate within the cell for the first hour, when RNA synthesis has come to a halt. It is evident, therefore, that

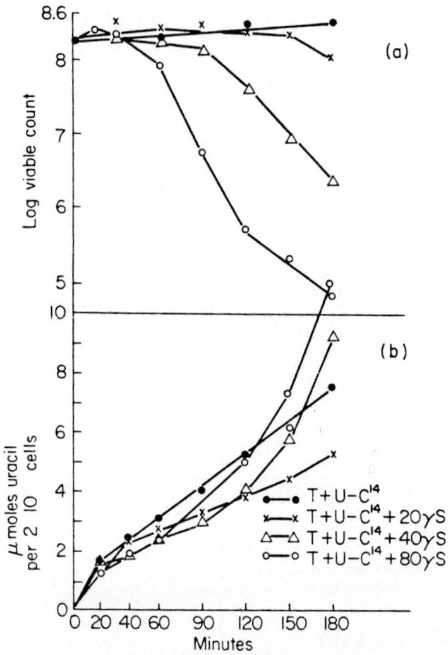

FIG. 4. The effects of different concentrations of streptomycin on (*a*) viability and (*b*) RNA synthesis in a random culture lacking arginine (32). U-C^{14} = uracil-2-C^{14}; T = thymine; S = streptomycin.

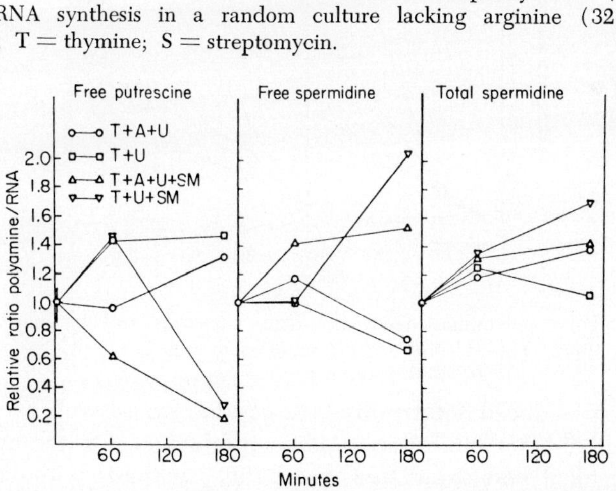

FIG. 5. The effect of streptomycin (60 μg/ml) on the relative ratio of polyamine to RNA in *E. coli* strain TAU. The experimental conditions were comparable to those indicated in Fig. 3, except that 3 mg of glucose (instead of 1 mg) per milliliter was added to the growth medium. The actual values for zero time—65.5 mμmoles of free putrescine, 11 mμmoles of free spermidine, and 26.4 mμmoles of total spermidine per micromole of RNA phosphorus—were taken to be 1.0. T = thymine; A = arginine; U = uracil; SM = streptomycin.

the two polyamines are in very different states within the cell; *i.e.*, either putrescine is mainly free and spermidine is essentially entirely bound, or a highly specific permeability effect has taken place. We lean toward the former interpretation.

In the absence of arginine, the cells do not begin to die for an hour, concomitant with the stimulation of RNA synthesis. At this time, loss of putrescine is precipitous and the accumulation of free spermidine is striking, far exceeding that obtained in the absence of the antibiotic. In both instances, then, in which an antibiotic relaxes the synthesis of ribosomal RNA—*i.e.*, chloramphenicol and streptomycin applied to TAU lacking arginine—we observe a marked increase in the relative ratio of spermidine to total polyamine and an absolute increase in the intracellular level of this compound.

As was indicated above and as can be seen in Fig. 6, these antibiotics act differently in producing these effects. In the absence of arginine,

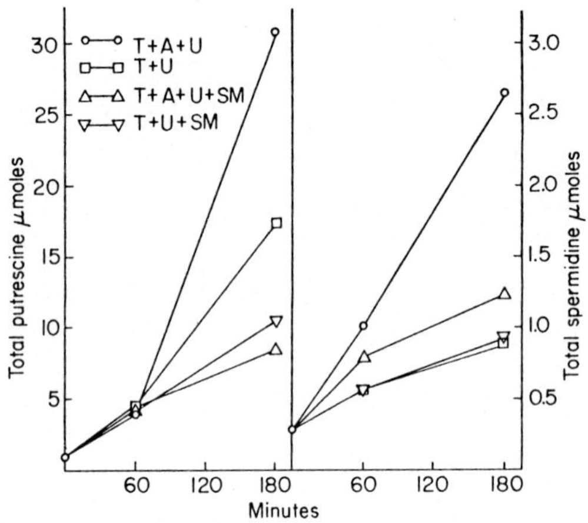

FIG. 6. Total polyamines recovered from the cells and medium of 150 ml aliquots of culture. The cultures are the same as in Fig. 5.

streptomycin reduced total putrescine production as well as putrescine content but did not inhibit spermidine production. On the other hand, chloramphenicol had stimulated spermidine synthesis without affecting putrescine content.

EFFECT OF EXOGENOUS POLYAMINES ON RNA SYNTHESIS IN THE ABSENCE OF ARGININE

Although antibiotics which relaxed RNA synthesis also concomitantly increased the cellular concentration of spermidine and decreased the ratio of putrescine to spermidine, it is not possible from these data to determine whether RNA synthesis controls polyamine level, or vice versa. We therefore studied the direct effect of the polyamines on RNA synthesis in the absence of the amino acid. It can be seen in Fig. 7 that putrescine at 150 m*M* is slightly stimulatory; at lower concentrations there is essentially no effect. Spermidine at a concentration of 20 m*M* or higher markedly increased uracil incorporation into RNA. This effect has been obtained with other strains—*e.g.*, a methionine-requiring strain B in the absence of methionine.

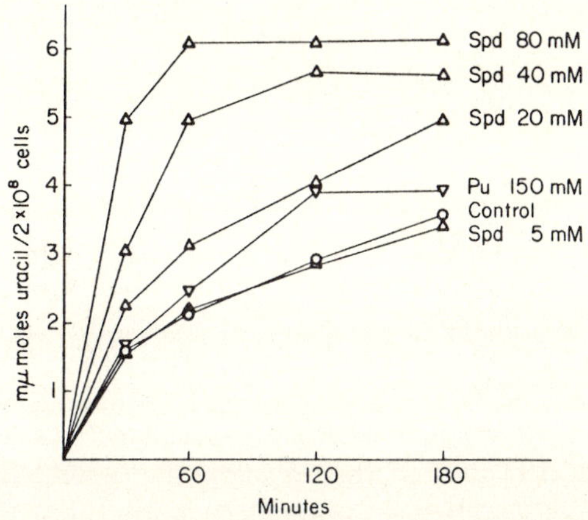

FIG. 7. The effect of polyamines on uracil incorporation in *E. coli* strain TAU in the absence of arginine. The cells grown in a complete medium were harvested at a concentration of 2×10^8 cells/ml and incubated in media containing uracil-C^{14} and lacking arginine. The concentrations of the polyamines added into the media are given in the graph. T = thymine; U = uracil-2-C^{14}; Pu = putrescine; Spd = spermidine. O-O-O = T + U; △-△-△ = T + U + Spd; ▽-▽-▽ = T + U + Pu.

When increasing amounts of putrescine are added to 80 m*M* spermidine in cultures of TAU minus arginine, the stimulatory effect of spermidine on RNA synthesis is reduced and essentially eliminated (see Fig. 8). High concentrations of spermidine are lethal; this effect, too, is prevented by putrescine. It can also be seen in Fig. 8 that the stimulation by spermidine is sustained for a longer period than that produced by chloramphenicol at 20 μg/ml.

We have studied the kinetics of stimulation of uracil incorporation more closely. In Fig. 9 we see that the stimulation of RNA synthesis by spermidine has a significant lag. At the moment we attribute this to a lag in penetration of spermidine into the cell. We have isolated a streptomycin-resistant strain of TAU, which is not stimulated to make RNA by the antibiotic in the absence of arginine. This organism has a normal content of polyamines and does not lose putrescine on exposure

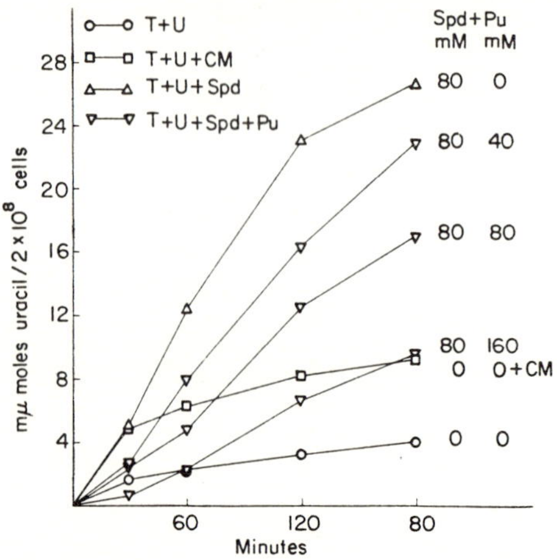

FIG. 8. The effect of putrescine on the stimulation of RNA synthesis caused by spermidine. The concentration of chloramphenicol was 20 μg/ml; the concentrations of putrescine and spermidine are given in the graph. T = thymine; U = uracil-2-C^{14}; CM = chloramphenicol; Pu = putrescine; Spd = spermidine.

to the antibiotic. Although selected for resistance to streptomycin, RNA synthesis in the organism in the absence of arginine is not stimulated by spermidine, although it is stimulated in the normal way by chloramphenicol. In this instance, then, resistance appears not to affect the

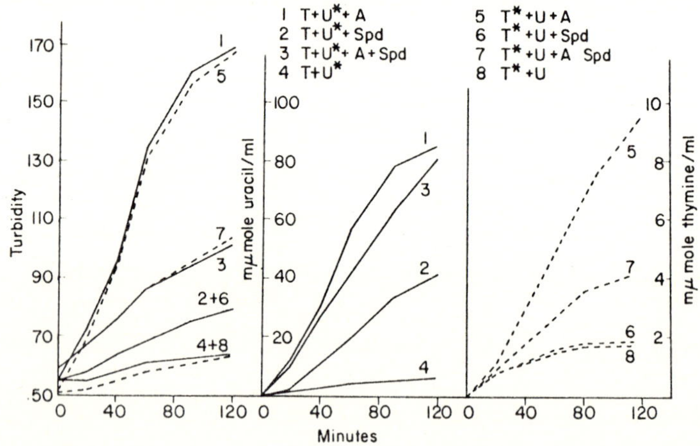

FIG. 9. The effects of spermidine on RNA and DNA synthesis in a culture of *E. coli* strain TAU at 2×10^8 cells/ml in a mineral medium containing glucose. T = thymine (2 μg/ml); T* = thymine-2-C^{14}; A = L-arginine (20 μg/ml); U = uracil (12 μg/ml); U* = uracil-2-C^{14}; Spd = spermidine (75 m*M*). The radioactivities of incorporated bases were converted to their molar equivalents (33).

relaxation mechanism but relates perhaps to the exclusion of basic compounds.

Under the same conditions as those presented in Fig. 9, the polyamine does not appear to affect the completion of a cycle of DNA synthesis. In the presence of arginine, spermidine inhibits cell multiplication, causing a slowly developing inhibition of DNA synthesis and growth, thereby preventing an increase in the rate of RNA synthesis, which is maintained at essentially the initial rate.

It has been shown by sucrose density gradient centrifugation that the RNA accumulated in the presence of spermidine consists largely of 16 S and 23 S RNA.

On RC^{rel} Organisms

Since an increase in the natural polyamine spermidine in stringent cells relaxes synthesis of ribosomal RNA, even as in strains carrying the mutant rc gene, we have asked whether relaxation in these strains does not arise from an excess of spermidine in these strains. A metabolic approach to this problem has suffered until recently from the lack of availability of isogenic pairs of stringent and relaxed organisms. We have analyzed a relaxed K12 strain requiring both leucine and methionine and have observed that on harvest from growth medium it had a relatively higher initial spermidine content than other strains we have examined. In the absence of either amino acid the organism did elaborate RNA at a significant rate for the first hour, but this was not maintained very long thereafter, as indeed is observed with other relaxed strains. The presence of methionine facilitated the continuation of RNA synthesis better than did leucine alone. Spermidine biosynthesis was blocked, as was anticipated, by withholding methionine. In the absence of leucine alone spermidine biosynthesis continued for an hour at almost the normal rate but then fell off, as did RNA synthesis. Although these results were not inconsistent with the notion that a high initial spermidine content caused the maintenance of synthesis of RNA in the absence of amino acid, we did not have the stringent parent to compare with this organism.

We now have a relaxed derivative of TAU but have not yet analyzed its polyamine content.[1] In the absence of arginine it makes RNA at an uninhibited rate for 40 minutes before slowing this function. Chlor-

[1] Note added after presentation of the paper: The cells of relaxed derivative of TAU have now been found to accumulate both free and acetyl spermidine in the absence of arginine. The production of free spermidine parallels the continued synthesis of RNA in this organism, as compared to the cessation of accumulation of free spermidine in the stringent strain, whose RNA synthesis is inhibited sharply by absence of arginine.

amphenicol has no effect on RNA synthesis in this culture, while spermidine extends the initial high rate of synthesis. In the absence of the amino acid—*i.e.*, while the organism is making ribosomal RNA—TAU rc^{rel} is very sensitive to streptomycin but does not show a stimulation of RNA synthesis. Our relaxed K12 is similarly very sensitive to streptomycin and dies far more quickly than do stringent strains in the absence of the amino acid; *i.e.*, the production of ribosomal RNA, even in the absence of an amino acid, sensitizes the cells to this antibiotic. In brief, then, our preliminary data on relaxed mutants are still consistent with our working hypothesis—namely, that the control of synthesis of ribosomal RNA is mediated to a significant degree through the polyamines.

Reversal by Spermidine of Levorphanol Action

Simon and Van Praag have shown that the basic narcotic, levorphanol, selectively inhibits the synthesis of ribosomal RNA in *E. coli* (28, 29). We are pleased to have had the opportunity to study in collaboration with Dr. Simon the effects of polyamines on the inhibition produced by levorphanol, as well as the effects of the drug on the polyamines (27). In Fig. 10, it can be seen that, on simultaneous application, spermidine but not putrescine almost completely reverses levorphanol inhibition of RNA synthesis in *E. coli* K13. This was also

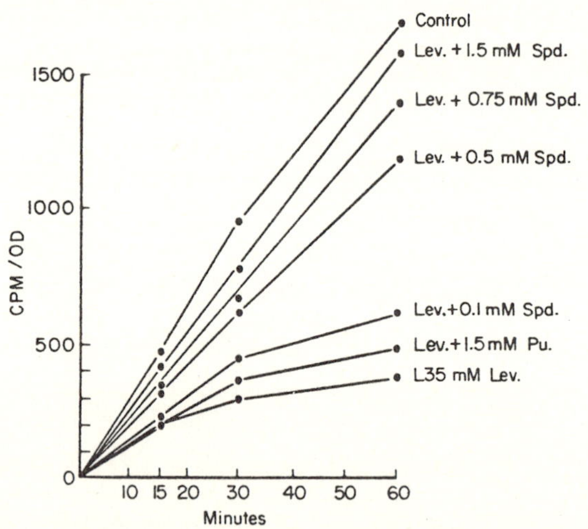

FIG. 10. Effect of polyamines on inhibition of RNA synthesis by levorphanol. Uracil-C^{14} (7.2 μg/ml, 0.06 μc/ml) was added to a culture of *E. coli* K13 in exponential growth 5 minutes after addition of levorphanol and polyamine. Samples were removed into 1 *M* perchloric acid at indicated intervals, filtered through Millipore membranes, and counted. The optical density at 550 mμ was measured in a Lumetron colorimeter. Spd = spermidine; Pu = putrescine; Lev = levorphanol.

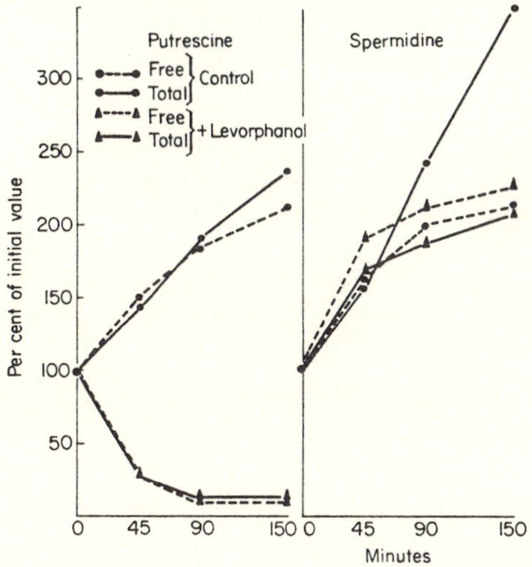

FIG. 11. Effect of levorphanol on accumulation of putrescine and spermidine in *E. coli* 15 TAU. Results are expressed as percentage of initial content. Aliquots (150 ml) of the cultures were removed at the indicated intervals and analyzed for polyamines. The actual zero time values for cellular polyamines in millimicromoles were: putrescine free—593, total—641; spermidine free—203, total—282.

observed in TAU and other organisms. Levorphanol alone, as presented in Fig. 11, causes the selective extrusion of putrescine in TAU, and spermidine synthesis continues only as long as synthesis of ribosomal RNA. Levorphanol is not lethal, and it appears then that lethality, as in treatment with streptomycin, cannot be attributed merely to the loss of putrescine. In Fig. 12, it can be seen that addition of spermidine to a

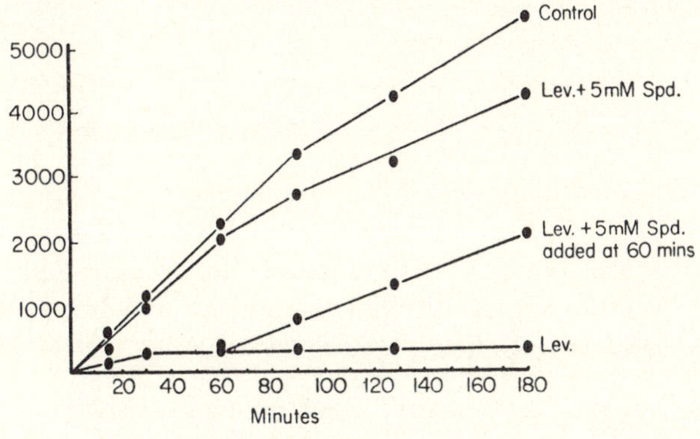

FIG. 12. Reversal of levorphanol inhibition of RNA synthesis by spermidine in *E. coli* K13.

levorphanol-inhibited culture permits the restoration in large part of RNA synthesis and of growth. The reversal of levorphanol inhibition required relatively low concentrations of spermidine, and we suspect that levorphanol has helped to increase permeability to the polyamine. It may well be that this combination of compounds will prove useful in the study of membrane permeability in *E. coli.* From time to time, we also wonder if these effects will prove relevant to the behavior of animal cells.

Summary on Relation of Polyamines to RNA Synthesis

We have observed that the content of polyamines present in *E. coli* bears a close relation to the content and synthesis of RNA in the bacteria. Although a higher molar ratio of Mg^{++} to RNA-P than of polyamine-N to RNA is found in bacteria grown in high Mg^{++} media, the polyamines, and particularly spermidine, can be synthesized and replace Mg^{++} in some measure in low Mg^{++} media. The absence of an amino acid permits an accumulation of putrescine and reduction of spermidine production, and this occurs concomitantly with an inhibition of synthesis of ribosomal RNA. In systems marked by a relaxation of RNA synthesis, as in the presence of chloramphenicol, streptomycin, the rc^{rel} gene, etc., the levels of these polyamines have been shifted to higher relative and absolute levels of spermidine *vis à vis* putrescine. With streptomycin and levorphanol, the cells selectively lose putrescine, which is possibly a pool component, and retain spermidine, which is probably associated firmly to ribosomes and other structures. The exogenous addition of spermidine totally relaxes synthesis of ribosomal RNA, and putrescine, which is unable to effect this relaxation, will markedly inhibit this effect. The inhibition of synthesis of ribosomal RNA by levorphanol occurs concomitantly with a curtailment of spermidine biosynthesis, following the loss of most of the cellular putrescine. Exogenous spermidine reverses the inhibition of growth and RNA synthesis by levorphanol. Although spermidine inhibits DNA synthesis and multiplication at high concentrations, it does not appear to affect DNA synthesis within a single cycle of multiplication. In uninfected bacteria, the polyamines appear to relate mainly to RNA synthesis. At this time, the data are consistent with the hypothesis that the absolute and relative levels of spermidine among the polyamines are important in the regulation of the synthesis of ribosomal RNA.

Polyamines in T-Even Virus Infection

Although we have many unsolved problems concerning the role of the polyamines in RNA synthesis, we have also begun work on the relation of these compounds to T-even phage infection. This infection is characterized by a relatively huge production of DNA, whose accumulation occurs, in a quantitative sense, at the expense of ribosomal synthesis. The biological situation provides an obvious test of the question of whether the polyamines relate to synthesis of RNA alone. In examining the limited literature on the polyamines in phage infection, it becomes evident that, although the T-even phages contain large amounts of putrescine and spermidine, which may derive in part during infection from an exogenous precursor (*i.e.*, arginine or polyamine), it has not been demonstrated that a significant net synthesis has in fact occurred (1, 34). Hershey (13) had shown that chloramphenicol did not reduce the conversion of arginine to polyamine in infection, but it was not clear

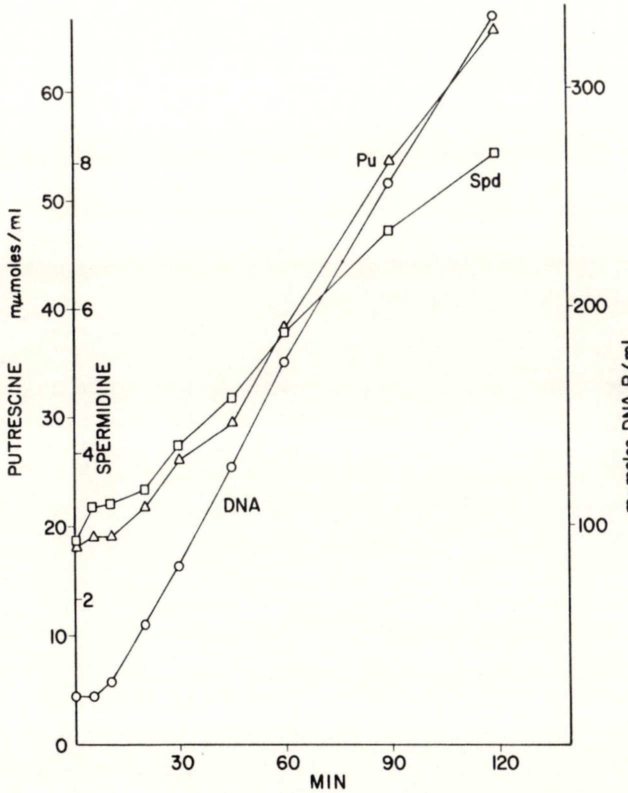

FIG. 13. The net synthesis of polyamines and DNA in T6r$^+$-infected *E. coli* strain B. The bacteria were grown in a mineral medium (M9) containing glucose and infected with a multiplicity of five virus particles per cell in the presence of 25 μg of L-tryptophan per milliliter. Aliquots of 150 ml were sedimented, and the pellets and supernatant fluids were analyzed for polyamines separately.

how much cellular polyamine had been shunted to virus, whether in infection the synthesis of polyamine was inhibited or stimulated, or if the relative ratio of syntheses of putrescine to spermidine was dramatically altered (1).

In Fig. 13 it can be seen that there is a continuing net synthesis of polyamine during T6r⁺ infection of *E. coli* strain B. It has appeared consistently that in the first 30 minutes of infection the increment in polyamine is less rapid than is DNA synthesis but that a significant increase in rate occurs at about this time—*i.e.,* after 30 minutes. The production of putrescine in the total culture eventually parallels DNA synthesis, whereas that of spermidine is not maintained at this high a rate. Breaking this down, as in Fig. 14, it appears that late in infection

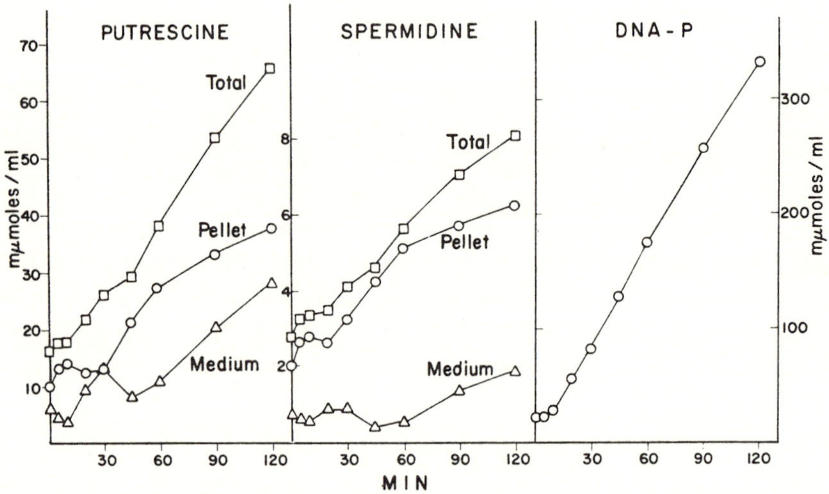

FIG. 14. The accumulation of polyamines in both cells and media of a T6r⁺-infected culture (see legend, Fig. 13). DNA was estimated in the entire culture.

putrescine excretion is very extensive and comprises more than half of the putrescine made. On the other hand, spermidine tends once again to be conserved within the cell. From about 20 to 60 minutes after infection, the rate of accumulation of spermidine within the cells does parallel that of DNA synthesis, though the molar ratio of DNA-P to spermidine-N and putrescine-N synthesized in that interval was about 13 and 4, for spermidine and putrescine, respectively. It can be seen, therefore, that in certain defined periods the rates of accumulation of polyamines in the cells are similar in both growth and virus multiplication.

It is evident here, however, that we are confronted with a number of complexities that will present difficulties for our interpretations and future experiments. The slow initial rate of polyamine synthesis suggests that infection does inhibit synthesis. Conceivably, enough internal polyamine is freed from cellular structure—*i.e.*, degradation of cell DNA, ribosomes, etc.—to fill early polyamine requirements and to effect a feedback inhibition on polyamine synthesis. Of course, other interpretations are possible. There is some loss of putrescine and spermidine to the medium, a reutilization of these polyamines from the medium, and eventually a large-scale release. Late in infection DNA synthesis is maintained, although polyamine is released. This may suggest either that some of the cells have lysed or that the need for polyamine cations may be reduced by packaging the DNA.

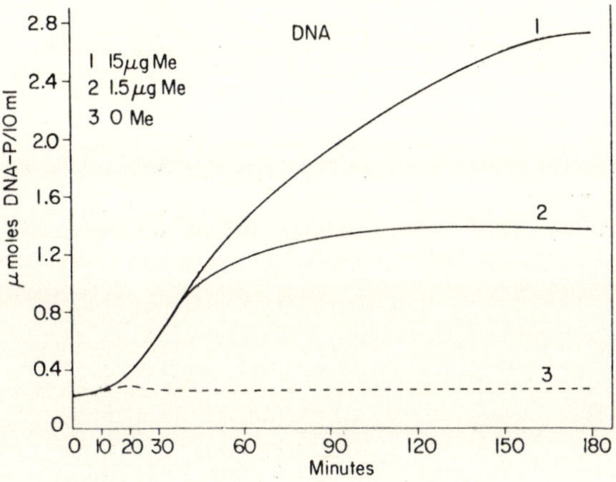

FIG. 15. The synthesis of DNA in T6r⁺-infected *E. coli* strain B_{45}, requiring methionine for growth. Me = L-methionine. The bacteria were grown to 2.5×10^8 cells/ml and infected at a multiplicity of five virus particles per cell in the presence of tryptophan.

In any case, we have begun to explore the question of whether the polyamine controls DNA synthesis or whether DNA synthesis evokes (or controls) polyamine synthesis. It is difficult to set up a rigorous experiment on this question in this system. As one approach, we have asked whether a methionine deficiency established after the production of early enzymes and an initial synthesis of DNA at a normal rate will not limit spermidine and DNA synthesis. In Fig. 15, we see that reducing the methionine content of the medium during $T6r^+$ infection of a methionine-requiring *E. coli* strain B does fulfill some of these conditions—*i.e.*, permit a normal synthesis of virus DNA for 30 minutes, followed by a slow increment for another 30 minutes. In Fig. 16, it can be seen that the

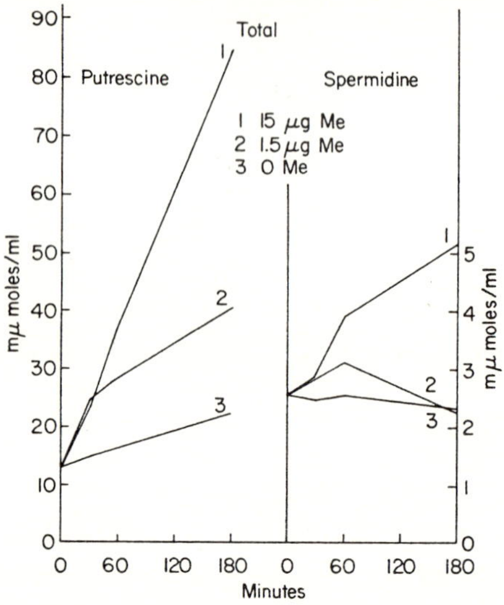

FIG. 16. The net synthesis of polyamines in T6r⁺-infected *E. coli* strain B_{45} requiring methionine for growth. Me = L-methionine. DNA synthesis in these cultures is presented in Fig. 15. Aliquots of 150 ml were sedimented, and the pellets and supernatant fluids were analyzed for polyamines separately.

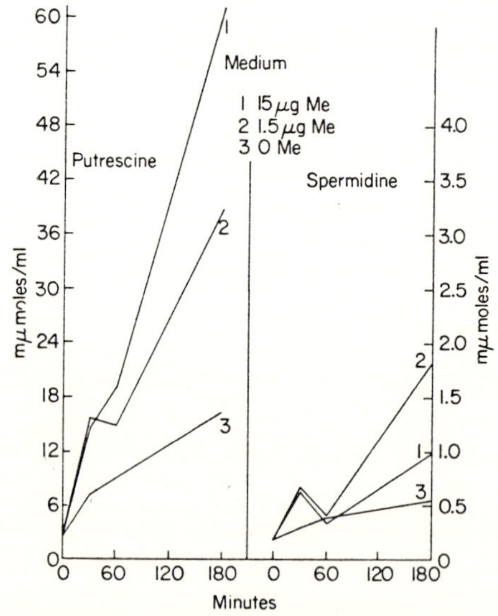

FIG. 17. The appearance of polyamines in medium in a T6r⁺-infected culture of *E. coli* strain B_{45}. Total polyamine production in the culture is described in Fig. 16.

establishment of methionine deficiency at 0 time or at 30 minutes sharply inhibits putrescine biosynthesis and completely prevents spermidine synthesis. It can be shown, as in Fig. 17, that an initial slight accumulation of spermidine in the medium at 30 minutes is actually reused during the slow synthesis of DNA occurring between 30 and 60 minutes in the depleted culture.

We do not consider that this experiment does say anything about control of DNA synthesis in infection. However, it does underline the existence of a correlation between the availability of polyamines and DNA synthesis not previously observed for any cell. Exogenous addition of spermidine to methionine-deficient infected cells does not stimulate DNA synthesis; on the contrary, at concentrations above 5 m*M* spermidine inhibits DNA synthesis in infection.

Conclusions

We feel that we have opened a new Pandora's box. It seems to us that much of our knowledge of the physiology of nucleate biosynthesis will bear reexamination, with the cations, and particularly the polyamines, in mind. Our experiments suggest that studies on the physiology of synthesis of polymeric anions, particularly that of the nucleates, will be woefully incomplete if they continue to neglect the cations; this remark refers to both inorganic and organic cations. We have little insight at present on the exact mechanisms by which the polyamines produce their physiological effects, and suspect that the determination of their role will be with us for a long time to come.

Acknowledgments

These investigations have been supported by U.S. Public Health Service Grant 7005 from the National Institute of Allergy and Infectious Diseases. A. Raina is a USPHS International Fellow F05-887. We are pleased to acknowledge the capable technical assistance of Miss Miekie Jansen, Mrs. Mary Moore, and Miss Nancy Hoffner in these studies.

References

1. Ames, B. N., and Dubin, D. T., *J. Biol. Chem.*, **235**, 769 (1960).
2. Ames, B. N., Dubin, D. T., and Rosenthal, S. M., *Science*, **127**, 814 (1958).
3. Caldarera, C. M., Barbiroli, B., and Moruzzi, G., *Biochem. J.*, **97**, 84 (1965).
4. Cohen, S. S., and Barner, H. D., *J. Biol. Chem.*, **237**, PC1376 (1962).
5. Cohen, S. S., and Lichtenstein, J., *J. Biol. Chem.*, **235**, 2112 (1960).
6. Doering, A. M., Jansen, M., and Cohen, S. S., *J. Bacteriol.*, in press.
7. Dubin, D. T., and Rosenthal, S. M., *J. Biol. Chem.*, **235**, 776 (1960).
8. Fox, C. F., Robinson, W. S., Haselkorn, R., and Weiss, S. B., *J. Biol. Chem.*, **239**, 186 (1964).

9. Fox, C. F., and Weiss, S. B., *J. Biol. Chem.*, **239**, 175 (1964).
10. Freda, C., and Cohen, S. S., *Bacteriol. Proc. 1966; J. Bacteriol.*, in press.
11. Herbst, E. J., Glinos, E. B., and Amundsen, L. H., *J. Biol. Chem.*, **214**, 175 (1955).
12. Herbst, E. J., Weaver, R. H., and Keister, D. L., *Arch. Biochem. Biophys.*, **75**, 171 (1958).
13. Hershey, A. D., *Virology*, **4**, 237 (1957).
14. Hurwitz, C., Paper presented at Los Angeles meeting of the American Society of Microbiology, May 1966.
15. Johnson, M. W., and Markham, R., *Virology*, **17**, 276 (1962).
16. Kanazir, D., Barner, H. D., Flaks, J. G., and Cohen, S. S., *Biochim. Biophys. Acta*, **34**, 341 (1959).
17. Kim, K., *J. Bacteriol.*, **91**, 193 (1966).
18. Klein, A., and Sauerbier, W., *Biochem. Biophys. Res. Commun.*, **18**, 440 (1965).
19. Krakow, J. S., *Biochim. Biophys. Acta*, **72**, 566 (1963).
20. Krakow, J. S., *J. Biol. Chem.*, in press.
21. Morris, D. R., and Pardee, A. B., *Biochem. Biophys. Res. Commun.*, **20**, 697 (1965).
22. Morris, D. R., and Pardee, A. B., *Bacteriol. Proc.* 1966.
23. Nathans, D., and Lipmann, F., *Proc. Natl. Acad. Sci. U.S.*, **47**, 497 (1961).
24. Raina, A., and Cohen, S. S., *Proc. Natl. Acad. Sci. U.S.*, **55**, 1587 (1966).
25. Raina, A., Jänne, J., and Siimes, M., *Biochim. Biophys. Acta*, **123**, 197 (1966).
26. Ron, E., and Davis, B., *J. Mol. Biol.*, in press.
27. Simon, E. J., Cohen, S. S., and Raina, A., *Biochem. Biophys. Res. Commun.*, **24**, 482 (1966).
28. Simon, E. J., and Van Praag, D., *Proc. Natl. Acad. Sci. U.S.*, **51**, 877 (1964).
29. Simon, E. J., and Van Praag, D., *Proc. Natl. Acad. Sci. U.S.*, **51**, 1151 (1964).
30. Smith, T. A., and Richards, F. J., *Biochem. J.*, **84**, 292 (1962).
31. Sneath, P. H. A., *Nature*, **175**, 818 (1955).
32. Stern, J. L., Barner, H. D., and Cohen, S. S., *J. Mol. Biol.*, **17**, 188 (1966).
33. Stern, J. L., Sekiguchi, M., Barner, H. D., and Cohen, S. S., *J. Mol. Biol.*, **8**, 629 (1964).
34. Tabor, H., and Tabor, C. W., *Pharmacol. Rev.*, **16**, 245 (1964).

Discussion of Part IV

R. W. Schlesinger, K. G. Lark, S. S. Cohen, D. Nakada, R. L. Metzenberg, K. R. Hanson, A. Nason

Chairman Schlesinger: Thank you very much, Dr. Cohen. Now I open this paper for discussion.

Dr. Lark: Have you tried chloramphenicol in the methionine phage studies to increase the DNA synthesis? Is there an effect of starvation for other amino acids?

Dr. S. S. Cohen: It is precisely because of such experiments that we are unwilling to conclude that there really is a direct correlation between spermidine and DNA. Now there are effects, but it is very difficult to say which is primary. One thing that you do see, as you deplete amino acids, is a lysis of infected cells and you can't be sure that this is not the basis of the arrest of DNA synthesis in methionine deficiency.

Dr. Nakada: Where is the ribosomal RNA of yours located during arginine starvation with the addition of spermidine? In other words, are the cells making ribosomes or ribosomal RNA which ends up in the so-called abnormal particles?

Dr. S. S. Cohen: One finds very small traces of the 16 and 23S RNA in the 30 and 50S ribosome peaks. You find a fairly large amount in the incomplete-particle region.

Dr. Metzenberg: Did I notice that the absence of methionine curtailed the production of putrescine as well as spermidine? Do you have any explanation for this or any speculations?

Dr. S. S. Cohen: No, it just seems obvious to me that we are going to have to look at all the enzymes for the biosynthesis of these components and study the feedback mechanisms and the repression mechanisms just as we go through any pathways. It has just been neglected. Since Tabor described these enzymes, no one has really studied the physiology of the enzymes, and this is the state we are in. It is clear that a methionine deficiency cuts off putrescine, as I showed in one of those last slides, but just how it does it, I am not in a position to say.

Dr. Hanson: I wonder if there is a possible structural role in the cell for a coordination shell with, say, magnesium at the center and with a hydrocarbon hydrophobic exterior about 10 A across, also, (this is perhaps a little wild) whether there could be any relationship between these materials and the adamantane derivatives which have some anti-virus activity.

Dr. S. S. Cohen: I know that the adamantane effect is reversed by ammonia, but I don't know that it was tested with the polyamines. Concerning the first part of your question, I don't know what the answer is to this. There has been virtually no study in the biochemical literature. I have undoubtedly neglected the organic chemical literature concerning coordination compounds of this type.

CHAIRMAN SCHLESINGER: I think the adamantane analogy might be weakened by the fact that, as far as I know, the adamantanes are supposed to prevent attachment of the viruses that they affect to the cells rather than replication.

DR. NASON: Are there any studies of the effects of polyamines in cell-free systems, such as DNA polymerase or RNA polymerase; can there be possible substitution of spermidine for magnesium in enzyme systems that require magnesium?

DR. S. S. COHEN: There have been many effects recorded. For example, Kalckar and group have recently stated or reported that spermidine affects one of the epimerases in the *gal* system. Perhaps Dr. Weiss would comment on spermidine in the RNA polymerase system. Many workers now add 1 to 5×10^{-3} *M* spermidine as part of the RNA polymerase assay routinely. But there are a wide variety of enzymes which seem to be stimulated in an as yet undetermined way by these compounds.

CHAIRMAN SCHLESINGER: Are there any questions? Well, thank you very much.

PART V

JUXTAPOSITION PHENOMENA I

Chairman's Remarks

Noboru Sueoka

Department of Biology,
Princeton University,
Princeton, New Jersey

This morning we shall hear three papers on regulatory mechanisms and their genetic basis in the pathways of synthesis of cysteine, histidine, and arginine. The juxtaposition of genes involved in a pathway is a major point of interest here.

The Biosynthesis of L-Cysteine in *Escherichia coli* and *Salmonella typhimurium* by a Multifunctional Enzyme Complex

NICHOLAS M. KREDICH AND GORDON M. TOMKINS

Laboratory of Molecular Biology,
National Institute of Arthritis and Metabolic Diseases,
National Institutes of Health,
Bethesda, Maryland

INTRODUCTION

The pathway of L-cysteine biosynthesis in *Salmonella typhimurium* and *Escherichia coli* is shown in Fig. 1. It is a branched pathway catalyzed by a group of enzymes which reduce sulfate to sulfide, together with the enzymes which lead from L-serine to O-acetyl-L-serine and then to L-cysteine.

The genetics of L-cysteine biosynthesis in S. *typhimurium* have been elucidated by the work of Clowes (1) and Dreyfuss and Monty (2), and are illustrated in Figs. 1 and 2. Some fourteen cistrons are located in five clusters on the bacterial chromosome. The A cluster is responsible for a sulfate permease, while the D and C cistrons in the C cluster appear to be the structural genes for the two kinases which lead to the synthesis of 3′-phosphoadenosine 5′-phosphosulfate. The H cistron together with the three cistrons in the B cluster are needed for the reduction of 3′-phosphoadenosine 5′-phosphosulfate to sulfite. The complex sulfite-to-sulfide reduction is also dependent on the B cluster as well as on the I and J cistrons of the C cluster and on the solitary G cistron. Mutants in the E region are cysteine auxotrophs unable to grow on sulfide, presumably because they lack the ability to synthesize cysteine from serine and sulfide.

It should be a point of interest to this Symposium that the sulfite reductase of this system seems to be a large multifunctional, multisite enzyme complex which is coded for by at least 6 cistrons (8).

We wish to describe the isolation of a multifunctional protein catalyz-

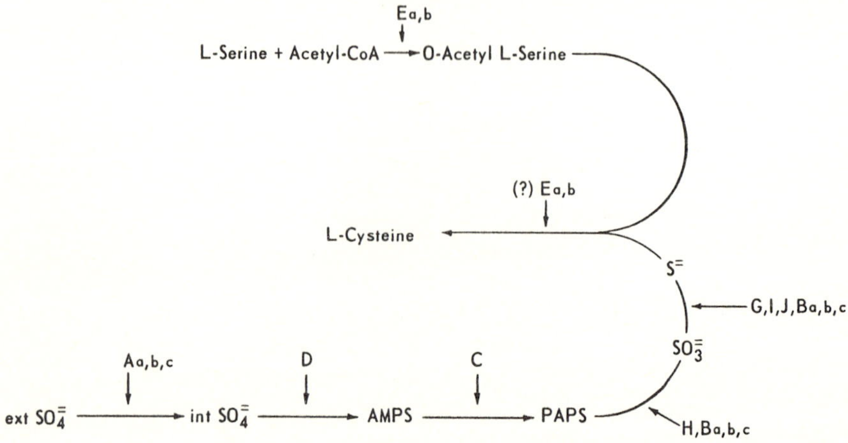

FIG. 1. Pathway of L-cysteine biosynthesis in S. *typhimurium* (2). The genetic loci concerned with each step are indicated as in Fig. 2. Abbreviations are: AMPS, adenosine 5′-phosphosulfate; PAPS, 3′-phosphoadenosine 5′-phosphosulfate.

ing the two steps between L-serine and L-cysteine, and to discuss some of its physical properties and the regulation of its activity (4).

SERINE TRANSACETYLASE

Serine transacetylase, the first enzyme in this pathway, catalyzes the acetylation of L-serine by acetyl-CoA to give O-acetyl-L-serine. The

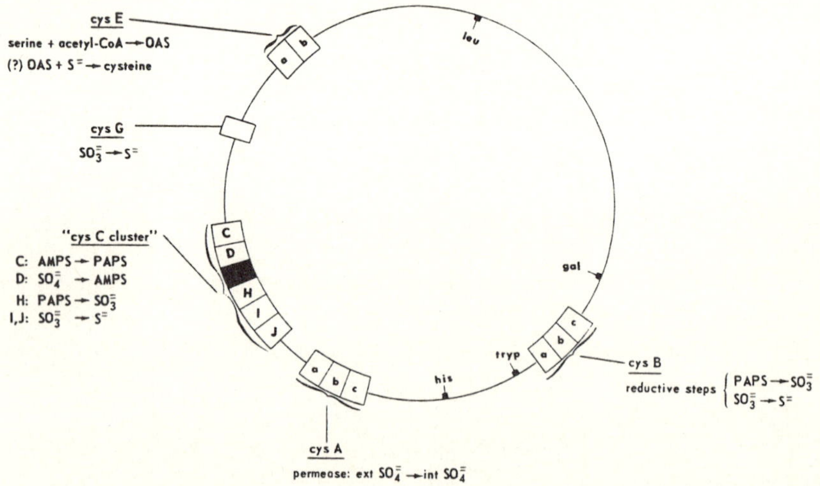

FIG. 2. Map of S. *typhimurium* chromosome showing the genetic areas concerned with the biosynthesis of L-cysteine. OAS, abbreviation for O-acetyl-L-serine.

assay is carried out in the presence of L-serine, acetyl-CoA, and dithionitrobenzoic acid (DTNB). The free thiol, CoA-SH, liberated by the acetylation reacts instantaneously with DTNB to give a chromophore with a strong absorbancy at 412 mμ, and the reaction is followed spectrophotometrically.

$$\text{L-Serine} + \text{acetyl-CoA} \longrightarrow \text{O-Acetyl-L-serine} + \text{CoA-SH}$$
$$\text{CoA-SH} + \text{Ar-S-S-Ar (DTNB)} \longrightarrow \text{CoA-S-S-Ar} + \text{Ar-S}^- \text{ (chromophore)}$$
$$\text{O-Acetyl-L-serine} \xrightarrow{\text{nonenzymatic}} \text{N-Acetyl-L-serine}$$

The product of the reaction has been identified as O-acetyl-L-serine by chromatographic techniques. It undergoes nonenzymatic acyl migration to give N-acetyl-L-serine with a first-order rate constant of 1% per minute at pH 7.6. Thus N-acetyl-L-serine is an indirect product of the enzyme reaction.

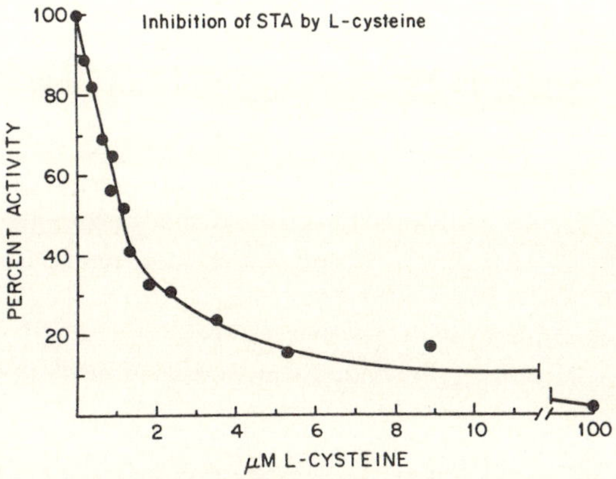

FIG. 3. Inhibition of serine transacetylase (STA) by L-cysteine. The K_i is 1×10^{-6} *M*.

The K_m's for L-serine and acetyl-CoA are 5×10^{-4} *M* and 1×10^{-4} *M*, respectively. Figure 3 shows that the reaction is strongly inhibited by low concentrations of L-cysteine. The K_i for this inhibition is 1×10^{-6} *M*. Other thiols, D-cysteine, and derivatives of L-cysteine have no appreciable effect on the rate of the reaction. Since this step is required for cysteine biosynthesis, this is a typical example of specific feedback inhibition.

Serine transacetylase has been purified approximately 700-fold from *E. coli* B to a state of apparent homogeneity. It has a sedimentation coefficient of 7.8 and a molecular weight of approximately 110,000. The pure enzyme has a slight yellow color and an absorption peak at 415 mμ similar to that of pyridoxal phosphate-containing enzymes. However, neither added pyridoxal phosphate nor carbonyl reagents known to react with pyridoxal affect the rate of the reaction.

Since an organism unable to synthesize O-acetyl-L-serine should be a cysteine auxotroph unable to grow on sulfide (a *cys* E mutant), a number of such mutants were assayed for their serine transacetylase activity; the results are shown in Table I. The bacteria were grown at 37°C in

TABLE I
SERINE TRANSACETYLASE AND O-ACETYLSERINE SULFHYDRYLASE ACTIVITIES IN *Cys* E MUTANTS OF *Salmonella typhimurium*[a]

Mutant	Serine transacetylase activity (%)	OASS activity (%)
Ea 2	3.0	510
Ea 6	<0.2	290
Eb 11	1.8	86
Eb 17	<0.2	33
Ea 30	<0.2	25
Eb 396	0.7	56

[a] Mutants were grown at 37°C on nutrient broth, and activities expressed as percentage of wild-type.

nutrient broth, and the activity of crude extracts is expressed as the percentage of that found in the wild-type organism grown in the same way. As can be seen, mutants in either the a or b cistrons of the *cys* E region contained 0 to 3% wild-type serine transacetylase activity. From these results it seems likely that the *cys* E region represents the structural gene for serine transacetylase, although conclusive evidence, such as the demonstration of an altered enzyme in a *cys* E mutant, is not available at this time.

O-ACETYLSERINE SULFHYDRYLASE

The second enzymatic activity in this two-step pathway is O-acetylserine sulfhydrylase (OASS). Since the assay procedure which we use is a considerable improvement over the nitroprusside reaction usually used to determine aliphatic thiols in the presence of sulfide, we shall briefly describe it here. In this assay, modified from a procedure published by Saville (7), O-acetyl-L-serine and sodium sulfide are incubated for several minutes with buffer and bacterial extract. The reaction is stopped by the addition of nitrous acid, which reacts with cysteine to give a stable S-nitrosothiol derivative. After removal of excess nitrous acid with ammonium sulfamate, a mixture of mercuric ion, sulfanilamide, and N-1-naphthylethylenediamine is added. In the presence of the mercuric ion the S-nitrosothiol decomposes to give nitrous acid, diazotizing the sulfanilamide, which then couples with the naphthylethylenediamine to give an azo dye as the chromophore.

$$\text{O-Acetyl-L-serine} + SH^- \longrightarrow \text{Cysteine-SH} + \text{acetate}^-$$
$$\text{Cysteine-SH} + HONO \longrightarrow \text{Cysteine-S-NO} + H_2O$$
$$\text{Cysteine-S-NO} + H_2O \xrightarrow{Hg^{++}} \text{Cysteine-SH} + HONO$$
$$\text{Sulfanilamide} + HONO \longrightarrow Ar\text{-}N_2{+} + H_2O$$
$$Ar\text{-}N_2{+} + \text{N-1-naphthylethylenediamine} \longrightarrow$$
$$AR\text{-}N = N\text{-}AR' \text{ (azo dye, chromophore)}$$

L-Cysteine, the product of the enzymatic reaction, has been crystallized in its oxidized form, L-cystine.

During its purification from S. *typhimurium* it was noted that most of the OASS activity precipitates at a higher ammonium sulfate concentration than serine transacetylase. When a crude extract is subjected to ion-exchange chromatography on DEAE-Sephadex, the pattern seen in Fig. 4

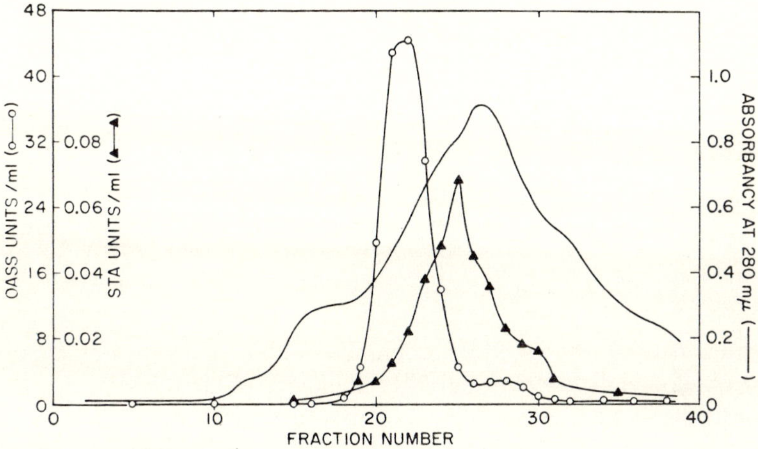

FIG. 4. DEAE-Sephadex chromatography of a crude extract of S. *typhimurium*, showing the relationship between serine transacetylase (STA) and O-acetylserine sulfhydrylase (OASS).

is obtained. Although the OASS and serine transacetylase activities overlap somewhat, it is clear that they elute differently on DEAE-Sephadex. When a crude extract is gel-filtered on Sephadex G-200 (Fig. 5), 98% of the OASS activity is readily separable from the more excluded serine transacetylase. However, the small amount of OASS activity which elutes with the serine transacetylase is interesting in that it coincides with the serine transacetylase with a reasonably constant ratio of activities throughout this peak. We refer to this fraction as OASS-STA and to the majority of the activity as OASS-A. When the OASS-STA peak is gel-filtered on a Sephadex G-150 column, the OASS activity runs with the serine transacetylase in the same manner, giving the same ratio of activities as in the first gel filtration.

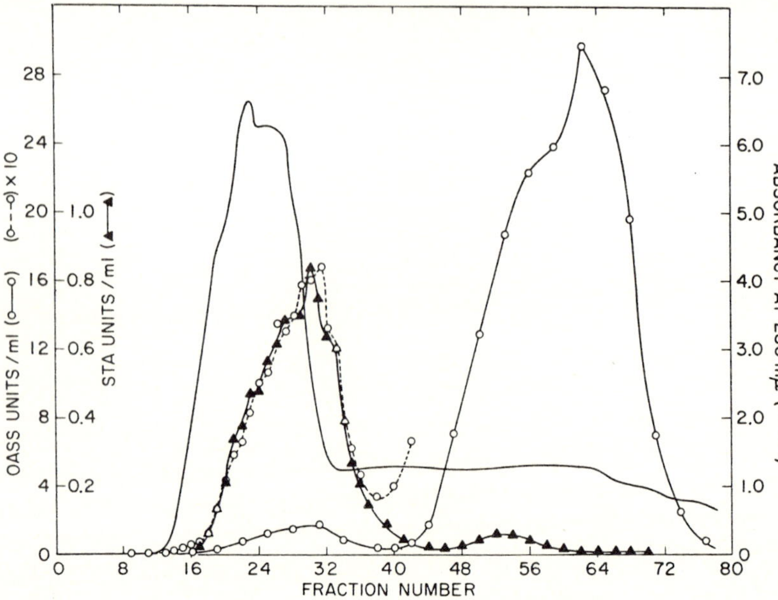

FIG. 5. Sephadex G-200 gel filtration of a crude extract of *S. typhimurium*. The small amount of O-acetylserine sulfhydrylase which elutes with serine transacetylase is designated OASS-STA, while remainder is OASS-A.

If this material is now chromatographed on DEAE-Sephadex (Fig. 6), we again find that the two activities coincide. (This relationship would obviously have been missed in the experiment starting with the crude extract, Fig. 4, since the OASS-STA is only a small fraction of the total OASS.) Sedimentation through a 5 to 20% sucrose gradient also fails to separate the two activities, as seen in Fig. 7.

When various stored fractions from the purification of the serine transacetylase from *E. coli* B were assayed for OASS activity, the results seen in Table II were obtained. Note that, in the first three steps listed in the table, the OASS-to-serine transacetylase ratios decreased from 120 to 2.0, and then remained relatively constant during DEAE-cellulose chromatography, Sephadex G-200 gel filtration, and preparative acrylamide gel electrophoresis, three procedures in which one generally obtains a high degree of protein resolution. The discrepancies between the ratios of 2.0, 6.4, and 4.5 are probably due to early difficulties with the OASS assay, now resolved, and to the fact that both OASS and serine transacetylase activities vary a great deal with storage under various conditions of temperature and salt and protein concentration. The ratios obtained in *S. typhimurium* under more carefully controlled conditions are also presented in Table II and are much more constant.

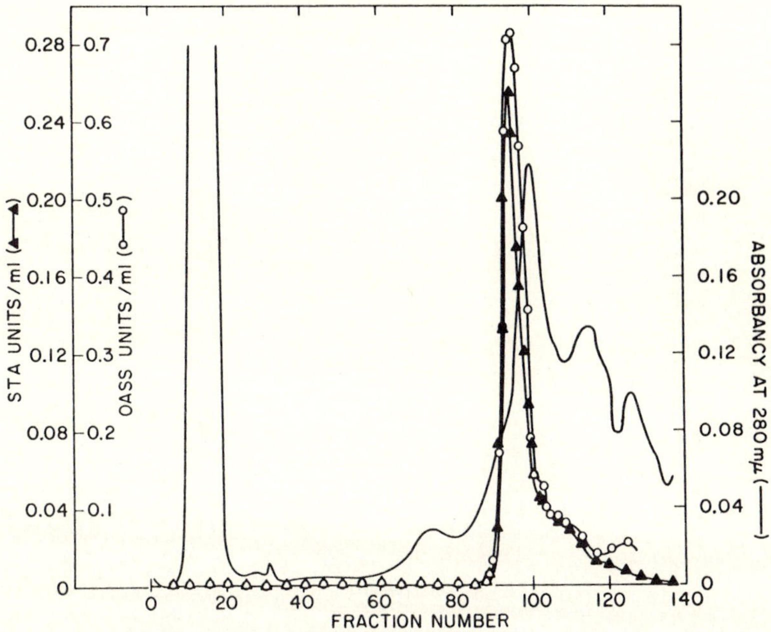

FIG. 6. DEAE-Sephadex chromatography of serine transacetylase and OASS-STA. The serine transacetylase-OASS-STA peak from a G-200 column was rechromatographed as in Fig. 4, showing the correspondence between the two activities.

TABLE II
COPURIFICATION OF SERINE TRANSACETYLASE AND O-ACETYLSERINE SULFHYDRYLASE ACTIVITIES IN *Escherichia coli* B AND *Salmonella typhimurium*

Fraction	Ratio of OASS to serine transacetylase
E. coli B	
Crude extract	120
Ammonium sulfate ppt.	62
Acid pH ppt.	25
Calcium phosphate gel	2.0
DEAE-cellulose	2.0
Sephadex G-200	6.4
Prep. gel electrophoresis	4.5
S. typhimurium	
Crude extract	117
Ammonium sulfate ppt.	60
Sephadex G-200	2.4
Sephadex G-150	2.3
DEAE-Sephadex	2.5
Sucrose gradient	2.8

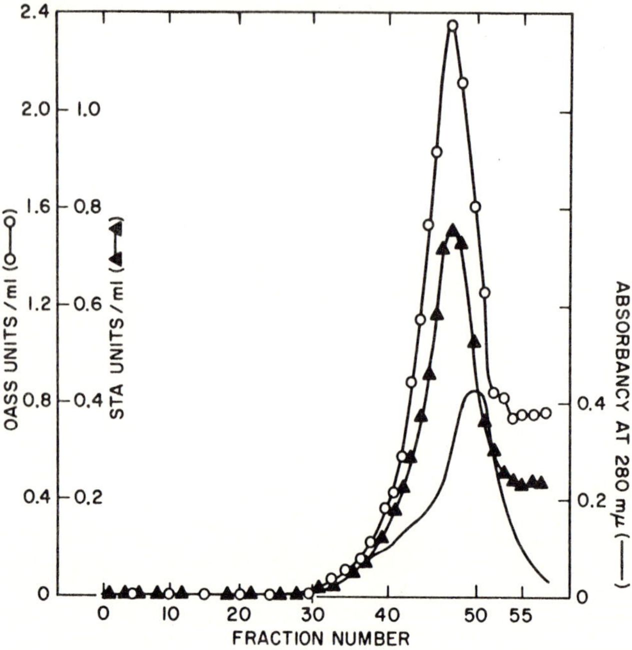

FIG. 7. Sucrose gradient centrifugation of the serine transacetylase-OASS-STA peak in Fig. 6. The two activities correspond in this 5 to 20% sucrose gradient, centrifuged at 22,000 rpm for 16 hours. The bottom of the gradient is to the left.

Since organisms lacking OASS activity should also be phenotypically *cys* E mutants unable to grow on sulfide, OASS activities in *cys* E mutants were examined (Table I). Although OASS activities in crude extracts varied considerably (between 25% and 500% wild-type levels), none of the six mutants examined had less than 25% wild-type activity. The OASS activities in mutants E_a 30 and E_b 17 have been examined and appear to be all of the OASS-A type, although we have not completely ruled out the possibility of a small amount of OASS-STA, minus the serine transacetylase, occurring in these organisms. The presence of OASS activity in these mutants correlates with the observation that all six will exhibit some growth about a crystal of O-acetyl-L-serine on a minimal agar plate. This growth, together with the absence of serine transacetylase in the mutants, suggests that O-acetyl-L-serine is the precursor of L-cysteine in S. *typhimurium.*

When wild-type S. *typhimurium* is grown on a sulfate-free medium in the presence of 0.5 m*M* L-cysteine or L-djenkolic acid (the latter presumably slowly releases L-cysteine and is known to derepress the enzymes of sulfate reduction in this organism), the OASS activities vary over a sevenfold range (Table III). Cells grown on L-cysteine have less than the wild-type level, and cells grown on L-djenkolic acid have about twice the wild-type value. Preliminary experiments have indicated that these changes are primarily, if not entirely, in OASS-A levels rather than in OASS-STA. It is interesting to note that under the same growth conditions lesser but reciprocal changes occur in serine transacetylase levels. The mechanism and significance of these changes are not presently understood.

TABLE III
EFFECT OF DIFFERENT SULFUR SOURCES ON THE LEVELS OF O-ACETYLSERINE SULFHYDRYLASE AND SERINE TRANSACETYLASE IN *Salmonella typhimurium*[a]

Sulfur source	OASS (units/mg)	Serine transacetylase (units/mg)
Sulfate	5.3	0.031
L-Cysteine	1.6	0.057
L-Djenkolic acid	10.9	0.023

[a] The wild-type organism, LT-2, was grown on a minimal salt medium (9) supplemented with 0.5% glucose and a 0.5 m*M* concentration of the indicated compound as the sole sulfur source.

The variation in OASS-A activity in *cys* E mutants and in the wild-type organism grown on different sulfur sources suggests the direct involvement of this enzyme activity in cysteine biosynthesis. On the other hand, the physical association between serine transacetylase and OASS-STA is also very suggestive of a role for this OASS activity in cysteine biosynthesis. The differences between these two OASS activities have not yet been systematically investigated, but those which have thus far been found are as follows: (1) They are physically different in regard to ammonium sulfate precipitation, gel filtration, and ion-exchange chromatography. (2) The K_m for O-acetyl-L-serine is 5 m*M* for OASS-A and 14 m*M* for OASS-STA. (3) Heating at 62°C for 5 minutes results in 10% inactivation of OASS-A and 90% inactivation of OASS-STA. (4) Preincubating OASS-A in 4 *M* urea for 5 minutes results in a slight decrease in activity, while the same treatment gives a threefold increase in OASS-STA activity.

If OASS-STA consists of OASS-A which is in some way associated with serine transacetylase, then the association must produce some conformational change in the OASS, to account for the different K_m's for O-acetyl-L-serine, and differences in sensitivity to heat denaturation and preincubation with urea. The existence of a large amount of free OASS, not associated with serine transacetylase, is, for the moment, problematic.

It may be that during the preparation of cell extracts a naturally occurring complex is dissociated. This and other possibilities are currently being investigated.

In any case it seems reasonable that the physiologically significant fraction of OASS activity is that which is bound to serine transacetylase, since the association of these two activities produces a particle capable of synthesizing L-cysteine from L-serine, acetyl-CoA, and sulfide. This has been demonstrated by using labeled L-serine or labeled sulfide as a precursor in the enzymatic reaction, followed by chromatography and autoradiography of the products. Furthermore, as would be expected, the serine transacetylase reaction, catalyzed by this complex, becomes rapidly feedback-inhibited by the L-cysteine formed in the reaction. Although these results are implicit in the data which we have presented, we feel that they are worth emphasizing, since it is a demonstration of end-product synthesis together with end-product inhibition occurring in a metabolic pathway mediated by a multifunctional enzyme complex.

Another interesting characteristic of this pathway is that, in the absence of sulfide, N-acetyl-L-serine is an indirect product of the serine transacetylase reaction. Since N-acetyl L-serine has been reported as the N-terminal residue in several pure proteins (3, 5, 6), we examined *E. coli* protein for the occurrence of this compound. Crude, soluble *E. coli* protein was exhaustively dialyzed and then subjected to pronase digestion. After ion-exchange and thin-layer chromatography, N-acetyl-L-serine was found as a product of the digest in amounts of 0.02 to 0.03 mmole per 40,000 mg of crude protein. We have not determined whether this compound exists as such or as O-acetyl-L-serine which is converted to N-acetyl-L-serine during the isolation and digestion of protein; indeed, the only evidence for its being bound in a peptide linkage is its release by pronase digestion. Nevertheless it is interesting to speculate that N-acetyl-L-serine may occur as an N-terminal residue in 2 to 3% of *E. coli* protein. This possibility and its relationship to serine transacetylase are presently under investigation.

References

1. Clowes, R. C., *J. Gen. Microbiol.*, **18**, 154 (1958).
2. Dreyfuss, J., and Monty, K. J., *J. Biol. Chem.*, **238**, 1019 (1963).
3. Harris, J. I., *Biochem. J.*, **71**, 451 (1959).
4. Kredich, N. M., and Tomkins, G. M., *J. Biol. Chem.*, **241**, 4955 (1966).
5. Narita, K., *Biochim. Biophys. Acta*, **30**, 352 (1958).
6. Reichmann, M. E., Chang, A. Y., Faiman, L., and Clark, J. M., Jr., *Cold Spring Harbor Symp. Quant. Biol.* **31**, 139 (1966).
7. Saville, B., *Analyst*, **83**, 670 (1958).
8. Siegel, L. M., personal communication.
9. Vogel, H. J., and Bonner, D. M., *J. Biol. Chem.*, **218**, 97 (1956).

Derepression and Repression of the Histidine Operon: Sequential and Simultaneous Modes

ROBERT F. GOLDBERGER AND MARY ANNE BERBERICH[1]

National Institute of Arthritis and Metabolic Diseases,
National Institutes of Health,
Bethesda, Maryland

The pathway for histidine biosynthesis in *Salmonella typhimurium*, elucidated largely through the efforts of Dr. B. N. Ames and co-workers (4, 5, 25, 29, 30), is shown in Fig. 1. The structural genes for the enzymes catalyzing the ten reactions are localized in a small segment of the bacterial chromosome, the *histidine operon* (6, 19). The isolation and characterization of over a thousand histidine-requiring mutants by Dr. P. E. Hartman and associates has provided the basis for a detailed map of the histidine operon (Fig. 2) (14, 15, 19). Through the studies of these groups, each of the genes has been identified with a specific enzyme. These studies showed that the relative positions of the genes in the operon do not correspond with the sequence in which the enzymes function in the metabolic pathway; however, the product of the first gene (at the operator end) does catalyze the first step of the pathway for histidine biosynthesis.

The phenomenon of coordinate derepression of the histidine enzymes was originally described by Ames and Garry (3), who found that the histidine operon functions as a single unit in response to the level of histidine available to the organism. They showed that, when histidine becomes the limiting growth factor, the levels of the enzymes for histidine biosynthesis increase (derepress) to the same extent. Later work (27, 28) indicated that histidyl tRNA charged with histidine is more directly involved in repression than is free histidine.

Evidence presented by Martin (21) indicated that the histidine operon is transcribed into a polycistronic messenger RNA, and studies on polarity in the histidine operon suggested that this messenger RNA is translated from the operator end (6). Thus, it appeared that utiliza-

[1] Research Fellow, The Helen Hay Whitney Foundation.

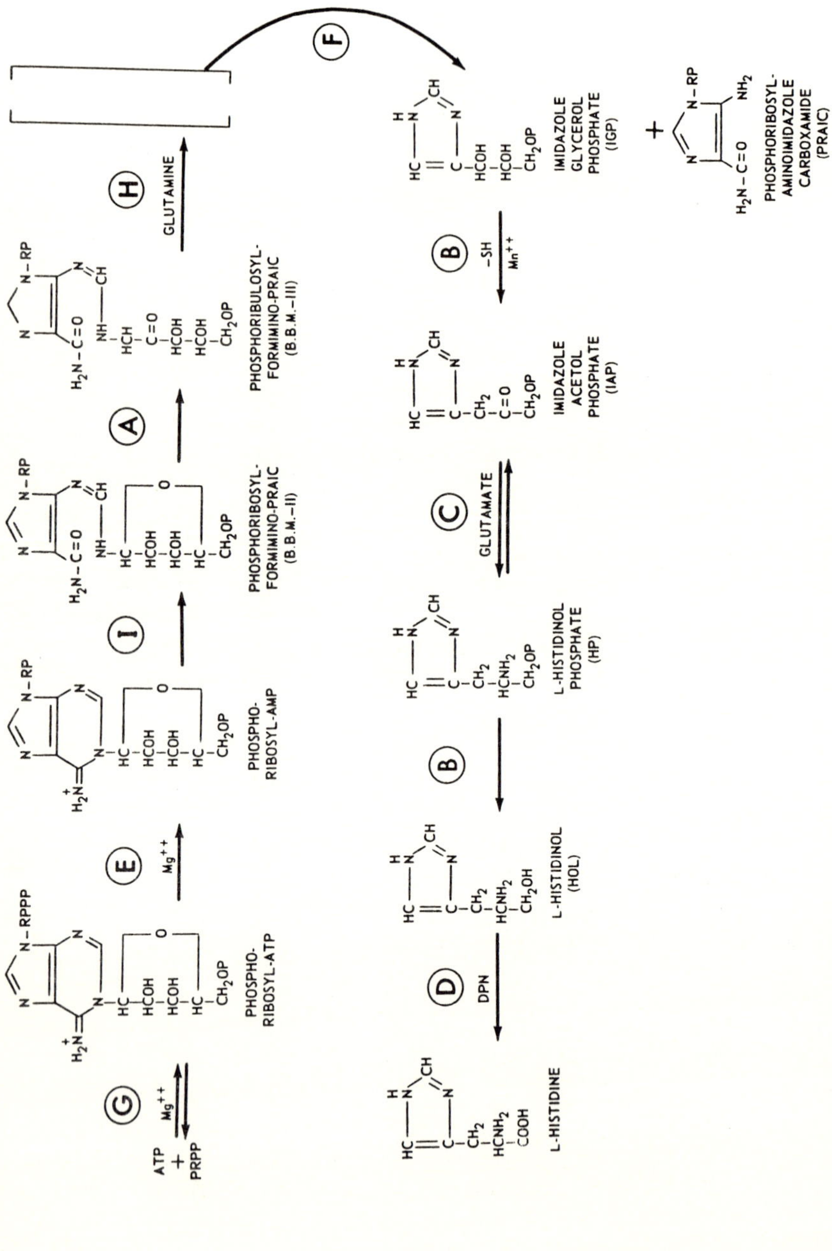

FIG. 1. Pathway for histidine biosynthesis in *Salmonella typhimurium*. Phosphate is designated by P; ribose, by R; and aminoimidazole carboxamide, by AIC. The brackets are used to indicate an intermediate of unknown structure.

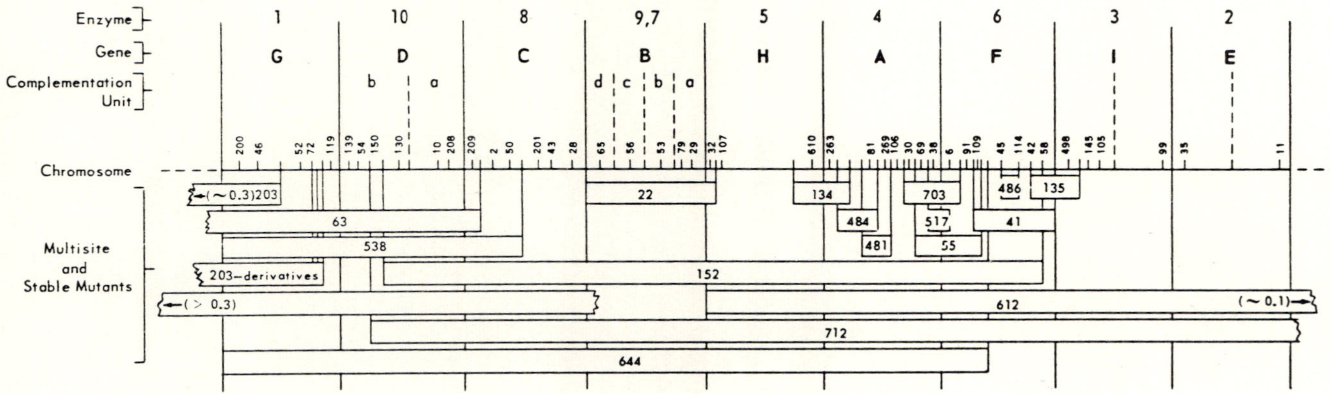

FIG. 2. Map of the histidine operon of *Salmonella typhimurium*, constructed by Hartman and co-workers. The numbers at the top of the diagram indicate the steps in the pathway for histidine biosynthesis catalyzed by the various enzymes. The structural genes for these enzymes, designated by capital letters, are shown in the second line of the diagram. Below this are shown, in descending order, complementation units (designated by small letters), some of the point mutants (designated by arabic numerals), and some of the multisite (deletion) mutants (designated by arabic numerals enclosed within boxes, the lengths of which indicate the sizes of the deletions).

tion of the information encoded in the histidine operon is oriented in a specific direction. To determine whether this orientation could be reflected in a temporal sequence of events, we undertook a study of the kinetics of derepression and repression of the histidine enzymes (8, 12). For this study, the specific activities of the enzymes were examined at frequent intervals during the periods immediately preceding and following the changes in growth rate that occur when histidine becomes a limiting growth factor (derepression) and when excess histidine is subsequently added back to the culture medium (repression). These studies served to demonstrate a temporal sequence in the derepression and repression of the histidine enzymes. We have also defined the conditions under which this sequence may be observed, and have been able to use the kinetic technique to study changes in the sequence which occur in the presence of 4-amino-5-imidazolecarboxamide ribonucleoside (ribosyl-AIC), adenine, and chloramphenicol (8, 12).

In this paper we present the results of our investigations. On the basis of these results, together with some of the pertinent information obtained in other laboratories, we shall discuss the various possibilities for the utilization of the information encoded in the histidine operon.

Methods and Results

Derepression of the histidine enzymes was studied in a large number of histidine auxotrophs, obtained from the collection of Dr. P. E. Hartman. In the presence of excess histidine the growth of these organisms proceeds with a generation time of approximately 50 minutes. As was previously shown (4, 5), the growth rate of these mutants can be controlled by limiting the amount of histidine. In our studies, cells were grown on sufficient histidine to allow normal growth to mid-log phase. Following histidine depletion, leaky[2] mutants were allowed to grow on endogenously produced histidine, which resulted in a doubling time of approximately 4 hours. In the case of complete[2] mutants, growth following depletion of histidine was supported by histidine precursors which were added to the medium (4, 5). The concentrations of these precursors were adjusted so that their slow conversion to histidine limited the growth to a doubling time of approximately 4 hours. As originally shown by Ames and Garry (3), the mutants elaborate large quantities of the enzymes for histidine biosynthesis (derepress) during the period of slow growth due to histidine limitation. We studied the kinetics of the derepression process by frequent measurements of the specific activities of six of the enzymes during the period immediately preceding and following histidine depletion in the growth medium.

[2] See footnote *b* for Table I.

Classification of Histidine Auxotrophs According to Mode of Derepression

When the kinetics of derepression were examined in one mutant, the enzymes for histidine biosynthesis were found to derepress simultaneously (8), as illustrated by the experiment shown in Fig. 3*A*. In another mutant, the enzymes derepressed in a temporal sequence which corresponds with the positional sequence of the genes in the histidine operon (8), as illustrated by the experiment shown in Fig. 3*B*. Approximately 20 minutes intervened between the times of derepression of the first and last enzymes studied. Alpers and Tomkins (1) observed that, immediately following induction, the order of appearance of two enzymes of the lactose operon of *Escherichia coli* corresponded with the positions of the genes in the operon. These workers observed an interval of approximately 3 minutes between the induction of β-galactosidase and thiogalactoside transacetylase, the enzymes of the first and third structural genes. It is interesting that the intervals between the times of derepression or induction of any two enzymes appear to be related to the distance between the corresponding genes, and are of the same order of magnitude in the lactose system of *E. coli* and the histidine system of *S. typhimurium*.

Having observed two different modes of derepression in two histidine auxotrophs, we undertook a survey of the mode of derepression in a large number of mutants. A summary of the results obtained with all mutants examined (8, 12, 24) is shown in Table I. No experiment gave equivocal results; the assignment of mutants to one of the two groups (simultaneous or sequential) could be made with a confidence of greater than 99%. When the modes of derepression (simultaneous or sequential) in the various auxotrophs are compared with the locations of the mutations in the histidine operon, no clear correlation can be made. However, when the modes of derepression are compared with the metabolic steps affected by the mutations, it becomes apparent that those mutants demonstrating the simultaneous mode of derepression have a nonfunctional enzyme involved in one of the first six steps of the metabolic pathway, whereas those mutants demonstrating the sequential mode of derepression have either a nonfunctional enzyme involved in one of the last four steps of the metabolic pathway or a partially defective enzyme involved in one of the first steps of the pathway. The metabolic step which divides the two groups is the reaction in which D-*erythro*-imidazoleglycerol phosphate (IGP) and phosphoribosyl-AIC are formed (cf. Fig. 4 and Table I). The mode of derepression is sequential in mutants which are capable of producing any amount of phosphoribosyl-AIC via the histidine pathway; the mode of derepression is simultaneous in

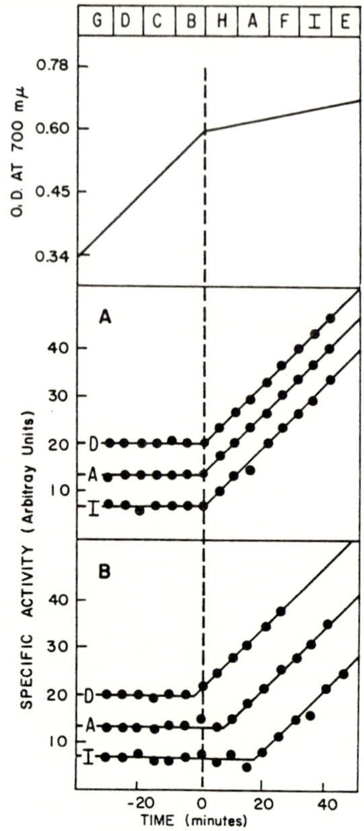

FIG. 3. Alternative modes of derepression of the histidine operon. At the top is shown the order of the genes in the operon. Below this, a growth curve is shown, which demonstrates the change in growth rate due to histidine limitation. The histidine auxotrophs were grown in the presence of an amount of L-histidine (20 μM) sufficient to support growth to an optical density at 700 mμ of 0.6 (5.6×10^8 cells/ml). Following histidine depletion, growth was supported by L-histidinol (10 μM), and the growth rate changed from a doubling time of approximately 50 minutes to one of approximately 4 hours. Aliquots were removed from the cultures periodically before and after the change in growth rate (derepression). Extracts prepared from these aliquots were assayed for protein and for five of the enzymes for histidine biosynthesis. *A*, the pattern obtained in a mutant (*hisG52*) characterized by the simultaneous mode of derepression (8). *B*, the pattern obtained in a mutant (*hisC2*) characterized by the sequential mode of derepression (8). Although not shown in this figure, the other two enzymes examined in each case followed the characteristic pattern. The vertical reference line (- - -) denotes the time of change in the growth rate due to limitation of histidine. For the sake of clarity, overlapping of the curves has been avoided by adding a constant factor to each set of points. If the base lines were at the same level, the positive slopes for all the enzymes would describe a single line, due to the fact that the rate at which all the enzyme doubled in quantity was identical.

TABLE I
MODE OF DEREPRESSION IN VARIOUS MUTANTS OF THE HISTIDINE OPERON

Organism	Type of defect	Defective enzyme(s)	Mode of derepression
LT-2[a]	Leaky[b]	*G*	Sequential
hisC2	Complete[b]	*C*	Sequential
hisB206	Complete	*B*	Sequential
hisB59	Leaky	*B*	Sequential
hisE11	Leaky	*E*	Sequential
hisI648	Leaky	*I*	Sequential
hisG52	Complete	*G*	Simultaneous
hisH, B22	Complete	*H, B*	Simultaneous
hisA30	Complete	*A*	Simultaneous
hisF41	Complete	*F*	Simultaneous
hisF486	Complete	*F*	Simultaneous

[a] Derepressed by the addition of DL-2-thiazole alanine, which is known to inhibit the *G* enzyme (4, 7).

[b] The term *leaky* designates an organism in which the mutation allows partial functioning of the metabolic pathway, as is evident from its slow growth in minimal medium; the term *complete* designates an organism in which the mutation causes a complete block in the metabolic pathway.

mutants which are not capable of producing phosphoribosyl-AIC via this pathway.

Effects of Ribosyl-AIC and Adenine

Because the sequential mode of derepression appeared to depend on the ability of the organism to produce phosphoribosyl-AIC via the histidine pathway, the effect of exogenous ribosyl-AIC on the mode of derepression was investigated (8). The results of a typical experiment are shown in Fig. 5. The simultaneous mode of derepression characteristic of the mutant used for this study is shown in Fig. 5*A*. The addition of ribosyl-AIC to the culture medium changed the mode of derepression to sequential (Fig. 5*B*). Since this shift in the mode of derepression was obtained for all mutants examined, we have concluded that phosphoribosyl-AIC is responsible, either directly or indirectly, for the phenomenon of sequential derepression.

The kinetics of derepression of the enzymes for histidine biosynthesis have also been examined in mutants grown in the presence of adenine (8). This metabolite had an effect opposite to that of ribosyl-AIC; in mutants ordinarily characterized by the sequential mode of derepression, adenine changed the mode of derepression to simultaneous, as illustrated in Figs. 6*A* and 6*B*. When adenine changed the mode of derepression from sequential to simultaneous, it delayed derepression of the early enzymes, whereas it advanced the time of derepression of the later enzymes. Thus, all the enzymes derepressed approximately 10 minutes after the change in the growth rate (cf. Figs. 6*A* and 6*B*).

The concentration of adenine required to elicit this effect in each mutant was directly related to the amount of phosphoribosyl-AIC produced via the histidine pathway in that mutant. Phosphoribosyl-AIC is not only a by-product of the histidine pathway but is also an intermediate in the *de novo* pathway for purine biosynthesis (13). In all histidine auxotrophs examined, the purine pathway is intact; therefore, phosphoribosyl-AIC produced in this pathway does not affect the mode of derepression of the histidine enzymes. The effect of adenine appears to be purine-specific, since guanine, but not uracil, could also shift the mode of derepression from sequential to simultaneous. However, guanine and adenine are interconvertible *in vivo* (10, 13, 18, 20).

The opposing effects of the two metabolites, adenine and ribosyl-AIC, are illustrated by the experiments shown in Fig. 7. In a mutant characterized by the simultaneous mode of derepression (Fig. 7*A*), the mode was changed to sequential when ribosyl-AIC was present in the growth

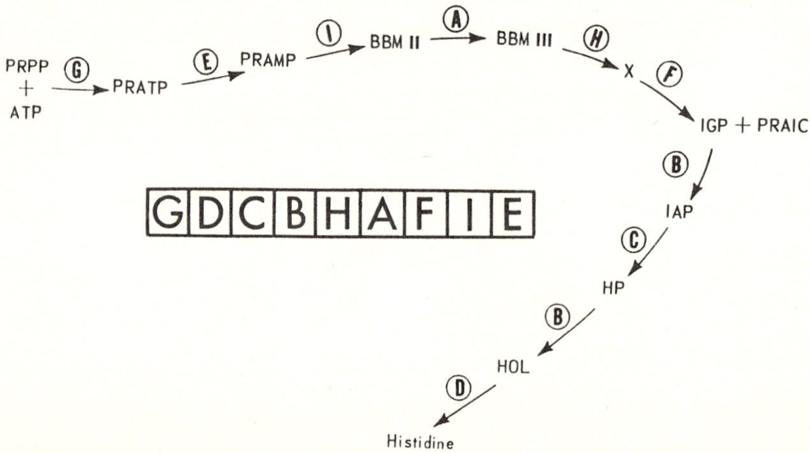

FIG. 4. Schematic representation of the pathway for histidine biosynthesis and the histidine operon. Abbreviations used: PRAIC, 4-amino-5-imidazolecarboxamide ribonucleotide; PRPP, 5-phosphoribosyl-1-pyrophosphate; PRATP, N-1-(5′-phosphoribosyl) adenosine triphosphate; PRAMP, N-1-(5′-phosphoribosyl) adenosine monophosphate; BBM II, N-(5′-phospho-D-ribosylformimino)-5-amino-1-(5″-phosphoribosyl)-4-imidazolecarboxamide; BBM III, N-(5′-phospho-D-1′-ribulosylformimino)-5-amino-1-(5″-phosphoribosyl)-4-imidazolecarboxamide; IGP, D-*erythro*-imidazoleglycerol phosphate; IAP, imidazole acetol phosphate; HP, L-histidinol phosphate; HOL, L-histidinol.

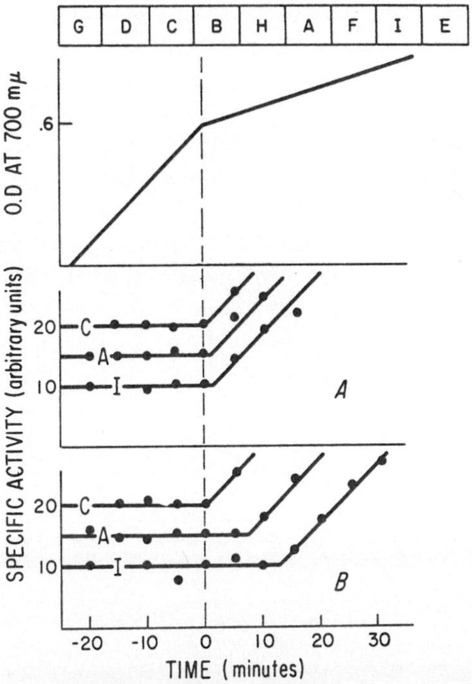

FIG. 5. The influence of ribosyl-AIC on the mode of derepression of *hisG52*. For methods of growth, sampling, and assay, see legend for Fig. 3. At the top of the figure is shown the order of the genes of the histidine operon, and below this is shown the growth curve. *A*, no additions; *B*, with ribosyl-AIC (4.5×10^{-5} *M*). The vertical reference line (- - -) denotes the time of change in the growth rate due to limitation of histidine. The lines are based on ten points before derepression and ten after. The figure focuses on the period relevant to the derepression times.

medium (Fig. 7*B*). When a sufficient amount of adenine was added to the medium in addition to the ribosyl-AIC, the mode was changed back to simultaneous (Fig. 7*C*).

Effect of Chloramphenicol

The effect of chloramphenicol on the kinetics of derepression was studied under conditions in which this antibiotic prolonged the doubling time of the derepressed organism to 8 to 12 hours (8). For these studies, the chloramphenicol was added approximately 10 minutes prior to histidine limitation in the growth medium. As the concentration of chloramphenicol was increased, the rate of increase in specific activity of all the enzymes was proportionately diminished and the intervals between the times of derepression of the enzymes were proportionately prolonged (Fig. 8). Whereas the usual time interval between derepression of the *D* and *A* enzymes was 12 minutes (Fig. 8*A*), this interval was extended to 35 minutes in the presence of chloramphenicol at a concentration of 0.5 μg/ml (Fig. 8*B*), and to 49 minutes in the presence of chloramphenicol at a concentration of 1.0 μg/ml (Fig. 8*C*). At the higher concentration employed in the present study, chloramphenicol totally inhibited derepression of the *I* enzyme within the period covered by the experiment. These results are in agreement with previous studies of Alpers and Tomkins (2), which showed that chloramphenicol increased the interval between the times of induction of β-galactosidase and thiogalactoside transacetylase. However, these workers observed

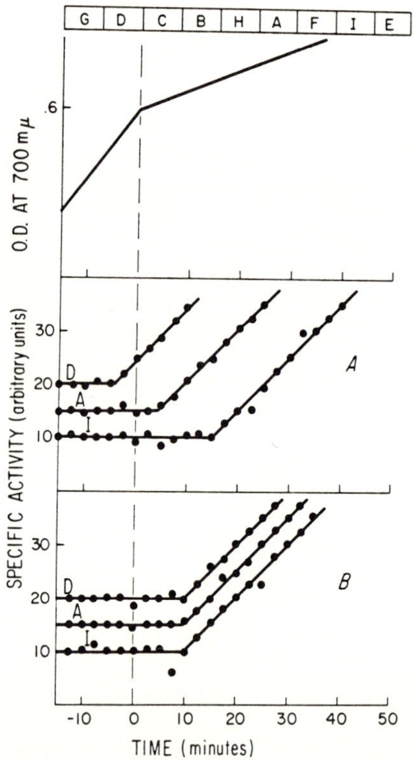

FIG. 6. The influence of adenine on the mode of derepression in *hisC2*. The methods of growth, sampling, and assay are described in the legend for Fig. 3. At the top of the figure is shown the order of the genes of the histidine operon, and below this is shown the growth curve. The sequential derepression of the enzymes characteristic of *hisC2* is shown in *A*. The effect of adenine ($7.5 \times 10^{-4}\ M$) is shown in *B*. The vertical reference line (- - -) denotes the time of change in the growth rate due to limitation of histidine.

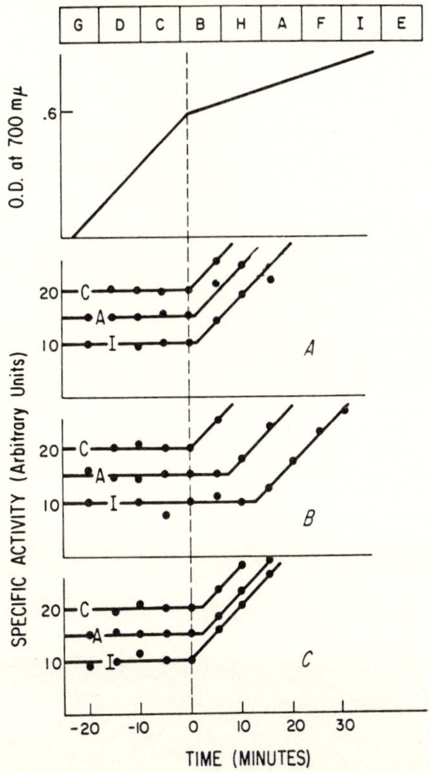

FIG. 7. The effects of ribosyl-AIC and adenine on the mode of derepression in *hisG52*. The methods of growth, sampling, and assay are described in the legend for Fig. 3. At the top of the figure is shown the growth curve. *A*, no additions; *B*, with ribosyl-AIC at 3.5×10^{-5} *M*; and *C*, with both adenine at 1.5×10^{-4} *M* and ribosyl-AIC at 3.5×10^{-5} *M*. The vertical reference line (- - -) denotes the time of change in the growth rate due to limitation of histidine. The lines are based on all the data: ten points before derepression and ten after. The figure focuses on the period relevant to the derepression times.

that chloramphenicol did not delay the induction of β-galactosidase, whereas in the present study, derepression of the first enzyme of the histidine operon was delayed. Alpers and Tomkins found that chloramphenicol inhibited the rate of increase in specific activity of thiogalactoside transacetylase more than that of β-galactosidase (2), whereas in the experiments discussed above, the rate of increase in specific activity for all the histidine enzymes tested was affected to the same degree (8).

In mutants characterized by simultaneous derepression, chloramphenicol had the same effect on the rate of increase in specific activity of all the enzymes, as illustrated by the experiment shown in Fig. 9. However, the mode of derepression remained simultaneous. Thus, chloramphenicol reduces the rate of increase in specific activity in both groups of mutants, sequential and simultaneous, but does not alter the mode of derepression in either.

Adenine exerts its effect on the mode of derepression, shifting sequential derepression to simultaneous, even when protein synthesis is partially inhibited by chloramphenicol, as shown in Fig. 10. Parts *A* and *B* of Fig. 10 show the effect of adenine in a mutant characterized by sequential derepression; parts *C* and *D* show that adenine has the same effect on

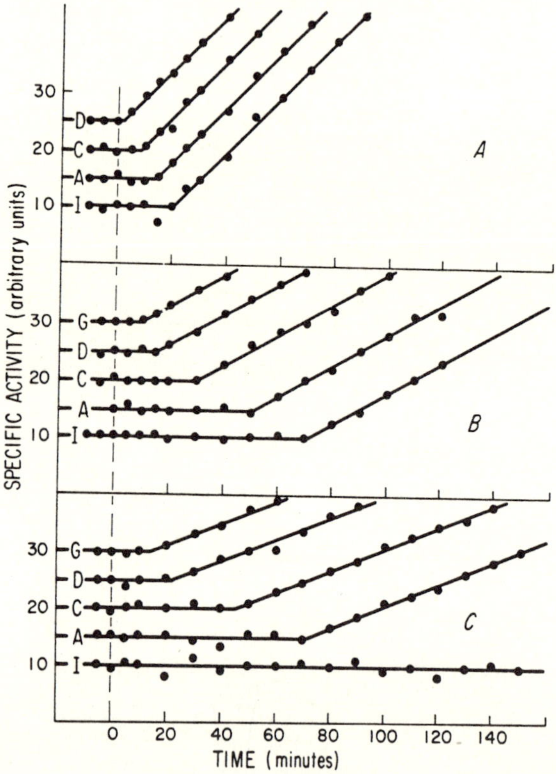

FIG. 8. The effect of chloramphenicol on the kinetics of derepression in *S. typhimurium* LT-2 (wild type). The specific activities for five of the enzymes for histidine biosynthesis are plotted against time. The vertical reference line (- - -) denotes the time of change in growth rate due to the addition of DL-2-thiazolealanine, which is known to cause derepression of the histidine operon (7). *A*, control experiment, without chloramphenicol; *B*, with chloramphenicol at 0.5 μg/ml; *C*, with chloramphenicol at 1.0 μg/ml.

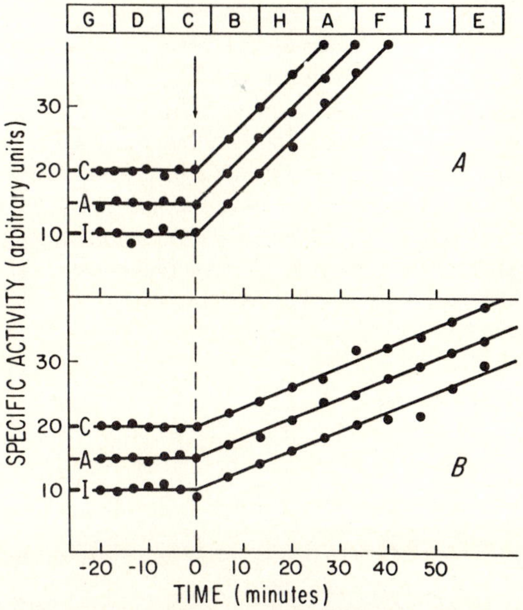

FIG. 9. The effect of chloramphenicol on the kinetics of derepression in *hisG52*. The specific activities for three of the enzymes for histidine biosynthesis are plotted against time. The vertical reference line (- - -) denotes the time of change in the growth rate due to limitation of histidine. At the top of the figure is shown the order of the genes of the histidine operon. *A*, control experiment, without chloramphenicol; *B*, with chloramphenicol at 1.0 μg/ml.

the mode of derepression when the same organism is grown in the presence of chloramphenicol at a concentration of 0.5 μg/ml.

Effect of Deletion Mutation

When sequential derepression occurs, the times of derepression of the various enzymes can be correlated with the positions of the structural genes in the operon (12). Thus, a significant shortening of the operon, by loss of a segment of its DNA, should give rise to a decreased interval between the times of derepression of the two enzymes for which the structural genes are located on either side of the deletion. However, if a genetically defined deletion were due to the inversion, rather than to the actual loss, of a segment of DNA, no alteration in the kinetic pattern of derepression would be expected for these two enzymes. This question was investigated with the use of the deletion mutant, *hisH,B22* (see Fig. 2), grown in the presence of ribosyl-AIC (24). The derepression pattern for three of the histidine enzymes in this mutant is shown in Fig. 11. The interval between the times of derepression of the *A* and *I* enzymes was approximately 10 minutes, whereas the *D* and *A* enzymes, the genes for which are located on either side of the deletion, derepressed essentially at the same time. When the operon is intact, the intervals between the times of derepression of the *D* and *A* enzymes and of the *A* and *I* enzymes are both approximately 10 minutes (see Fig. 3*B*). The absence of an interval between the times of derepression of the *D* and *A* enzymes in the deletion mutant suggests that there is an actual physical absence of genetic material ordinarily situated between the two corresponding structural genes in the histidine operon.

Repression

Kinetic analyses have also been applied to repression of the enzymes for histidine biosynthesis (12). The specific activities of the histidine enzymes were measured under conditions of histidine limitation and following the addition of excess histidine to the culture. When excess

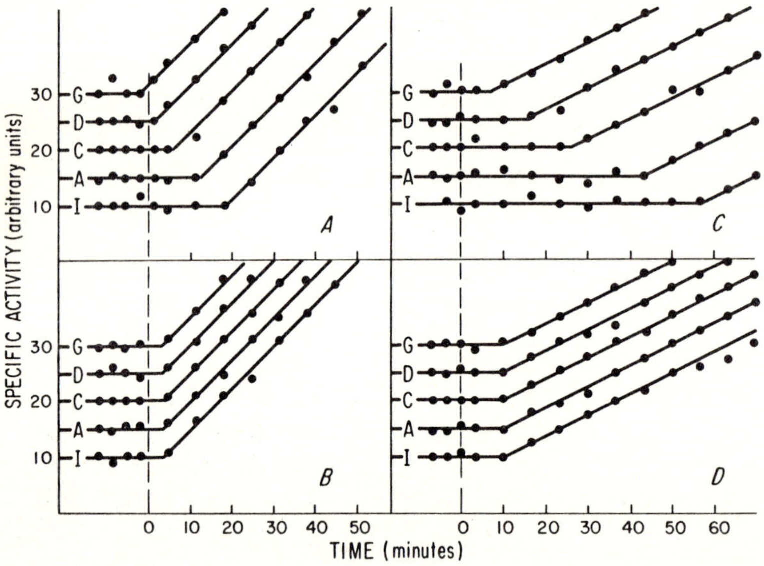

FIG. 10. The effect of chloramphenicol plus adenine on the kinetics of derepression in *hisE11*. The specific activities for five of the enzymes for histidine biosynthesis are plotted against time. The vertical reference line (- - -) denotes the time of change in the growth rate due to limitation of histidine. *A*, control experiment, with no additions; *B*, adenine alone, at 1.5 × 10^{-4} *M*; *C*, chloramphenicol alone, at 0.5 μg/ml; *D*, both adenine, at 1.5 × 10^{-4} *M*, and chloramphenicol, at 0.5 μg/ml.

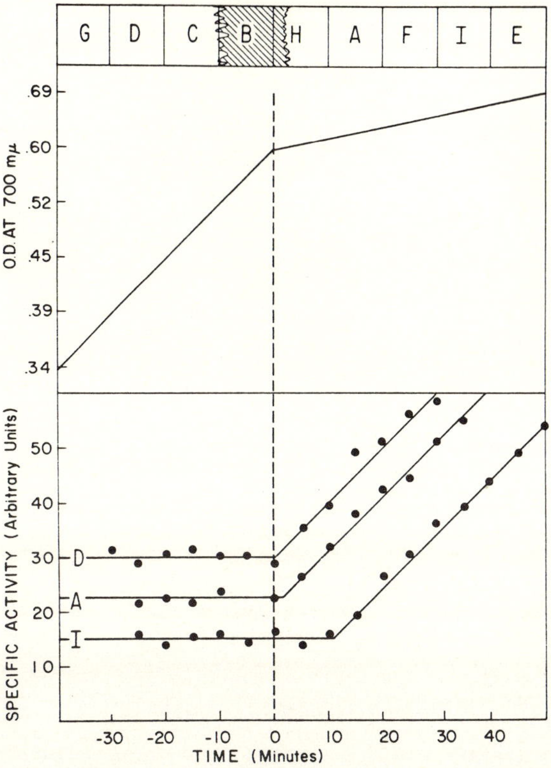

FIG. 11. Sequential mode of derepression observed in the deletion mutant, *hisH,B22,* when ribosyl-AIC was present in the culture medium at a concentration of 3.5×10^{-5} *M*. For details of sampling and assay, see legend for Fig. 3.

histidine was added 1 hour after derepression had occurred, the doubling time of the cells immediately changed from 4 hours to 50 minutes (the growth rate of repressed cells). The results obtained with a mutant characterized by the sequential mode of derepression are shown in Fig. 12. When histidine was added, the increases in the specific activities of the histidine enzymes came to a halt in the same temporal sequence in which derepression had occurred, again corresponding to the positional sequence of the structural genes in the operon. The time intervals observed for repression were of the same order of magnitude as those for derepression, and were similar to those observed by Alpers and Tomkins (1) during deinduction of two of the enzymes of the lactose operon in *E. coli*. The same experiment was performed, using a mutant characterized by the simultaneous mode of derepression (*hisG52*, see

Table I). The results were identical with those described above, demonstrating that the kinetics of repression are independent of the mode of derepression.

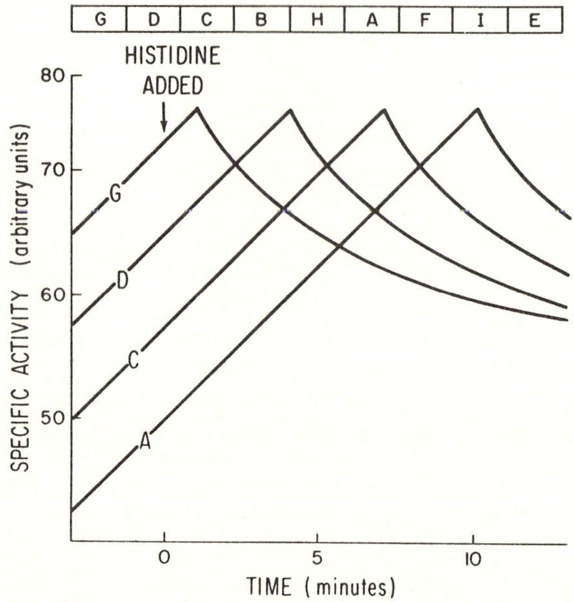

FIG. 12. Repression of enzymes for histidine biosynthesis. The specific activities shown above were measured during derepression and following the addition of excess histidine to the culture. When excess histidine was added, a generation time comparable to repressed growth was restored. The time of addition of histidine is indicated by the arrow. In order to clearly represent the repression times, only the central portions of the experiments are shown here; the lines were drawn, however, in accordance with all the data obtained in the 2 to 3 hours of the experiments.

INTERPRETATIONS

There are many theoretical possibilities for the mechanism of derepression and repression of the histidine operon. Some of these possibilities are discussed below in relation to the kinetic studies we have presented.

In some mutants, the information encoded in the genes of the histidine operon is utilized (ultimately translated) in a temporal sequence, whereas in other mutants this information is utilized simultaneously.

The effects of adenine and ribosyl-AIC cannot be interpreted without knowledge of the mechanisms responsible for determining the sequential and simultaneous modes of derepression. We have considered the possibility that the difference in mechanism between the two modes of derepression involves a difference in the rate of the rate-limiting step of the derepression process, considering only transcription or translation. That is, if sequential derepression were due to a slow sequential utilization of the genetic information of the histidine operon, then, conceivably, a great increase in the rate would be essentially equivalent to a simultaneous utilization of that information. Conversely, if simultaneous derepression were due to a rapid utilization of the genetic information of the histidine operon, then a reduction in the rate might reveal an underlying sequence in the mechanism of that utilization. Therefore, if the two different modes of derepression observed in our studies were due to a difference in the rate of translation or transcription, they would result in two different rates of enzyme synthesis—in the case of simultaneous derepression, the rate of increase in specific activity of the histidine enzymes would be greater than it would be in the case of sequential derepression. Since the rate of increase in specific activity of the histidine enzymes was found to be identical in mutants characterized by simultaneous derepression and in those characterized by sequential derepression, we consider it unlikely that the difference in the two modes of derepression involves a difference in the rate of the rate-limiting step of the derepression process. It follows, therefore, that the basis for the effects of adenine and ribosyl-AIC on the mode of derepression is not an acceleration or deceleration of the rate-limiting step, but, rather, a change in some *mechanism* involved in the derepression process.

In this discussion we have not yet considered the possibility that the rate-limiting step in the derepression process involves the attachment of ribosomes at the beginning of the polycistronic histidine messenger RNA. In this case, the difference between simultaneous and sequential derepression could be a difference in the rate of translation, but the rate-limiting step, attachment of ribosomes, would remain identical in both cases. Although this possibility must be kept in mind, we have tentatively put it aside because of the many difficulties encountered in fitting this model to the data from studies with chloramphicol discussed below.

Chloramphenicol has no effect on either mode of derepression. For example, in a mutant characterized by the simultaneous mode, derepression remained simultaneous even when chloramphenicol was present in sufficient concentration to greatly diminish the rate of translation, as evidenced by the diminished slopes of the curves shown in Fig. 10. Because chloramphenicol is known to inhibit translation of messenger RNA (11, 26, 33), the prolongation of the intervals between the times of derepression of the enzymes for histidine biosynthesis observed in the presence of this antibiotic (see Fig. 8) is probably due to a reduction in the rate of translation. The fact that adenine, in the presence of chloramphenicol, eliminates the intervals but does not alter the rate of enzymic increase (see Fig. 9) further supports the conclusion that the effect of adenine and ribosyl-AIC is to change the *mechanism* of derepression.

Sequential utilization of the information encoded in the histidine operon could involve either *the sequential translation of a polycistronic message* (as depicted in Fig. 13*A*) or *the sequential synthesis of individual messenger RNA molecules, one for each cistron* (as depicted in Fig. 13*B*). The latter alternative is excluded by the demonstration of Martin (21) that the histidine operon is transcribed into a polycistronic messenger RNA and by the subsequent verification of this finding under conditions of sequential derepression (32). Although the sequence observed must be ultimately due to a sequential translation of the polycistronic message, the rate of this translation may be restricted by the rate at which the message is transcribed. In other words, the polycistronic message may be translated as rapidly as it is being made.

It is interesting that, at the higher concentrations of chloramphenicol tested, derepression of the *I* enzyme did not occur (see Fig. 8*C*). If transcription were the rate-limiting step in the process of sequential derepression, one would have to envision a mechanism by which tran-

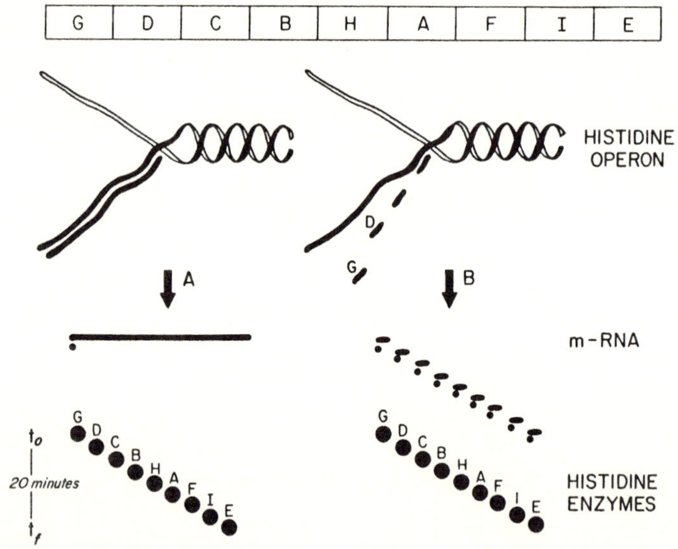

FIG. 13. Models for the sequential derepression of the enzymes for histidine biosynthesis. The time scale is represented at the left of the figure, where t_o is the time of derepression of the *G* enzyme, and t_f is the time of derepression of the *E* enzyme. As depicted here, the sequence could result either from sequential translation of a polycistronic message (*A*) or from sequential transcription of individual messages (*B*).

scription stops just short of the *I* gene when it has been progressing normally for almost 2 hours. On the other hand, if translation were the rate-limiting step in the process of sequential derepression, the failure of the *I* enzyme to derepress could be explained in the following way. The end of the message distal to the operator, unprotected by association with ribosomes for a relatively long time in the presence of chloramphenicol, might be vulnerable to degradation by nucleases. This mechanism is consistent with the concept suggested previously (9, 31) that transcription and translation are coupled processes, and is only dependent on the idea that the polymerase can operate some distance ahead of the most advanced ribosome. Alpers and Tomkins have presented evidence that the polymerase does, indeed, continue to function for at least the distance of one cistron in the absence of protein synthesis (2).

In those mutants characterized by the simultaneous mode of derepression, simultaneous utilization of the information encoded in all the genes of the histidine operon could involve either *the simultaneous translation of all cistrons of a polycistronic message* or *the simultaneous synthesis and release of individual messenger RNA molecules, one for each cistron.* These two possibilities are represented schematically in Figs. 14*A* and 14*B*, respectively.

Our studies with adenine and ribosyl-AIC demonstrate that all the organisms tested can exhibit both sequential and simultaneous derepression, depending on the conditions of the experiment. The shift from sequential to simultaneous derepression could involve a shift from the sequential translation of the polycistronic message to either the simultaneous translation of all the cistrons of this message or the simultaneous synthesis and release of many individual messenger RNA molecules. The nature of the histidine messenger RNA in mutants characterized by simultaneous derepression has not yet been ascertained. However, it seems reasonable to us to assume that this message is also polycistronic. Any other assumption would require two different mechanisms of transcription in every organism, one for the synthesis of a polycistronic message and the other for the synthesis of many messages.

The most probable explanation for the shift from the sequential to the simultaneous mode of derepression is a shift in the mechanism of translation of the polycistronic messenger RNA of the histidine operon. In the case of sequential derepression, translation of the message is initiated only at one end and proceeds by the movement of ribosomes toward the other end. The rate of transcription may or may not be dependent on the rate of translation. Ultimately, however, the relatively slow sequential translation is responsible for the sequential synthesis of the enzymes for histidine biosynthesis, translation of the whole message

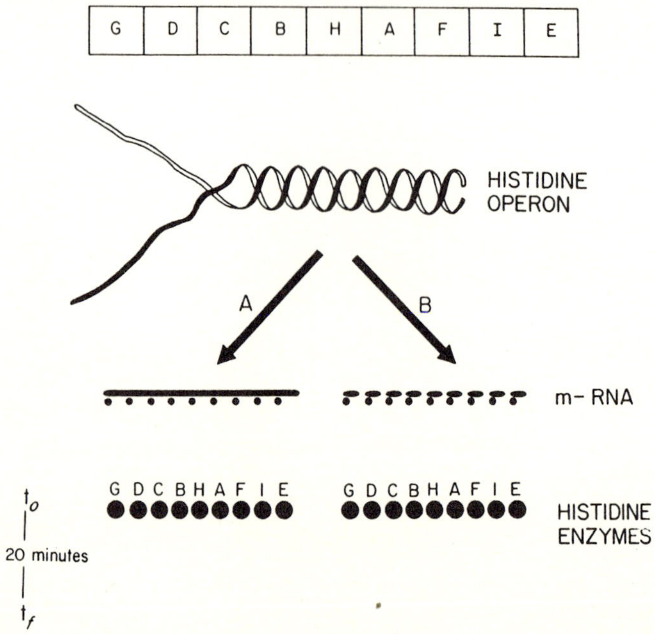

FIG. 14. Models for the simultaneous derepression of the enzymes for histidine biosynthesis. The time scale is represented at the left of the figure, where t_o is the time of derepression of all the enzymes. As depicted here, simultaneous derepression could result either from simultaneous translation of all parts of a polycistronic message (*A*) or from simultaneous synthesis and release of individual messages (*B*).

requiring approximately 20 minutes. The rate-limiting step in this translation may be the process by which each ribosome negotiates the borders between cistrons—that is, the initiation and termination of polypeptide chains. This is suggested by the kinetics of derepression in a deletion mutant, where the time eliminated might be more closely related to the *number of cistrons,* than to the *length of genetic material,* deleted (24). Sequential utilization of the information in the histidine operon involves recognition of the beginnings and ends of approximately fifteen cistrons (19).

In the case of simultaneous derepression, all the enzymes for histidine biosynthesis derepress at the same time. Therefore, it appears that all portions of the polycistronic message are translated simultaneously. This suggests that the entire message is available for translation before a significant amount of translation occurs. We conclude that simultaneous derepression reflects the attachment of ribosomes at multiple sites along the polycistronic message. However, this conclusion does not necessarily exclude the possibility that transcription and translation are still coupled. The effect of adenine, according to the hypothesis presented here, is to allow the attachment of ribosomes at multiple sites along the polycistronic message, whereas phosphoribosyl-AIC acts to prevent such attachment.

On the basis of this hypothesis one might make the prediction that adenine would eliminate the polar effect of certain classes of mutations of the histidine operon, which have been the subject of extensive studies by Ames and Hartman (6) and by Martin *et al.* (22, 23). These mutants have been shown to have reduced levels of all the enzymes encoded in the DNA more distal to the operator than the site of mutation (6, 22). Ames and Hartman (6) and Martin *et al.* (23) have proposed that the polarity is due to a reduction in the number of ribosomes which may pass the codon altered by the mutation and successfully complete the translation of the polycistronic message. One might expect, then, that attachment of ribosomes at multiple sites along a polycistronic message would, in effect, bypass the altered codon and eliminate the effect of the mutation on the levels of the subsequent enzymes. According to this reasoning, adenine should "cure" polarity, whereas ribosyl-AIC might magnify the polar effect. However, recent studies of Imamoto *et al.* (16) on the tryptophan operon in *E. coli* demonstrate that the decreased levels of the enzymes affected by polar mutations are associated with decreased levels of the specific messenger RNA for those enzymes. Thus polar mutations prevent, to various degrees, completion of a polycistronic message, owing, perhaps, to a dependence of transcription on translation (2, 9, 31). If polar mutations in the histidine operon have similar effects, the hypothesis discussed above for the mechanism of action of adenine on the mode of derepression should not necessarily be extended to the case of polar mutants. Studies on the kinetics of derepression in polar mutants of the histidine operon are in progress. Preliminary results show that the sequence of derepression of the histidine enzymes is different from that ordinarily observed (17). Therefore, because there appear to be fundamental differences between polar and nonpolar mutants under conditions of derepression, it is impossible at the present time to predict the effect of adenine or ribosyl-AIC on the relative levels of the histidine enzymes in polar mutants.

Interpretation of the meaning of sequential repression of the enzymes for histidine biosynthesis is complicated by the uncertainty concerning the level of protein synthesis at which the repressor acts. One possibility for the sequential cessation of enzyme synthesis is a cessation of message initiation at the operator end of the operon, nascent messenger molecules proceeding to completion. In addition, the polycistronic messenger RNA may be degraded from the operator end by an exonuclease (1). However, a more likely possibility, previously proposed (27), is that repression involves a cessation of ribosomal attachment to the operator end of the polycistronic message. For the latter hypothesis it is not necessary to invoke the action of a 5′-exonuclease, which as yet has not been shown to be involved in the degradation of messenger RNA.

Summary

The information encoded in all the genes of the histidine operon may be utilized in two different ways—sequentially from the operator end or simultaneously. Phosphoribosyl-AIC compels the organism to utilize this information sequentially, whereas adenine has the reverse effect. It is important to determine whether phosphoribosyl-AIC and adenine have the same effects in other systems or whether the specific substrates, intermediates, or by-products of the other metabolite pathways have analogous effects on their respective operons.

The significance of two alternative modes of derepression is not limited to the histidine operon. If, as we have proposed, a polycistronic messenger RNA can be translated not only sequentially (by ribosomes attaching only at the operator end), but also simultaneously (by ribosomes attaching at multiple sites), then organisms must be equipped with a system for selecting between the two alternative, as well as systems for implementing both.

Acknowledgments

We are indebted to Dr. C. B. Anfinsen for his encouragement and support throughout the course of these investigations. We also wish to thank Drs. B. N. Ames, P. E. Hartman, R. G. Martin, G. M. Tomkins, and P. Venetianer for their many helpful discussions and critical reading of the manuscript. It is again a pleasure to acknowledge the expert technical assistance of Mrs. Diana Marver and Mrs. Marilyn Meyers.

Note added in proof:

Recent studies on the mechanism involved in simultaneous versus sequential derepression (Berberich, M. A., Kovach, J. S., and Goldberger, R. F., *Proc. Natl. Acad. Sci. U.S.*, **57**, 1857, 1967) substantiate the hypothesis presented here. It appears that the mode of derepression depends upon the capacity of the cell to initiate polypeptide chains.

In mutants characterized by sequential derepression, the phosphoribosyl-AIC formed via the histidine pathway enters the purine pathway by conversion to formylphosphoribosyl-AIC. This conversion requires N^{10}-formyltetrahydrofolate, and therefore causes a shortage of formyl groups for the formylation of methionyl-tRNA. Under such conditions the ability of the cell to initiate polypeptide chains is limited, and translation of the polycistronic histidine message is initiated preferentially at the operator end. In mutants characterized by simultaneous derepression (those which do not produce phosphoribosyl-AIC via the histidine pathway) translation of the polycistronic histidine message can be initiated at multiple points because there is no deficit of N^{10}-formyltetrahydrofolate. Compounds such as adenine favor the simultaneous mode of derepression because they act as one-carbon donors and/or spare formyl groups which are ordinarily required for their own biosynthesis, whereas ribosyl-AIC and inhibitors of folic acid biosynthesis favor the sequential mode of derepression because they limit the formylating ability of the cell.

Recent studies on the repression process (J. S. Kovach, M. A. Berberich, P. Venetianer, and R. F. Goldberger, in preparation) have suggested that the feed-back-sensitive site of the enzyme encoded in the first structural gene of the histidine operon is required for the sequential mode of repression described above.

References

1. Alpers, D. H., and Tomkins, G. M., *Proc. Natl. Acad. Sci. U.S.*, **53**, 797 (1965).
2. Alpers, D. H., and Tomkins, G. M., *J. Biol. Chem.*, in press.
3. Ames, B. N., and Garry, B. J., *Proc. Natl. Acad. Sci. U.S.*, **45**, 1453 (1959).
4. Ames, B. N., Garry, B., and Herzenberg, L. A., *J. Gen. Microbiol.*, **22**, 369 (1960).
5. Ames, B. N., Martin, R. G., and Garry, B. J., *J. Biol. Chem.*, **236**, 2019 (1961).
6. Ames, B. N., and Hartman, P. E., *Cold Spring Harbor Symp. Quant. Biol.*, **28**, 349 (1963).
7. Ames, B. N., Hartman, P. E., and Jacob, F., *J. Mol. Biol.*, **7**, 23 (1963).
8. Berberich, M. A., Venetianer, P., and Goldberger, R. F., *J. Biol. Chem.*, **241**, 4426 (1966).
9. Byrne, R., Levin, J. G., Bladen, H. A., and Nirenberg, M. W., *Proc. Natl. Acad. Sci. U.S.*, **52**, 140 (1964).
10. Carter, C. E., and Cohen, L. H., *J. Biol. Chem.*, **222**, 17 (1956).
11. Das, H., Goldstein, A., and Kanner, L., *Mol. Pharm.*, **2**, 158 (1966).
12. Goldberger, R. F., and Berberich, M. A., *Proc. Natl. Acad. Sci. U.S.*, **54**, 279 (1965).
13. Gots, J. S., and Gollub, E. G., *Proc. Natl. Acad. Sci. U.S.*, **43**, 826 (1957).
14. Hartman, P. E., Loper, J. C., and Serman, D., *J. Gen. Microbiol.*, **22**, 323 (1960).
15. Hartman, P. E., Hartman, Z., and Serman, D., *J. Gen. Microbiol.*, **22**, 354 (1960).
16. Imamoto, F., Ito, J., and Yanofsky, C., *Cold Spring Harbor Symp. Quant. Biol.*, **31**, (1966).
17. Kovach, J., unpublished results.
18. Lieberman, I., *J. Biol. Chem.*, **223**, 327 (1956).
19. Loper, J. C., Grabner, M., Stahl, R. C., Hartman, Z., and Hartman, P. E. *Brookhaven Symp. Biol.*, **17**, 15 (1964).
20. Magasanik, B., *Ann. Rev. Microbiol.*, **11**, 221 (1957).
21. Martin, R. G., *Cold Spring Harbor Symp. Quant. Biol.*, **28**, 357 (1963).
22. Martin, R. G., Silbert, D., Smith, D. W. E., and Whitfield, H., Jr., *J. Mol. Biol.*, **21**, 357 (1966).
23. Martin, R. G., Whitfield, H., Jr., Berkowitz, D., and Voll, M. J., *Cold Spring Harbor Symp. Quant. Biol.*, **31**, 215 (1966).
24. Marver, D., Berberich, M. A., and Goldberger, R. F., *Science*, 153, 1655 (1966).

25. Moyed, H. S., and Magasanik, B., *J. Biol. Chem.*, **235**, 149 (1960).
26. Rendi, R., and Ochoa, S., *J. Biol. Chem.*, **237**, 3711 (1962).
27. Roth, J. R., Silbert, D. F., Fink, G. R., Voll, M. J., Anton, D., Hartman, P. E., and Ames, B. N., *Cold Spring Harbor Symp. Quant. Biol.*, **31**, 383 (1966).
28. Schlesinger, S., and Magasanik, B., *J. Mol. Biol.*, **9**, 670 (1964).
29. Smith, D. W. E., and Ames, B. N., *J. Biol. Chem.*, **239**, 1848 (1964).
30. Smith, D. W. E., and Ames, B. N., *J. Biol. Chem.*, **240**, 3056 (1965).
31. Stent, G. S., *Science*, **144**, 816 (1964).
32. Venetianer, P., unpublished results.
33. Weber, M. J., and DeMoss, J. A., *Federation Proc.*, **25**, 582 (1966).

Gene-Ribosome-Enzyme Organization in the Arginine System of *Escherichia coli*

H. J. VOGEL, S. BAUMBERG, D. F. BACON, E. E. JONES, L. UNGER, AND R. H. VOGEL

*Institute of Microbiology,
Rutgers, The State University,
New Brunswick, New Jersey*

The arginine system of *Escherichia coli*, with its genes partly clustered and partly dispersed, is well adapted to the study of various aspects of organizational biosynthesis (18, 22). The eight steps leading from glutamate to arginine, as well as the enzymes catalyzing them, are given in Fig. 1. The arrangement of the corresponding structural genes and of the system-specific regulatory gene (*R*) are indicated in Fig. 2 (22). The order (7) of genes in the four-gene cluster has been determined in *E. coli* K-12. Only two of the enzymes (Enzyme 2 and Enzyme 3) specified by the four genes actually are in biosynthetic sequence, and their genes are flanked, apparently contiguously, by those for Enzyme 8 and Enzyme 5.

NONCOORDINATE REPRESSION

One can now inquire whether the repression control for these four enzymes corresponds to a single genetic site or, possibly, to more than one site. On the basis of the unmodified operon model (9), a single control site governing the formation of the four enzymes would be expected. A suggestion that there may at least be two such sites was obtained some time ago, when it was found that the repression of Enzymes 5 and 8 is noncoordinate (3). These experiments, carried out in strain K-12, have now been extended, as summarized in Table I (2). It can be seen that not only do Enzymes 5 and 8 fail to repress coordinately, but also Enzymes 2 and 3, taken as a pair, do not repress coordinately. This behavior seems all the more remarkable in view of the findings mentioned in the following section, which are in line with separate repressor recognition sites corresponding to Enzymes 5 and 8,

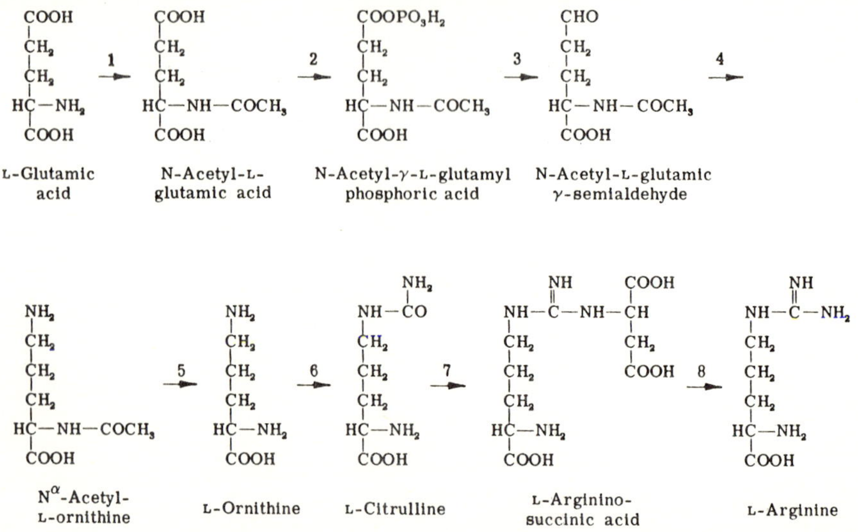

FIG. 1. Path of arginine synthesis in *Escherichia coli.* The enzymes corresponding to the eight steps shown are: 1, N-acetylglutamate synthetase; 2, N-acetyl-γ-glutamokinase; 3, N-acetylglutamic γ-semialdehyde dehydrogenase; 4, acetylornithine δ-transaminase; 5, acetylornithinase; 6, ornithine transcarbamylase; 7, argininosuccinate synthetase; 8, argininosuccinase.

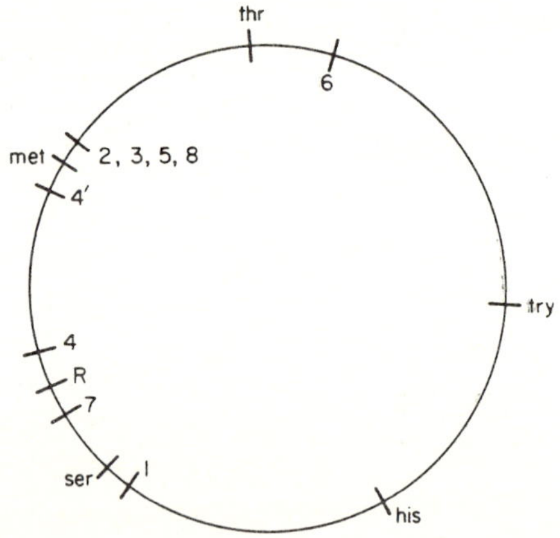

FIG. 2. Genes of the arginine system in *E. coli,* and reference markers (22). The arginine genes are labeled inside the schematic circular chromosome. Genes 1 through 8 correspond to Enzymes 1 through 8, respectively (see Fig. 1). In the nomenclature of Sanderson and Demerec (14), Genes 1 through 8 are *argB*, *argC*, *argH*, *argG*, *argA*, *argD*, *argE*, and *argF*, respectively. Gene 4′ (or *M*) corresponds to arginine-inducible acetylornithine δ-transaminase, and *R* is the system-specific regulatory gene.

TABLE I
NONCOORDINATE REPRESSION OF ENZYMES SPECIFIED BY CLUSTERED ARGININE GENES (AS RELATIVE DIFFERENTIAL RATES)[a]

Regulation	Enzyme 8	Enzyme 2	Enzyme 3	Enzyme 5
Derepression	100	100	100	100
Repression	3	4	17	12
Ratio[b]	33	25	6	8

[a] Two derivatives of *E. coli* K-12, both wild-type with respect to the structural genes of the arigine system, were used (cf. 20). The values for derepression were obtained with an R^- strain (which gives the same results for cultivation in the presence or in the absence of added arginine). The values for repression were obtained with an R^+ strain grown in the presence of L-arginine hydrochloride (0.1 mg per ml). The results are expressed as differential rates of enzyme formation relative to the respective values for derepression taken as 100 (2).

[b] Ratio of derepression value to repression value.

although we would conclude that noncoordination alone need not imply separate sites.

A Pleiotropic Mutant

Support for the notion that the four-gene cluster contains information for at least two repressor recognition sites came from the isolation of a K-12 mutant in which the repression-derepression behavior, *as well as the level,* of three out of the four enzymes specified by the clustered genes are affected (4). The three enzymes (Enzymes 8, 2, and 3), correspond to a subcluster of genes (see Fig. 3). The fourth enzyme (Enzyme 5) is normal in level and in repressibility properties. The mutant behaves as an arginine auxotroph. It does not grow on unsupplemented minimal medium, but grows at wild-type rate in the presence of arginine. On ornithine, the mutant gives slow growth reflecting a restrictive rate of arginine formation. The pleiotropic mutation of this organism maps in close association with the gene cluster. Genetic tests indicate that a single-point mutation is involved. In reversion studies, a positive response was obtained to 2-aminopurine but not to the acri-

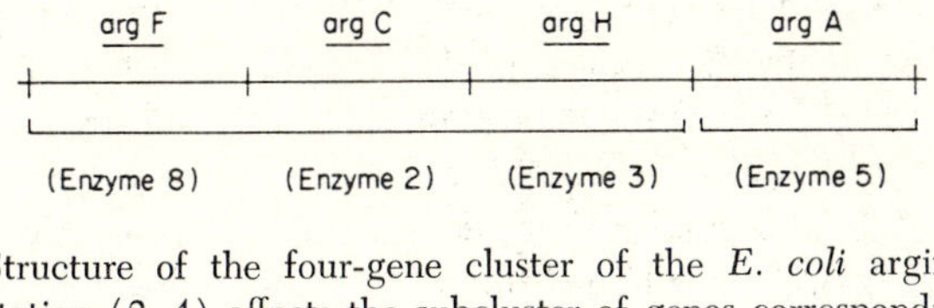

FIG. 3. Structure of the four-gene cluster of the *E. coli* arginine system. A pleiotropic mutation (2, 4) affects the subcluster of genes corresponding to Enzymes 8, 2, and 3. (See the text for details.)

dine-like compound ICR 170; single-step revertants to full wild-type character were isolated (2, 4).

Enzyme levels were examined after cultivation of the pleiotropic mutant in the presence of ornithine. Under these conditions, Enzyme 5 as well as the arginine enzymes specified by genes outside of the cluster are normally derepressed. However, Enzymes 8, 2, and 3 are all produced at relatively very low levels (2, 4).

It then seemed particularly interesting to determine if the three enzymes which are produced at such low levels are still responsive to arginine as a repressor. This indeed was indicated to be the case for all three enzymes (2, 4). The results obtained with Enzyme 8, for which the best assay is available, are given in Table II. It is seen that, in the presence of arginine, Enzyme 8 in the mutant is repressed by a factor of about 3. In a strain carrying the pleiotropic mutation and also a mutant allele (R^-) of the regulatory gene, Enzyme 8 is not repressed by arginine (2, 4). Enzymes 2 and 3 appear to behave similarly. It is concluded that the repressibility of the three enzymes that are affected in the (R^+) pleiotropic mutant is still governed by the system-specific regulatory gene (2). The finding that the ratio of the derepressed level of Enzyme 8 to its repressed level (Table II) amounts to about 3 is noteworthy, since this ratio is approximately 33 in normal strains (Table I). The usual experience is that the repressibility is unchanged even when the level of enzyme activity is drastically lowered (17).

TABLE II
RELATIVE DIFFERENTIAL RATES OF ENZYME 8 FORMATION IN STRAINS CARRYING THE PLEIOTROPIC MUTATION[a]

	Cultivation in the presence of	
Strain	Ornithine	Arginine
961 (R^+)	5.3	1.8
977 (R^-)	4.7	4.7

[a] When the pleiotropic mutant 961 is grown in the presence of ornithine, the rate of arginine synthesis is restrictive (because of the low level of Enzyme 8); the arginine system is thus generally under derepressive conditions. Strain 977, which carries R^- as well as the pleiotropic mutation, is seen to give values comparable to that obtained under arginine restriction in the repressible strain 961. Ornithine or arginine were used as the hydrochlorides of the L forms at 0.1 mg per ml. The results are expressed as differential rates of enzyme formation relative to the value of 100 for Enzyme 8 in Table I (2).

The results with the pleiotropic mutant indicate that there are at least two repressor recognition sites for the enzyme-forming systems corresponding to the four-gene cluster (presumably, one site for Enzyme 5 and at least one more for Enzymes 8, 2, and 3); that what appears to be a single-point mutation can simultaneously affect the level and the repression control of three enzymes; and that even for the three enzymes thus related, the regulation need not be coordinate.

The Problem of the Gene Cluster

The indicated occurrence of at least two repressor recognition sites for the enzyme-forming system corresponding to the four-gene cluster is not in harmony with an unmodified form of the operon model. If we do not have the economy of a single repressor recognition site, what then is the selective advantage (that must be assumed) of having the four genes clustered together (cf. 6, 15)? One possible explanation is in terms of the compartmentalization of metabolic systems. Presumably, in *E. coli* and related organisms, enzymes are produced in close proximity to their structural genes (16, 20). The clustering of genes would thus lead to the side-by-side deposition of the corresponding enzymes. On the assumption of a nonuniform distribution of small-molecular intermediates within the cell, we can then envisage a kind of compartmentalization (cf. 5, 19) in which substrates tend to be readily accessible to their respective enzymes, with advantages for biosynthetic flow. If so, how are we to view the fact that only some of the arginine genes are clustered and others are well dispersed in the *E. coli* genome?

Gene Aggregation

Some evidence is available that certain of the arginine genes, which occur at considerable map distances from one another in the *E. coli* genome, actually lie close together in the cell (22). Such a coming together of functionally related, not closely linked genes, or gene aggregation, would have an effect similar to that of the clustering of genes, as far as compartmentalization is concerned. Gene aggregation would imply that the bacterial chromosome has a higher-order structure which might be regarded not as completely fixed or rigid but rather as a fairly regular or preferred disposition of the chromosome.

The evidence for gene aggregation is based on the demonstration of appropriate concomitant double mutations. The underlying thought was that such double mutations might emerge from an interaction of a strand segment from one portion of DNA helix with a strand segment from another portion of DNA helix. For instance, when ultraviolet radiation is used as mutagenic agent, two pyrimidine residues, each swung outward from its portion of helix, may yield a pyrimidine dimer photoproduct which conceivably could lead to mutations in both portions of helix. Thus, the isolation, at sufficiently high frequency, of double mutants with mutations in not closely linked arginine genes would be taken as evidence for a juxtaposition of the affected genes during mutagenesis and, hence, for gene aggregation (22).

Of several hundred arginine auxotrophs examined, a minimum of 0.7% proved to be double mutants of the type contemplated here (22). The nature of the concomitant mutations found is given in Table III.

TABLE III
CONCOMITANT MUTATIONS IN NOT CLOSELY LINKED GENES OF THE ARGININE SYSTEM[a]

Strain number	Mutation selected for		Additional mutation found	
	Gene	Product	Gene	Product
250	*argB*	Enzyme 1	*R*	*R* product
514	*argB*	Enzyme 1	*P*	AcO permease
55	*argC*	Enzyme 2	*argG*	Enzyme 4
1429	*M*	Enzyme 4′	*argF*	Enzyme 8

[a] From Vogel and Bacon (22). See Figs. 1 and 2. AcO: N^{α}-acetylornithine.

We are, therefore, led to surmise that the clustering of genes, on the one hand, and the aggregation of unclustered genes, on the other, are complementary phenomena. The relative positions of genes in the chromosomes of such organisms as *E. coli* probably depend on a number of factors, of which compartmentalization of metabolic systems appears to be a significant one. That the genic arrangement, far from being haphazard, is very conservatively handled in evolution can be inferred from the close similarity between the genetic maps of *E. coli* and of *Salmonella typhimurium* (14, cf. 6).

Repression and Induction Governed by a Single Regulatory Gene

Among the genes concluded to be aggregated are those which are concerned with Enzymes 4 and 4′. Enzyme 4 is the normal, *arginine-repressible* transaminase of the *E. coli* arginine pathway. Enzyme 4′ is an *arginine-inducible* transaminase. Both enzymes catalyze the same reaction. The inducible transaminase has been demonstrated in two-step mutants deficient in the repressible enzyme (24). The two transaminases behave as different proteins (11) and have been obtained (10) in ultracentrifugally homogeneous form (see Fig. 4). Quite remarkably, the repressibility of Enzyme 4 and the inducibility of Enzyme 4′ are governed by the system-specific regulatory gene (1). Thus, in the same

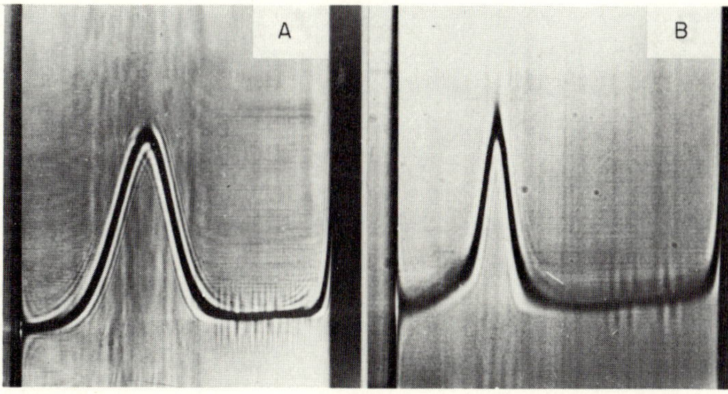

Fig. 4. Sedimentation pattern of arginine-induced (A) and derepressed (B) acetylornithine δ-transaminases in buffer containing 0.2 *M* KCl, at 59,780 rpm and 20° (10). Sedimentation proceeded from left to right. The buffers were 0.2 *M* Tris-HCl at pH 8.0 (A) and 0.01 *M* potassium phosphate at pH 7.0 (B). Protein concentrations of 16 mg per ml (A) and 9 mg per ml (B) were used. The pictures were taken at phase plate angles of 60° (A) and 65° (B), 45 min (A) and 32 min (B) after the stated rpm value was reached. Sedimentation coefficients ($s_{20,w}$) of 4.1 and 9.2 S were calculated for A and B, respectively.

culture and at the same time, arginine in conjunction with the same regulatory-gene product can bring about the induction of Enzyme 4′ and the repression of the other arginine enzymes.

This versatile functioning of a regulatory-gene product led us to consider, in a new light, the regulation of enzyme synthesis, as effected by arginine and perhaps by other amino acids. It was thought that the induction of Enzyme 4′ need not be typical of inductions by amino acids and quite likely proceeds by a different mechanism from that of the induction of, say, carbohydrate-degrading enzymes. Nevertheless, the fact that one and the same regulatory-gene product can be either inducive or repressive (depending presumably on the properties of the target enzyme-forming system) required an explanation and seemed especially interesting from the point of view of repression.

The tRNA Analogue Model of Enzyme Repression and Induction

The simplest model (21) that we could construct then to fit these findings and to be in harmony with results from other laboratories (for references, see 21, 23) placed the primary effect of repression and induction at the level of the ribosome and invoked transfer RNA's (tRNA's).

In this model, repression is attributed to the interaction of two suitable tRNA's positioned on the messenger-ribosome complex. The tRNA's are (*a*) an analogue tRNA, e.g., an analogue arginyl-tRNA, which would be specified by the system-specific regulatory gene (*R*) and would resemble normal arginyl-tRNA in having a standard arginine anticodon and in being chargeable with arginine by the usual arginine-activating enzyme, although the analogue tRNA would have a lesser affinity for the enzyme than has the normal tRNA, and (*b*) a repression trigger tRNA which would be so constructed that it can interact with the analogue tRNA when they are in association with the messenger-ribosome complex; the corresponding normal tRNA would not give a repressive interaction (21). In bacteria, where gene-messenger-ribosome complexes may be prevalent, the primary repressive action at the ribosome may have a secondary effect on the transcription of the gene involved (21).

Derepression is readily accounted for in terms of a replacement of the analogue tRNA by the corresponding normal tRNA species. For induction, the basic feature of the model would be that, under appropriate circumstances, the analogue tRNA functions in protein synthesis more effectively than does the normal tRNA (21).

The part of the message directly concerned with repression would be an arginine codon and the codon for the repression trigger tRNA. A favorable place for such a codon pair would be near the beginning of the message, in which case repression would be associated with the initiation of the protein chain involved (21). Since the repression trigger tRNA would have some special properties and would yet be used rather generally in repressions by amino acids, the repression trigger might be none other than the chain-initiating N-formylmethionyl-tRNA.

Thus, the desirability of postulating a DNA operator would have vanished, and the "RNA operator" for an arginine enzyme-forming system would simply be a standard arginine codon.

In the message, the N-formylmethionine codon and the repression-signaling arginine codon are thought to be adjacent. Possibly depending on which of these two codons precedes the other, the arginine codon may or may not be translated. The proposed mechanism would be particularly simple if the arginine codon did not direct the incorporation of arginine into the nascent peptide chain; otherwise, there may or may not be an enzymatic modification of the early part of the chain. (For references regarding peptide chain initiation, see 12, 23.)

One way in which the repression-producing effect involving the analogue tRNA and the trigger tRNA could be pictured is that the two tRNA's, when positioned side by side on the messenger-ribosome complex, interact so as to impede the relative motion of messenger and ribosome (21). However, the primary repression event can be viewed in a somewhat modified way, for which a measure of support is provided by the properties of the pleiotropic mutant discussed above.

A Possible Chain Initiation Mutant: Implications for tRNA Models of Repression

The above-mentioned results and conclusions regarding the pleiotropic mutant can be summarized as follows:

(*a*) the single-point mutation involved maps in the region of the four-gene cluster;

(*b*) the mutation affects the formation of three enzymes corresponding to a subcluster of genes;

(*c*) under derepressive conditions, the levels of the three enzymes are drastically lowered relative to the normal derepressed levels;

(*d*) under repressive conditions, arginine represses the three enzymes to levels comparable to the normal repressed ones.

A salient feature of the pleiotropic mutation is that it simultaneously affects enzyme levels and repression behavior. Since, in the tRNA model of repression, a close connection between repression by amino acids and the initiation of polypeptide chains is envisaged, the existence of the pleiotropic mutant supports elements of this model.

Presumably, we are dealing with a polycistronic message for Enzymes 8, 2, and 3. The normal message would be headed by the peptide-chain-initiating codon and a neighboring arginine codon, and the pleiotropic mutation would, therefore, have affected one of the six nucleotides of these two codons.

One possibility is that the arginine codon is mutationally altered. If so, one would attribute the low levels of the three enzymes to repression by an amino acid other than arginine. This possibility, however, is not encouraged because arginine still gives a pronounced repression.

The other possibility is that the pleiotropic mutation has changed the peptide-chain-initiating codon. This possibility is particularly attractive since it allows the construction of a harmonious picture. We now assume that the mutation has largely, but not completely, abolished the chain-initiating property of the altered codon. The normal peptide-chain-initiating mechanism would function to a minor extent because of some residual recognition, by the mutant codon, of the relevant tRNA (probably N-formyl-methionyl-tRNA). To the extent that the chain is properly initiated, repression by arginine would still occur. However, the bulk of the translation of the message would proceed without proper initiation, at a relatively low rate. Such enzyme formation, without proper initiation, is thought *not* to be subject to repression. The rate of enzyme synthesis attained under these conditions might depend somewhat on the particular mutation that has occurred. All known properties of the pleiotropic mutant can be accounted for in these terms.

An extension, to enzyme repression in general, of the conclusions drawn for the pleiotropic mutant is suggested by the finding that the repressed levels of the three enzymes are not very different in the mutant and in normal strains. Enzyme repression could then be attributed specifically to an interference with peptide chain initiation; under derepression, the interference would be minimal or absent, and under full repression, the interference would be maximal, with the enzyme level similar to that obtained if the usual peptide chain initiation machinery were greatly impaired or not functional. In this way, the very reproducible repressed levels would find an explanation.

As for the molecular mechanism of the inferred repressive interference with peptide chain initiation, we can contemplate, among several possibilities, the involvement of an interaction between N-formylmethionyl-tRNA and (for the arginine system) a species of arginyl-tRNA, somewhat as discussed above.

A Gradation Phenomenon in the Repression of Polycistronic-Message Enzymes

Further support for a close connection between repression and peptide chain initiation can be seen in the striking deviation from coordinate repression for Enzymes 8, 2, and 3 in the wild-type strain. Table I indicates that the ratio of the derepressed level to the repressed level is approximately 33 for Enzyme 8, 25 for Enzyme 2, and 6 for Enzyme 3. It will be noted that this ratio decreases in line with the sequence of the corresponding genes in the cluster, i.e., the repressibility of the enzymes forms a gradation. Another way of characterizing such graded repressibility is to say that the translation rate of an earlier part of the message relative to that of a later part is a function of degree of repression.

In their studies of noncoordinate repression in the tryptophan case, Margolin and Bauerle postulated that, under repressive conditions, ribosomes can attach themselves to points within the bulk of a polycistronic message, which would lead to a relatively more efficient translation of part of the message; under derepressive conditions, this kind of ribosomal attachment would not occur (13). While there might be a contribution from some such effect, results with the histidine system suggest that ribosomal attachment at points within a message actually can occur during derepression (see Goldberger and Berberich, 8).

The present model, in which repression is viewed as antagonizing peptide chain initiation, provides a natural explanation for noncoordinate repression of enzymes specified by a polycistronic message. Thus, if the rate of enzyme synthesis is heterogeneous along the length of the message (which it seems necessary to assume) then proper initiation at the head of the message *versus* the lack of such initiation could provide the basis for noncoordination of the repression of the first enzyme, or of early enzymes, with respect to the repression of later enzymes. For example, let us assume that translation begins with Enzyme 8 and, for the later enzymes, increases in efficiency because of a certain number of added initiations within the message; let us also assume that, for a message under derepression and for one under repression, the number of added (intra-message) initiations is about the same. Then, the repressibility of Enzyme 8 (whose formation would not involve intra-message initiations) would be greatest and the repressibility of Enzyme 3 (formed with the largest number of added initiations) would be least —which is in line with the observations. In connection with the present model, it is suggested that the derepression of the three arginine enzymes reflects predominantly an increase in the rate of initiation and hence *translation,* and at most a relatively modest increase in the rate of *transcription.*

Summary

Studies of the *E. coli* arginine system have indicated that at least two repressor recognition sites are involved in the regulation of the synthesis of four enzymes specified by a gene cluster: a pleiotropic, probably single-point mutation mapping in the region of the cluster affects the level *and* repressibility of three of the enzymes corresponding to a gene subcluster, the formation of the fourth enzyme being entirely unaffected. The pleiotropic mutation is thought to have led to an altered chain initiation codon heading a polycistronic message.

Enzyme repression is viewed as occurring at the ribosomal level and involving transfer RNA's. Specifically, repression (in the arginine system) would be exerted by a species of arginyl-tRNA antagonizing peptide chain initiation (probably involving N-formylmethionyl-tRNA) at the beginning of the message. The repressor recognition site ("RNA operator") would be a standard arginine codon adjacent to the chain initiation codon.

The repressibility of the three enzymes specified by the gene subcluster is noncoordinate and graded: the ratios for the derepressed level to the repressed level of the enzymes are 33, 25, and 6. This gradation is accounted for in terms of the present model of enzyme repression taken together with postulated intra-message initiations under both repressive and derepressive conditions.

Gene clustering and gene aggregation (a coming together of functionally related, not closely linked genes) are thought to complement each other in leading to the localized deposition of metabolically related enzymes.

Acknowledgment

This work was aided by grants from the U. S. Public Health Service, the National Science Foundation, and the Damon Runyon Memorial Fund.

References

1. Bacon, D. F., and Vogel, H. J., *Cold Spring Harbor Symp. Quant. Biol.*, **28**, 437 (1963).
2. Baumberg, S., and Vogel, H. J., manuscript in preparation.
3. Baumberg, S., Bacon, D. F., and Vogel, H. J., *Proc. Natl. Acad. Sci. U.S.*, **53**, 1029 (1965).
4. Baumberg, S., Bacon, D. F., and Vogel, H. J., *Genetics*, **54**, 322 (1966).
5. Davis, R. H., this volume.
6. Demerec, M., *in* "Evolving Genes and Proteins" (V. Bryson and H. J. Vogel, eds.), p. 505, Academic Press, New York, 1965.
7. Glansdorff, N., *Genetics*, **51**, 167 (1965).
8. Goldberger, R. F., and Berberich, M. A., this volume.
9. Jacob, F., and Monod, J., *J. Mol. Biol.*, **3**, 318 (1961).
10. Jones, E. E., and Vogel, H. J., manuscript in preparation.
11. Jones, E. E., Ortigoza-Ferado, J. A., and Vogel, H. J., *Federation Proc.*, **24**, 416 (1965).
12. Lengyel, P., *J. Gen. Physiol.*, **49**, 305 (1966).
13. Margolin, P., and Bauerle, R. H., *Cold Spring Harbor Symp. Quant. Biol.*, **31**, 311 (1967).
14. Sanderson, K. E., and Demerec, M., *Genetics*, **51**, 897 (1965).
15. Stahl, F. W., and Murray, N. E., *Genetics*, **53**, 569 (1966).
16. Stent, G. S., this volume.
17. Unger, L., Bacon, D. F., and Vogel, H. J., *Genetics*, **50**, 292 (1964).
18. Vogel, H. J., *Proc. Natl. Acad. Sci. U.S.*, **39**, 578 (1953).
19. Vogel, H. J., *in* "Amino Acid Metabolism" (W. D. McElroy and B. Glass, eds.), p. 335, The Johns Hopkins Press, Baltimore, 1955.
20. Vogel, H. J., *Cold Spring Harbor Symp. Quant. Biol.*, **26**, 163 (1961).
21. Vogel, H. J., *Proc. Seventh Canadian Cancer Res. Conf. 1966*, p. 133, Pergamon Press of Canada, Toronto, 1967.
22. Vogel, H. J., and Bacon, D. F., *Proc. Natl. Acad. Sci. U.S.*, **55**, 1456 (1966).
23. Vogel, H. J., and Vogel, R. H., *Ann. Rev. Biochem.*, **36**, 519 (1967).
24. Vogel, H. J., Bacon, D. F., and Baich, A., *in* "Informational Macromolecules" (H. J. Vogel, V. Bryson, and J. O. Lampen, eds.), p. 293, Academic Press, New York, 1963.

Discussion of Part V

F. Lynen, N. M. Kredich, E. Racker, R. F. Goldberger, R. O. Burns, H. J. Vogel, F. Lipmann, M. Riley, P. Rogers, J. R. Mattoon, M. Schaechter, N. Sueoka

Dr. Lynen: Dr. Kredich, is your enzyme able to use free serine for the formation of cysteine? You may remember that Schlossmann, in our laboratory, did a screening experiment with various organisms. He found that a crude extract of *E. coli* is very active in incorporating H_2S into serine with the formation of cysteine.

Dr. Kredich: We have tried this. It does not work with serine, Dr. Lynen. The activity in the crude extract is unbelievable. The crude extract has an activity, when you use O-acetylserine and sulphide in saturating concentrations, of 5 units per mg. Now, this unit is 1 micromole per minute of cysteine formed. In our assays, using a crude extract, we assay with 5 micrograms of protein per ml. A much more purified enzyme preparation using much more protein has no effect with serine. Actually, we find a non-enzymic formation of thiol under these conditions. It is very, very slight. Obviously, it isn't what you reported, but we find this using O-acetylserine and sulphide. If we add a little pyridoxal phosphate, it goes just a little bit faster, but it is nothing compared to the activity we get with enzyme. It doesn't work with the serine at all.

Dr. Racker: Dr. Goldberger, what is the significance and meaning of the derepression before zero time?

Dr. Goldberger: We defined zero time by the break in the growth curve. The finding that derepression, as determined by enzyme assays, occurs before this time simply indicates that the change in growth rate, as determined from the turbidity of the culture, is not the earliest event which can be observed upon depletion of excess histidine.

Dr. Burns: Dr. Vogel, your observations concerning the 5, 3, 2, 8 gene cluster are very interesting as you recognize, if viewed in an evolutionary sense. The situation becomes even more so if you consider the recent work of Bauerle and Margolin on the tryptophan cluster in *Salmonella typhimurium.* There is one question I would like to ask concerning the cluster in the arginine pathway, and it is a question that also arose while reading the original paper, that is whether or not you have ever demonstrated the presence of, or attempted to find, polarity mutants in this cluster. This would constitute very good evidence, that is, the absence of polarity effects would constitute very good evidence that the two regions are in fact acting independently when they are clustered.

DR. VOGEL: This is an interesting point. We do not have polarity mutants affecting the two regions. I might mention that Dr. Baumberg used an acridine half-mustard to examine the nature of the pleiotropic mutant, and in this case, the half-mustard had no effect.

DR. LIPMANN: I have a very short question. If your idea that repression is due to an altered sRNA is really right, couldn't you try to do an *in vitro* experiment by crossing the two—taking the sRNA's from one strain and putting them with the other and see if you repress?

DR. VOGEL: Yes, I think I understand your question. We would very much like to try this kind of *in vitro* experiment, although we expect that it would not be easy technically.

DR. RILEY: I thought it might be interesting to throw into the pot the fact that the catabolic enzyme, tryptophanase, inducible by tryptophan, is also inducible by 5-methyltryptophan which, as far as I know, is not activated and does not attach to tryptophanyl transfer RNA.

DR. VOGEL: I would agree that one must consider the possibility that not all amino acids act via aminoacyl-tRNA's, even if it should turn out that 5-methyltryptophan is activable after all.

DR. ROGERS: In support of Dr. Vogel's work and his model, we have isolated R^-_{arg} strains from the normal R^+_{arg} *E. coli* K12, and then we have partially purified the arginyl-RNA synthetase from both of these strains. Measuring attachment of arginine to transfer RNA, in support of your idea, the affinity for arginine is identical for the two enzymes. You would not expect any difference. Also, at low arginine concentrations, the isolated transfer RNA's from R^+_{arg} and R^-_{arg} organisms give the same kinetics of arginyl-RNA formation. So there doesn't seem to be any difference. However, there is one difference we did note. In crude extracts, Coles and I found [R. Coles and P. Rogers, *Bacteriol. Proc.* p. 87 (1964)] a difference in rate and in final amount of arginine attached. We still don't know why this is, because as soon as we purify the enzyme and add back transfer RNA from the R^+_{arg} or R^-_{arg} strain, the amount of arginine attached is the same. I think there may be another factor which we are removing in these extracts, once we start purifying the system.

DR. MATTOON: I would like to ask Dr. Kredich if he has tested the effect of methionine on his two OASS enzymes?

DR. KREDICH: You mean *in vitro*? Does it have any effect on the reaction rate? We have tried no *in vivo* experiments, but *in vitro*, methionine has no effect on either of the two OASS enzymes or on the serine transacetylase.

DR. SCHAECHTER: I would like to ask a question of Dr. Goldberger.

Since 20 minutes seem like a long time even for making 10 enzymes, I wonder if you have considered experiments to show that this is not just the time for finding the activity of the enzyme but that it is the actual time required for translation itself.

DR. GOLDBERGER: I agree that 20 minutes is a long time for translation of the histidine message, especially in view of the fact that the assembly of amino acids into polypeptide chains has been shown to occur very rapidly in bacterial cells. The experiments of Dr. Ray Byrne, for example, show that the time required for the assembly of alkaline phosphatase in *E. coli,* as determined by the nonuniform labeling technique, is extremely short. But you must remember that the histidine message is *polycistronic.* We have considered the possibility that the time it takes for each ribosome to get from one end of the polycistronic histidine message to the other is not primarily involved in the translation of the individual cistrons but rather in crossing from one cistron to the next. The question of how the 20 minutes are utilized would thus boil down to a question of how ribosomes negotiate the borders between cistrons, initiating and terminating polypeptide chains. Our experiments with the deletion mutant *hisH, B22* lend support to this possibility, as I mentioned in the body of the talk.

CHAIRMAN SUEOKA: Before closing, I would like to ask one question of Dr. Goldberger. When you add adenine and eliminate the sequential synthesis, do you eliminate the polarity of polar mutants?

DR. GOLDBERGER: We have just begun our study of polar mutants and are finding all sorts of interesting things, but I'm afraid they are too preliminary to discuss as yet.

CHAIRMAN SUEOKA: Thank you.

PART VI

JUXTAPOSITION PHENOMENA II

Chairman's Remarks

CHRISTIAN B. ANFINSEN

National Institute of Arthritis and Metabolic Diseases,
National Institutes of Health,
Bethesda, Maryland

Dr. Vogel mentioned that if we can finish not more than 15 minutes later than it says on the program, we can eat. So the speakers have all agreed to keep their talks within the time allotted, and the discussion must be very clever and quick. Now, I would like to ask Dr. Lynen to come and present his lecture.

Multienzyme Complex of Fatty Acid Synthetase

FEODOR LYNEN

In Experimental Collaboration with:

ALEXANDER HAGEN, INGRID HOPPER-KESSEL,
DIETER OESTERHELT, HENRY C. REEVES,
ECKHART SCHWEIZER, KLAUS WILLECKE,
AND MOHAMED YALPANI

Max-Planck-Institut für Zellchemie, München, Germany

Biochemistry and particularly enzymology have arrived at a point where they must probe into the structural complexity of the living cell. The improvement of microscopic techniques, especially the introduction of the electron microscope, and the development of methods for cell fractionation by differential centrifugation have steadily advanced the frontiers of cell physiology. The realization has grown that subdividing the cell into cell membrane, nucleus, and cytoplasm is inadequate because the highly ordered structural organization of the cell requires a much more textured description. Many cytoplasmic organelles, such as mitochondria, ribosomes, and lysosomes, were discovered, as well as membranous fragments accumulating in the microsome fraction. Biochemical analysis of the various structural elements revealed a broad spectrum of specific functions which generally result from the coordinated interplay of several enzymes. Therefore, the problem became one of gaining insight into the molecular architecture of these structural elements and of elucidating the functional coordination of the various enzyme activities, depending not only on their direct interaction but also on specific regulatory mechanisms.

Biochemical investigations in this field may be facilitated by experience from studies on far simpler multienzyme complexes which are characterized by the exclusive participation of proteins with different reaction specificities in a functional, coordinated unit. The various protein components possess both (a) catalytic sites, responsible for enzyme activity, and (b) binding sites, responsible for interaction with other proteins to form specific spatial structures.

It is the purpose of this paper to discuss results which we achieved during studies on fatty acid synthesis from malonyl CoA catalyzed by an enzyme system from yeast. These studies were of general importance because they also provided the first insight into the chemical details of this biosynthesis (11, 12).

From yeast cells ruptured by vigorous shaking with glass beads, we were able to isolate a protein fraction 150 times as active as the crude extract. The purified enzyme, which we named fatty acid synthetase, proved to be homogeneous in the Tiselius apparatus and in the ultracentrifuge. Its molecular weight was estimated to be 2.3 million. It was found that 1 mole of freshly prepared fatty acid synthetase at 25°C and pH 6.5 incorporates between 3200 and 5800 moles of malonyl CoA into fatty acids per minute. The synthesis of fatty acids from malonyl CoA requires TPNH as a reducing agent and small amounts of acetyl CoA, in analogy with the avian and mammalian enzymes first studied by Wakil and Ganguly (29) and by Brady (2). The yeast enzyme synthesizes a mixture of palmityl and stearyl CoA (18), according to equation 1 ($n = 7$ or 8). Acetyl CoA serves as "primer" of the process. Its C_2 unit is

$$\text{Acetyl CoA} + n \text{ malonyl CoA} + 2n \text{ TPNH} + 2n \text{ H}^+ \rightarrow \text{CH}_3(\text{CH}_2\text{CH}_2)_n\text{CO-CoA} + n \text{ CO}_2 + n \text{ CoA} + 2n \text{ TPN}^+ + n \text{ H}_2\text{O} \quad (1)$$

recovered only in the methyl end of the fatty acid produced (18), indicating that C_2 units from malonyl CoA are added to the acetyl residue during the synthetic reaction. In its function as "primer," acetyl CoA can be replaced by homologous saturated acyl CoA compounds but not by their oxidation products, identified in studies on fatty acid oxidation. Furthermore, all attempts to find low-molecular-weight intermediates of the synthesis were unsuccessful. The explanation of these puzzling observations was our discovery that the transformation of malonyl CoA into fatty acids is achieved through intermediates which are covalently bound to sulfhydryl groups of the synthetase (12). We found that two different types of sulfhydryl groups have carrier functions in the synthetic process. We denoted them as "central" and "peripheral" sulfhydryl groups for purposes of differentiation. In the scheme of fatty acid synthesis shown in Fig. 1, first presented in 1961 (12, 13), they are distinguished by bold-faced and normal print.

The synthetic process is initiated by the transfer of an acetyl residue from acetyl CoA to the "peripheral" sulfhydryl group, designated the "priming reaction." It is followed by the transfer of a malonyl residue from malonyl CoA to the "central" sulfhydryl group. The next step is a condensation between the enzyme-bound acetyl and malonyl groups resulting in the formation of acetoacetyl-enzyme with the concomitant liberation of CO_2. The stepwise conversion of the β-keto acid into the

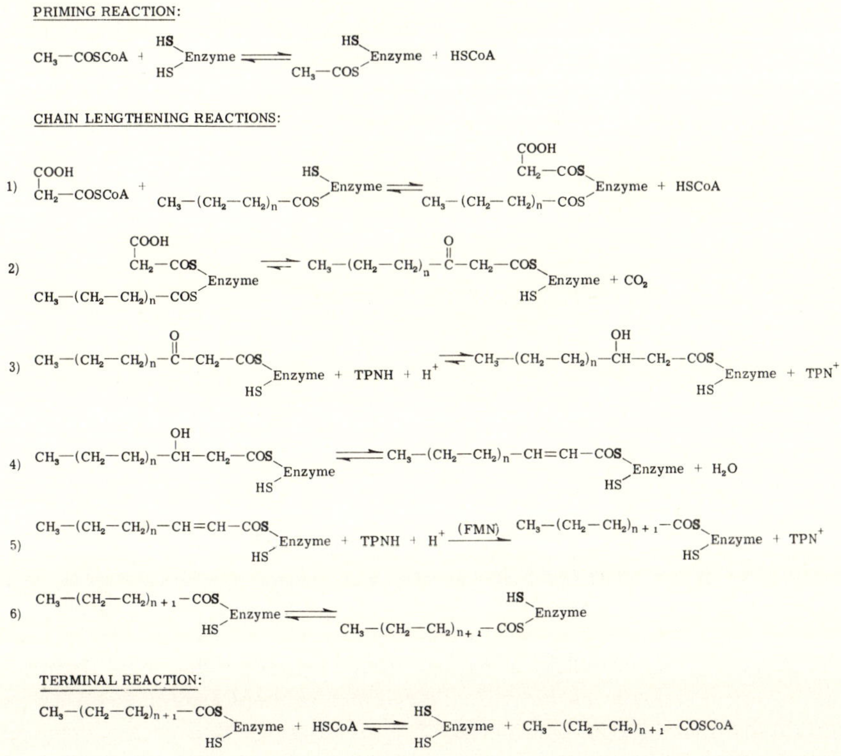

FIG. 1. The mechanism of fatty acid synthesis.

saturated acid is accomplished by way of its reduction by TPNH to D(—)-β-hydroxybutyryl-enzyme, followed by dehydration to crotonyl-enzyme and another TPNH-linked reduction to form the saturated butyryl-enzyme. In the second reduction step flavin mononucleotide (FMN) serves as hydrogen carrier. All the acyl residues involved in these chemical transformations are bound to the "central" sulfhydryl group. At the stage of the saturated acid, the butyryl group finally is transferred to the "peripheral" sulfhydryl group, thus liberating the "central" sulfhydryl group for introduction of the next malonyl residue. The reaction cycle can then proceed again, starting with butyryl malonyl-enzyme, and it is repeated until long-chain saturated fatty acids with 16 or 18 carbon atoms are formed. In the terminal reaction step, the acyl residue of palmityl- or stearyl-enzyme is transferred from the "central" sulfhydryl group to coenzyme A with the formation of palmityl or stearyl CoA and the regenerated enzyme. The free enzyme can again react with acetyl and malonyl CoA, thereby reinitiating the entire process.

The whole sequence of reactions is accomplished by a multienzyme complex. Its functional unit was proposed to consist of a combination of seven different enzymes arranged around the "central" sulfhydryl group in such a manner that the intermediates bound covalently to this group can come in close contact with the active sites of the participating enzymes (Fig. 2).

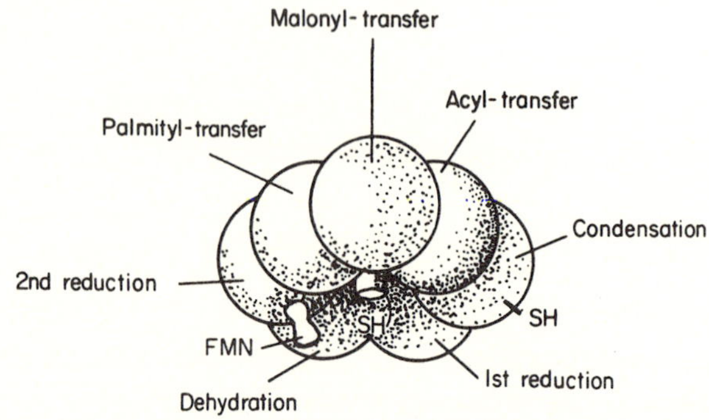

FIG. 2. Hypothetical structure of the multienzyme complex of fatty acid synthetase. The seven enzyme units shown refer to the seven reactions of Fig. 1.

We have now obtained experimental evidence that the "central" SH group belongs to an individual structural element (see p. 254). As we originally assumed in our model, it is not part of one of the enzymic components. This means that the architecture of the fatty acid synthetase from yeast resembles the structure of the analogous enzyme system from bacteria. As the elegant experiments of Vagelos and his associates (20) with the *Escherichia coli* system have demonstrated, the "central" sulfhydryl group, which carries the fatty acid intermediates, is bound to a readily dissociable protein of molecular weight about 9500, designated "acyl carrier protein." The bacterial enzyme system is not arranged in a stable multienzyme complex. By using standard methods of protein fractionation, it was possible to separate this enzyme system into enzymatically active individual components (19, 30). A similar "acyl carrier protein" has been reported to be present in the corresponding enzyme system from plants (23). Our repeated attempts to split the multienzyme complex of yeast into its subunits with retention of the individual enzyme activities were without success. To split the complex, it was necessary to use such drastic conditions as 0.2 *M* sodium deoxycholate or 6 *M* urea, so that most of the individual enzyme activities of fatty acid synthesis were lost (7). With respect to its stability the multienzyme complex of yeast fatty acid synthetase resembles the avian and mammalian synthetase complexes (3, 4, 10).

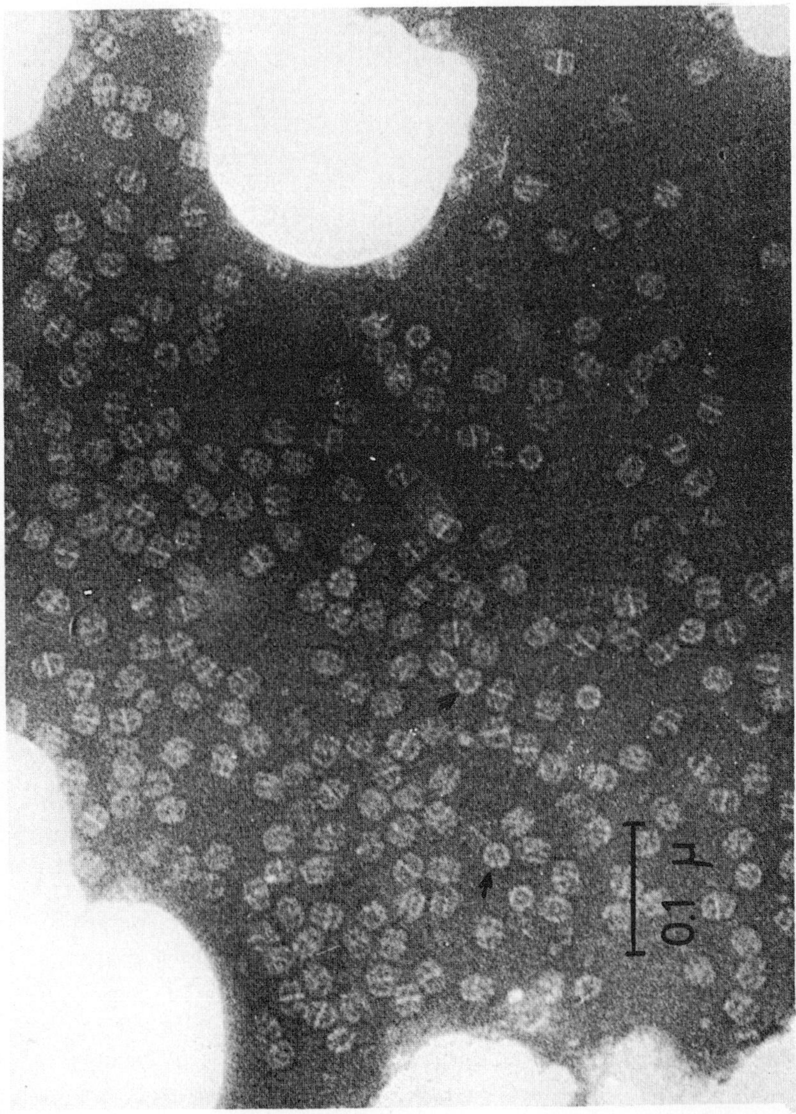

FIG. 3. Electron micrograph of the purified fatty acid synthetase from yeast.

From the standpoint of chemical mechanism, all fatty acid synthetases studied so far seem to catalyze the same reaction sequence. One minor dissimilarity concerns the terminal reaction, which yields palmityl and stearyl CoA with the yeast synthetase but free palmitate with the animal and bacterial enzyme systems. This difference may be due to the intervention of a hydrolytic enzyme which replaces the acyl transferase in the terminal reaction.

The structural organization of the fatty synthetase from yeast was verified by electron microscopy (Fig. 3). By use of the negative staining technique with phosphotungstic acid, Hofschneider (8) was able to recognize single particles of oval shape surrounded by an equatorial ring. The longitudinal diameter of the particles is 250 A, their cross diameter 210 A. Occasionally ring structures also can be seen (indicated by arrows in Fig. 3). This structure is built from about seven subunits and, according to its diameter, might be a top view of the equatorial ring of the particles. As can be seen, the oval-shaped particles are also composed of subunits. Unfortunately we cannot yet translate the information from electron micrographs into known schemes of structure. However, a structure composed of three circular subunits fitting together seems possible. From the results of the chemical studies, we have some evidence which supports the concept that each particle of molecular weight 2.3 million is composed of three functional assemblies.

To demonstrate the manifold catalytic activities attributed to the synthetase, we used model substrates in which the carboxylic acid intermediates of fatty acid synthesis were bound to pantetheine or N-acetylcysteamine (12, 13) (Fig. 4). Lacking the strong covalent bond to the "central" sulfhydryl group of the natural substrates, the affinity of these model substrates for the component enzymes is rather small. The defect can be circumvented, however, by employing high concentrations of the model substrates. In studying the bacterial enzyme system, the natural substrates—that is, the carboxylic acid intermediates bound to the "acyl carrier protein"—could be used (1, 19, 26, 30).

In addition to the experiments with model substrates, we used stoichiometric amounts of the yeast enzyme and demonstrated the enzyme-bound intermediates directly (13). For example, short incubation of labeled 1-^{14}C-acetyl CoA with synthetase led to the formation of the radioactive acetyl-enzyme (see Fig. 1). It could be precipitated with trichloroacetic acid with retention of radioactivity, indicating that the labeled acetyl group was covalently linked to the protein. When the radioactive acetyl-enzyme was separated from excess substrate by passing the reaction mixture over Sephadex and then incubated with malonyl

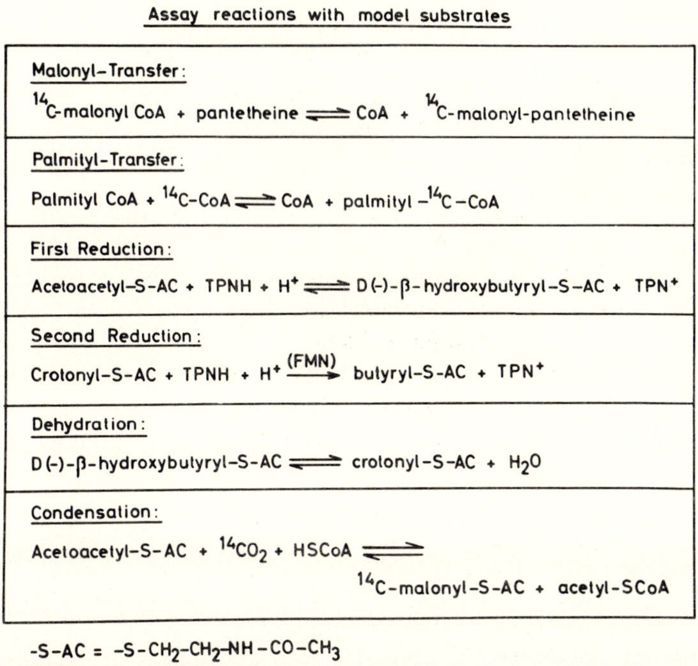

FIG. 4. Summary of assay reactions with model substrates for the determination of the enzymic components of fatty acid synthetase.

CoA and TPNH, more than 95% of the protein-bound acetic acid could be recovered in the fatty acids formed.

If the incubation mixture of labeled 1-^{14}C-acetyl CoA and synthetase was supplemented with malonyl CoA, a radioactive acetoacetyl-enzyme was formed (see Fig. 1). This could also be precipitated with trichloroacetic acid and yielded 3-^{14}C-acetoacetate after mild alkaline hydrolysis (18). Our attempts to isolate the native acetoacetyl-enzyme, employing the Sephadex technique, failed, because acetoacetyl-enzyme is slowly decomposed spontaneously to free acetoacetate (13).

In these experiments we also measured the equilibrium constant of the formation of acetoacetyl-enzyme and found the value (16)

$$K_{eq} = \frac{[\text{Acetoacetyl-enzyme}] \times [\text{CoA}]^2 \times [\text{CO}_2]}{[\text{acetyl CoA}] \times [\text{malonyl CoA}] \times [\text{enzyme}] \times [\text{H}^+]} = 2 \times 10^5 \ (0°\text{C})$$

Eliminating the H^+ concentration in this equation, the equilibrium constant at pH 7.0 equals

$$K'_{eq} = \frac{[\text{Acetoacetyl-enzyme}] \times [\text{CoA}]^2 \times [\text{CO}_2]}{[\text{acetyl CoA}] \times [\text{malonyl CoA}] \times [\text{enzyme}]} = 2 \times 10^{-2}$$

If we compare this equilibrium constant (condensation with malonyl CoA) with the equilibrium constant of the thiolase reaction (condensation with acetyl CoA),

$$K_{eq}' = \frac{[\text{Acetoacetyl CoA}] \times [\text{CoA}]}{[\text{acetyl CoA}]^2} = 1.6 \times 10^{-5} \text{ (pH 7.0; 25°C)}$$

which also generates an acetoacetyl thioester, the great thermodynamic advantage gained by the decarboxylation accompanying the condensation with malonyl CoA becomes evident.

The favorable shift in the equilibrium position of the condensation reaction is ultimately due to the delivery of energy by ATP. At the expense of one pyrophosphate bond, acetyl CoA is first bound to CO_2 as a carrier (equation 2).

$$CH_3CO\text{-}SCoA + HCO_3^- + ATP^{4-} \rightleftharpoons \underset{}{\overset{COO^-}{\overset{|}{C}}}H_2\text{-}CO\text{-}SCoA + ADP^{3-} + HPO_4^{2-} + H^+ \quad (2)$$

After replacement of coenzyme A by the "central" sulfhydryl group of the enzyme complex, condensation occurs with the acetyl residue bound to the "peripheral" sulfur. The condensation is accompanied by the release of CO_2 and the cleavage of one thioester bond. Because both reactants are in an "activated" state, the condensation reaction becomes very efficient.

From a chemical standpoint, the condensation reaction may be classified as an acylation of a malonic ester. The methylene group of the malonyl thioester, which is known to be more nucleophilic than the methyl group of an acetyl thioester, adds to the electrophilic carbonyl carbon of the sulfur-bound carboxylic acid, as shown in the upper half of Fig. 5. The intermediate formed is converted to the β-keto acyl derivative by the subsequent elimination of mercaptan and CO_2. An alternative reaction mechanism might be the concerted process shown on the lower half of Fig. 5. A decision between the two mechanisms may be reached by tracer experiments with heavy water.

When we studied the formation of acetoacetyl-enzyme and its spontaneous decomposition, in addition to free acetoacetate we found another compound characterized by an absorption band at 278 mμ. This compound was also observed by Bressler and Wakil (4) in similar experiments with the purified fatty acid synthetase from chicken liver. Its chemical nature remained unknown. Reports of the isolation of triacetic acid lactone by Brodie *et al.* (6) and by Brock and Bloch (5) in experiments with enzyme systems of avian and bacterial origin prompted the

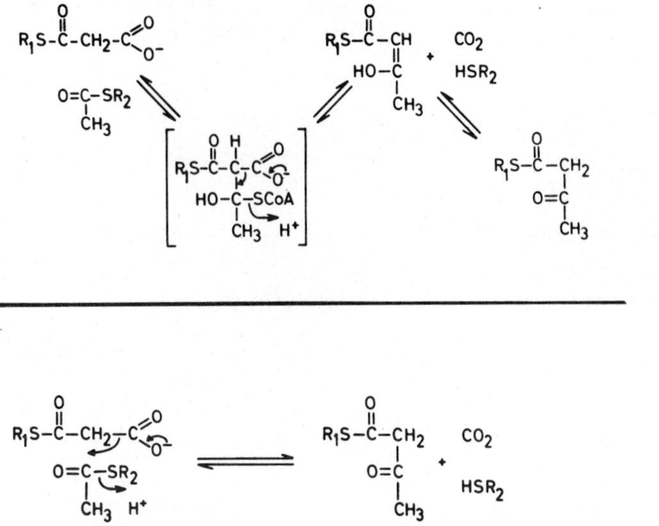

FIG. 5. Mechanism of the formation of β-ketoacids from malonyl thioester.

reinvestigation of the 278 mμ compound produced by the fatty acid synthetase of yeast. Yalpani and Willecke (33) incubated the purified yeast enzyme with 1-^{14}C-acetyl CoA and malonyl CoA in the absence of TPNH, isolated the radioactive reaction product, and identified it as triacetic acid lactone by comparison with the authentic compound. The identification was based on paper chromatography, paper electrophoresis, and recrystallization to constant specific radioactivity after the addition of carrier material. Furthermore, the crystalline triacetic acid lactone was oxidized with chromic acid following the procedure of Kuhn-Roth. Eighty-six percent of the original radioactivity was recovered in the acetic acid produced, indicating that the triacetic acid lactone carries the radioactive carbon in position 5.

To explain the formation of the lactone labeled in this position, we assume that the acetoacetyl residue of acetoacetyl-enzyme, bound to the "central" SH group, can be slowly transferred to the "peripheral" sulfhydryl group, thus liberating the "central" acceptor group for introduction of another malonyl residue. The next condensation leads to the formation of enzyme-bound triacetic acid which can easily be released as lactone, resulting from the nucleophilic attack of the neighboring enolate anion on the thioester bond (Fig. 6).

It seems to us that the formation of triacetic acid is a derailment, dependent on the absence of TPNH, which otherwise would guarantee the fast reduction of acetoacetyl-enzyme to produce hydroxybutyryl-

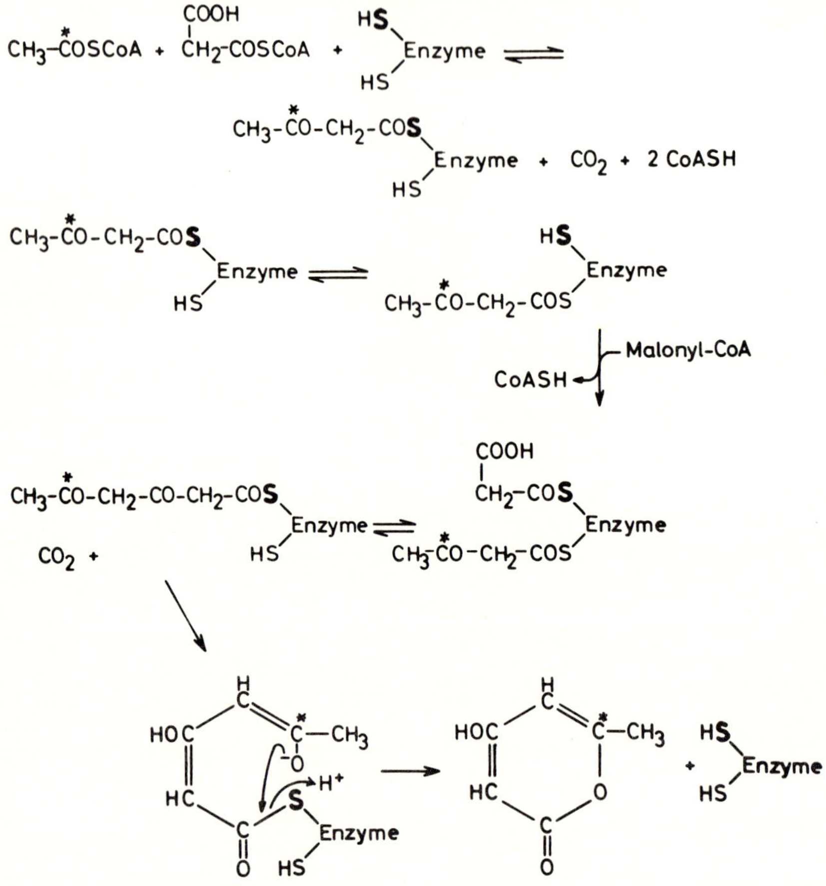

FIG. 6. Hypothetical mechanism of the synthesis of triacetic acid lactone.

enzyme. On the other hand, what must be considered as a derailment in fatty acid synthesis is supposed to be an essential reaction step in the biosynthesis of phenolic polyacetate compounds like orsellinic acid or acetylphloroglucinol and their derivatives, catalyzed by other enzyme systems (17).

Investigation of the chemical nature of the two types of sulfhydryl groups involved in fatty acid synthesis first led to the identification of the "peripheral" one as belonging to cysteine (14). In addition, we have obtained experimental evidence that this sulfhydryl group is located in the condensing enzyme components of the multienzyme complex.

The carrier of the "central" sulfhydryl group of the yeast fatty acid synthetase was identified as 4′-phosphopantetheine bound through a phosphodiester linkage with the hydroxyl group of a serine residue of the polypeptide (27). This kind of attachment was discovered by Vagelos (21) and by Wakil (24) in studies on the chemical structure of the "acyl carrier protein" of *E. coli*, and was also found in the mammalian fatty acid synthetase complex (10).

Our evidence for the occurrence of 4′-phosphopantetheine stems from experiments in which the purified fatty acid synthetase of yeast was heated at pH 12 at 98°C for one hour, following the procedure described by Vagelos (21). Under these conditions the protein released a low-molecular-weight compound, which after benzoylation and further purification was identified as S-benzoyl-4′-phosphopantetheine by chemical analysis and by comparison with the authentic compound (Fig. 7). The release of 4′-phosphopantetheine by mild alkaline treatment is in accord with the concept of an elimination reaction in the β-position of a polypeptide-bound serine.

The presence of 4′-phosphopantetheine is further supported by experiments with yeast cells grown in a medium containing ^{14}C-labeled pantothenic acid. Wells *et al.* (31) isolated the fatty acid synthetase from these

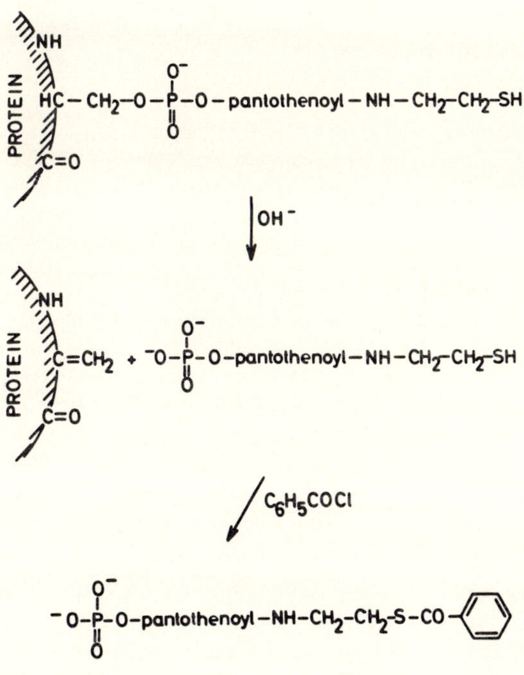

Fig. 7. Schematic representation of the identification of protein-bound 4′-phosphopantetheine.

cells and found the purified enzyme to contain three moles of radioactive pantothenate covalently bound per mole of enzyme.

Using the labeled fatty acid synthetase, Willecke (32) was able to demonstrate that 4′-phosphopantetheine belongs to an individual structural element of the multienzyme complex. It appears to be smaller than the rest of the proteins because dissociation of the complex by treatment with 6 *M* guanidine hydrochloride at pH 8.6 in the presence of dithiothreitol released a ^{14}C-labeled protein, which moved more slowly through a column of Sephadex G 200 (Fig. 8). Preliminary experiments have

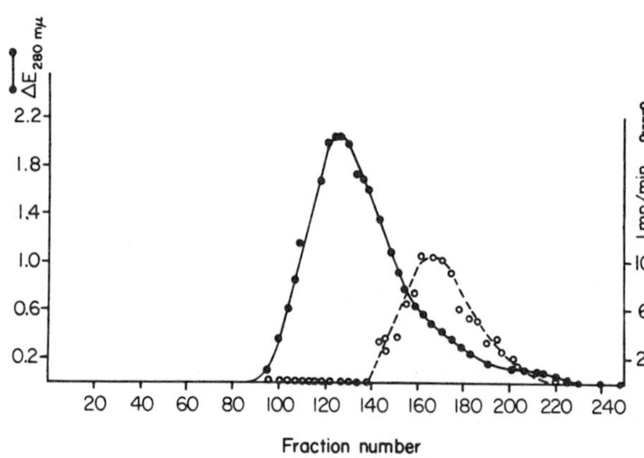

FIG. 8. Chromatography of the ^{14}C-pantothenate-labeled fatty acid synthetase, dissociated by 6 *M* guanidine hydrochloride, on Sephadex G 200. The mixture of 630 mg of purified ^{14}C-pantothenate-labeled yeast synthetase (specific activity 1.2 enzyme units/mg; 12×10^4 cpm), 3.0 ml of 2.8 *M* Tris buffer, pH 8.6, 0.5 ml of 0.5 *M* EDTA, 460 mg of dithiothreitol, and 17 gm of guanidine hydrochloride was diluted with water to a total volume of 30 ml and incubated for 12 hours at 32°C. The reaction mixture was then chromatographed on a column of Sephadex G 200 (5.1×100 cm), equilibrated with 6 *M* guanidine hydrochloride, 0.15 *M* K-phosphate, pH 7.1, and 0.15 *M* thioglycol. The eluting solvent had the same composition. Volume of fraction: 5 ml. The protein content of the fractions (filled circles) was determined by measuring ultraviolet-absorption at 280 mμ, their radioactivity (open circles) by liquid scintillation counting.

shown that in the absence of the reducing agent separation cannot be achieved, which may indicate that the "acyl carrier protein" of yeast synthetase is covalently bound to one of the component enzymes by a disulfide bridge. To verify this hypothesis, further experiments are needed.

The attachment through 4′-phosphopantetheine seems to be very appropriate for the functioning of the multienzyme complex. 4′-Phosphopantetheine provides a flexible arm of more than 15 A in length carrying the "central" sulfhydryl group and conceivably permitting rotation of the latter between the various enzymes. In this manner it is easily possible to bring the fatty acid intermediates, bound covalently to this sulfhydryl group, into close contact with the active site of each component enzyme which has only limited freedom of motion in the stable multienzyme complex. This is schematically illustrated in Fig. 9, where the circles should indicate the active sites of the participating enzymes.

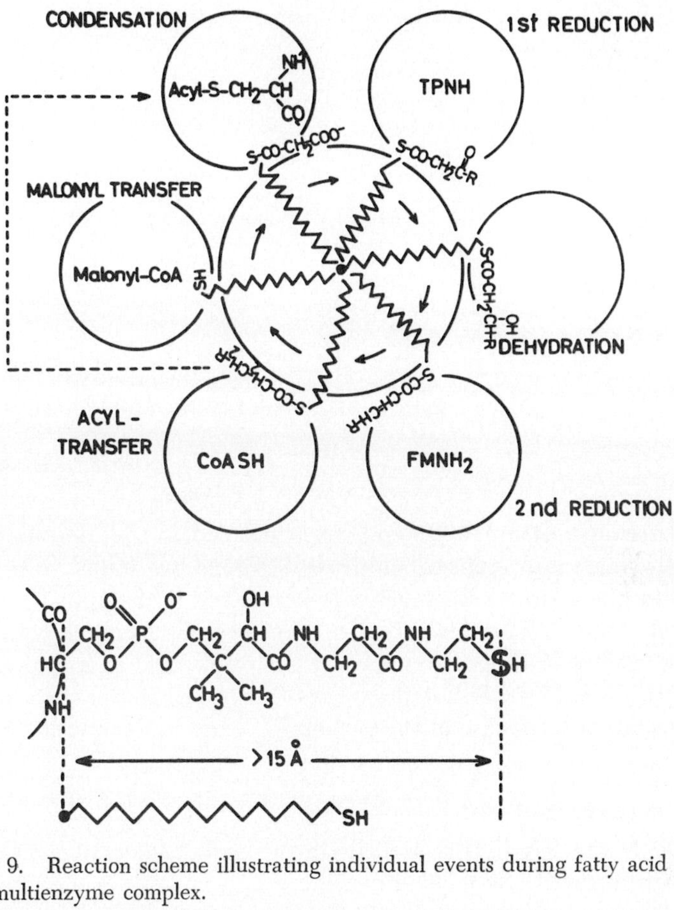

FIG. 9. Reaction scheme illustrating individual events during fatty acid synthesis on the multienzyme complex.

With this picture in mind one realizes how important it is that a larger molecule like pantothenic acid is inserted as a bridge between the protein and the acyl-carrying cysteamine. Possibly this fact may be the clue to the old problem of why coenzyme A is synthesized in the living organisms with incorporation of the rare and valuable vitamin pantothenic acid, which, by itself, is not involved in the function of coenzyme A as acyl carrier. It is tempting to suggest that coenzyme A or its precursor dephosphocoenzyme A functions in cellular metabolism not only as acyl-transferring coenzyme, well documented by numerous investigations since Lipmann's pioneering work, but also as donor of 4′-phosphopantetheine in the biosynthesis of fatty acid synthetase. A chemical reaction, as formulated in equation 3, would be analogous to the synthesis of lecithin, from diglyceride and CDP-choline (equation 4). In both cases, phosphodiesters are formed starting with primary alcohols.

$$\left.\begin{matrix}-\mathrm{CO} \\ -\mathrm{NH}\end{matrix}\right\rangle \mathrm{CH\text{-}CH_2OH} + \underset{\mathrm{P(?)}}{\text{adenine-ribose-P-P-pantetheine}} \rightarrow$$

$$\left.\begin{matrix}-\mathrm{CO} \\ -\mathrm{NH}\end{matrix}\right\rangle \mathrm{CH\text{-}CH_2\text{-}O\text{-}P\text{-}pantetheine} + \underset{\mathrm{P(?)}}{\text{adenine-ribose-P}} \quad (3)$$

$$\begin{matrix}\mathrm{H_2C{-}O{-}CO{-}R_1} \\ | \\ \mathrm{HC{-}O{-}CO{-}R_2} \\ | \\ \mathrm{H_2C{-}OH}\end{matrix} + \text{cytosine-ribose-P-P-choline} \rightarrow$$

$$\begin{matrix}\mathrm{H_2C{-}O{-}CO{-}R_1} \\ | \\ \mathrm{HC{-}O{-}CO{-}R_2} \\ | \\ \mathrm{H_2C{-}O{-}P{-}choline}\end{matrix} + \text{cytosine-ribose-P} \quad (4)$$

In the course of our investigations which led to the identification of the "peripheral" and "central" sulfhydryl groups, we found to our great surprise that acetate and malonate are not bound to the enzyme complex exclusively via sulfur atoms (15). In these experiments, Schweizer (27) used the lability of thiol esters toward performic acid as a tool for their characterization. Performic acid oxidizes thiol esters to the corresponding sulfonic acids with release of the carboxylic acids (9) (equation 5).

$$\mathrm{R{-}S{-}\overset{\overset{\displaystyle O}{\|}}{C}{-}R'} \xrightarrow{\mathrm{H{-}\overset{\overset{O}{\|}}{C}{-}OOH}} \mathrm{R{-}SO_3H + R'{-}COOH} \quad (5)$$

In Table I, the results of an experiment with radioactive ^{14}C-acetyl-enzyme are shown. The labeled enzyme was prepared by the interaction of constant amounts of enzyme with varying concentrations of 1-^{14}C-acetyl CoA, followed by precipitation with trichloroacetic acid and measurement of the protein-bound radioactivity. From this table it can be seen that the transfer of radioactive acetate to the enzyme depends on the concentration of acetyl CoA used, as expected, if the acetyl transfer is a reversible process. However, in the whole concentration

TABLE I

FORMATION OF ^{14}C-ACETYL-ENZYME AND RELEASE OF ^{14}C-ACETIC ACID BY OXIDATION WITH PERFORMIC ACID[a]

^{14}C-Acetyl CoA ($\times 10^{-6}$ *M*)	Protein-bound radioactivity (cpm)		Stable radioactivity (%)
	Direct	After oxidation	
2	15623	7825	50
4	19582	9100	46.5
6	23847	10045	42
10	29325	13260	45
15	38590	17270	45
20	39830	17731	45
30	40817	18921	46
40	45437	20741	45.5
50	50113	22921	45.5

[a] Each sample contained, in 1.0 ml of 0.1 *M* K-phosphate, pH 6.5, 2 mg of yeast fatty acid synthetase (specific activity 850 milliunits), 10 μmoles of cysteine, and ^{14}C-acetyl CoA (24.8×10^6 cpm/μmole) in the concentrations listed in the table. The mixture was incubated for 5 minutes at 0°C, and the reaction was stopped by the addition of 0.3 ml of 3 *M* trichloroacetic acid. After careful washing the precipitated protein was dissolved in 0.5 ml of 98% formic acid, 0.2-ml aliquots were dried on strips of Whatman I paper, and the radioactivity was measured before and after 12 hours of exposure to an atmosphere of performic acid. Before measurement in the scintillation counter the paper strips were dried in vacuum over KOH. The radioactivity values listed in the table are based on 2 mg of protein.

range studied, only about 50% of the radioactive acetyl groups bound to the protein were released by treatment with performic acid.

From this and other experiments, we came to the conclusion that the acyl transfer to the multienzyme complex is initiated with the transfer to some nonsulfhydryl acceptor group X in the protein. As illustrated in Fig. 10, malonyl and acetyl residues are transferred from group X to the

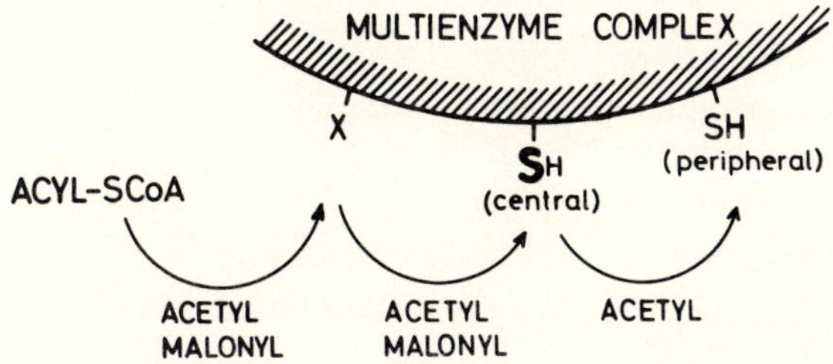

FIG. 10. Specificity of acyl transfer to the different acceptor groups on the multienzyme complex.

"central" sulfhydryl group. The further transfer to the "peripheral" sulfhydryl group is specific for acetate and its higher homologs.

According to our scheme, the acyl residue of malonyl-enzyme is bound to both group X and the "central" sulfhydryl group. In agreement with this assumption, more than one radioactive acyl peptides were isolated from the peptic hydrolysate of 3-^{14}C-malonyl-enzyme by chromatography on DEAE-Sephadex. This is illustrated in Fig. 11. The fastest moving fraction, designated A, was split by performic acid, whereas fractions B and C were stable to the same treatment. From the acid

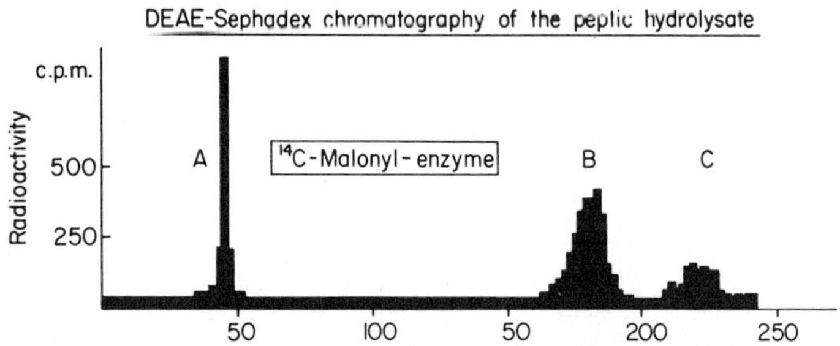

FIG. 11. Chromatography of the peptic hydrolysate of ^{14}C-malonyl-enzyme on DEAE-Sephadex. For the preparation of ^{14}C-malonyl-enzyme, 1.6 gm of synthetase (specific activity 550 milliunits) and 28 μmoles of 3-^{14}C-malonyl CoA (1.2×10^6 cpm/μmole) in 120 ml of 0.05 *M* K-phosphate buffer, pH 6.5, were incubated for 2 minutes at 22°C. The reaction was stopped by the addition of 6 ml of 3 *M* trichloroacetic acid. After careful washing, the precipitated protein was suspended in 2 liters of 0.01 *M* HCl and digested by incubation with 160 mg of crystalline pepsin at room temperature. After evaporation of the water in vacuum, the residue was dissolved in 7 ml of pyridine acetate buffer, pH 6.25. Insoluble material was removed by centrifugation, and the clear solution was chromatographed on a column of DEAE-Sephadex (2.1×150 cm). The eluting solvent until fraction 70 (volume of fraction: 6 ml) was 0.05 *M* pyridine acetate buffer, pH 6.25; thereafter a gradient was employed. For this purpose 0.2 *M* acetic acid was continuously introduced into a mixing chamber containing 200 ml of 0.05 *M* pyridine acetate buffer, pH 6.25.

hydrolysate of malonyl peptide A, cysteamine and β-alanine were identified by ion exchange chromatography (15, 27). Since cysteine was absent, this peptide is derived from the polypeptide area around the "central" sulfhydryl group.

The radioactivity of fractions B and C was due to the presence of a ^{14}C-malonyl heptapeptide and a ^{14}C-malonyl pentapeptide which could be purified by chromatographic procedures. The acyl pentapeptide contained serine besides histidine, glycine, alanine, and leucine (C-terminal position). The acyl heptapeptide was similar but contained, in addition, glycine and glutamic acid. These results seem to indicate that group X is identical with the hydroxyl group of serine. Further experiments should indicate whether the carrier group X is identical for acetate and malonate, and whether it is connected with the enzyme components that catalyze acyl transfer reactions.

That acylation of the serine residue is not an unphysiological process, but is part of the enzyme function, is indicated by the experiment with the isolated 1-^{14}C-acetyl-enzyme, mentioned earlier. Although only half of the protein-bound acetic acid was bound to sulfur, more than 95% was incorporated into long-chain fatty acids, when the system was supplemented with malonyl CoA and TPNH.

In this connection, the problem is raised whether the acetyl transfer to the "peripheral" sulfhydryl group involves the "central" sulfhydryl group as acyl carrier. Schweizer (27) tried to answer this question experimentally by measuring the amount of protein-bound 3-^{14}C-methylmalonate in the presence of increasing amounts of acetyl CoA. In these studies he used methylmalonyl CoA instead of malonyl CoA because it can transfer its acyl residue to the same acceptor groups on the synthetase ("central" SH group and serine hydroxyl), but does not condense with the enzyme-bound acetyl residue.

The results of these competition experiments were surprising. Contrary to expectations, the amount of protein-bound radioactive methylmalonate did increase and did not decrease with increasing amounts of acetyl CoA added (Fig. 12, left half). Chromatographic separation of the radioactive methylmalonyl peptides, as formed by peptic digestion then showed that the effect of acetyl CoA addition was different with different acceptor groups (Fig. 12, right half). The transfer of methylmalonate to the "central" sulfhydryl group (peptide A) was decreased with the addition of acetyl CoA, as is expected if both acyl residues compete for the acceptor. On the other hand, the methylmalonyl transfer to the serine site (peptides B and C) was increased with the addition of acetyl CoA to such an extent that the decrease at the sulfhydryl group was overcompensated. It appears that this phenomenon might be related to mutual interactions of the various acceptor groups in the multienzyme structure. The transfer of methylmalonyl to the serine site seems to be aggravated as long as the "peripheral" sulfhydryl group is free, and to be facilitated when this group carries the acetyl residue. This phenomenon might be the result of changes in the molecular architecture of the multienzyme complex, thereby affecting the energetics of methylmalonyl transfer.

In good agreement with this assumption it was found that interaction of the enzyme complex with iodoacetamide or N-ethylmaleimide induces an analogous effect. Iodoacetamide reacts preferentially with the "periph-

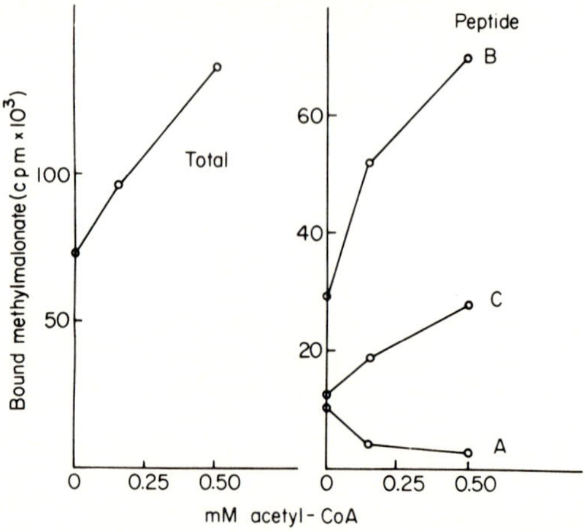

FIG. 12. Effect of acetyl CoA on the transfer of ^{14}C-methylmalonate to various acceptor sites of fatty acid synthetase. Each of the samples contained in a volume of 6.0 ml (in micromoles): K phosphate, pH 6.5, 100; 3-^{14}C-methylmalonyl CoA (13.5×10^6 cpm/μmole), 0.25; acetyl CoA, as indicated on the abscissa, and 100 mg of yeast synthetase (specific activity 1.3 enzyme units/mg). The mixtures were incubated for 5 minutes at 0°C. The reaction was stopped by the addition of 0.1 ml of 3 *M* trichloroacetic acid. After careful washing of the protein, the total protein-bound radioactivity was determined in aliquots (left diagram). Other aliquots were digested by incubation with pepsin, the radioactive methylmalonyl peptides A, B, and C were separated by chromatography on DEAE-Sephadex (see legend of Fig. 11), and their radioactivity was determined (right diagram).

eral" SH group and leaves the "central" one nearly unaffected, whereas N-ethylmaleimide reacts with both types of SH groups simultaneously. As a result, fatty acid synthetase, inhibited by iodoacetamide, can accept 3-^{14}C-methylmalonyl from the CoA derivative on both the pantetheine site and the serine site, whereas the enzyme, pretreated with N-ethylmaleimide, can accept acyl residues only at the serine site. This could be demonstrated by chromatographic separation of the radioactive peptides obtained by pepsin hydrolysis. Further evidence is given by the experiments of Fig. 13, in which total protein-bound radioactivity (curve I) and the radioactivity stable to oxidation (curve II) were measured simultaneously. Two facts are striking: (1) the nearly complete disappearance of sulfur-bound radioactivity (curve III) in the enzyme preparation pretreated with N-ethylmaleimide; and (2) the increased transfer of methylmalonyl residues to the hydroxyl group of serine (curve II) when the "peripheral" sulfhydryl group was substituted either with iodoacetamide or with N-ethylmaleimide. This observation was in line with the previously discussed experiments.

It should be emphasized that substitution of the "peripheral" sulfhydryl with iodoacetamide increases also the transfer of methylmalonate to the "central" sulfhydryl group (curve III), indicating some mutual interactions between "central" and "peripheral" acceptor sites.

Some interactions were also indicated by observations related to the kinetics of our enzyme assay, in which the rate of TPNH oxidation was determined. In our routine assay procedure, all reactants except malonyl CoA are placed in a cuvette and, after 1 or 2 minutes of preincubation, the reaction is initiated by the addition of malonyl CoA. Under these conditions the oxidation of TPNH is linear for several minutes. We

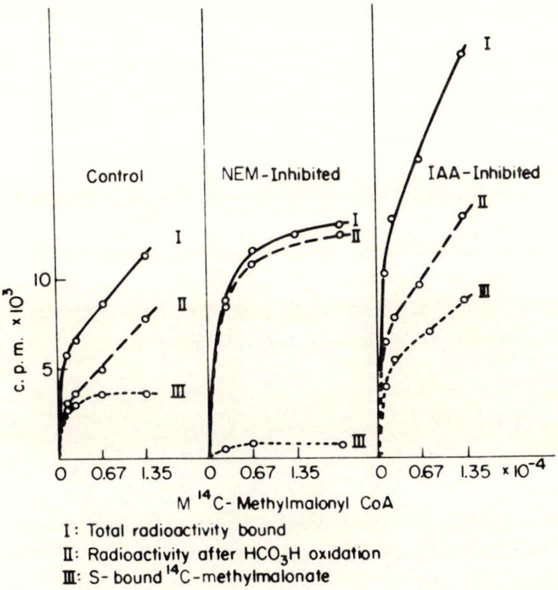

FIG. 13. Effect of pretreatment of the synthetase with N-ethylmaleimide (NEM) or iodoacetamide (IAA) on the formation of 3-^{14}C-methylmalonyl-enzyme. Each of the samples contained in a volume of 0.35 ml (in micromoles): K phosphate, pH 6.5, 32; cysteine, 2.25; 3-^{14}C-methylmalonyl CoA (13.5×10^6 cpm/μmole), as indicated on the abscissa, and 1.2 mg of yeast synthetase (specific activity 1.7 enzyme units/mg), either untreated (left diagram), pretreated with 0.01 *M* N-ethylmaleimide (central diagram), or pretreated with 0.001 *M* iodoacetamide (right diagram). The mixtures were incubated for 5 minutes at 0°C. The reaction was stopped by the addition of 1.0 ml of 0.3 *M* trichloroacetic acid, and the precipitated protein was carefully washed and dissolved in 1.0 ml of formic acid. One-half milliliter of the protein solution, without further addition or after addition of 0.05 ml of 30% aqueous H_2O_2, was incubated for 10 hours at 4°C. After the addition of 5 to 10 ml of water, protein was again precipitated by the addition of 0.1 ml of 3 *M* trichloroacetic acid, carefully washed, and its radioactivity measured.

observed that when the assay is initiated with acetyl CoA, rather than malonyl CoA, there is a lag period prior to the linear oxidation of TPNH. Further, under these conditions the rate of TPNH oxidation never reaches that observed when the reaction is started with malonyl CoA. Experiments of Reeves (25) are shown in Fig. 14.

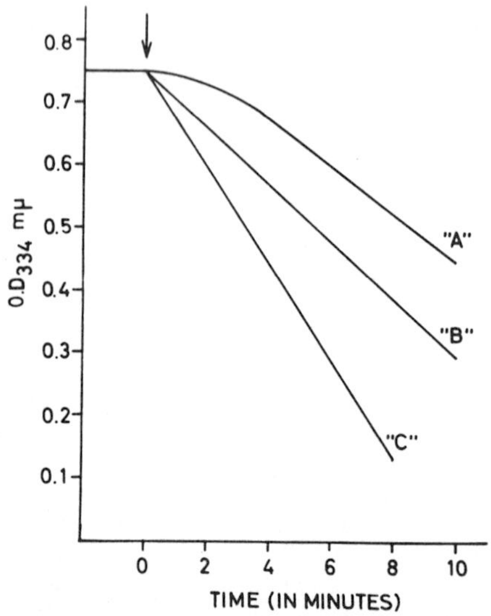

FIG. 14. Kinetics of the optical assay of fatty acid synthetase. The cuvette contained 100 m*M* K phosphate buffer, pH 6.5; 20 m*M* cysteine; 0.03 m*M* TPNH; 5 mg of serum albumin; 15 μg of enzyme, and either 0.04 m*M* acetyl CoA or 0.05 m*M* malonyl CoA. Temperature: 25°C. The reaction was started at the point indicated by the arrow by either acetyl CoA (curve "A") or malonyl CoA (curve "C"). Curve "B" represents a reaction mixture containing, in addition to the above, 100 m*M* EDTA and was initiated by the addition of acetyl CoA.

When the enzyme is preincubated with malonyl CoA, the "central" sulfhydryl group may become saturated with malonyl residue in such a manner as to block the movement of the acetyl residue to the peripheral sulfhydryl group. The lag period observed prior to TPNH oxidation under these circumstances may be a reflection of the displacement of malonyl residues by acetyl residues which must occur before the condensation and subsequent reduction steps can start.

The observation that the activity of fatty acid synthetase is always less when the enzyme has been preincubated with malonyl CoA may suggest that two forms of the enzyme complex exist. One form could be envisioned as an "inactive complex" which, in the presence of acetyl CoA, is converted to an "active complex":

$$\text{Enzyme}_{(\text{inactive complex})} \xrightarrow{\text{Acetyl CoA}} \text{Acetyl-enzyme}_{(\text{active complex})}$$

It is the "active complex" which would then accept malonyl CoA and, following the subsequent reactions, lead to the formation of long-chain fatty acids. Under conditions where a mixture of the two complexes is preincubated with malonyl CoA, the malonate unit may become bound in such a manner as to cause a partial denaturation or inactivation so that the "active complex" could not be formed following the subsequent addition of acetyl CoA. The observation that the inhibition of fatty acid synthetase is independent of the preincubation time in the presence of malonyl CoA lends support to the suggestion that two forms of the enzyme complex may exist and that it is only the "inactive complex" which is affected by malonyl CoA.

Finally, let me briefly discuss some experiments which support our differentiation between "central" and "peripheral" sulfhydryl groups. When we studied the inactivation of fatty acid synthetase by N-ethylmaleimide, we found that preincubation of the enzyme with acetyl CoA or homologous saturated acyl CoA derivatives, leading to the formation of the fully acylated enzyme, protected the synthetase against the action of N-ethylmaleimide (Table II). In contrast, however, preincubation of the synthetase with malonyl CoA or methylmalonyl CoA, leading to the acylation of the "central" sulfhydryl group but not to that of the "peripheral" one, had no protective effect. Furthermore the peptide spectrum, obtained by DEAE-Sephadex chromatography of the peptic

TABLE II

Protection of Synthetase from Inactivation with N-Ethylmaleimide by Preincubation with Acyl CoA Derivatives[a]

Acyl CoA during preincubation (0.28×10^{-3} *M*)	Relative enzyme activity
—	5.8
Malonyl CoA	4.8
Acetyl CoA	84.6
Butyryl CoA	74.1
Decanoyl CoA	66.9
Myristyl CoA	48.4
Palmityl CoA	33.7
Untreated enzyme	100

[a] Equal amounts of synthetase were preincubated with the CoA derivatives listed in the table. N-Ethylmaleimide (final concentration 0.005 *M*) was added, and the mixture was incubated for 5 minutes at 0°C. Excess inhibitor was then removed by the addition of cysteine, after which the enzyme activity was measured and compared with the activity of the untreated enzyme (= 100).

hydrolysate of 1-^{14}C-acetyl-enzyme, was much more complex than the peptide spectrum of malonyl- or methylmalonyl-enzyme (Fig. 15). At least six radioactive peaks were found, three of them disappearing, when fatty acid synthetase was inactivated by pretreatment with iodoacetamide. Using chromatographic procedures, Oesterhelt (22) purified the

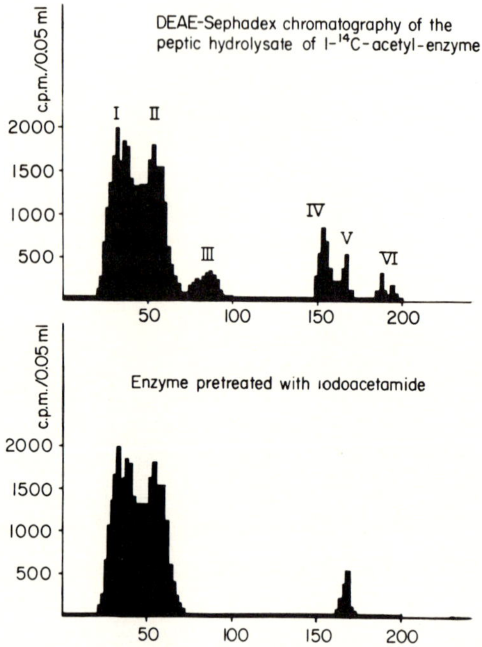

FIG. 15. Chromatography of the peptic hydrolysate of ^{14}C-acetyl-enzyme on DEAE-Sephadex. Each of the two samples contained in a volume of 3.0 ml: 180 mg of yeast synthetase (specific activity 0.8 enzyme units/mg), 0.1 *M* K phosphate, pH 6.5, and either water (upper diagram) or 0.02 *M* iodoacetamide (lower diagram). After incubation for 10 minutes at 0°C, 0.3 ml of *M* cysteine was added to both reaction mixtures, thereby stopping the action of iodoacetamide on the enzyme. For the preparation of ^{14}C-acetyl-enzyme, 1 ml of 1.73 m*M* 1-^{14}C-acetyl CoA (25×10^6 cpm/μmole) was added to each sample 10 minutes later. After another incubation for 5 minutes at 0°C, the reaction was stopped by the addition of 0.4 ml of 3 *M* trichloroacetic acid. The precipitated and carefully washed protein was suspended in 400 ml of 0.01 *M* HCl and digested by incubation with 30 mg of crystalline pepsin at room temperature. After evaporation of the water in vacuum the residue was dissolved in a little water, and the solution was brought to pH 6.5 by the addition of pyridine. Insoluble material was removed by centrifugation, and the clear solution was chromatographed on a column of DEAE-Sephadex (1.5×200 cm). The eluting solvent until fraction 89 (volume of fractions: 5 ml) was 0.05 *M* pyridine formate buffer, pH 6.25; thereafter a gradient was employed. For this purpose 2 *M* pyridine formate buffer, pH 4.5, was continuously introduced into a mixing chamber containing 500 ml of 0.05 *M* pyridine formate buffer, pH 6.25.

radioactive peptides further and demonstrated that peaks III and IV (Fig. 15), which are not present in the iodoacetamide-pretreated enzyme, contain exclusively sulfur-bound acetyl peptides. The thiol present in these peptides was found to be cysteine. Cysteamine was absent and was found to be present only in the peptides of peak II, confirming our previous results with the peptic hydrolysates of radioactive malonyl- and methylmalonyl-enzyme.

The experiments on the protection of fatty acid synthetase against the inactivation by N-ethylmaleimide may also help to explain why the synthetic process terminates at the stage of palmityl and stearyl CoA. Earlier we assumed that this might be due to the specificity of the enzymic component which stopped the process by transferring the fatty acid radical from the enzyme to coenzyme A (12). However, when Schweizer studied (28) the relationship between rate of acyl transfer and chain length of the acid, it was found to our great surprise that the saturated acids from C_6 to C_{16} were transferred at nearly equal rates. Thus we are forced to the conclusion that other factors must be responsible for stopping the synthesis at the stage of the C_{16} and C_{18} acids.

When we compared the effect of equal concentrations of the homologous saturated acyl CoA derivatives, it was found that protection decreases with increasing chain length of the acyl residue. Now, if it is true that the protective effect of acyl CoA derivatives against the action of N-ethylmaleimide is due to the degree of acylation of the "peripheral" sulfhydryl groups, we must conclude from Table II that acetyl radicals can acylate to a much greater extent than palmityl radicals. This would be possible if the "peripheral" sulfhydryl group of the multienzyme complex were located in a hydrophilic environment, while the "central" one was in a lipophilic environment. Thus with increasing chain length of the fatty acids the tendency to move from the "central" group to the "peripheral" one would gradually become smaller. Palmitic and stearic acid would preferentially remain on the "central" sulfhydryl group, thereby blocking the uptake of a new malonyl radical until the fatty acid is released from the enzyme complex by transfer to coenzyme A.

References

1. Alberts, A. W., Majerus, P. W., Talamo, B., and Vagelos, P. R., *Biochemistry,* 3, 1563 (1964).
2. Brady, R. O., *Proc. Natl. Acad. Sci. U.S.,* **44,** 993 (1959).
3. Brady, R. O., *J. Biol. Chem.,* **235,** 3099 (1960).
4. Bressler, R., and Wakil, S. J., *J. Biol. Chem.,* **236,** 1643 (1961).
5. Brock, D. J. H., and Bloch, K., *Biochem. Biophys. Res. Commun.,* 23, 775 (1966).

6. Brodie, J. D., Wasson, G., and Porter, J. W., *J. Biol. Chem.*, **239**, 1346 (1964).
7. Hagen, A., Thesis, University of Munich, 1963.
8. Hagen, A., and Hofschneider, P. H., "Proceedings of the Third European Regional Conference on Electron Microscopy," Vol. B, p. 69. Czechoslovak Academy of Sciences, Prague, 1964.
9. Harris, I., Meriwether, B. P., and Harting-Park, J., *Nature,* **197,** 154 (1963).
10. Larrabee, A. R., McDaniel, E. G., Bakerman, H. A., and Vagelos, P. R., *Proc. Natl. Acad. Sci. U.S.,* **54,** 267 (1965).
11. Lynen, F., *Sitzber. Bayer. Akad. Wiss. Math.-Naturw. Kl.* (March 4, 1960).
12. Lynen, F., *Federation Proc.,* **20,** 941 (1961).
13. Lynen, F., *in* "Proceedings of the Robert A. Welch Foundation Conferences on Chemical Research. V. Molecular Structure and Biochemical Reactions," p. 293. Welsh Foundation, Houston, Texas, 1962.
14. Lynen, F., *in* "New Perspectives in Biology" (M. Sela, ed.), (B.B.A. Library Volume 4), p. 132. Elsevier Publishing Company, Amsterdam, 1964.
15. Lynen, F., *in* "Aspects of Yeast Metabolism, Guiness Symposium," Dublin, 1965, in press.
16. Lynen, F., *Angew. Chem.,* **77,** 929 (1965).
17. Lynen, F., and Tada, M., *Angew. Chem.,* **73,** 513 (1961).
18. Lynen, F., Hopper-Kessel, I., and Eggerer, H., *Biochem. Z.,* **340,** 95 (1964).
19. Majerus, P. W., Alberts, A. W., and Vagelos, P. R., *Proc. Natl. Acad. Sci. U.S.,* **51,** 1231 (1964).
20. Majerus, P. W., Alberts, A. W., and Vagelos, P. R., *Proc. Natl. Acad. Sci. U.S.,* **53,** 410 (1965).
21. Majerus, P. W., Alberts, A. W., and Vagelos, P. R., *J. Biol. Chem.,* **240,** 4723 (1965).
22. Oesterhelt, D., unpublished experiments.
23. Overath, P., and Stumpf, P. K., *J. Biol. Chem.,* **239,** 4103 (1964).
24. Pugh, E. L., and Wakil, S. J., *J. Biol. Chem.,* **240,** 4727 (1965).
25. Reeves, H. C., unpublished experiments.
26. Sauer, F., Pugh, E. L., Wakil, S. J., Delaney, R., and Hill, R. L., *Proc. Natl. Acad. Sci. U.S.,* **52,** 1360 (1964).
27. Schweizer, E., unpublished experiments.
28. Schweizer, E., Thesis, University of Munich, 1963.
29. Wakil, S. J., and Ganguly, J., *J. Am. Chem. Soc.,* **81,** 2597 (1959).
30. Wakil, S. J., Pugh, E. L., and Sauer, F., *Proc. Natl. Acad. Sci. U.S.,* **52,** 106 (1964).
31. Wells, W. W., Schultz, J., and Lynen, F., *Proc. Natl. Acad. Sci. U.S.,* **56,** 633 (1966).
32. Willecke, K., unpublished experiments.
33. Yalpani, M., and Willecke, K., unpublished experiments.

The Synthesis of Amino Acids by Organized Enzyme Systems

R. P. WAGNER, A. BERGQUIST, B. BROTZMAN,
E. A. EAKIN, C. H. CLARKE, AND R. N. LE PAGE[1]

Genetics Foundation, Department of Zoology,
University of Texas, Austin, Texas

INTRODUCTION

For many years two quite antithetical schools of thought have existed with regard to the organization of those parts of the cell that carry out the catalytic role. One of these, as held by many biologists, is well expressed by the following quotation from L. Monné written about twenty years ago (22).

"It is indisputable that the protoplasm is not a random mixture of a great variety of enzyme molecules. Metabolism is a regular sequence of chemical reactions which occur at the right time and the right place, and which are appropriately modified in adaptation to changing environmental conditions. Therefore, the protoplasm must be regarded as an organized system of enzymes whose activity is strictly controlled. The enzymes are activated and inactivated according to the needs of the living cell. The enzymes must be concatenated with each other and arranged in certain patterns. The harmonious collaboration of enzymes is maintained by the ordered structure of protoplasm."

The other point of view was represented by many biochemists or at least enzymologists and bacteriologists who looked upon cells primarily as bags of enzymes and did not at all agree that it was indisputable that enzymes were organized, especially those enzymes that were easily solubilized after breaking open the cell.

At the present day it is apparent that Monné's statement is probably either all true or not very far from it (including the perceptive observation about metabolic control), and there remains the task now of working out the details of this organization. This is not easily done. For while it is a relatively simple matter to isolate enzymes and determine their physical and catalytic properties, such data as are found are not neces-

[1] C. H. Clark contributed to the results presented in the first section only. B. Brotzman, E. A. Eakin, and R. N. Le Page contributed to the second section.

sarily informative about the enzyme's actual mode of action in the cell. On the other hand, dealing with groups of enzymes contained in specific cell fractions also leaves much to be desired, since one seldom knows all of what is in a given fraction, and further one is always faced with the problem of dealing with possible artifacts of preparation. The biological problem of dissecting the cell to find out how it works at the molecular level in metabolism bears some relation to the physical problem of investigating the atom. The uncertainty principle enters, because, in manipulating the object in order to study it, it is impossible not to change it.

However, despite these difficulties, it has been possible to demonstrate that some enzymes are sequestered in such things as mitochondria, chloroplastids, lyzosomes, and the like, and some beginnings have been made in the direction of showing how enzymes are organized within such entities, particularly the mitochondria. Thus, according to Green and co-workers (2, 3, 7), the outer membrane of beef heart mitochondria can be separated into two components, the S fraction, which contains a high proportion of the readily soluble enzymes that implement the citric acid cycle and the oxidation and elongation of fatty acids, and the K fraction, which is rich in the particulate macromolecular dehydrogenating complexes of the citric acid cycle. Neither of these fractions contains elements of the electron transport system, which appears to be confined to the inner membrane of the mitochondrion. What is of particular interest here is that even the so-called "soluble" enzymes may also be bound *in vivo* and function as part of an organized entity.

Not only is organization found in such easily discernible organelles as mitochondria, but also in the soluble so-called "ground substance" of the cytoplasm which gives no evidence of ordered structure under the electron microscope. Reed and others (28), for example, have found that the oxidative decarboxylation of pyruvate is carried out in *Escherichia coli* by a set of three enzymes which are bound together in a definite molecular ratio to give unit complexes consisting of 50 molecules each; these are demonstrable under the electron microscope, and it is doubtful that they are artifacts. It can be readily assumed that the efficient decarboxylation of pyruvate (and of α-ketoglutarate by a related similar complex) is accomplished by an ordered, structured enzyme complex in the intact *E. coli* cell. Almost identical complexes with the same specificity have been found in mammalian cells.

If enzyme organization is a fundamental attribute of cells, its study becomes of considerable significance in the analysis of the effects of gene mutation, for mutations should be expected to affect parts of a protein in addition to its active catalytic site. Such modifications can be expected to lead to a nonfunctioning overall pathway even though the activity of the changed enzyme for its specific reaction may remain unaltered or only slightly altered when tested *in vitro. In situ, in vivo* an altered enzyme may be completely inactive, but *in vitro* in solution it may be quite active because of allosteric effects (20). A possible example of this was described some years ago, although not then so interpreted. Mutants of *Neurospora crassa* requiring pantothenate, and blocked at the step immediate to pantothenate synthesis (the union of pantoate and β-alanine), were found to have an active enzyme for this synthesis when reduced to acetone powders (32, 37). Additionally, it was found that, if the mutants were grown under constant vigorous aeration in the presence of pantothenate, they began to synthesize their own pantothenate after several days of growth (38). At present this mutant phenotype is probably best interpreted as being the result of an altered pantothenate synthetase which functions in a biosynthetic pathway leading from α-ketoisovalerate only under certain environmental conditions. *In vitro* it readily carries out its specific function, the synthesis of pantothenate from β-alanine and pantoate. It is not necessary, in other words, that it be integrated into a multienzyme system for its specific function to be demonstrated *in vitro,* but it is important *in vivo.* The reason for this may be (as pointed out by Munkres and Woodward, 24) that a mutant allosteric enzyme attached at its proper site may not be active by reason of changes in its active site. These workers have presented evidence for the existence of such a situation in the *Neurospora* mutants which have altered, but active *in vitro* malic dehydrogenases.

Our own interest in the problem of cellular organization grew out of the study of a group of mutants of *Neurospora crassa* requiring isoleucine and valine; a number of these were found, like the malic dehydrogenase mutants of Munkres *et al.* (23, 24), to have enzyme activities which were unexpected in view of their known requirements for added growth factors (39). This prompted a search for an organized system involving at least those enzymes required for isoleucine and valine synthesis from pyruvate. As has been shown for a number of organisms, both these amino acids are synthesized from pyruvate via a series of five steps requiring four enzymes which are common to both pathways, as shown in Fig. 1.

The Overall Synthesis of Valine and Isoleucine from Pyruvate

If a homogenate of *Neurospora* mycelium is prepared by grinding with sand in the presence of sucrose using a mortar and pestle (or some other similar gentle method), and the large remaining particles such as

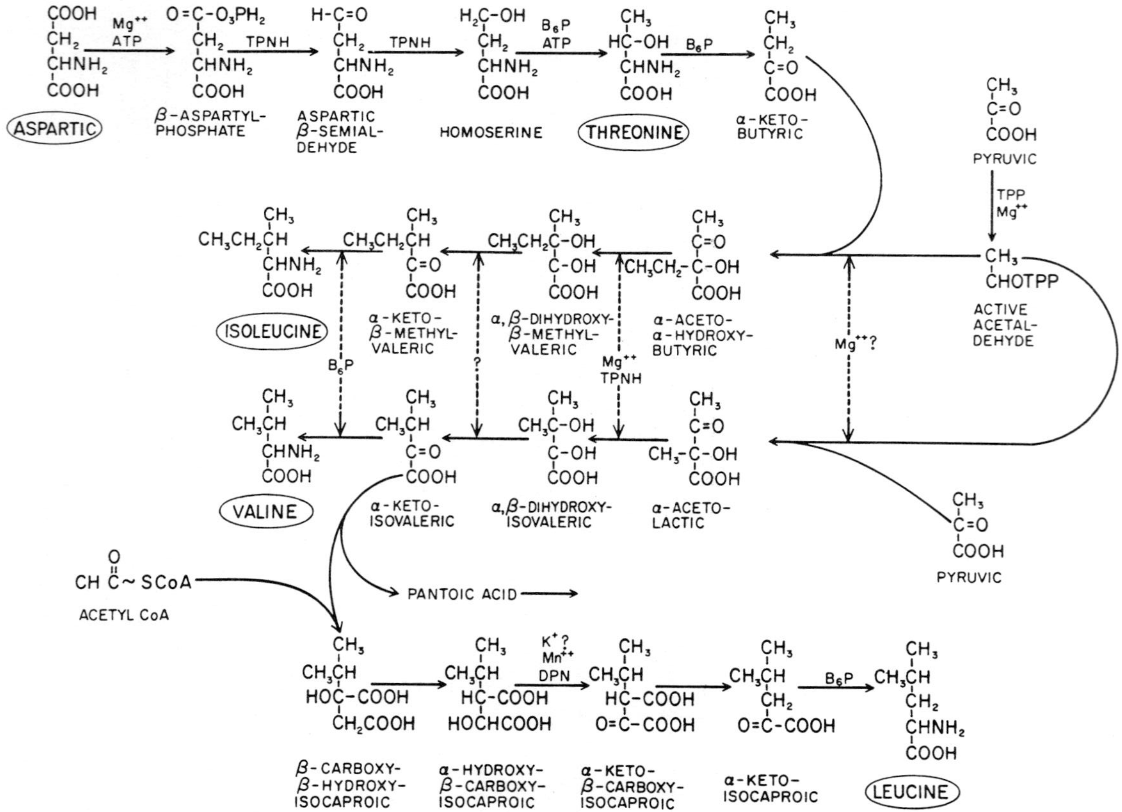

FIG. 1. The biosynthetic pathways leading to isoleucine, valine, and leucine. The condensing enzyme acts at the step at which the α-aceto acids are synthesized; the reductoisomerase converts these to the dihydroxy acids, which are in turn converted to the α-keto acids by the dehydratase.

nuclei and cell wall fragments are removed by low-speed (500 *g*) centrifugation, the supernatant will show some activity for the synthesis of valine from exogenously supplied pyruvate plus the required cofactors and phenylalanine. The pellet from centrifugation at higher speed (10,000 *g* to 35,000 *g*) shows the major portion of the overall activity for pyruvate to valine, provided it is supplemented with coenzymes and other necessary substrates, as given in Table I. The specific activity may be tenfold as high in the pellet as in the supernatant at pH 8, the approximate pH optimum (17). It has been shown by centrifugation in a sucrose density gradient that the activity of the pellet material is associated with the mitochondrial fraction (34). This activity may be removed from the pellet by a variety of methods such as sonication (Table I) or the use of various detergents (Table II). However, merely suspending the pellet in buffer in the presence of sucrose leads to the solubilization of some activity as shown in Tables I and II.

TABLE I
EFFECT OF SONIC IRRADIATION ON THE ACTIVITY OF 39,000 *g* CRUDE PELLET[a]

Cell fraction	Micromoles valine/mg protein/3 hr
Crude pellet	0.80
Supernatant after irradiation	1.17
Pellet after irradiation	0.02
Supernatant control	0.21
Pellet control	0.63

[a] Mycelium grown 24 hours. Ground in sand in presence of 0.1 *M* sucrose and 0.05 *M* Tris buffer, pH 8.5. Filtered through glass wool. Filtrate centrifuged at 500 *g* for 10 minutes. Supernatant from this centrifuged at 39,000 *g* to give crude pellet. Pellet taken up in 0.1 *M* sucrose plus 0.05 *M* Tris and treated with sonic irradiation for 30 seconds with a Sonifier, Model S75, at setting No. 4. Irradiated material then centrifuged at 150,000 *g* for 60 minutes to give supernatant and pellet. Control represents supernatant and pellet from centrifuged crude pellet at 150,000 *g* without prior sonic irradiation. Assayed for valine-synthesizing activity in the presence of the following mixture: 10 m*M* $MgSO_4$, 0.28 m*M* NADPH, 0.30 m*M* thiamine pyrophosphate, 0.40 m*M* pyridoxal phosphate, 25 m*M* glucose-6-phosphate sodium salt, 40 m*M* L-phenylalanine, 50 m*M* sodium pyruvate, 50 m*M* Tris buffer at pH 8.5, 0.1 Kornberg unit glucose-6-phosphate dehydrogenase (actually not necessary) and 4 mg of protein to be assayed in a total volume of 2 ml. Incubation at 35°C.

The overall activity obtained in either the pellet or the supernatant is unstable; to prepare fractions with good activity, sucrose or pyruvate must be present during the homogenization and subsequent steps, as shown by Table III and by Kiritani *et al.* (17). This observation is of considerable significance, because neither sucrose nor pyruvate is required to stabilize the activities of the four enzymes involved in the

TABLE II
SOLUBILIZATION OF OVERALL ACTIVITY BY VARIOUS DETERGENTS[a]

Treatment	Micromoles valine/mg protein/3 hr	
	Supernatant	Pellet
None	0.30	0.27
Tween 80	0.20	0.39
Triton X100	0.70	0.00
Brij 58	0.56	0.09
Deoxycholate	1.03	0.00

[a] Mycelial fractions prepared as in Table I without sonic irradiation. The original crude pellet treated before centrifugation at 150,000 g had a specific activity of 0.67 except for that treated with sodium deoxycholate, which had an activity of 1.27. The final concentration of detergents was 0.33%. These were added to the crude pellet 5 minutes before centrifugation at 150,000 g for 60 minutes. Supernatants and pellets obtained were then assayed for activity by using the mixture described in Table I.

transformation of pyruvate to valine (17). Indeed the presence of sucrose during the preparation of the cell fractions may result in a lower specific activity for the condensing enzyme and the reductoisomerase and dehydratase in both the supernatant and pellet than is the case when sucrose is absent, as shown by Table IV. These data also show that when the homogenate is prepared in the presence of sucrose a greater percentage of total enzyme activity is in the pellet.

We have concluded from this that the four enzymes, at least, are bound together in some fashion to form an active system for carrying out the overall synthesis of valine from pyruvate. The further testing of

TABLE III
THE EFFECT OF SUCROSE AND PYRUVATE ON OVERALL ACTIVITY[a]

Cell fraction	Micromoles valine/mg protein/3 hr			
	0.25 M sucrose	0.10 M sucrose	No sucrose	50 mM pyruvate, no sucrose
Crude pellet	0.67	0.68	0.16	1.06
Crude pellet after sonic irradiation	0.64	0.64	0.06	0.65
Supernatant after sonic irradiation	1.02	1.13	0.15	2.57

[a] Extraction procedure was as follows: 24-hour-old mycelium ground with sand in 0.05 M Tris buffer at pH 8.5 plus indicated additives. Homogenate filtered through glass wool, centrifuged at 500 g for 10 minutes, and supernatant of this centrifuged at 39,000 g for 60 minutes. Pellet from this taken up in Tris buffer plus indicated additive. Sonic irradiation and other procedures as in Table I.

TABLE IV

INDIVIDUAL ENZYME ACTIVITIES AFTER PREPARATION OF FRACTIONS IN THE PRESENCE AND IN THE ABSENCE OF SUCROSE[a]

	Specific activities (μmoles/mg protein/hr)					
	No sucrose			0.25 *M* sucrose		
Enzyme	Pellet	Super-natant	Percent total activity in pellet	Pellet	Super-natant	Percent total activity in pellet
Condensing	4.2	1.9	32	4.0	0.19	75
Dehydratase	12.3	4.4	38	7.0	0.51	66
Reductoisomerase	13.7	3.9	41	7.8	0.54	65

[a] Procedure: 24-hour-old mycelium ground in 0.1 *M* Tris buffer at pH 6.5 or in 0.25 *M* sucrose+0.15% bovine serum albumin plus buffer. Centrifuged for preparation of pellet and supernatant as previously described at 39,000 *g*. Enzyme assays as described previously using these substrates: pyruvate for condensing enzyme, α,β-dihydroxyisovalerate for the dehydratase, and α-aceto-α-hydroxybutyrate for the reductoisomerase. Bovine serum albumin had previously been found not to affect the activities of these enzymes.

this hypothesis has proved to be quite difficult. Attempts to destroy the overall activity with various hydrolytic enzymes have given inconclusive results. The effects of lecithinase C, phospholipase A from *Crotalus adamanteus* venom, a phospholipase from *Naja naja* venom, and purified pancreatic lipase I and III (Worthington) were negative. However, wheat germ lipase Type I (Sigma Chemical Co.) destroyed the overall activity of soluble fractions which had been solubilized either by sonication or by detergents. But since this is not a highly purified enzyme preparation, and is known to contain acid phosphatase activity, nothing specific can be ascribed to this result at present.

Both the mitochondrial pellet fraction and the soluble fraction derived from it have activity for the overall synthesis of isoleucine from pyruvate plus α-ketobutyrate, and under proper conditions both valine and isoleucine are formed simultaneously (35). A definite competitive inhibition exists between pyruvate and α-ketobutyrate which is not unexpected, since they depend on the same enzymes for their conversion. Both threonine and aspartate can replace α-ketobutyrate to a certain extent with the formation of isoleucine in the presence of pyruvate. Figure 2 shows the effect of adding L-threonine at different levels in the presence of a constant level of pyruvate to a suspension of the mitochondrial pellet fraction. An inhibition of valine formation is evident, just as in the case of α-ketobutyrate. However, threonine is not as inhibitory to valine produc-

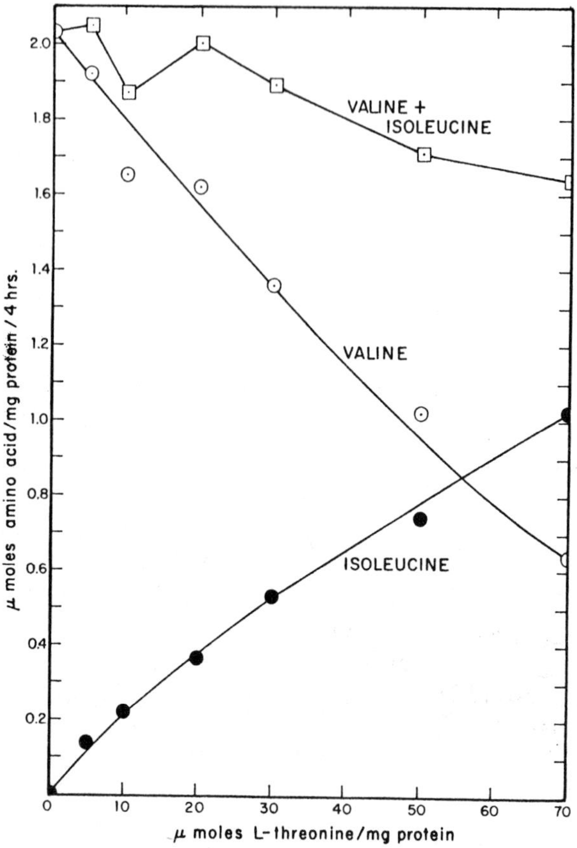

FIG. 2. The simultaneous synthesis of valine and isoleucine in the presence of threonine and pyruvate. Assays for valine and isoleucine activity made in presence of mixture given in Table I. Two milligrams of protein present per milliliter. Pyruvate concentration held constant at 25 μmoles/mg protein.

tion as is α-ketobutyrate. To obtain equimolar synthesis of isoleucine and valine, one must have a molar ratio of α-ketobutyrate to pyruvate of 0.03, compared to a ratio of threonine to pyruvate of 2.3, in the presence of 2 mg of protein per milliliter.

When aspartate is used as the partial isoleucine precursor in the presence of pyruvate, results such as those shown in Fig. 3 are obtained. It is evident that not as much isoleucine relative to valine is made as in the presence of threonine. However, isoleucine is produced in considerable quantity, and there is again an inhibitory effect on the valine production. The reason for the inhibition of isoleucine synthesis in the presence of excess aspartate is not understood.

Since both threonine and aspartate are active substrates for isoleucine synthesis with the system described here, it could mean that the organized system which we hypothesize incorporates more than four enzymes, and has at least as many as nine. The fact that one gets a definite synthesis of isoleucine in 4 hours in the presence of aspartate might be taken as support for the existence of some sort of organization. However, the differences in the kinetics of systems of free soluble enzymes versus systems of bound, organized enzymes have not been worked out, and one cannot draw any firm conclusions from the above observations.

In a previous communication, we have used the fact that the dilution of the pyruvate to valine system, in either the pellet or the soluble form, does not change the specific activity of the system so far as valine is concerned until a dilution of about 2 mg of protein per milliliter or below is achieved, as evidence in agreement with the existence of an organized particulate entity (17). Other explanations for this behavior can be advanced, but this dilution effect would be expected of an organized system which fell apart upon dilution. The results of a further test of the system's capacity to withstand dilution are given in Fig. 4, which presents data from an experiment in which the production of both isoleucine and

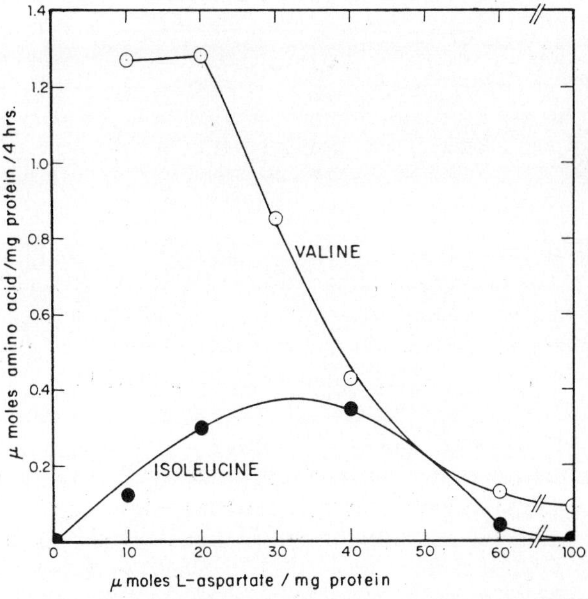

FIG. 3. The simultaneous synthesis of valine and isoleucine in the presence of aspartate and pyruvate. In addition to the assay mixture given in Table I, there were present 20 μmoles of ATP and 5 mg of NADH per 8 mg of protein. Pyruvate concentration held constant at 10 μmoles/mg of protein.

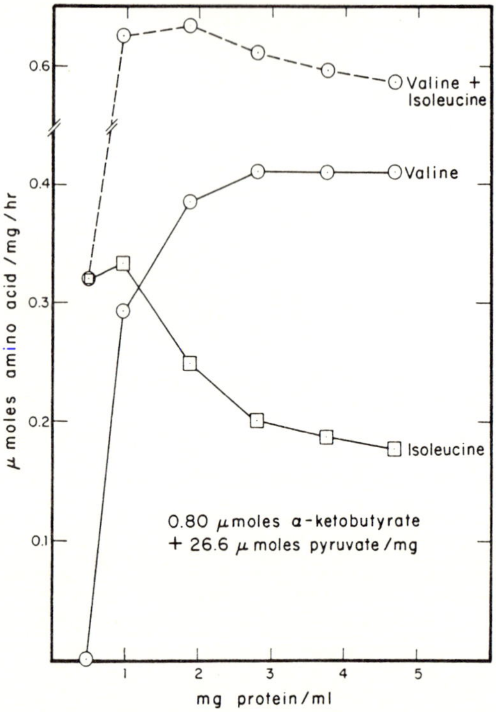

FIG. 4. Effect of dilution on the synthesis of isoleucine and valine simultaneously from pyruvate and α-ketobutyrate. Assay mixture as in Table I.

valine was followed. It will be noted that, although valine production falls off rapidly with protein below 2 mg/ml, isoleucine production does not, but appears to increase almost till the lowest dilution at which it was possible to make measurements of amino acid concentration was reached. There appears to be no simple explanation for this, but again the evidence is in favor of some sort of organized particle.

Results from Mixing Mutant Fractions

If an organized system is involved in the synthesis of isoleucine and valine, one might expect that cell fractions of mutants which cannot synthesize isoleucine and valine from pyruvate should not be able to do so in mixtures even if they have different blocks in the isoleucine–valine pathway, or at most should be much less effective than the wild-type cell fractions.

We had previously shown that the isoleucine–valine-requiring mutants, 327, 304, 318, 311, 321 and 331, have a similar phenotype in so far as they have somewhat lower dihydroxy acid dehydratase activities than the wild type, and cell fractions from them have little or no activity for the synthesis of valine from α-acetolactate, some activity from α,β-dihydroxyisovalerate, and none from pyruvate (36). They do, however, have active condensing enzyme, reductoisomerase, and transaminase activities at wild-type level or better. The mutant, 305 is deficient in reductoisomerase activity and accordingly cannot convert either pyruvate or acetolactate to valine. It has wild type or higher levels of the other three enzymes. Cell fractions of mutant 364 cannot convert pyruvate to valine but can do so by using the succeeding precursors. Cell-free extracts from it do, however, contain activities for all four enzymes (36). On the basis of these observations, one should not expect mixtures of those mutants apparently blocked at the dehydratase step (327, 304, 318, 311, 321, and 331) to produce valine from pyruvate, but they might or might not when mixed with 305 or 364, which are blocked at two other different steps. Fractions of all eight mutants were tested in mixtures of 305, 327, and 364, alternatively, these representing each of the three different types of mutants. Mixtures of pellet with pellet, as well as supernatant with supernatant, were made as shown in Table V. Mixtures of pellet with supernatant were also made, but the results of these are not recorded here, since they are essentially the same as those shown in the table.

TABLE V

SYNTHESIS OF VALINE BY COMBINATIONS OF CELL FRACTIONS OF *Neurospora* STRAINS WITH A REQUIREMENT FOR ISOLEUCINE AND VALINE[a]

	Pellet mixtures			Supernatant mixtures		
Mutant	327	305	364	327	305	364
327	—	0.26	0.20	—	0.23	0.28
304	0.00	0.25	0.11	0.02	0.23	0.21
318	0.00	0.33	0.16	0.01	0.22	0.13
311	0.00	0.01	0.01	0.00	0.01	0.00
321	0.00	0.06	0.02	0.00	0.23	0.06
331	0.00	0.01	0.03	0.00	0.06	0.02
305	0.30	—	0.10	0.19	—	0.26
364	0.19	0.13	—	0.28	0.15	—

[a] Specific activities given as micromoles of valine per milligram of protein per 4 hours. Mycelia were grown in presence of isoleucine and valine for 24 hours and ground in 0.5 *M* sucrose plus 0.05 *M* Tris buffer at pH 8.5 with sand. The homogenates were filtered through glass wool and centrifuged at 39,000 *g* for 60 minutes. The pellets and supernatants were combined as indicated and incubated in the presence of the mixture given in Table I for 3.5 hours. All mixtures contained 2 to 3 mg of protein from each strain.

It will be seen from the data that some of the pellet-with-pellet combinations did give activity for valine production from pyruvate, but not at the same level as the wild-type pellet which generally showed activity two to four times as high as the highest of these. As expected, those with the same type of block—327, 304, 318, 311, 321, and 331—did not show activity with 327. On the other hand, 327, 304, and 318 did show fairly good activity with 305 and 364, which are presumably blocked in different positions. The most interesting ones are 311, 321, and 331, which showed barely measurable activity with 305 and 364. In general, the supernatant mixtures gave approximately the same results as the pellet. One marked exception was the activity of the combination 321 + 305, which was barely significant in the pellet but quite high in the supernatant combinations. It should be noted here that it was difficult to make a very clean separation between the pellet and supernatant fractions because the mutant pellets did not pack tightly as did the wild type.

These results are in general agreement with the hypothesis that an organized, attached system is being dealt with. In the first place, the

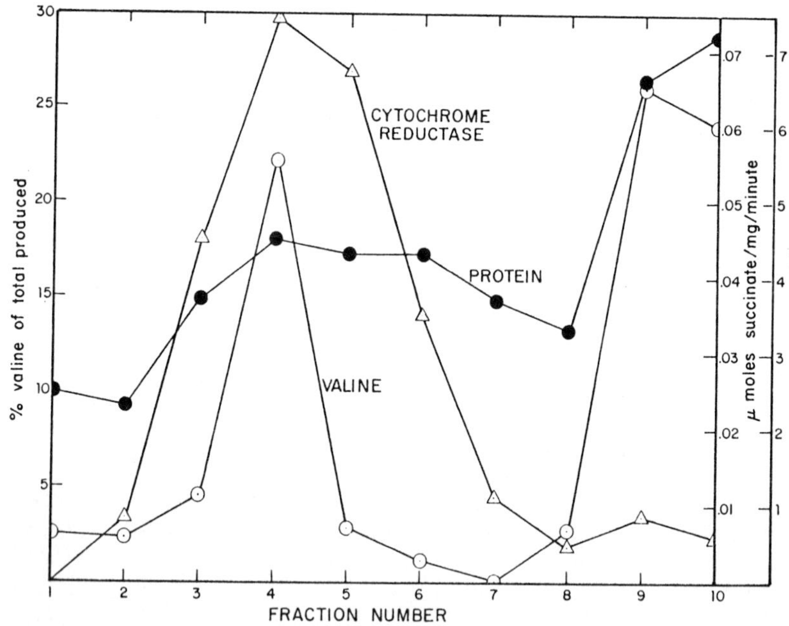

FIG. 5. Mutant pellet combination, 305 + 327, centrifuged in a linear sucrose gradient. Gradient 0.58 *M* to 1.9 *M*. Centrifuged in Spinco SW 25 swinging bucket for 4.5 hours at 25,000 rpm. Valine-synthesizing activity of each fraction determined by assay given in Table I. Fractions removed from bottom of tube. A total of 5.0 ml was divided into ten equal fractions.

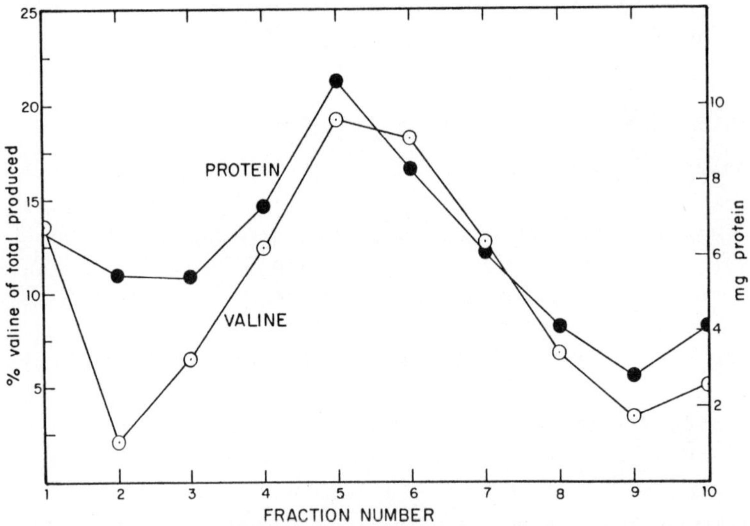

FIG. 6. Mutant supernatant combinations, 305 + 327, centrifuged in linear sucrose gradient. Gradient 0.25 *M* to 0.58 *M*. Other conditions as in Fig. 5.

activities are in general lower than would be expected from wild type and, secondly, some combinations in which all four wild-type enzymes are presumably present give practically no activity.

The mutant combination 305 + 327 was· investigated further to determine the behavior of the valine-synthesizing system in a sucrose gradient. Figure 5 depicts the results from centrifuging in a linear sucrose gradient ranging from 0.58 *M* to 1.9 *M* sucrose. Pellet preparations from each mutant were mixed and centrifuged as indicated. A main peak of activity at 1.46 *M* sucrose was obtained. This corresponds to the peak for cytochrome c succinate reductase and also to the density of wild-type mitochondria. An additional zone of activity was found at the top of the gradient. This presumably corresponds to the soluble activity in the mixed supernatants. Figure 6 presents the results of centrifuging the supernatant combination of 305 and 327 in a sucrose gradient ranging from 0.25 *M* to 0.58 *M*. The peak for this activity comes in the region of 0.44 *M* sucrose. Again this corresponds to the density of the soluble activity in the supernatant. A pellet was formed at the bottom of the tube after centrifugation which had activity for valine synthesis. Presumably this is material with a density equivalent to 1.46 *M* sucrose which was not centrifuged down when the supernatants were prepared.

These results make it difficult to decide whether new functional complexes are actually being formed out of good parts of old ones, or whether the activity that is noted is merely due to a sort of complementation in which one set of enzymes of one mutant carries the sequence to one point and the other finishes it, and nothing new is really formed.

Isoleucine Synthesis in Salmonella

A system resembling the *Neurospora* isoleucine–valine system has been demonstrated in *Salmonella typhimurium* by Cronenwett and Wagner (5). Sonic irradiation of *Salmonella* spheroplasts prepared in a glycine medium results in a fraction which is capable of synthesizing isoleucine on the addition of α-ketobutyrate, and, after the removal of unbroken spheroplasts, further centrifugation at 39,000 *g* segregates this activity in the pellet or membrane fraction, leaving little or no overall activity in the supernatant except from α-aceto-α-hydroxybutyrate and the succeeding precursors. It resembles the *Neurospora* system in its requirement for sucrose as a stabilizer. Surprisingly, the membrane fraction has no demonstrable activity for the synthesis of valine from either pyruvate or α-acetolactate. An endogenous source of a two-carbon substrate seems to be present in the membrane fraction however, because isoleucine is synthesized in the absence of pyruvate. Also noteworthy is the fact that 30 to 45% of the activity remains in the absence of exogenous NADPH which is necessary for the reduction of α-aceto-α-hydroxybutyrate, and 100% of the activity remains in the absence of exogenous thiamine pyrophosphate and Mg^{++}. The system is further different from that found in *Neurospora* in that it has two pH optima, one between pH 6.5 and 7.0 and the other between pH 8.0 and 8.6.

Despite these differences, however, and the inability for synthesis of valine in *Salmonella* fractions, the finding that the isoleucine pathway is associated with the membrane fraction in *Salmonella* is in general agreement with what has been found in *Neurospora*. The inactivity for the synthesis of valine is inexplicable, since it is certain that in *Salmonella* both amino acids are synthesized through the same biosynthetic pathway just as in *Neurospora* (33).

Amino Acid Production by "Intact" Mitochondria

It was assumed from the results described above that, if a system for the synthesis of valine and isoleucine could be demonstrated, it should also be possible to demonstrate synthesis of leucine from pyruvate or α-ketoisovalerate (see Fig. 1). But it was found that no leucine was produced from either substrate even when all the necessary cofactors were added to the *Neurospora* pellet or supernatant preparations described in the foregoing section (17). On the other hand, if the homogenate was prepared from the mycelium by the method of Hall and

Greenawalt (9) for isolating *Neurospora* mitochondria capable of oxidative phosphorylation, amino acids did appear after incubation of the particulate fraction of the homogenate centrifuged down at 10,000 g (14). This method differs from the preceding one in the use of a medium containing 0.15% bovine serum albumin and 0.005 *M* EDTA plus 0.25 *M* sucrose. Mitochondrial fractions prepared in this way do not respond to substrates such as pyruvate or α-ketoisovalerate. But on incubation they do produce a number of amino acids, including valine and leucine, provided ATP and Mn^{++} ion are present. ATP must be present at low levels, however, as increasing its concentration results in drastic inhibition in the free amino acid-producing capacity. Additionally, enhancement of leucine production occurs in the presence of β-carboxy-β-hydroxyisocaproate, which is a precursor of leucine (see Fig. 1). Although the presence of added pyruvate does not enhance the production of either leucine or valine during the first 4 hours of incubation, the addition of pyruvate labeled with ^{14}C at carbon-2 resulted in the incorporation of ^{14}C into valine and leucine–isoleucine. This was determined by paper chromatography by which means leucine was not separated from isoleucine.

The evidence points to the origin of these amino acids, at least in part, by synthesis in the incubating mitochondria presumably from endogenous substrates already present in the mitochondria. The significant difference between this mitochondrial preparation and that described in the previous section is that bovine serum albumin was present. The presence of EDTA alone does not produce mitochondrial preparations with an ATP requirement for the production of free amino acids (13). On the other hand, the pH optimum for production of amino acids was in the basic range in each type of preparation—that is, pH 7.8 for the BSA (bovine serum albumin) treated material, and 8.0 to 8.5 for that prepared in the absence of BSA. The BSA-treated preparations show, however, a sharp drop in activity between pH 7.8 and 8.2, so that the optima cannot be considered the same.

The foregoing observations (as well as others which are described in the references to the work done in this laboratory) make it quite obvious that the results obtained with any given preparation may be drastically altered by even minor modifications in the method of preparation. Under ordinary circumstances during the purification of individual enzymes these changes may have little or no effect, but they may greatly alter the behavior of the overall synthetic system. This is made especially evident by the effect of serum albumin on the ability of the mitochondrial fractions to utilize pyruvate readily in the synthesis of valine and isoleucine. If we assume that the mitochondria approximate the intact condition when isolated in the presence of serum albumin, then we must assume that the mitochondria can act only in the production of amino acids from endogenous substrates (if they are indeed synthesizing them *de novo* in the first place), or the endogenous substrates are present in excess.

The requirement for ATP and Mn^{++} is also not readily explained. It is known that ATP together with ADP and P_i controls the rate of respiration of mitochondria, and that excess ATP upsets the ratio of [ATP] to [ADP] [P_i] and leads to inhibition of respiration (18, 19). Since the free amino acid production by the mitochondria is inhibited by ATP in excess, this may indicate that the endogenous process is dependent on respiration in the mitochondria. In agreement with this is the finding that ADP is as effective as ATP in stimulating amino acid production. ADP has been shown by Chance and Williams (4) to drastically increase the O_2 uptake of "tightly bound" mitochondria.

Mn^{++} can be replaced by Mg^{++} or Ca^{++}, but these are only about 60% as effective as Mn^{++}. The requirement for these remains unexplained.

Purification of Enzymes Involved in the Isoleucine–Valine Pathway

As part of the process of trying to understand the structure of the indicated enzyme complex, attempts were made to purify certain of the enzymes in *Neurospora crassa* which are involved jointly in the synthesis of isoleucine and valine. Partial success was obtained with the dehydratase and reductoisomerase by using standard methods of enzyme purification (15, 16). A purification of 100- to 150-fold was achieved with the dehydratase, and from 200 to 300 with the reductoisomerase. Extraction of both purified enzyme preparations with a chloroform–methanol mixture yielded a considerable quantity of lipid-soluble material ranging from 40 to 50% by weight of the total. The exact nature of this lipid material has not been determined, but it must be rather firmly bound to the enzyme protein, since it persisted with it through Sephadex and DEAE-Sephadex columns used in the purification. The phosphorus content of the lipid material is either zero or close to it. Presumably it is all neutral lipid. We have taken the presence of the lipid to indicate the attachment of these enzymes *in situ* to membranes. Admittedly, however, the absence of phosphorus does not support this. But there remains, of course, the possibility that the phosphorus-containing moeity of the lipid was eliminated during the purification procedure.

One of the *Neurospora* isoleucine–valine mutants, 330, which appears to be blocked at the dehydratase step, but has relatively high concentrations of the dehydratase enzyme (36 to 50% of wild-type activity), has been used as a source of mutant dehydratase. Preliminary results show rather conclusively that the dehydratase from mutant 330 is different from the wild type. Its behavior is quite different during elution from Sephadex gels, and it is more unstable under certain conditions (M. Jenkins, unpublished results). This finding suggests that the dehydratase mutants in general cannot convert the dihydroxy acids *in vivo* because they produce an altered dehydratase which either does not attach or, if it does attach, upon doing so has its active site altered to an inactive condition.

Attempts were also made to purify the condensing enzyme. It was found that standard methods of purification such as ammonium sulfate precipitation and gel column chromatography with such agents as Sephadex or DEAE columns were not effective in producing a reproducible purification much beyond what had originally been obtained by Radhakrishnan and Snell (26), who by their procedure, however, had separated the condensing enzyme activity from the acetolactate decarboxylating enzyme. Nonetheless, attempts to purify the condensing enzyme led to an interesting finding which is discussed in the next section.

Amino Acids from Supernatant Fractions "Free" of Mitochondria

As a consequence of attempting to purify the condensing enzyme, it was found that supernatant from homogenate centrifuged with a force of 10,000 g or greater to remove mitochondria formed two visibly distinct bands on a Sephadex G100 column. These bands were obtained only when the temperature was above 10°C and when the column was used before it became packed in the form normally recommended for its use.

In general, we have used columns of about 1 cm inside diameter which were prepared by pouring in a suspension of Sephadex in distilled water to a height of about 10 cm. The column was then washed with 10^{-3} *M* $MnSO_4$. Before the Sephadex columns became packed they were charged with about 2 ml of *Neurospora* supernatant. Soon after the supernatant was added two quite distinct bands formed, the lower of which was white opaque and the upper of which was translucent and tinged with pink. The bands grow more distinct as they move down the column and remain separated provided the temperature is maintained between about 15° and 30°C. Long columns (15 cm and longer) may result in the lower band's becoming excessively dense and sticking to the gel so that it can no longer be eluted. Elution is best accomplished with distilled water containing Mn^{++} as described above. However, the Mn^{++} is not essential for the formation of the two bands. The whitish band coming off the column contains the majority of the protein and lipid of the supernatant (see Fig. 7). What follows in the colored band consists of free amino acids, carbohydrate, and presumably a variety of other smaller molecules.

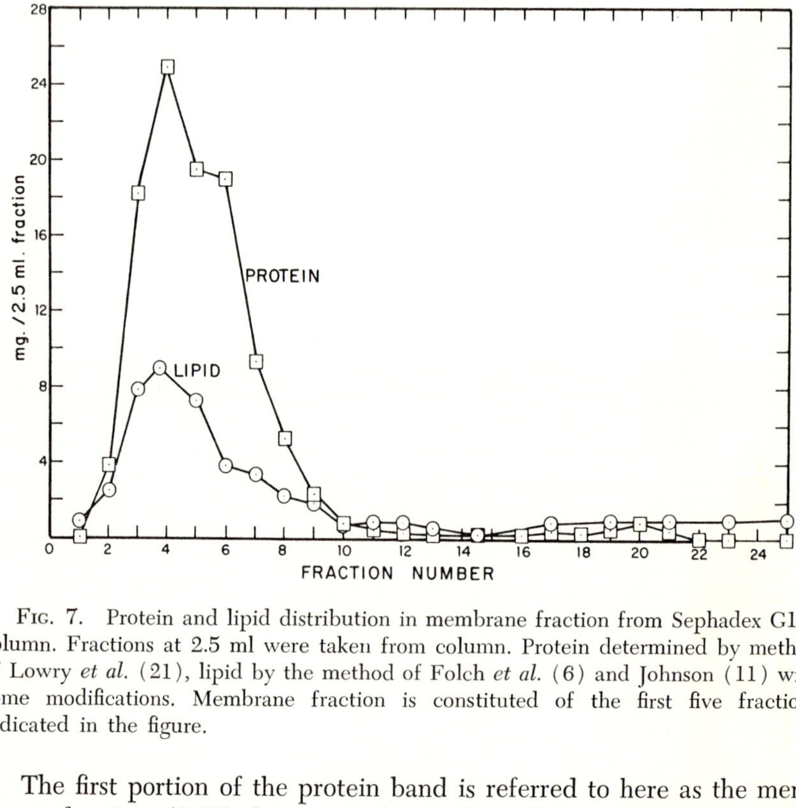

FIG. 7. Protein and lipid distribution in membrane fraction from Sephadex G100 column. Fractions at 2.5 ml were taken from column. Protein determined by method of Lowry *et al.* (21), lipid by the method of Folch *et al.* (6) and Johnson (11) with some modifications. Membrane fraction is constituted of the first five fractions indicated in the figure.

The first portion of the protein band is referred to here as the membrane fraction (MF) because when collected from the column and allowed to sit either in distilled water or in buffer at pH 6.0 to 8.0 it flocculates and presents the appearance of membrane material extracted from mammalian cells. Additionally, chemical analysis shows MF to have an average composition of about 57% protein, 28% lipid, 9% carbohydrate, 2.5% RNA, 0.03% DNA, and 0.4% phosphorus. These values are similar to those found for membrane material from *Mycoplasma* (27) and rabbit liver cells (31). Furthermore, MF was extracted by the method of Woodward and Munkres (40), and structural protein apparently identical to the *Neurospora* mitochondrial protein described by them was identified (Wagner and Riggs, unpublished results). We found that in our hands lipid-free mitochondrial material yielded 3.8% by dry weight of structural protein, while the lipid-free MF material yielded 2.8% by dry weight. Structural protein is presently considered to be a constituent of all cell membranes in *Neurospora* (Woodward, unpublished results). Finally, the cloudy flocculant is easily dissolved, making the solution quite clear, with the addition of sodium dodecyl sulfate. About 5 μmoles of SDS is required per milligram of MF protein to obtain a minimum absorbance in the spectrophotometer. This is again a characteristic of membrane material (27).

A study of MF was begun because it was found to contain a considerable number of enzymatic activities—some of them with higher specific activity than originally present in the supernatant from which the MF was derived. Table VI presents some representative results

TABLE VI
ACTIVITY OF FOUR ISOLEUCINE–VALINE ENZYMES IN MEMBRANE FRACTIONS AND SUPERNATANTS FROM WHICH DERIVED[a]

	Micromoles/mg protein/hr			
Fraction	Condensing enzyme	Reducto-isomerase	Dehydratase	Trans-aminase
Supernatant	0.4	1.1	0.34	4.7
MF	12.2	5.1	2.22	14.0

[a] Supernatants from 39,000 *g* centrifugation of homogenates previously centrifuged at 500 *g*. Homogenates prepared initially in 0.1 *M* Tris-maleate buffer at pH 6.5. Enzyme activities determined as described by Wagner *et al.* (36).

obtained and compares the activity of the supernatants and the MF derived from them. In addition, a tenfold increase in the malic dehydrogenase activity was found in the MF as compared to the supernatant from which it was derived.

These findings prompted us to investigate the activity of MF for the synthesis of valine and isoleucine. Not unexpectedly, incubation of the MF in the presence of pyruvate led to the appearance of valine, but, unexpectedly, isoleucine and other free amino acids were formed. In a period of 3 to 4 hours of incubation at 35°C, 2 to 3 μmoles of amino acids appear per milligram of protein. Actually, incubation of MF in buffer at pH 4.0 to 8.0 in the *absence* of pyruvate and/or other exogenous factors and cofactors results in the appearance of the same free amino acids although in somewhat different ratios.

If the supernatant from which MF is derived is incubated at 35°C there is but a slight increase, or none at all, in the total amount of amino acids after incubation, but a considerable increase occurs after incubation of MF. Table VII presents data from an experiment in which supernatant from centrifugation at 150,000 *g* was used to produce an MF fraction on the Sephadex column. Unincubated and incubated supernatant and MF were then compared for free amino acid content as determined on a Spinco amino acid analyzer. It will be noted that there is a difference in the amount of free amino acids produced in the MF at pH 6.0 and 8.0. We have consistently found that quantitatively more free amino acids are produced in the acid than in the basic range. The addition of SDS to the MF sufficient to dissolve it results in the absence of free amino acids after incubation.

TABLE VII
TOTAL AMINO ACIDS IN UNINCUBATED AND INCUBATED FRACTIONS[a]

	Micromoles amino acid/mg protein			
	Supernatant		Membrane Fraction	
Incubation period	pH 6.0	pH 8.0	pH 6.0	pH 8.0
None	2.40	3.00	0.00	0.00
3 hours	2.55	2.68	2.16	0.38

[a] Supernatant from 150,000 *g* centrifugation of homogenate prepared initially in Tris-maleate buffer, 0.1 *M*, at pH 6.5. Temperature of incubation 37°C. Total amino acids determined by summing amounts of detectable amino acids as determined by a Beckman Amino Acid Analyzer. MF incubated in presence of 0.5 *M* Tris-maleate at pH 6.0 and 0.5 *M* Tris at pH 8.0.

The appearance of these amino acids during the incubation of the MF in the absence of any exogenous substrate or cofactors naturally leads one to the initial conclusion that protein of the MF is hydrolyzing and the free amino acids are products of hydrolysis. It may be further postulated that the protein of the supernatant from which the MF is derived is not hydrolyzed because of the presence of an inhibitor(s) which is removed on the Sephadex column.

However, the data in Table VIII show that not all the amino acid derived from MF can be the result of hydrolysis alone, since aspartate, glycine plus alanine, and valine incorporated ^{14}C from pyruvate uniformly labeled with ^{14}C. The specific activities for the aspartate and the valine formed are given in Table IX. Except for the free aspartate formed in the presence of supernatant at pH 6.0, the specific activities are quite low compared to the specific activity of the added pyruvate which was 6.5 mC/mmole. Additionally, the total incorporation of ^{14}C into aspartate and valine was quite low, considering that about 20×10^6 cpm was present initially.

A number of points need to be stressed concerning the data in Table

TABLE VIII

DISTRIBUTION OF RADIOACTIVITY AFTER INCUBATION OF MEMBRANE FRACTION AND SUPERNATANT WITH ^{14}C-LABELED PYRUVATE[a]

Fraction	pH	Counts per minutes $\times 10^6$								
		Before incubation	After incubation	Protein-free fraction	Total "organic acids"	Total "amino acids"	Aspartate	Glycine + alanine	Valine	"Amino acids recovered" (%)
MF	6.0	18.78	4.84	4.66	1.79	2.00	0.38	0.64	—	50
MF	8.0	22.06	20.97	20.29	17.50	1.96	0.02	1.15	0.04	62
Supernatant	6.0	18.15	16.52	18.24	1.87	12.15	0.15	2.78	0.06	25
Supernatant	8.0	21.00	21.95	21.89	2.61	12.71	0.06	1.44	—	12

[a] Pyruvate uniformly labeled with ^{14}C and with a specific activity of 6.5 mC/mmole. Each fraction contained 10 mg of protein in 4.4 ml. MF from supernatant after centrifugation at 150,000 *g* as in Table VII. Protein after 3 hours of incubation precipitated with hot TCA. Total "organic acids" represents effluent from Dowex 50 column in acid form removed with distilled water; total "amino acids" represents effluent from same column removed with 10% ammonia. Amino acids determined with a Beckman Amino Acid Analyzer to which was attached an Ansitron Flow Counter to determine radioactivity. Radioactivity in other fractions determined in an Ansitron Scintillation Counter.

VIII in addition to those in the foregoing. First it should be noted that the membrane fraction, MF, incubated at pH 6.0 lost considerably more radioactivity after the 3-hour incubation period than that incubated at pH 8.0, or the supernatant fractions. It was established by passing the atmosphere over the incubating mixtures through $Ba(OH)_2$ that radioactive CO_2 was given off during incubation in large quantity (about 5×10^6 cpm) from the MF fraction at pH 6.0.

The protein-free fraction remaining after TCA precipitation was added to a Dowex 50 column in the acid form. Elution was begun first with distilled water. Material eluted with water is designated as "organic acid." It is apparent from Table VIII that the main part of the radioactivity in the MF at pH 8.0 was in the organic acid fraction. This is in contrast to the low amount in the MF at pH 6.0 or the supernatant fractions. After elution with water a solution of 10% ammonia was passed through the column. The effluent from the column resulting from this

TABLE IX
TOTAL ASPARTATE AND VALINE FORMED AND SPECIFIC RADIOACTIVITY IN PRESENCE OF MEMBRANE FRACTION AND SUPERNATANT[a]

		Aspartate		Valine	
Fraction	pH	Micro-moles	Specific activity	Micro-moles	Specific activity
MF	6.0	1.97	195,000	0.23	—
MF	8.0	0.10	150,000	0.23	188,000
Supernatant	6.0	<0.001	1.5×10^6	0.44	136,000
Supernatant	8.0	0.25	225,000	0.00	—

[a] Conditions as in Table VIII. Specific activity given as counts per minute per micromole.

treatment is designated as the amino acid fraction. It was this fraction after concentration which was placed in the amino acid analyzer. It will be noted that the MF incubate contained much less radioactivity than the supernatant incubates in this Dowex fraction. However, the last column in the table shows that a larger percentage of identifiable amino acid was present in the MF incubate than in the supernatant.

These data again show that the supernatant and MF fractions are quite different metabolically, and that the pH of the incubation mixture is also of importance.

An additional significant observation about the MF is that ATP stimulates the production of ninhydrin-positive substances among which are isoleucine and valine, as shown in Fig. 8. This is reminiscent of the

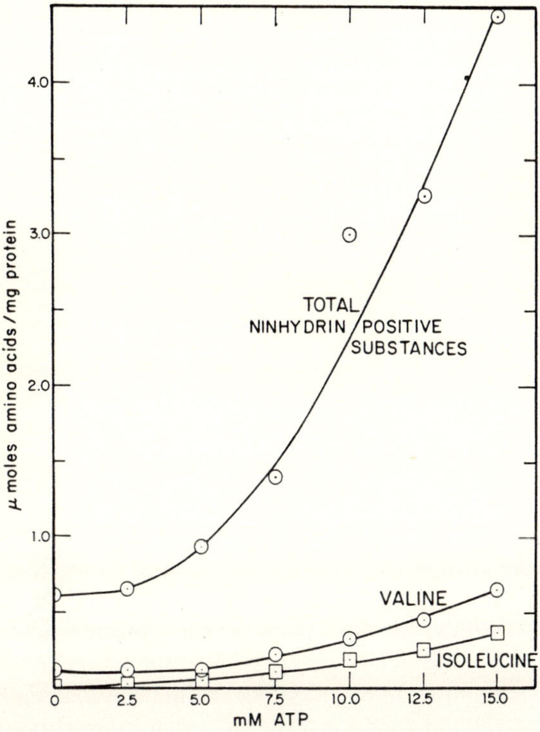

FIG. 8. Effect of ATP on the appearance of ninhydrin-positive substances in the presence of the membrane fraction. Total ninhydrin-positive substances determined by the method described by Spies (30). Valine and isoleucine determined by bioassay (10).

ATP effect of the mitochondrial fraction discussed in the preceding section. However, we have ascertained that the major part of the increase in the ninhydrin-positive material is due to the appearance of something other than free amino acids, but have as yet not established the nature of the compound(s).

DISCUSSION

The original hypothesis proposed, that the isoleucine–valine mutants of *Neurospora* are incapable of synthesizing these amino acids because of the disorganization of the enzymes catalyzing the necessary biosynthetic steps, has not been proved correct by the results so far accumulated, but neither has it been disproved. What is perhaps as important is that a number of findings have been made which do indicate that some of the enzymes, if not all, involved in the biosynthesis of isoleucine and valine are associated with the mitochondria or particles that have the same density as mitochondria. In general, mitochondria have in the past been implicated primarily as the seats of the Krebs cycle, the electron transport apparatus, and the synthesis and breakdown of fatty acids. However, it is not unreasonable that they should also be involved in the biosynthesis of amino acids, particularly those derived from aspartate and glutamate, since these two amino acids are themselves generated by the Krebs cycle. Indeed, more than twenty years ago Reichard (29) presented evidence to show that the synthesis of ureidosuccinate, a precursor of arginine, from aspartate, CO_2, and ammonia occurred in the mitochondria of rat liver. This may be taken as the first evidence for the role of mitochondria in general amino acid synthesis.

If the mitochondria are indeed centers of synthesis for at least certain amino acids, then we should expect the enzymes involved to be organized in some fashion for maximum efficiency. The "stripping" of mitochondria by divers techniques such as by the use of phospholipase (2) enables one to separate groups of enzymes with related functions together into the same cell fractions, thus indicating a considerable degree of compartmentalization, presumably by attachment to ground substances such as structural protein and membranes within the mitochondrion. In view of such compartmentalization, it is hard to believe that there is no reason for enzymes of related function to be together. Unfortunately, very little analysis has been made of the kinetic advantages of compartmentalization and organization versus free floating enzymes. Such as has been done has, however, raised some interesting points indicating that the matter should be more vigorously pursued despite the inherent difficulties (8, 12, 25). The suggestion that most enzymatic reactions are vectorial rather than scalar has been considered seriously by a number of workers (20).

The two most important observations which point toward the biosynthetic mechanism for the synthesis of isoleucine and valine as being an organized entity are (1) the stabilizing effect of sucrose on the overall activity and (2) the inability of certain mutant homogenate mixtures to "complement," or at best to complement at a relatively low level of activity. The ability of pyruvate, one of the initial substrates, to replace sucrose (the stabilizing effect of which does not itself appear to be understood) also may be taken to indicate an ordered entity, but again for unknown reasons. The general conclusion seems in order that the system hangs together so long as it is active—that is, has substrate. This is true whether the system is attached to the particulate fraction, or has been freed of it by sonication or the use of detergents. This observation may be of considerable importance in the study of biosynthetic systems in general in cells.

Whether all the enzymes required for synthesis of isoleucine and valine are attached to mitochondria and/or other particles *in vivo,* or whether some are also free floating as organized entities in the extra-particulate cytoplasm, is presently a moot question. However, the finding that the preparation of *Neurospora* homogenate in the presence of bovine serum albumin results in the presence of a greater amount of three of the four valine-synthesizing enzymes in the pellet fraction than in the supernatant (as determined by their individual total activity) points toward one site for these to be in the particulate fraction, probably primarily or totally the mitochondria. In addition, as was already pointed out, mitochondrial preparations made in the presence of serum albumin plus sucrose act quite differently from those made in the presence of sucrose only, since the former produce valine in the absence of exogenous pyruvate, whereas the latter either do not produce it or produce it at a very low level. In addition the BSA-treated mitochondria produce other amino acids in the absence of exogenous substrate.

The reason for this result may be that the BSA-treated mitochondria are relatively intact (since they carry out oxidative phosphorylation) and as a consequence synthesize amino acids, as they possibly do *in vivo,* by the utilization of their structural substances. It has been shown that the mitochondria of spermatozoa, when respiring in the absence of substrate, have a reduction in the cristae of the mitochondrial midpiece and a parallel decrease in endogenous phospholipids (1). We suggest, therefore, that under *in vivo* conditions amino acids are synthesized in mitochondria from endogenous substrates present in the substance of the mitochondria. Under these conditions a flux equilibrium exists in which the endogenous material is replaced from outside as it is used. Mitochondria prepared in the absence of BSA do not utilize endogenous substances in the same way and are "open" to exogenous substrate utilization at least in part—in part, because they can utilize pyruvate for valine, but not for leucine synthesis.

The behavior of BSA-treated mitochondria resembles in certain respects the behavior of the membrane fraction, MF, since it also produces amino acids in the absence of endogenous substrate. The origin of these in incubating MF remains unknown, but it is quite possible that they derive in part from the lipid content of the fraction. The nature of the nitrogen source also remains to be considered. We suggest, tentatively, that the reason MF produces amino acids in large amounts compared to the supernatant from which it is derived is that it is released from feedback inhibition. As it forms on the Sephadex gel the smaller molecules such as the amino acids are removed, permitting the enzyme systems to become active in synthesis. This removal role is played by protein synthesis in the intact cell.

Acknowledgments

The work reported here has been supported in part by grants from the National Institutes of Health (GM 12323) and The Robert A. Welch Foundation. C. H. Clarke and R. N. Le Page were recipients of Rosalie B. Hite postdoctoral fellowships.

References

1. Afzelius, B. A., and Mohri, H., *Exptl. Cell Research,* **42,** 10 (1966).
2. Allmann, D., Bachmann, E., and Green, D., in press.
3. Bachmann, E., Allmann, D., and Green, D., in press.
4. Chance, B., and Williams, G. R., *Advan. Enzymol.,* **17,** 65 (1956).
5. Cronenwett, C. S., and Wagner, R. P., *Proc. Natl. Acad. Sci. U.S.,* **54,** 1643 (1965).
6. Folch, J., Lees, M., and Sloane-Stanley, G. H., *J. Biol. Chem.,* **226,** 497 (1957).
7. Green, D., Bachmann, E., Allmann, D., and Perdue, J., in press.
8. Gutfreund, H., and Jones, E. A., *Biochem. J.,* **90,** 208 (1964).
9. Hall, D. O., and Greenawalt, J. W., *Biochem. Biophys. Res. Commun.,* **17,** 565 (1964).
10. Henderson, L. M., and Snell, E. E., *J. Biol. Chem.,* **172,** 15 (1948).
11. Johnson, M. J., *J. Biol. Chem.,* **181,** 707 (1949).
12. Jones, E. A., and Gutfreund, H., *Biochem. J.,* **91,** 1c (1964).
13. Kinsey, J. A., Dissertation, University of Texas, 1965.
14. Kinsey, J. A., and Wagner, R. P., *Proc. Natl. Acad. Sci. U.S.,* **55,** 404 (1966).
15. Kiritani, K., Narise, S., and Wagner, R. P., *J. Biol. Chem.,* **241,** 2042 (1966).
16. Kiritani, K., Narise, S., and Wagner, R. P., *J. Biol. Chem.,* **241,** 2047 (1966).
17. Kiritani, K., Narise, S., Bergquist, A., and Wagner, R. P., *Biochim. Biophys. Acta,* **100,** 432 (1965).
18. Klingenberg, M., and Schollmeyer, P., *Biochem. Z.,* **333,** 335 (1960).
19. Klingenberg, M., and Schollmeyer, P., *Biochem. Z.,* **335,** 231 (1961).
20. Lehninger, A. L., *Naturwissenschaften,* **53,** 57 (1966).
21. Lowry, O. H., Rosebrough, N. J., Farr, A. L., and Randall, R. J., *J. Biol. Chem.,* **193,** 265 (1951).
22. Monné, L., *Advan. Enzymol.,* **8,** 1 (1948).
23. Munkres, K. D., and Richards, F. M., *Arch. Biochem. Biophys.,* **109,** 466 (1965).
24. Munkres, K. D., and Woodward, D. O., *Proc. Natl. Acad. Sci. U.S.,* **55,** 1217 (1966).
25. Pollard, E., *J. Theoret. Biol.,* **4,** 98 (1963).
26. Radhakrishnan, A. N., and Snell, E. E., *J. Biol. Chem.,* **235,** 2316 (1960).
27. Razin, S., Morowitz, H. J., and Terry, T. M., *Proc. Natl. Acad. Sci. U.S.,* **54,** 219 (1965).
28. Reed, L. J., and Cox, D. J., *Ann. Rev. Biochem.,* **35,** 57 (1966).
29. Reichard, P., *Acta Chem. Scand.,* **8,** 795 (1954).
30. Spies, J., "Methods in Enzymology," Vol. III, p. 467, Academic Press, 1957.

31. Takeuchi, M., and Terayama, H., *Exptl. Cell Res.,* **40,** 32 (1965).
32. Wagner, R. P., *Proc. Natl. Acad. Sci. U.S.,* **35,** 185 (1949).
33. Wagner, R. P., and Bergquist, A., *Genetics,* **45,** 1375 (1960).
34. Wagner, R. P., and Bergquist, A., *Proc. Natl. Acad. Sci. U.S.,* **49,** 892 (1963).
35. Wagner, R. P., Bergquist, A., and Barbee, T., *Biochim. Biophys. Acta,* **100,** 444 (1965).
36. Wagner, R. P., Bergquist, A., Barbee, T., and Kiritani, K., *Genetics,* **49,** 865 (1964).
37. Wagner, R. P., and Guirard, B. M., *Proc. Natl. Acad. Sci. U.S.,* **34,** 398 (1948).
38. Wagner, R. P., and Haddox, C. H., *Am. Naturalist,* **85,** 319 (1951).
39. Wagner, R. P., Kiritani, K., and Bergquist, A., "Molecular Basis of Neoplasia," p. 346, University of Texas Press, 1962.
40. Woodward, D. O., and Munkres, K. D., *Proc. Natl. Acad. Sci. U.S.,* **55,** 872 (1966).

On the Role of Glucose-6-phosphate Dehydrogenase in the Morphology of *Neurospora*

STUART BRODY AND E. L. TATUM

The Rockefeller University,
New York, New York

One of the facets of cellular organization of mutual interest to both biochemists and developmental biologists is the regulation of a particular area of metabolism and its relation to cellular growth and differentiation. More precisely, it is information about the spatial and temporal organization of metabolism which is of current interest.

In order to obtain this type of information about multicellular organisms, it would probably be interesting to study any organism which undergoes cytologically visible changes in shape and structure during its development, and which possesses an organized nuclear structure containing distinct chromosomes. It would probably be more fruitful, however, to be able to apply the techniques of microbial genetics and biochemistry toward the understanding of this type of organism and to the description of its particular developmental changes. *Neurospora crassa* is an organism which will not only satisfy all these operational requirements but, in addition, is capable of mutation to many bizarre morphologically distinct forms in both its sexual and asexual cycles.

One particular type of morphological mutant, known as a colonial mutant, can be described as a strain which grows as a dense, compact colony rather than the spreading filamentous form typical of the wild-type *Neurospora.* Figure 1 indicates these characteristics. This change in the external morphology of the culture appears to be due to a change in the growth pattern—i.e., an increase in the amount of branching. This general phenotype is shown by forty to fifty mutants, all genetically distinct (3). In addition, the altered phenotype is associated with changes in the amount of the various carbohydrate polymers in the cell wall (4). However, it is not known whether these cell wall changes are responsible for the different morphology or are just another symptom of an altered metabolism. In any event, investigation of carbohydrate

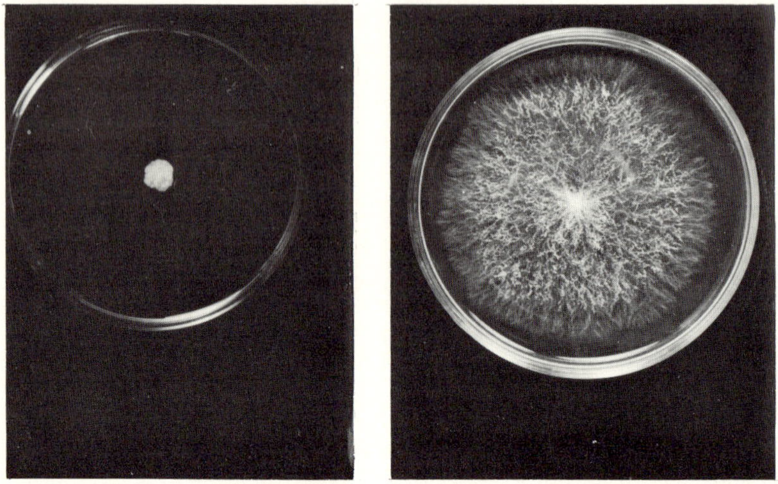

FIG. 1. Growth on minimal agar after 48 hours at 30°C. The wild-type strain is on the right, *col-2* strain on the left.

metabolism in colonial mutants was initiated to gain more knowledge about the genetic control of cell wall structure, growth patterns, and external morphology.

Little is known about the synthesis of the *Neurospora* cell wall or about the underlying organization of the cytoplasm and cell wall which allows extension of the mycelia mainly at the tip. Since some of the morphological mutants appeared to be altered in their ability to achieve rapid linear extension of the cell wall, it was thought that these mutations might affect those enzymes involved in cell wall synthesis and assembly. The relationships between these enzymes, the cell wall, and the genetic material are outlined in a simplified and schematic way in Fig. 2. The genetic material could be directly responsible for the structure and regulation of the three classes of enzymes and possibly indirectly for the structure of existing templates. Of the three classes of enzymes, the class of enzymes described as assembly enzymes are not yet known in *Neurospora,* and, in fact, this study was undertaken to detect altered assembly enzymes. However, this paper obviously describes a different type of enzymatic change.

Extracts of one particular mutant, *col-2,* were found to contain eight times as much glucose-6-P as did extracts of the wild-type strain. The glucose-6-P content was estimated by chromatographing ethanol extracts, eluting the glucose-6-P, and quantitatively assaying with highly purified glucose-6-P dehydrogenase. This increase in the steady-state level was characteristic of this particular colonial mutant, since the high glucose-6-P content segregated with the *col-2* marker in all cases tested and since many other different colonial mutants did not have an elevated glucose-6-P level. Therefore, it would appear that that state of the *col-2* locus directly or indirectly determines the level of glucose-6-P. Unfortunately, this clue did not indicate whether the glucose-6-P accumulation was the cause of the altered morphology (or closely related to the primary cause), or whether it was just part of the description of the effects of some distantly related cause.

After extensive analysis, it was found that the glucose-6-P accumulation in the *col-2* mutant was apparently due to a change in the structure of the glucose-6-P dehydrogenase. The evidence for this is as follows: Highly purified preparations of this enzyme from the colonial strain differed in substrate affinities, heat lability, and general stability from similar preparations obtained from the wild-type strain (2). Since stringent proof in the form of peptide pattern differences has not yet been obtained, it was important to purify the glucose-6-P dehydrogenase preparations until they were free from measurable activities of those enzymes which might have interfered with this type of measurement. Also, numerous experiments were performed which indicated that the assay conditions and the state of the purified enzymes were not responsible for the observed differences (1). There are seven differences between the two enzyme preparations; as an example of these differences, Fig. 3 indicates the effect of temperature on the glucose-6-P K_m of the enzyme

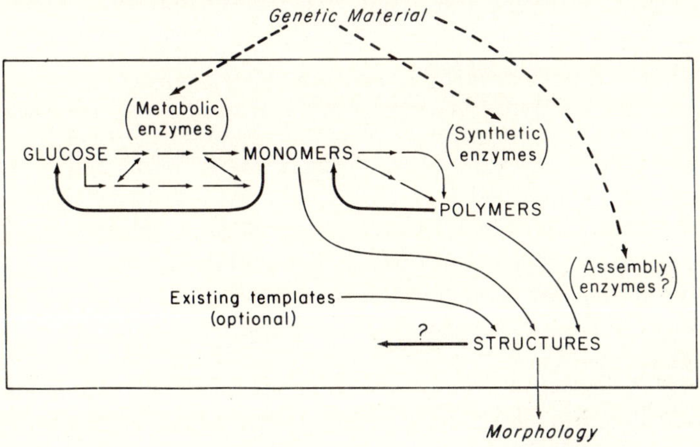

FIG. 2. Schematic representation of the possible relationships between the genetic material and morphology. Heavy unbroken lines denote probable modes of feedback regulation.

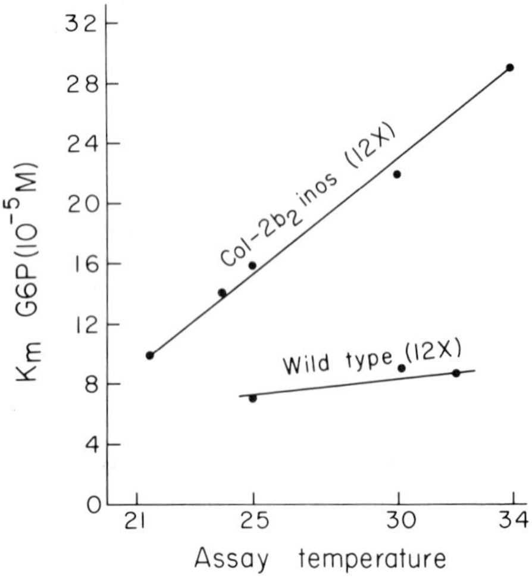

FIG. 3. Effect of temperature on glucose-6-P affinity. The numbers in parentheses (12×) indicate twelve times the TPN K_m concentration.

from these two strains. It is clear that the colonial enzyme differs from the wild-type enzyme at all temperatures and, in addition, is strongly affected by temperature, as might be expected of a mutationally altered enzyme. It is also interesting that this change in K_m with temperature can be nicely correlated with an increased level of glucose-6-P in the *col-2* strain at higher temperatures (2). In other words, it would appear that, under conditions where the affinity for the substrate is lowered, more substrate is actually found in the cell. Both of these parameters showed little variation in the wild type.

The altered glucose-6-P dehydrogenase in the *col-2* strain, which probably differs in its primary structure from the wild-type enzyme, segregates with the *col-2* marker through many crosses and is found in all *col-2* segregants tested. Wild-type-like revertants, selected on the basis of morphology, had the wild-type enzyme, illustrating that the alteration of the enzyme is not irreversible. Other mutants with a colonial phenotype, which are genetically distinct, do not show any alterations in the glucose-6-P dehydrogenase, indicating that colonial growth alone does not cause these changes in this enzyme. These facts and arguments, published in detail elsewhere (2), indicate that the *col-2* locus is the structural gene for glucose-6-P dehydrogenase and that a subtle change in the structure of this enzyme leads to the accumulation of glucose-6-P in the *col-2* strain, and eventually to the altered morphology.

It is one thing to indicate the primary biochemical cause of the altered morphology and another thing to describe the ensuing pleiotropic effects that lead to the altered growth pattern. Among the many possible secondary effects of this mutation are a reduction in the overall rate of the pentose phosphate shunt, leading to a possible lowered TPNH steady-state level. Employing specifically labeled C^{14}-glucose, it was found that, in the *col-2* strain, the ratio of $C^{14}O_2$ produced from C-1 labeled versus C-6 labeled glucose was one-half that of the ratio for the wild-type strain, as shown in Table I. Although numerous supporting measurements are

TABLE I

ESTIMATES OF *in Vivo* METABOLIC PATHWAYS[a]

Strain	Description	Recovery of counts as $C^{14}O_2$ from C^{14}-glucose (mmoles CO_2/mole glucose used)		
		UL	C-1	C-6
RL 3-8A	Wild type	690 (100%)	218 (24%)	55 (8.%)
col-2b₂, inos	Colonial	245 (100%)	100 (18%)	55 (22%)

[a] A tiny mycelial inoculum was added to 20 ml of minimal medium containing 0.5% glucose and 1 μC of the C^{14}-glucose listed above. The flasks were incubated in a 30°C water bath, and air was vigorously bubbled through them for 16 to 24 hours. The exhaust was bubbled through 100 ml of 1 *M* NaOH, and the CO_2 that was trapped was precipitated as $BaCO_3$. The $BaCO_3$ was filtered and washed, and measured amounts of a slurry were plated onto planchets. Radioactivity was determined with a windowless Geiger counter. All values have been corrected for self-adsorption and are the numerical averages from six to eight separate determinations on log-phase cultures.

needed for a better estimate of the *in vivo* rate of the shunt, it appears that a significant reduction had taken place. This reduction would imply that, in *Neurospora,* one of the rate-determining factors in the shunt is the affinity of this enzyme for glucose-6-P.

It is interesting to point out that the level of this enzyme, as judged by activity measurements, does not seem to play an important role in the regulation of this pathway. Crude extracts of both wild-type and *col-2* strains appeared to have the same amount of this enzyme. Secondly, there is enough glucose-6-P dehydrogenase present (0.1 to 0.2% of the total protein) to sustain a theoretical maximum reaction rate approximately 100 times as great as the estimated rate of the *in vivo* pentose phosphate shunt. This indicates that factors other than the amount of the enzyme are probably responsible for the regulation of this pathway.

The role of other possible controlling factors such as the steady-state level of TPN, as well as feedback inhibitors, has not yet been conclusively established. Growth of both wild-type and *col-2* strains on ribose or xylose as a carbon source did not appear to affect either strain. In other words, the reduction in the activity of the initial steps of the shunt due to the *col-2* mutation cannot be mimicked or reversed by the increased availability of some of the end products of this pathway. Still remaining is the possibility that the steady-state level of TPNH has been reduced, producing many pleiotropic effects. This is currently under investigation.

The increased glucose-6-P level in the *col-2* strain is not matched by a proportional increase in the level of fructose-6-P, as shown in Table II.

TABLE II
GLUCOSE-6-P AND FRUCTOSE-6-P CONTENT[a]

Strain	G-6-P (μmoles/gm dry weight)	F-6-P (μmoles/gm dry weight)	G-6-P/F-6-P
Wild type	0.3–0.5	0.10–0.15	3.2
col-2b$_2$	3.0–4.0	0.50–0.75	5.6

[a] Glucose-6-P was enzymatically assayed by employing highly purified glucose-6-P dehydrogenase (some details given in text). After the glucose-6-P was completely converted to 6-phosphogluconic acid, phosphohexose isomerase (Boehringer preparation) was added to the assay system for the determination of fructose-6-P.

These two intermediates are supposedly kept in equilibrium by hexosephosphate isomerase, and, in fact, many studies of other organisms have shown that the levels of the two compounds do rise and fall together. Since both strains have the same amount of hexosephosphate isomerase activity, it is possible that the lack of equilibration is due to some type of strong inhibition of this enzyme. The discrepancy does not appear to be due to any change in the isomerase itself, since purified preparations of this enzyme from both strains showed identical $K_m(s)$ for glucose-6-P, were identical with respect to their fructose-6-P $K_m(s)$, and at equilibrium, the ratio of glucose-6-P to fructose-6-P was 3.3–1.

One intriguing possibility is that there are two separate pools of glucose-6-P, but only one in equilibrium with fructose-6-P. The pool not in equilibrium is either formed anew or greatly increased in the *col-2* strain and is not drained off by the action of the isomerase and the Embden-Meyerhof pathway. If the isomerase were not present everywhere in the mycelial mass, or were inhibited in certain regions, this could provide a possible explanation for the lack of equilibrium.

Although localization or restriction of certain enzymes to particular regions may seem strange in an organism like *Neurospora,* it has already been histologically demonstrated that certain enzymes are found mainly in the growing tip regions, and certain enzymes primarily in other regions (5).

The question of possible localization of enzymes and substrates adds another dimension (literally and figuratively) to the interpretation of these results and to subsequent findings. It is conceivable that certain enzymatic reactions will be rate-limiting only at the tip of the growing mycelia and therefore will have a great effect on the balance between growth at the tip, and branching in back of the tip. It is this delicate balance which determines the growth pattern of *Neurospora* and thereby its external morphology. Therefore, the accumulation of glucose-6-P may have to be viewed as a local accumulation which either inhibits or activates certain enzymes (like glycogen synthetase or glucan synthetase) at or near the site of active cellular extensions. Similarly, the possible reduction in the TPNH level may also be only of importance at certain regions.

It should be pointed out that, in spite of all these complications, at least it is possible to describe the primary effect of this particular mutation. One should also note that the description so far has involved levels of substrates and properties of enzymes, and not gene activation or inactivation. This may not seem quite as exciting or glamorous as those analyses which propose control of differentiation solely at the transcription and translation level. However, focusing only at the RNA level may be somewhat one-sided, since a great deal of the description and mechanism of development may be more involved with certain crucial rate-limiting enzymatic reactions and their regulation by metabolic conditions. In this regard, it is possible that the morphological changes in some of the other colonial mutants are due to slight defects in key reactions, rather than to a whole operon being turned on or off.

In conclusion, this biological phenomenon of tip and lateral growth (in many ways similar to apical dominance) probably involves a complex balance between the levels of many intermediates and coenzymes, synthetic and degradative enzymes, and the existing cell wall matrix, as well as the localization of all or some of these factors. As to the underlying organization of all these factors, at present we have a few hard facts, many soft facts, and lots of interesting possibilities to investigate.

References

1. Brody, S., unpublished data.
2. Brody, S., and Tatum, E. L., *Proc. Natl. Acad. Sci. U.S.*, **56**, 1290 (1966).
3. Garnjobst, L., and Tatum, E. L., manuscript in preparation.
4. Mahadevan, P. R., and Tatum, E. L., *J. Bacteriol.*, **90**, 1073 (1965).
5. Zalokar, M., *Am. J. Botany*, **46**, 602 (1959).

Channeling in *Neurospora* Metabolism

ROWLAND H. DAVIS

Department of Botany,
University of Michigan,
Ann Arbor, Michigan

In metabolism, the synthesis of diverse end products from few starting materials inevitably entails branch points in the flow of intermediates. Aside from many accessible intermediates, often provided by quickly equilibrating sets of reactions such as the Krebs cycle, there are other intermediates, involved in only a few pathways, for which there is a potential for severe competition between two or more enzymes. The study of branch points in bacterial metabolism has revealed mechanisms which dampen competition largely by variations in enzyme production or activity (48, 34). There is another way of handling the problem of competition between enzymes which may be more highly developed in eucaryotic organisms than in bacteria. This mechanism depends on the spatial segregation of a common intermediate into multiple pools, each of which is specific for a given pathway. Localization of intermediate pools in cells, when seen in the context of metabolite flow, is referred to as *channeling* (44, 48, 33). The purpose of this paper is to analyze channeling of intermediates in relation to enzyme competition as it has been inferred to take place in the metabolism of the eucaryote *Neurospora*. I hope to revive thereby certain considerations of cell organization which have been neglected in much of the previous work in biochemical genetics. In all cases, discussion will center around enzymes normally considered "soluble." The topographic and kinetic features of a very simple model of channeling, the "surface" model, will be presented, followed by data on carbamyl phosphate metabolism which suggests the model. Alternative channeling mechanisms will then be considered in light of ornithine synthesis and utilization. In neither case have the mechanisms suggested been definitively tested, but quite specific questions may be posed. An excellent review of enzyme organization, covering many points considered below, has been published recently (39).

The Surface Model

The first requirement of the "surface" model, by which two pools of a common intermediate are maintained and channeled to two pathways, is the existence of two genetically distinct enzymes catalyzing the synthesis of this intermediate, each specific for one of the pathways. The second requirement is that each path-specific enzyme of synthesis of the common intermediate also catalyze the formation of the following compound, unique to the pathway in question, or be physically aggregated to the protein which does so. The effect of such relationships is to allow (but not necessarily enforce) segregation of the intermediate into two pools bound to the surface of two multifunctional units. It should be noted that the synthesis of the common intermediate might be catalyzed by the product of the same gene participating in two path-specific aggregates. This may be a common relationship, and difficult to discriminate from the case in which there is one pool of the common intermediate. The increasing number of multiple enzymes for a common biochemical step (42), however, together with their involvement with bifunctional ("polycephalic") proteins (6) or enzyme aggregates (20), suggests that the first alternative is quite general.

The important kinetic features of the surface model are quite simple. For even partial pool segregation to occur, the common intermediate must be physically bound. The simplest means of accomplishing this is by binding the intermediate to the catalytic site of the enzymes utilizing it. In a two-step sequence taking place on the surface of a single enzyme or aggregate, efficient binding of the product of the first reaction will result from (*a*) a high affinity of the (contiguous) second reaction site for this intermediate, and (*b*) incomplete saturation of the second active site with the intermediate, as will be the case if the first enzyme is rate-limiting to the two-step sequence. If the intermediate is actually released from the first site, the K_m's and the V_{max}'s of the sites, and the resultant local substrate concentration will determine the extent to which the intermediate escapes the vicinity of the protein and is thereby "unchanneled." If the intermediate remains truly (covalently) bound to the enzyme protein, the distance between the two active sites in question should not significantly exceed the dimensions of the metabolite produced by the first site without a compensating flexibility of the protein itself. The question of the "distance" between competing enzymes, however, cannot be considered in the general sense, but only as a formal parameter. The surface model is in essence based on the behavior of "enzyme-bound" intermediates often postulated in complex reactions (39). A classic case of such considerations is that of indole as an "enzyme-bound" intermediate of the reaction indole glycerol phosphate to tryptophan, catalyzed by tryptophan synthetase (51, 19).

A consequence of the surface model is that the cellular concentration of intermediates efficiently localized in this way be very low—that is, similar to or less than the concentration of active sites which bind it in succeeding reactions. This allows us, in a later example, to judge the model inadequate where a significant pool of an apparently channeled intermediate is found. A second consequence, related to the first, is that factors which elevate intermediate concentrations to saturating levels, or mutations which alter the velocities or affinities of enzymes, will in many cases lead to "overflow" to alternate pathways. This is not a natural expectation of a "compartment" model of channeling, to be discussed in a later section.

Genetics of Carbamyl Phosphate Metabolism in *Neurospora*

The model given above has been developed from genetic, enzymological, and nutritional studies of *Neurospora* mutants affecting carbamyl phosphate (CAP) metabolism. CAP, an extremely labile compound, is required in both arginine and pyrimidine synthesis: In arginine synthesis, it is used in the ornithine transcarbamylase (OTCase) reaction to form citrulline; in pyrimidine synthesis, it is used in the aspartate transcarbamylase (ATCase) reaction to form ureidosuccinic acid (US). While *Escherichia coli* has one enzyme catalyzing the formation of CAP (38), it is clear now that fungi such as *Neurospora* (12), yeast (30), and *Coprinus* (4), and possibly many higher organisms, have an enzyme more or less specific for each of the two pathways. This possibility was first recognized in studies of arginine–pyrimidine gene interaction in *Neurospora* (11); the mechanisms governing the distribution of the carbamyl group to the two pathways, however, are still not entirely clear.

The nutritional and enzymatic phenotypes of mutants affecting CAP metabolism are shown in Table I. In regard to the two transcarbamylases, the data are quite simple. The *arg-12* locus is the structural gene for OTCase (16). This locus was for many years undefined by auxotrophic mutants. The *arg-12*^{s} allele, discovered as a suppressor of certain pyrimidine and proline mutants (35), reduces OTCase to 5% normal activity without imposing a detectable growth requirement. The OTCase of the *arg-12*^{s} strain has a five- to ten-fold higher K_m for ornithine, and a lower K_m for CAP (10). From this strain, it is possible to select OTCase-less mutants with ease, by further mutation of the *arg-12* gene (16, 50). An OTCase-less mutant has also been recovered directly after irradiation of wild type (16).

The *pyr-3d* mutation is analogous to *arg-12* in that it leads to an

TABLE I
CAP SYNTHESIS AND TRANSCARBAMYLASES IN EXTRACTS OF *Neurospora* STRAINS[a]

Strain	*CAP Synthesis*[b]		*Transcarbamylases*		Requirement
	Glutamine-dependent	Ammonium-dependent	OTCase	ATCase	
Wild type	31	12	+	+	None
arg-2	8	6	+	+	Arginine
arg-3	4	0	+	+	Arginine
pyr-3a	19	11	+	+	Uridine
arg-2, pyr-3a	0	11	+	+	Arginine + uridine
arg-3, pyr-3a	0	0	+	+	Arginine + uridine
pyr-3d	(+)[c]	(+)	+	−	Uridine
arg-12	(+)	(+)	−	+	Arginine
arg-12[s]	(+)	(+)	±	+	None

[a] For CAP synthesis, extracts of acetone powders made with 0.05 *M* K^+ phosphate buffer, pH 7.0, contained 10 to 15 mg of protein per milliliter, and were passed once through a short column of Sephadex G-25 to remove small-molecular-weight materials.

[b] Reaction mixtures (0.5 ml) contained ATP (12 m*M*); $MgCl_2$ (12 m*M*); L-ornithine-HCl (6 m*M*) K^+ phosphate, pH 7.8 (85 m*M*) $KHCO_3$-C^{14} (30 m*M*); partially purified OTCase (10) from *Neurospora* (18 units); and either L-glutamine (6 m*M*) or NH_4Cl (6 m*M*) for the two types of activity. The temperature was 25°C. The concentration of glutamine used is saturating; the concentration of NH_4Cl used is about 1/12 K_m. Assays were done otherwise by the method of Davis (13). All assays were performed within 30 minutes after the removal of extracts from Sephadex columns. Activities are expressed as millimicromoles of radioactive bicarbonate rendered acid-stable per milligram of protein per hour, assuming a specific radioactivity of bicarbonate added to be 42,000 counts per minute per micromole.

[c] Both substrates supported CAP synthesis in *pyr-3d, arg-12,* and *arg-12*[s] extracts prepared at a different time and in a different manner than those for which figures are given, so the assays are not quantitatively comparable.

elimination of ATCase activity (9). Alleles leading to a partial deficiency are also known (41). These mutations are therefore presumed to lie in the structural gene for ATCase. The allelism of *pyr-3d* with *pyr-3* mutants of other types will be discussed below.

The genetic determination of CAP synthesis is extremely complex. It has been reported previously that the arginine-requiring mutant, *arg-3,* lacks completely an ammonium-dependent carbamyl phosphokinase activity (12, 14). Further work showed that a glutamine-dependent reaction (32) could be detected in wild type. This activity is also missing in *arg-3,* and, further, it is missing in *arg-2,* another arginine mutant unlinked to *arg-3* (15). Recent preliminary work, however, done with more concentrated, fresher extracts, shows that a very unstable, glutamine-dependent activity is present in *arg-2* and *arg-3* mutants. This activity is eliminated if the *pyr-3a* mutation is also present in the genome (Table I). We may, by analysis of these data, derive certain conclusions, all consistent with previous work, and similar in most respects to the same system in yeast and *Coprinus*.

First, the glutamine-dependent activity is referable to two enzymes. One, requiring the participation of both *arg-2* and *arg-3* genes, is specific for the arginine pathway. The other, determined by the *pyr-3* locus, is specific for pyrimidine synthesis. The former, while unstable, is not nearly as labile as the latter to heat and to elevation of pH. Only when double mutants (*arg-2, pyr-3a* and *arg-3, pyr-3a*) are tested does one

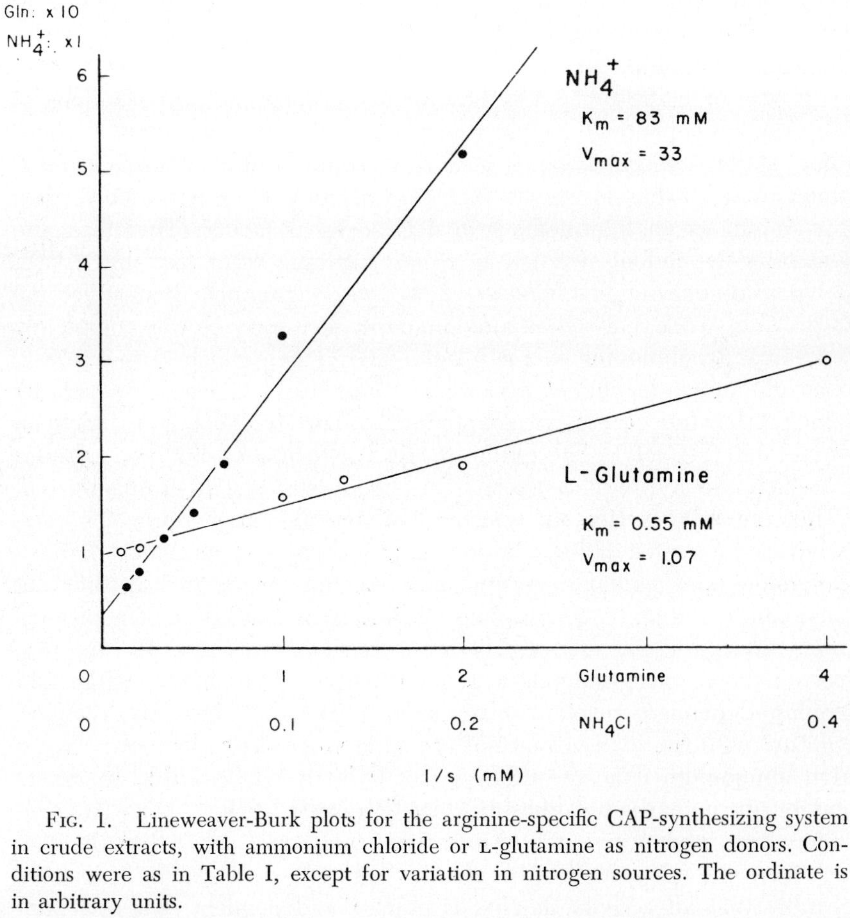

FIG. 1. Lineweaver-Burk plots for the arginine-specific CAP-synthesizing system in crude extracts, with ammonium chloride or L-glutamine as nitrogen donors. Conditions were as in Table I, except for variation in nitrogen sources. The ordinate is in arbitrary units.

note a complete deficiency of glutamine-dependent CAP synthesis (Table I). While more needs to be done with the pyrimidine-specific system to extend the enzymological conclusions adequately, a large amount of indirect evidence accumulated previously leaves no doubt of the existence of pyrimidine-specific CAP synthesis (14).

Second, the ammonium-dependent reaction appears to be a function solely of the *arg-3* gene, since mutations at this locus unconditionally eliminate it. Despite the presence of this activity in *arg-2* strains, they require arginine. This shows that the glutamine-dependent activity is probably the biological one. Some indication of why this might be so is provided by affinity measurements for ammonium and glutamine as substrates of the arginine-dependent system (Fig. 1). While the maximal velocity of the ammonium-dependent reaction is very much greater (especially at the pH used here) than that of the glutamine-dependent reaction, the system has an affinity for glutamine over 100-fold greater than for ammonium.

It was of interest to know how many proteins the two reactions of the arginine-dependent system represented. The observation that glutamine, at low concentration, is a more efficient substrate than equimolar ammonium (Table I) shows that this reaction does not involve free ammonium as an intermediate in a two-step sequence. Therefore, the glutamine-dependent reaction is almost certainly catalyzed by a two-polypeptide enzyme involving *arg-2*$^+$ and *arg-3*$^+$ products. It troubled me in previous work that the ammonium-dependent enzyme was completely indifferent to glutamine as a substrate (13). This problem was resolved, once glutamine-dependent activity in crude extracts was detected, by taking arbitrary, successive ammonium sulfate fractions from such an extract (Fig. 2, left). The crude extract was prepared under conditions in which the pyrimidine-specific CAP-producing system is inactivated. Using the assay system in which glutamine and ammonium are compared as substrates at low, equimolar concentrations, one finds that the glutamine- and ammonium-dependent systems precipitate at different ammonium sulfate concentrations. Mixed fractions show no unusual interactions, and the recovery of both activities is 60 to 80%. (In some fractionations, the glutamine-dependent activity is higher than the ammonium-dependent reaction in the same fractions.) These observations, together with the high velocity of the ammonium-dependent reactions at high ammonium concentration, explain why it was possible to detect and purify a "carbamyl phosphokinase" from wild type which was indifferent to glutamine (13). More important, they show that the two activities are not referable to identical protein moieties. It is entirely possible that all activity *in vivo* is glutamine-dependent, and that am-

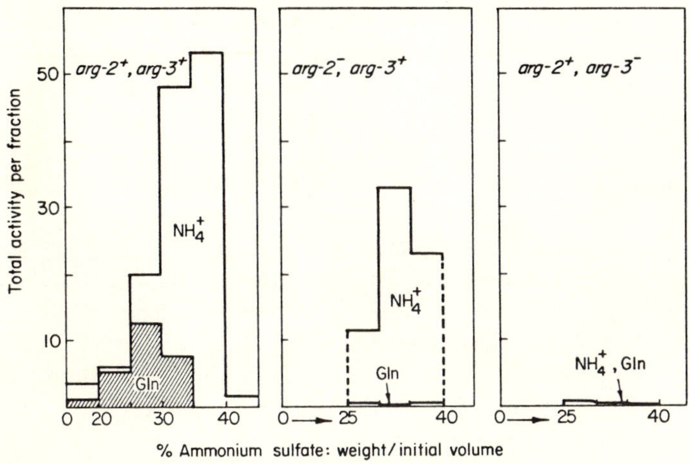

FIG. 2. Fractionation of crude extracts of wild type (left), *arg-2* (middle), and *arg-3* (right) with ammonium sulfate. Fractions, after centrifugation, were redissolved in a small volume and passed twice through a Sephadex G-25 column to remove small-molecular-weight materials. Assays were done by the method of Davis (13) as modified in footnote *b* to Table I, to determine ammonium-dependent (NH_4^+) and glutamine-dependent (Gln) activities.

monium-dependent activity appears only on extraction and dissociation of the polypeptides.

That the system being described here is arginine-specific is very nicely shown by similar fractionation of *arg-2* and *arg-3* extracts (Fig. 2). It supports the previous indications (Table I) that the *arg-2* mutation eliminates the glutamine-dependent reaction, while the *arg-3* mutation eliminates both activities of the arginine-specific system. In this fashion, we see again that the glutamine-dependent reaction for arginine is probably catalyzed by a heteromultimer constituted of *arg-2*$^+$ and *arg-3*$^+$ products. (Complementation for this reaction is seen between *arg-2* and *arg-3* mutations both *in vivo* and in mixed crude extracts.) The *arg-3*$^+$ polypeptide, moreover, when separated from the *arg-2*$^+$ polypeptide, is clearly capable of ammonium-dependent activity. To what extent the heteromultimer is capable of ammonium-dependent activity cannot be judged with the available data. The pyrimidine-specific system appears incapable of using ammonium under the conditions used (Table I, note *arg-3*).

The *pyr-3* locus requires further comment. The locus has three main types of alleles. The first, represented by *pyr-3d*, is deficient in ATCase activity. The second, represented by *pyr-3a*, is deficient in the pyrimidine-specific enzyme of CAP synthesis. In addition, an extensive series of mutants appear to lack both of these functions as the result of a single mutation (17). From some of these it is possible to obtain, by "partial reversion," derivatives lacking only one function (49, 21). The singly-deficient types of *pyr-3* mutant, whether obtained directly as mutants from wild, or as partial revertants of the doubly-deficient class, complement well despite the fact that the genetic map of the locus shows no clear grouping of alleles (17). The doubly-deficient class are actually defined best by their noncomplementing behavior. Because alleles of the *pyr-3a* type have kinetically altered ATCase activity (26), and because they have a unique genetic and mutational relationship with strains lacking this enzyme, it has been concluded that the "locus" specifies one or more polypeptides which constitute a single, bifunctional enzyme protein (17). With the discovery of CAP_{pyr} synthesis *in vitro,* this conclusion is now open to direct test.[1]

The Mechanism of CAP Channeling

We may now consider the channeling of CAP. Figure 3 summarizes the metabolism of CAP and the genetic determination of the enzymes

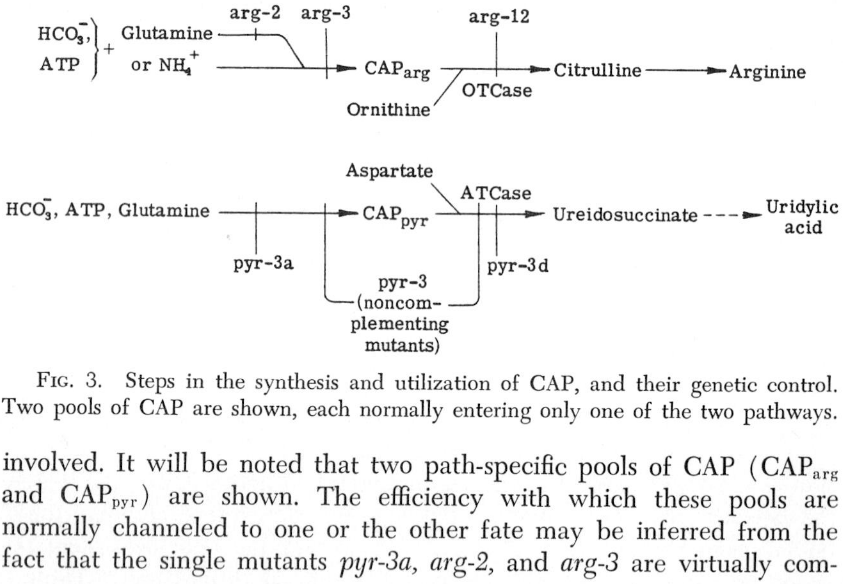

FIG. 3. Steps in the synthesis and utilization of CAP, and their genetic control. Two pools of CAP are shown, each normally entering only one of the two pathways.

involved. It will be noted that two path-specific pools of CAP (CAP_{arg} and CAP_{pyr}) are shown. The efficiency with which these pools are normally channeled to one or the other fate may be inferred from the fact that the single mutants *pyr-3a, arg-2,* and *arg-3* are virtually com-

[1] After the manuscript of this paper was submitted, I became aware that Reissig had simultaneously discovered the pyrimidine-specific mode of CAP synthesis in extracts of *Neurospora.* He reports this in a recent paper (J. L. Reissig, A. S. Issaly, and I. M. de Issaly, *National Cancer Institute Monographs,* in press), with a discussion of channeling of CAP similar to that presented here. It is a pleasure to acknowledge his contributions to this problem in the last seven years.

plete auxotrophs under normal conditions. Their requirements are not spared by the end product of the other pathway (14). Despite the fact that extracts of *pyr-3a* can catalyze formation of US from CAP produced by the arginine-specific system (see below), this mutant is unable, as mentioned, to synthesize US *in vivo* for growth, and 2-C^{14}-uracil is not diluted, during growth of a similar mutant on this substrate, by endogenous pyrimidine synthesis (3).

Because channeling is difficult to see below the level of the whole cell, there is an unsatisfactory distance between experimental fact and the molecular basis for channeling of CAP. A number of important observations may be understood in terms of the "surface" model, but they do not constitute a rigorous test of it. Nevertheless, it is useful to compare what we know with the predictions of the model.

The structural requirements of the surface model are at least partially satisfied by the data given. First, there are two enzymes of CAP synthesis, each specific for a single pathway. Second, the genetics of the *pyr-3* locus indicates a bifunctional protein both synthesizing and utilizing CAP_{pyr}. There is no satisfactory evidence yet, however, that OTCase and the arginine-specific system of CAP synthesis are aggregated *in vivo*, as was assumed in the surface model. This is a point to be discussed further below.

Conditions which disturb channeling, and thereby lead to overflow of CAP between pathways, are consistent with kinetic features of the surface model. It has been found that strains carrying both *arg-2* and *pyr-3d* (40) or *arg-3* and *pyr-3d* (14) display little or no arginine requirement, though their uridine requirement is characteristic of *pyr-3d* (Fig. 4, top). In a symmetrical fashion, a strain carrying both *pyr-3a* and *arg-12*^{s} has no uridine requirement (Fig. 4, bottom), and, since the latter mutation imposes no requirement, the double mutant is prototrophic (11). An added piece of evidence that the arginine pathway is supplying CAP for pyrimidine synthesis in *pyr-3a, arg-12*^{s} is that a uridine requirement reappears when the strain is grown in arginine (35, 11). It is probable that this reflects a normal feedback action of arginine upon the synthesis of CAP_{arg}, a hypothesis supported and extended by Thwaites (43). Thus *arg-12*^{s} is sufficient to "unchannel" an arginine-specific source of CAP such that it is now available to both pathways. The two sets of gene interaction may be summarized by emphasizing that a transcarbamylase deficiency in one pathway relieves a CAP deficiency in the other. In the case of *arg-12*^{s}, the low activity of OTCase and the resultant derepression of CAP_{arg} synthesis (Davis, unpublished observations) would lead to both an accumulation of CAP_{arg} and a saturation of the mutant OTCase for this substrate. In this fashion,

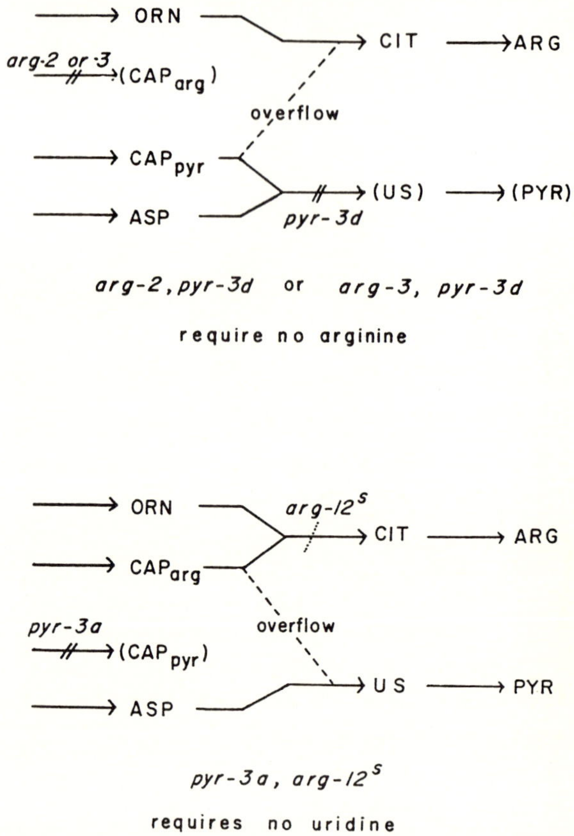

FIG. 4. Top: Diagram of the route of CAP metabolism in double mutants *arg-2, pyr-3d,* and *arg-3, pyr-3d.* Bottom: Diagram of the route of CAP metabolism in the double mutant, *pyr-3a, arg-12*^{s}. Complete genetic blocks are indicated by double lines, an incomplete block (*arg-12*^{s}) by a dotted line, and compounds not synthesized are parenthesized.

CAP_{arg} becomes available for pyrimidine synthesis. That the escape of CAP from one path to another can be accomplished by simple enzymatic mutations suggests that there is no barrier, such as a membrane, which intervenes between each CAP pool and the transcarbamylase of the other pathway. The combination of complete channeling in the normal cell with the evident ease of overflow in abnormal conditions is most simply explained by the surface model. It is therefore unnecessary to adopt at this point a "compartment" model, involving more elaborate structural features.

To the extent that CAP_{arg} overflow may be attributed to the effects of *arg-12*^{s} on OTCase (rather than the concomitant derepression of CAP_{arg} synthesis), the overflow is reminiscent of the behavior of indole in *E. coli* mutants with a defective tryptophan synthetase B protein. The mutant protein, derived by a novel selection method (7), allows, in the course of the indole glycerol phosphate (InGP)-to-tryptophan reaction, the production of twice as much indole as tryptophan. Because indole is normally an "enzyme-bound" intermediate, presumed to be transferred between contiguous reaction sites in the reaction beginning with InGP (19), it is probable that the altered B polypeptide studied allows escape of this intermediate and its accumulation as free indole. This could come about by an alteration of K_m or position of the (second) binding site for indole, or of the velocity of the second component of the InGP-to-tryptophan reaction.

Further evidence of the disturbance of channeling of CAP is available. Charles (5) found that the single mutants *arg-2*, *arg-3*, and *pyr-3a*, deficient in the synthesis of CAP_{arg} or CAP_{pyr}, will grow in minimal medium if the atmosphere is enriched in CO_2. The arginine mutants occasionally grow at lower temperature in CO_2 concentrations found in air (0.03%), not at all in CO_2-free air, and all mutants grow rather well at 30% CO_2. The response of the arginine mutants to CO_2 (but not to arginine) is inhibited by uridine, as though the CO_2-supported growth took place by overflow from the remaining, pyrimidine-regulated source of CAP. The symmetrical situation for *pyr-3a* was also observed; CO_2-supported growth is inhibited by arginine. The CO_2 responses suggest, inasmuch as CO_2 is a substrate of CAP synthesis, that it leads to greater rates of synthesis of this compound, by mass action or by induction of enzyme (37). If this were the case, the resulting concentrations of CAP might saturate the transcarbamylase of a given pathway, and overflow would take place. This sustains the possibility that a requirement for channeling (or localization) of CAP is that the enzyme utilizing it remain unsaturated. The CO_2 responses are important, because they indicate that a mutational transcarbamylase deficiency is not an obligatory condition for overflow of CAP to another pathway. We need be less concerned, therefore, with ancillary structural consequences of mutant enzymes.

Early work on the pyrimidine mutants by Fairley and his co-workers has shown that α-amino-*n*-butyric acid, among other compounds, is able to support the growth of a mutant, 1298, similar to *pyr-3a* (22, 24, 3). It appears to do so by alteration of arginine metabolism in such a way that CAP_{arg} enters the CAP-deficient pyrimidine pathway. Part of the action of α-aminobutyrate is an inhibition of OTCase activity in a manner which apparently duplicates the effect of the *arg-12*^{s} mutation on CAP_{arg}

channeling (23, 25). These data once again indicate that overflow need not be explained on the basis of organizational alterations brought about by mutation. It would be valuable to know the pool size of CAP in the cell under normal and abnormal circumstances. Although the assumption is often made, there is no good evidence that it is very low, as predicted by the surface model; the determination has not been attempted because of the lability of the compound.

The final aspect of CAP channeling is whether physical aggregation of the enzymes of CAP synthesis and utilization must obtain in both pathways, or merely in one. As mentioned above, there is no evidence for or against the aggregation of OTCase and the enzyme of CAP_{arg} synthesis, and it might be asked whether this is in fact essential. If we assume no aggregation, CAP_{arg} would, on the simplest assumption, be freely diffused. Its exclusive consumption by OTCase in the normal cell might be explained merely by an extreme competitive advantage of OTCase over ATCase (14). The ATCase protein, of course, would produce its own specific source of CAP, which would be bound tightly and consumed by the ATCase site. These postulates, however, lead to a contradiction, since, if ATCase is to bind CAP_{pyr} so well that it fails to enter the OTCase reaction, the enzyme is expected, especially when derepressed seven- to ten-fold in CAP_{pyr}-deficient mutants, to be able to use CAP_{arg} to some finite extent. This appears not to happen, according to growth and isotope dilution studies mentioned above. To explore this point somewhat more thoroughly, a set of experiments on transcarbamylase competition was performed in which ATCase and OTCase were allowed to compete with one another in free solution for very small amounts of CAP generated by the ammonium-dependent CAP-forming enzyme. Reaction mixtures contained one or both transcarbamylases, saturating concentrations of their amino acid substrates, and the requirements of the ammonium-dependent carbamyl phosphate-forming system; the bicarbonate substrate was labeled. The total radioactivity fixed into an acid-stable form was determined by plating an aliquot into acid on a planchet, which was dried and counted. The proportion of this which was citrulline was found by removing it from another aliquot with a short Dowex-50-H^+ column. In the first experiment, a constant amount of ATCase was allowed to compete at pH 8.5 with successively greater amounts of OTCase. In the normal range of OTCase activity in wild-type extracts relative to the ATCase used in the reaction mixtures, only 60 to 65% of the label was found in citrulline, and increasing the OTCase led only to a slow, asymptotic approach to complete channeling of label to citrulline. In a second experiment (Fig. 5), a crude extract of wild type, with a "normal" transcarbamylase ratio, was allowed to form ureidosuccinate and citrulline

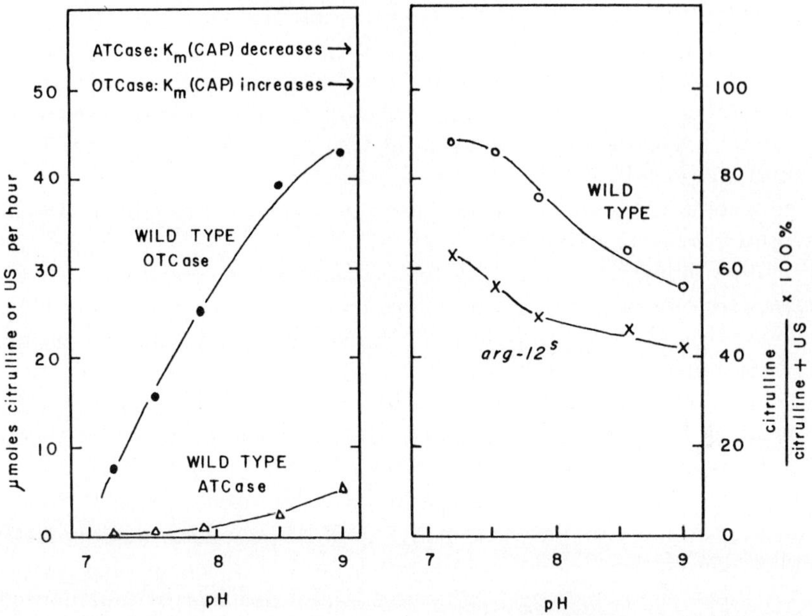

FIG. 5. (Left) Change of OTCase and ATCase activity (with saturating substrate concentrations) with pH. The ratios of the two enzyme activities represent those normally found in wild-type mycelia. The activity of the ammonium-dependent reaction of CAP synthesis is approximately 0.1 μmole per hour in the same extract. (Right) Citrulline formation, expressed as percent citrulline plus US, as pH is varied in reaction mixtures generating C^{14}-CAP by the ammonium-dependent reaction and utilizing it via ATCase and OTCase, both in the presence of saturating ornithine and aspartate. The wild-type extract was that used in the experiment shown at left. The *arg-12*^{s} extract was similar except for the mutant OTCase, present in activity similar to that of ATCase, but having abnormally high affinity for CAP.

from labeled bicarbonate at different pH's. The optima for both transcarbamylases are pH 9.1, and decreasing the pH to 7.2 leads to no more than a two-fold variation in the OTCase-to-ATCase ratio. In the same range of pH, however, the affinity of ATCase for CAP rises with increasing pH; that of OTCase declines. The effect of these changes on the competition is to decrease citrulline formation with increasing pH. At no pH, however, does the soluble system satisfy the requirements of absolute channeling. Use of *arg-12*^{s} extracts (the transcarbamylase activity ratio of which is more favorable to ATCase, but the OTCase of which has a higher affinity for CAP) leads to an expected decrease in citrulline formation (Fig. 5). Because the K_m values of the transcarbamylases at most pH's are 10^{-5} M or less, and because the steady-state concentration of CAP in these experiments is not known, detailed interpretation is not warranted. Moreover, experiments of this sort will never clearly duplicate an *in vivo* situation. But the experiment gives no support to the idea that channeling of CAP_{arg} is explained wholly in terms of transcarbamylase competition in free solution. It will not, therefore, be surprising if a physical link between the enzyme of CAP_{arg} production and OTCase is found to prevail *in vivo*. In this connection, it is encouraging that a "mitochondrial" preparation has been found which is capable of arginine synthesis in *Neurospora* (27).

Other information relevant to the channeling mechanism for CAP comes from yeast. This organism appears to have a complex locus homologous to *pyr-3* in *Neurospora* (29), as well as OTCase and an arginine-specific enzyme of carbamyl phosphate synthesis. The path-specific enzymes of CAP synthesis are products of different genes, and both appear to be glutamine-dependent (31). Despite the fact that OTCase has forty times the specific activity of ATCase (in assays reported very briefly in refs. 28 and 2), segregation of either source of CAP from the transcarbamylase of the other pathway is so poor that mutants lacking either one of the CAP-forming systems grow well on minimal medium. Only when the remaining CAP source is also deficient by mutation or is repressed or inhibited by the end product of its pathway do requirements for the CAP-deficient pathway appear. This evidence thus suggests that CAP_{pyr} localization and high OTCase-to-ATCase ratios are not sufficient conditions for channeling.

It is clear from the information presented that the mechanism of CAP channeling is only poorly understood. However, the surface model, as the simplest one, will guide further work on CAP pool size, on the molar ratios of enzymes and their aggregation, and on isolation or reconstitution of CAP channeling systems. The importance of the surface model of pool localization is probably greater, now that more enzyme aggregates and membrane localizations of coordinated pathways are being discovered. Alternative models of channeling which are slightly more complex may now be considered briefly.

The Compartment Model

Where the pool size of an intermediate exceeds the available catalytic sites of enzymes utilizing it, more familiar means of localization may be considered. One mechanism is that a catalytic system, composed, for instance, of the aggregated proteins of a pathway, constitutes a self-assembled phase in the cytoplasm to which the intermediates of the pathway are restricted. This might take the form of a hydrophobic envelope, internally differentiated perhaps, whose integrity might be difficult to maintain *in vitro*. This type of system would be similar to the enzyme aggregates referred to in the surface model, with another property emphasized to account for greater capacity for small molecules. The second alternative extends this type of thinking, and explains channeling in terms of unit-membrane-bound compartments. The alternatives are formally indistinguishable in reference to the techniques used, and may be grouped as "compartment" models of channeling.

Channeling of Ornithine

A metabolic relationship which may ultimately be understood in terms of a compartment model is the arginine–proline interaction in *Neurospora.* The biosynthetic pathways are shown in Fig. 6. In the first segment of the arginine pathway, glutamate is transformed, by way of acetylated intermediates, to ornithine (47, 18). These reactions actually form a

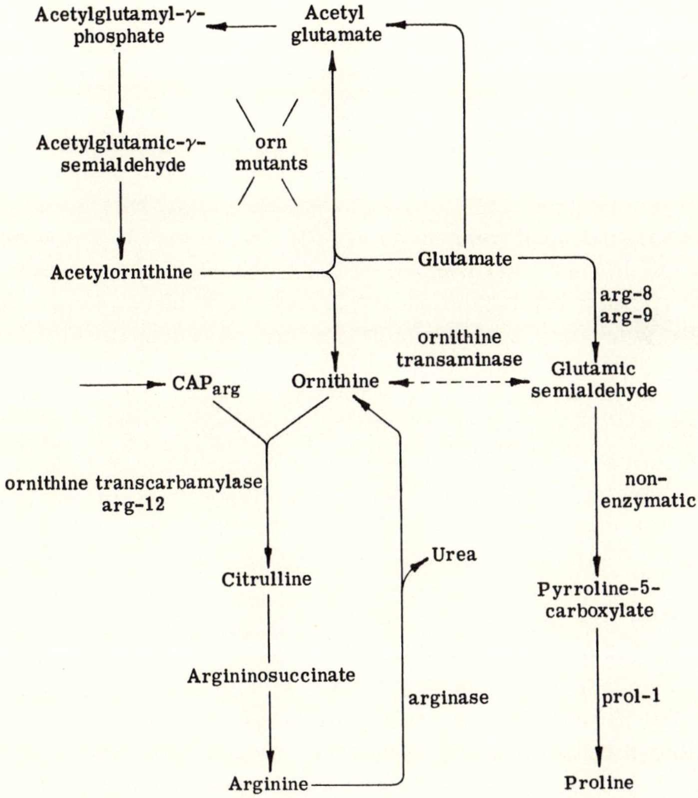

FIG. 6. Diagram of arginine and proline synthesis in *Neurospora.* Certain enzymes are named, and the genes affecting some steps are given. Most reactions are reversible.

cycle, since the acetyl group is conserved in a transacetylase reaction, in which ornithine is liberated. Mutants for these reactions, called *orn* mutants, are known at three loci. Their growth is supported by ornithine, citrulline, or arginine. Ornithine is transformed to arginine, in the last segment of the pathway, which begins with the OTCase reaction, controlled genetically by the *arg-12* locus.

Proline synthesis is initiated by the reduction of glutamate to glutamic γ-semialdehyde. The nature of the enzymatic reaction(s) is not known, but there are mutants at two loci, *arg-8* and *arg-9*, which block the formation of glutamic semialdehyde (45). Glutamic semialdehyde cyclizes nonenzymatically to Δ^1-pyrroline-5-carboxylate, which is subsequently reduced to proline. The last reaction is specified by the *prol-1* locus (52).

Two other enzymes are of great importance: The first is the inducible arginase, capable of hydrolyzing arginine to ornithine and urea (36). The second is ornithine transaminase, capable of converting ornithine to glutamic semialdehyde and its cyclized derivative (46). One or both of these enzymes, neither of which is in the normal route of biosynthesis, allow the use of arginine and its precursors as alternate sources of proline (Fig. 6). In line with this is the fact that *arg-8* and *arg-9* grow in medium supplemented either with compounds of the arginine pathway or with proline alone.

The original channeling hypothesis in *Neurospora* was suggested by Vogel and Bonner (44) when they realized that, despite normal endogenous ornithine production in *arg-8* and *arg-9* for arginine synthesis, virtually none of it was available for proline synthesis. Paradoxically, exogenous ornithine *could* be used for proline synthesis in these mutants, and in wild type, low levels of C^{14}-ornithine appeared to be used *preferentially* for proline synthesis (1, 45). While some debate prevailed at the time over channeling of arginine intermediates, it became rather confused by the question of whether ornithine arose in an acetylated pathway, or from glutamic semialdehyde via the ornithine transaminase reaction (8). At this point, it is sufficient to note that the fundamental problem still exists: endogenous ornithine does not flow to proline. The reason that this is seen in terms of a compartment model is that the ornithine pool is easily detectable, being 30 μmoles per gram dry weight during logarithmic growth. This is far in excess of the expected concentration of catalytic sites of OTCase capable of binding, and thereby restricting it to arginine synthesis. The fact that ornithine transaminase is present only in low levels in a minimal-grown culture might suggest that this alternative reaction is simply not available for endogenous ornithine. However, this does not explain the preferential use of low levels of exogenous ornithine for proline synthesis. This situation can be clarified only by experiments with logarithmic cultures, carefully sampled for pools and enzymes in conditions in which channeling is thought to occur.

More persuasive evidence for channeling in the arginine pathway has been accumulated recently in the laboratory of J. Mora. Under conditions in which L-citrulline is used as the sole source of nitrogen and proline for *arg-8*, arginine made endogenously from the citrulline provided accumulates to high levels in the mycelium, and high levels of arginase, urease, and ornithine transaminase are seen. However, the arginine accumulated is not catabolized, and the mycelium becomes starved for nitrogen and proline. Wild type, dependent on L-citrulline as a sole nitrogen source, becomes similarly nitrogen-deficient. The addition of exogenous arginine leads to growth; further, the addition of ammonium nitrate alone leads to the catabolism of the endogenous arginine such that the proline requirement of *arg-8* is satisfied (M. Castañeda, J. Martuscelli, and J. Mora, unpublished observations). It is likely that the exploration of the segregation of catabolic and biosynthetic functions implied by these phenomena will be a good test of the compartment model of channeling.

Two rather intriguing gene interactions are seen in connection with ornithine metabolism. The first is that the *arg-12*^{s} mutation (low OTCase activity), when combined with the *arg-8* or *arg-9* mutations, eliminates the nutritional requirement imposed by the latter (35). It does so presumably by making endogenous ornithine available for proline synthesis, via ornithine transaminase. It has also been found that *arg-12*^{s} alone leads to an elevated (two- to three-fold) ornithine pool during logarithmic growth on minimal medium (Davis, unpublished observations). It is of interest, then, that *arg-12*^{s} is able to "unchannel" ornithine as well as CAP_{arg}. If it is ultimately shown that the ease of CAP_{arg} overflow reflects the lack of a significant intracellular barrier between arginine and pyrimidine pathways, it is necessary to admit that a similarly labile confinement applies to ornithine. How any confinement of the large pool of ornithine is accomplished is of considerable interest.

The second gene interaction is the effect of *arg-12*^{s} on the utilization of exogenous ornithine by *orn* mutants. Because *arg-12*^{s} grows almost as well as wild type on minimal medium, it is clear that the OTCase found in this strain is sufficient for arginine synthesis from endogenous ornithine. However, when combined with any *orn* mutant such as *orn-1* (= *arg-5*), normally capable of growth on ornithine, *arg-12*^{s} completely prevents the use of exogenous ornithine (35) until the concentration reaches 50 mg/ml (Table II). Even at this concentration, growth is variable and sparse. These data seem to bear out the indications (44, 45) that the access of exogenous ornithine to the arginine synthetic pathway is poor. The situation is merely magnified here by the *arg-12*^{s} mutation, which lowers the affinity of OTCase for ornithine somewhat as well as lowering greatly its specific activity.

Recently, in collaboration with Dr. J. Mora, we have used the *orn-1*, *arg-12*^{s} double mutant strain as starting material for selection of mutants able to grow on 100 μg of ornithine per milliliter. A number have been isolated, one of which, UM-728, has been studied further. In Table II,

TABLE II
UTILIZATION OF ORNITHINE BY WILD-TYPE AND MUTANT STRAINS

Medium contains (per ml)	Mg dry weight per 10 ml medium				
	Wild type (74 A)	*arg-12*^{s}	*orn-1*	*orn-1*, *arg-12*^{s}	*orn-1*, *arg-12*^{s}, UM-728
No additions (minimal)	34	30	0	0	0
50 μg L-ornithine HCl	—	—	9	0	12
100 μg L-ornithine HCl	—	—	14	0	21
1 mg L-ornithine HCl	—	—	34	0	20
50 mg L-ornithine HCl	—	—	40	13	2
200 μg L-arginine HCl	34	34	25	19	25

the growth on ornithine of the derived triple mutant is shown. Its ability to utilize ornithine at low concentration is, if anything, better than that of the *orn-1* single mutant. It is inhibited by the highest concentration of ornithine, however, unlike the *orn-1* and *orn-1, arg-12*^{s} strains. The OTCase of the triple mutant has the same specific activity and the same affinity for ornithine as the double mutant from which it came (Table III). This is consistent with the finding that UM-728 is a single gene

TABLE III
ENZYME CHARACTERISTICS IN WILD TYPE, *orn-1*, *arg-12*^{s}, AND *orn-1*, *arg-12*^{s}, UM-728

Strain	OTCase		Ornithine transaminase activity[b]
	Specific activity[a]	K_m for ornithine	
Wild type	45.0	1–2 × 10^{-3} *M*	—
orn-1, arg-12^s	1.0	1–2 × 10^{-2} *M*	480
orn-1, arg-12^s, UM-728	1.4	1–2 × 10^{-2} *M*	37

[a] Micromoles of citrulline per milligram of protein per hour at 35°C.
[b] Arbitrary units, per milligram of protein.

unlinked to *arg-12*. The triple mutant has a large deficiency in ornithine transaminase, however (Table III). This indicates that the loss of this competitive enzyme is the reason that the OTCase reaction now proceeds in the triple mutant. The phenotype must be studied further before it is known whether it is a structural or a regulatory gene, or an indirect effect of some other change. Nevertheless, we are left with the impression that ornithine concentration and/or its physical location is normally well-controlled by several factors. (In fact, it has been observed that the administration of arginine to cells leads to rather small changes of ornithine pool size, and that ornithine itself is quickly metabolized.) That this control is advantageous is suggested by the apparent toxicity of high levels of ornithine in *orn-1*, *arg-12*^{s}, UM-728, the very strain in which exogenous ornithine apparently can gain free access to sites of endogenous arginine synthesis.

It is hoped that further work on the basis for observed distributions of CAP and ornithine among competitive pathways will bring out systems in metabolic organization applicable to many other cellular activities. It will be of great interest to note the relative importance of kinetic and structural features of these systems, and, among the latter, of simple molecular aggregates versus large, highly structured organelles.

Acknowledgments

This work has been supported, over the last eight years, by fellowships and grants from the National Science Foundation and the University of Michigan Cancer Research Institute. I gratefully acknowledge the long-term technical assistance of Mrs. Jane E. Zimmerman in all phases of the work. I also thank Drs. J. Mora and I. P. Crawford for valuable discussion, and Dr. Samson R. Gross for suggesting the "phase" model of pool segregation to me.

References

1. Abelson, P. H., and Vogel, H. J., *J. Biol. Chem.*, **213**, 355 (1955).
2. Bechet, J., and Wiame, J. M., *Biochem. Biophys. Res. Commun.*, **21**, 226 (1965).
3. Boyd, J. M., and Fairley, J. L., *J. Biol. Chem.*, **234**, 3232 (1959).
4. Cabet, D., Gans, M., Hirsch, M., and Prevost, G., *Compt. Rend.*, **261**, 5191 (1965).
5. Charles, H. P., *J. Gen. Microbiol.*, **34**, 131 (1964).
6. Cotton, R. G. H., and Gibson, F., *Biochim. Biophys. Acta*, **100**, 76 (1965).
7. Crawford, I. P., and Johnson, L. M., *Genetics*, **48**, 725 (1963).
8. Davis, B. D., *Advan. Enzymol.*, **16**, 247 (1955).
9. Davis, R. H., *Proc. Natl. Acad. Sci. U.S.*, **46**, 677 (1960).
10. Davis, R. H., *Arch. Biochem. Biophys.*, **97**, 185 (1962).
11. Davis, R. H., *Genetics*, **47**, 351 (1962).
12. Davis, R. H., *Science*, **142**, 1652 (1963).
13. Davis, R. H., *Biochim. Biophys. Acta*, **107**, 44 (1965).

14. Davis, R. H., *Biochim. Biophys. Acta,* **107,** 54 (1965).
15. Davis, R. H., *Genetics,* **54,** 330 (1966).
16. Davis, R. H., and Thwaites, W. M., *Genetics,* **48,** 1551 (1963).
17. Davis, R. H., and Woodward, V. W., *Genetics,* **47,** 1075 (1962).
18. De Deken, R. H., *Biochim. Biophys. Acta,* **78,** 606 (1963).
19. DeMoss, J. A., *Biochim. Biophys. Acta,* **62,** 279 (1962).
20. DeMoss, J. A., and Wegman, J., *Proc. Natl. Acad. Sci. U.S.,* **54,** 241 (1965).
21. Dutta, S. K., and Woodward, V. W., *Genetics,* **52,** 391 (1965).
22. Fairley, J. L., *J. Biol. Chem.,* **210,** 347 (1954).
23. Fairley, J. L., and Wampler, D. E., *Arch. Biochem. Biophys.,* **106,** 153 (1964).
24. Fairley, J. L., Herrmann, R. L., and Boyd, J. M., *J. Biol. Chem.,* **234,** 3229 (1959).
25. Herrmann, R. L., Lou, M. F., and White, C. W., *Biochim. Biophys. Acta,* **121,** 79 (1966).
26. Hill, J. M., and Woodward, V. W., *Genetics,* **52,** 448 (1965).
27. Kinsey, J. A., and Wagner, R. P., *Proc. Natl. Acad. Sci. U.S.,* **55,** 404 (1966).
28. Lacroute, F., *Compt. rend.,* **258,** 2884 (1964).
29. Lacroute, F., *Compt. rend.,* **259,** 1357 (1964).
30. Lacroute, F., Piérard, A., Grenson, M., and Wiame, J. M., *Arch. Intern. Phys. Biochim.,* **72,** 687 (1964).
31. Lacroute, F., Piérard, A., Grenson, M., and Wiame, J. M., *J. Gen. Microbiol.,* **40,** 127 (1965).
32. Levenberg, B., *J. Biol. Chem.,* **237,** 2590 (1962).
33. Matchett, W. H., and DeMoss, J. A., *Biochim. Biophys. Acta,* **86,** 91 (1964).
34. McFadden, B. A., and Howes, W. V., *Arch. Biochem. Biophys.,* **109,** 415 (1965).
35. Mitchell, M. B., and Mitchell, H. K., *Proc. Natl. Acad. Sci. U.S.,* **38,** 205 (1952).
36. Mora, J., Tarrab, R., and Bojalil, L. F., *Biochim. Biophys. Acta,* **118,** 206 (1966).
37. Nazario, M., and Reissig, J. L., *Biochem. Biophys. Res. Commun.,* **16,** 42 (1964).
38. Piérard, A., Glansdorff, N., Mergeay, M., and Wiame, J. M., *J. Mol. Biol.,* **14,** 23 (1965).
39. Reed, L. J., and Cox, D. J., *Ann. Rev. Biochem.,* **35,** 57 (1966).
40. Reissig, J. L., *Genet. Res.,* **1,** 356 (1960).
41. Reissig, J. L., *J. Gen. Microbiol.,* **30,** 327 (1963).
42. Stadtman, E. R., *Bacteriol. Revs.,* **27,** 170 (1963).
43. Thwaites, W. M., *Genetics* **55,** 769 (1967).
44. Vogel, H. J., and Bonner, D. M., *Proc. Natl. Acad. Sci. U.S.,* **40,** 688 (1954).
45. Vogel, R. H., and Kopac, M. J., *Biochim. Biophys. Acta,* **36,** 505 (1959).
46. Vogel, R. H., and Kopac, M. J., *Biochim. Biophys. Acta,* **37,** 539 (1960).
47. Vogel, R. H., and Vogel, H. J., *Genetics,* **48,** 914 (1963).
48. Wiame, J. M., *in* "Colloques Internationaux du C. N. R. S., Marseilles" (J. Senez, ed.), p. 381, Éditions du C.N.R.S., Paris, 1965.
49. Woodward, V. W., and Davis, R. H., *Heredity,* **18,** 21 (1963).
50. Woodward, V. W., and Schwarz, P., *Genetics,* **49,** 845 (1964).
51. Yanofsky, C., and Rachmeler, M., *Biochim. Biophys. Acta,* **28,** 640 (1958).
52. Yura, T., *Proc. Natl. Acad. Sci. U.S.,* **45,** 197 (1959).

Discussion of Part VI

C. B. Anfinsen, F. Lipmann, F. Lynen, K. Bloch, R. Y. Hsu, M. Levinthal, R. P. Wagner, D. Pramer, S. Brody, P. Siekevitz, J. R. Mattoon, J. K. Bhattacharjee

Chairman Anfinsen: So, we will now have discussion of the papers.

Dr. Lipmann: Listening to Dr. Lynen's wonderful talk, I have been again impressed by the similarity between peptide synthesis and fatty-acid synthesis. This has been a hobby of mine, but I think that particularly the transfer of the newly-elongated chain from one enzyme site to the other to make room for a newly incoming single unit to elongate further is very similar to what I tried to discuss yesterday for protein synthesis. How do you propose to get this translocation, as we call it in the case of polypeptide synthesis where we try to invoke GTP as a possible agent; how does it occur in your case?

Dr. Lynen: I would like to emphasize what is so nice about fatty-acid synthesis is that you have both reactants in an activated state and then you have these "arms." That is what I tried to demonstrate with this one slide where I had the "carousel." That is the reason you have to have a long arm—in order to swing it around.

Dr. Bloch: I would like to know, Dr. Lynen, whether you have any evidence for the intracellular localization of the complex in yeast, and related to this question, what your thoughts are why the yeast and the animal liver enzymes occur as tightly aggregated complexes whereas the bacterial and the plant systems are composed of easily separable enzymes which show no tendency to interact.

Dr. Lynen: With respect to the second question, I think that the state of multienzyme complexes is more efficient than a complex composed of separated units and, of course, this is a question of evolution. The bacterial system is on the lower level, and so maybe nature discovered evolutionary advantages in having a multienzyme complex. With respect to your first question, I must say no. We have no idea where the fatty-acid synthetase is located. We plan to do some experiments with protoplasts. Of course, in order to break the yeast cells, we use an ultrashake or some machine like the ultrashake, and this is a very drastic procedure; therefore, we can't say where the enzyme is located in the living cell.

Dr. Hsu: Upon hearing this interesting talk by Dr. Lynen, I have a couple of questions that Dr. Lynen might have some thoughts on. First, what is the mechanism of transfer of the butyryl moiety from the central to the peripheral sulphydryl group?

DR. LYNEN: It's just like a transacylation reaction. One and then another group comes in and you have a transfer reaction.

DR. HSU: So in other words, what you are saying is that actually the sulfhydryl groups have been moved to the vicinity so that there is probably a conformational change occurring so that these sulfhydryl groups are in close proximity.

DR. LYNEN: We have some evidence that there are conformation changes. I had it in mind to discuss it, but I had no time.

DR. HSU: My second question concerns the finding that, in the pigeon liver enzyme complex, there is no FMN but in the yeast complex, there is. What do you think of this difference?

DR. LYNEN: I don't know, but I can tell you that Auer, who was in my laboratory about three or four years ago, also studied the pigeon liver enzyme. We found the same as you published recently, namely, that there is no flavin in the enzyme.

DR. LEVINTHAL: I have two questions for Dr. Wagner. I wondered whether you were able to isolate any reductoisomerase-less mutants and show that they mapped in the same place where the group I mutants mapped, thereby identifying that gene as the structural gene for reductoisomerase.

DR. WAGNER: The group I mutants *are* reductoisomerase-less mutants having no activity for reductoisomerase that we can measure. This doesn't mean that they don't form a protein which is inactive. The indications are that they are the result of mutations at the reductoisomerase structural gene locus.

DR. LEVINTHAL: Will these mutants of the group 1, 2, 3, and 4 mutants complement other *ilva* mutants that are blocked closer in the pathway?

DR. WAGNER: These are the only loci that we know in the first place that are concerned with isoleucine-valine in *Neurospora,* unless you have found others. Complementation does not occur among the group 1 mutants we have so far. However, they complement with 2 and 3. We get a complementation pattern that shows a gradation going from group 1 to group 2 and 3. The 2's are sort of in the middle. We also get complementation in the group 4 mutants.

DR. PRAMER: Dr. Brody, I am wondering if your morphological effects are all under environmental or nutritional control, that is, if you can induce this type of colonial change by supplying glucose-6-phosphate exogeneously to the wild type.

DR. BRODY: We tried that, and it didn't work. As a matter of fact, we have tried extensive experiments involving different media and different temperatures and a whole series of things to see if we could affect the glucose-6-P level in the wild type, let alone the level in the colonial mutant. With the exception of the effect of temperature, they both seem to be pretty much steady-state levels.

DR. SIEKEVITZ: I was wondering if it is rather difficult to measure the shunt pathway the way you did with radioactivity, because in one case you have a high glucose-6-phosphate content and in the other case you have a low one. But this isn't matched by high and low levels of the ribose-5-phosphate. So I don't see how you can measure a good C-1 to C-6 radioactivity ratio in the presence of different ratios of concentrations of glucose-6-phosphate and ribose-5-phosphate.

DR. BRODY: Well, we never measured ribose-5-phosphate in any of the cells. And I think I did put in a disclaimer that I really don't know how quantitative the radiorespirometry methods are for something with an altered metabolism. I think it indicates that something is clearly different in this mutant, but exactly what it is one can't say from this type of gross measurement.

DR. MATTOON: You may be interested to know that we have a somewhat similar situation with mannose-requiring yeast mutants isolated in our laboratory. This mannose requirement also bestows on the mutant an increased sensitivity to snail enzyme which is used to make protoplasts.

DR. BHATTACHARJEE: I would like to add a small comment to Dr. Wagner's talk. We do have a handful of mutants for the complex pathway of isoleucine-valine in yeast which fall into more than four complementing groups. We also have complex mutants which require simultaneously leucine and valine, and leucine and isoleucine. Being a newcomer to the field, I foresee no unemployment problem in the field for the future young scientists.

CHAIRMAN ANFINSEN: No comment from Dr. Wagner.

PART VII

MEMBRANE SYNTHESIS AND ASSOCIATED PHENOMENA

Chairman's Remarks

F. Lynen

Max-Planck-Institut für Zellchemie,
Munich, Germany

May I call the meeting to order. Looking through the titles of the various sessions, I had a feeling that we are expecting to have a special session this afternoon, because you may realize that, generally, the titles were rather short. But this afternoon, we have the title "Membrane Synthesis and Associated Phenomena." I don't know whether it is just an expression of our ignorance about membranes. As you already know, there is still some discussion about the basic structure of membranes, and much interest developed in recent years regarding the function of membranes in ion transport and so on. So I personally feel that the study of membranes must become intensified in the following years and, with this in mind, I think it was a very good selection of the organizers of the symposium to have a special session on the membrane.

The Biogenesis of Intracellular Membranes

PHILIP SIEKEVITZ, GEORGE E. PALADE, GUSTAV DALLNER,[1]
ITHAK OHAD,[2] AND TSUNEO OMURA[3]

The Rockefeller University,
New York, New York

INTRODUCTION

During the last two years, our laboratory has been engaged in an examination of the biological mechanism whereby membrane constituents, particularly enzyme proteins and phospholipids, are assembled into complete membranes. Specifically, we have attempted to exclude the genetic control mechanisms involved in protein and lipid synthesis and have focused our attention on the means whereby already synthesized, specific proteins, and already synthesized lipids are brought together to form specific membrane structures. For this reason, we have been using two systems in which we know that a net synthesis of specific membranes does occur—namely, the rapid accumulation of smooth (that is, devoid of ribosomes) endoplasmic reticulum (ER) membranes in the parenchymal cells of the liver of just-born rats (3, 4), and the appearance of chloroplast lamellae upon exposure to light of the etiolated cells of a mutant *Chlamydomonas* (15). These cells are unable to synthesize chlorophyll in the dark, but can do so in the light. While genetic factors must be involved indirectly, and possibly directly, in the production of any membrane, there are differences in this regard between the two systems studied. In the production of the ER membranes, most of the genes controlling the synthesis of constitutive membrane proteins are activated (derepressed) at the onset of membrane production, while in the chloroplast lamellae, most of the corresponding genes are already active, and hence some constitutive proteins are already available for membrane assembly. The difference could be quantitative, for in the

[1] Dr. Dallner's present address is Department of Pathology, Karolinska Institutet, Stockholm, Sweden.

[2] Dr. Ohad's present address is Department of Biological Chemistry, The Hebrew University, Jerusalem, Israel.

[3] Dr. Omura's present address is Institute for Protein Research, Osaka University, Osaka, Japan.

former case only a few of the membrane proteins are present in the cell at the onset of membrane assembly (3, 4), while in the latter it appears that most of the proteins are existent in the cell at this time (15). However, I think that we have isolated, in both instances, the parameters having to deal only with the assemblying of membranes, and this will form the subject of this paper.

What are the possible mechanisms of membrane formation? A priori, there are two possible basic processes which can be involved—namely, a single-step process or a multiple-step process. The first hypothesis implies that all constituents of the membrane—enzyme proteins, structural protein (if this exists in the membranes under consideration), and lipids—are simultaneously assembled in a single operation. This idea has a correlate, that the genes involved are so geared to each other that the appropriate protein and lipids are synthesized at the same time, and moreover (and this applies also to the second postulate) that the protein, and perhaps the lipid, have built-in "recognition" sites which determine the specificity of the final assemblying. The second hypothesis implies that a functional membrane is produced stepwise, the first step in the process consisting in the assembly of certain lipids with a "structural" protein to form a primary or "ur-membrane," and that the constitutive enzymes are later tagged on at appropriate places along this membrane.

We have, up to now, assumed that specific cellular membranes as a whole—as, for example, the ER membranes—are composed of a randomization of their constituent enzyme molecules, but we really do not know if this is the case. Thus, except for the necessary proximity and positioning of enzymes acting in teams, as in an electron transport chain, it could well be that the membranes we are experimenting with, the ER membrane and the chloroplast lamellae, are mosaics consisting of a number of types of functionally different patches, or tesserae, each having a characteristic enzyme or enzyme set. If we could so pick out, we might find, for example, that the glucose-6-phosphatase molecules of the ER membrane are not randomly situated on the ER membrane throughout the cell, but are positioned at certain places, so that 100 molecules are in one patch of membrane and none in the next patch; at present we have no data about this at all.[4]

However, in the experiments the results of which are to be mentioned below, we think that we have come up with certain possible mechanisms, exclusive of others, and the reason for this is that we have deliberately

[4] This idea is amenable to experimental verification by histochemical electron microscopy, and is being worked on in our laboratory.

looked at situations in which (a) new membrane enzymes and new membrane structures appear simultaneously, as in the case of the ER membranes of the developing rat liver cell, or (b) new membrane structures appear without the appearance of new enzymes, most of the constitutive enzymes being already present, as in the case of the developing lamellae of the mutant-alga chloroplast. This approach to an elucidation of membrane structure is quite different from that used by others—the isolation of membrane subunits and the reconstruction of the whole membrane therefrom (9). However, we believe that some information about the validity of certain proposed structures can be derived from these data. For example, is a membrane a "unit-membrane" structure, a continuous layer with no repeating subunit in the plane of the membrane, or is it a "globular" structure with no continuous lipid layer, but made up of planar aggregates of subunits, small in two dimensions but as thick as the membrane? Thus, though we have no comprehensive answer to the above question, we believe that certain possibilities suggest themselves as a result of the data to be presented below. These data will omit the experimental procedure used to obtain them, for these are either already in press (3, 4, 16) or to be published shortly (15), and indeed are by now routine procedures of the laboratory.

For the purposes of this paper, and in grateful deference to the request of Dr. E. Racker, the following definitions are proposed: A biological membrane is a complex composed of different types of molecules, mainly lipids and proteins, which are held together by noncovalent bonds. Because of the nature of the lipids, the complex is so arranged as to give a separation of hydrophobic regions (within the membrane) from hydrophilic regions (on the outside of the membrane). Morphologically, the membrane appears as a stratified structure whose layers have been tentatively correlated with its hydrophilic and hydrophobic regions. Because of the nature of the proteins, the complex is so constituted as to give functional specificity to the membrane with regard to its type and its location within the cell. We say nothing about the intimate organization of the molecules within the membrane complex, only that the resultant biological membrane functions both structurally as a separator of compartments within the cell, and metabolically as a separator of biochemical functions within the cell. Finally, a membrane can be isolated from the cell as an entity with preservation of structure and function. The constitutive components of a membrane are those molecules which are an intimate part of the complete, functioning membrane structure; this definition does not imply that all the component molecules are necessary for the membrane structure to exist as such, to be visualized in the microscope or separated from the cell. It does mean that, in the living cell, these specific constituents are associated, wholly or in part, with various kinds of membrane structures, and as such partake of the functioning structural entity which is a membrane.

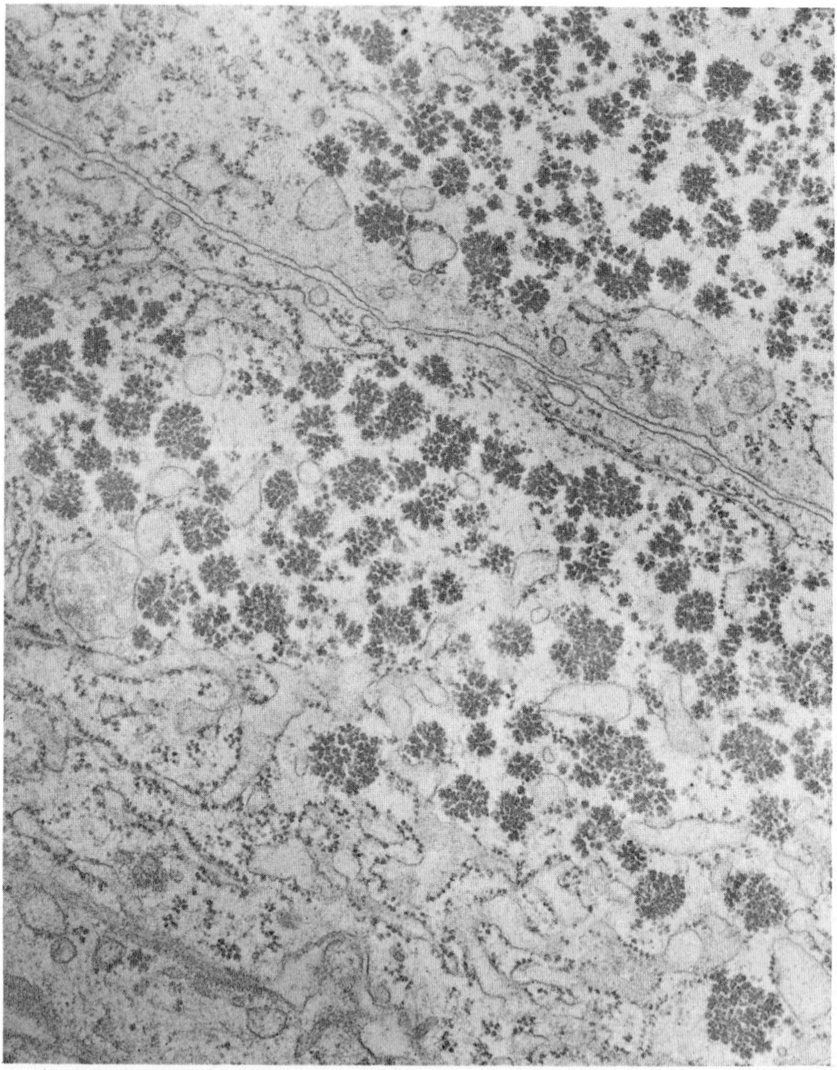

FIG. 1. Electron micrographs of glycogen-containing areas of hepatic parenchymal cells (3). (*a*) ~3 days before birth; ×46,000. (*b*) ~3 days after birth; ×49,000.

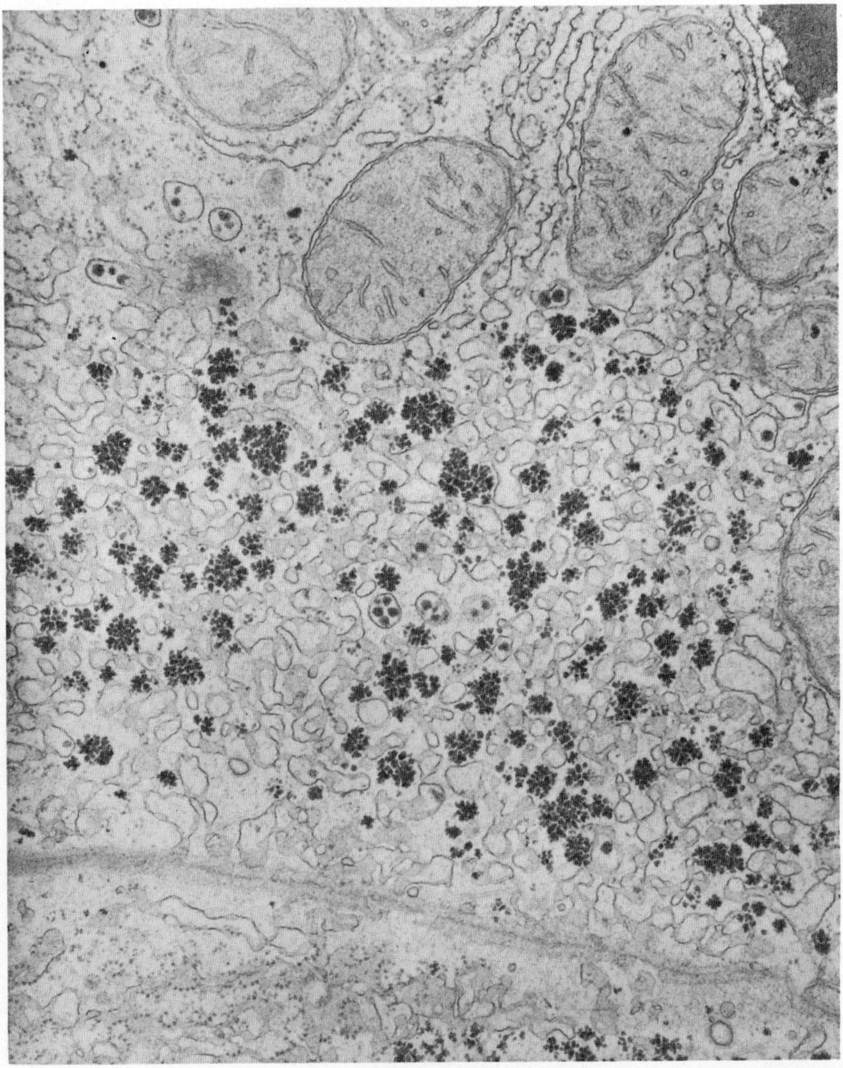

FIG. 1*b*. For legend see opposite page.

RESULTS AND DISCUSSION

The systems we are dealing with are illustrated in Figs. 1 and 2. Figure 1*a* shows an electron micrograph of the glycogen-particle-containing area of the hepatic parenchymal cell about 3 days before the birth of the animal, while Fig. 1*b* shows the same area 3 days after birth; it is in this area that the new membranes are finally located, and

Fig. 1*b* indicates the extent of the proliferation. It should be mentioned that the cells at birth have already an extensive amount of rough ER (membrane-bounded cisternae and vesicles with ribosomes attached), but little smooth ER; it is the latter which increases in amount after birth.

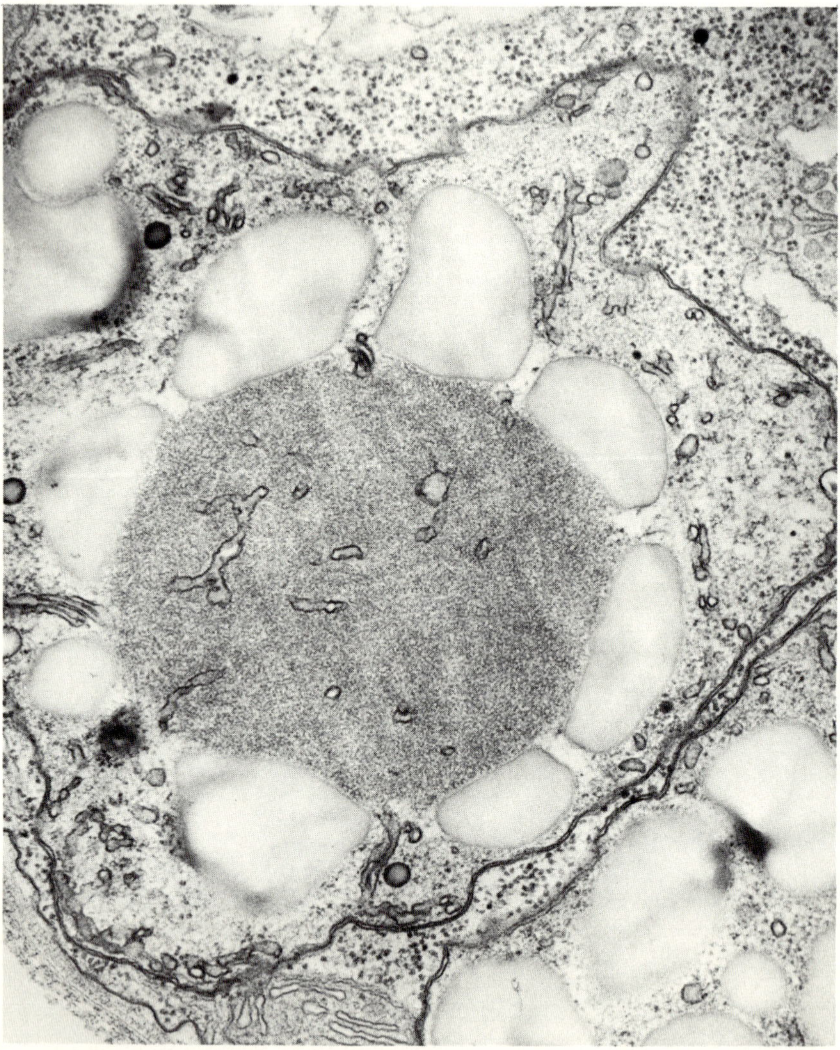

FIG. 2. Electron micrographs of chloroplast region in the y-1 mutant of *Chlamydomonas reinhardi* (15). (*a*) Cells grown in the dark for six generations; chlorophyll content below 0.5 μg per 10^7 cells. The central body is the pyrenoid, surrounded by starch granules; $\times$57,000. (*b*) Cells grown in the light for six generations; chlorophyll content $\sim$18 μg per 10^7 cells; $\times$23,000.

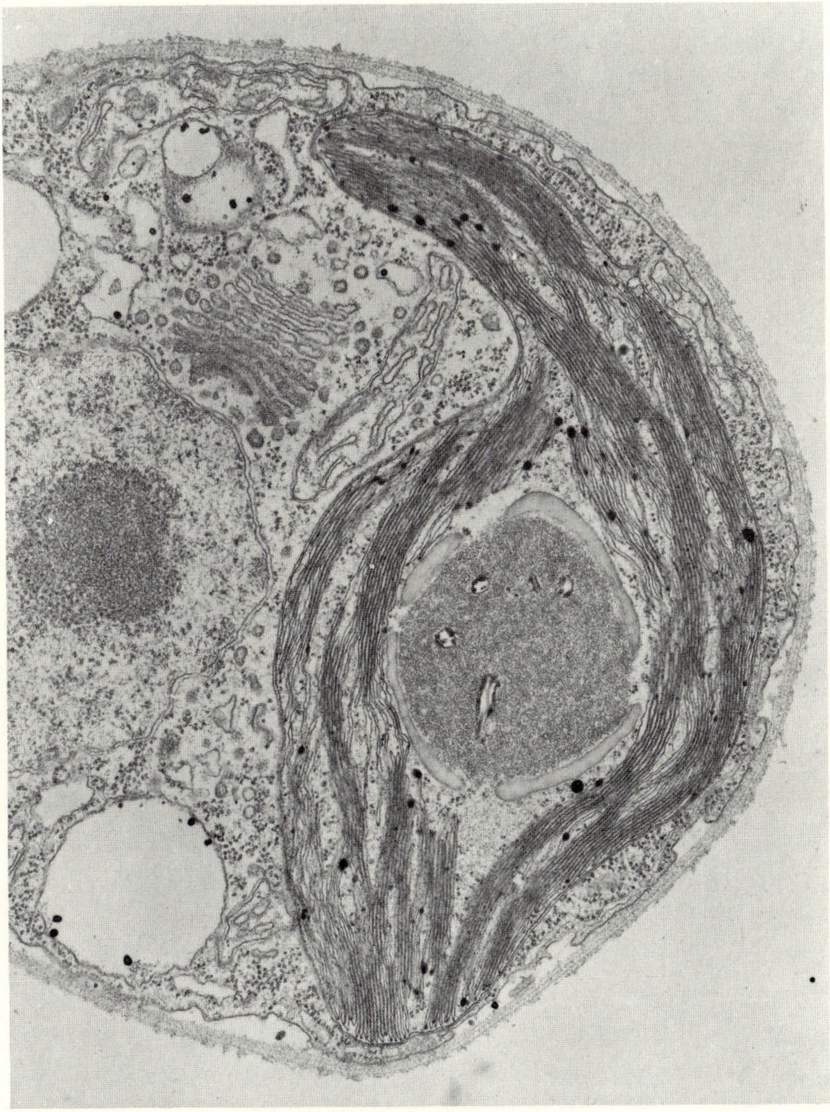

FIG. 2*b*. For legend see opposite page.

Figure 2*a* indicates the paucity of lamellae in the chloroplast of the dark-grown mutant *Chlamydomonas* cell, while Fig. 2*b* shows the full complement of lamellae found after exposure of this form for about 6 hours to light, with no cell division taking place during this period. It must be emphasized that, outside of the disappearance of the lamellae during growth in the dark, nothing else changes drastically in the morphology of the chloroplasts. Thus we have in these two systems examples of the morphological parameters of membrane formation. Details of these events are in press (3, 4, 16) or in the process of being submitted (15) for publication.

The biochemical events taking place at the same time are illustrated in Figs. 3 and 4 for the ER system, and in Figs. 6 and 7 for the chloroplast system. In the former, microsomes, fragments of the ER, were examined, while in the latter a homogenate (made in the French press) of the whole cell was examined. In the former system it was clear that, of the eleven enzyme activities examined here, only one, an ATPase, is present

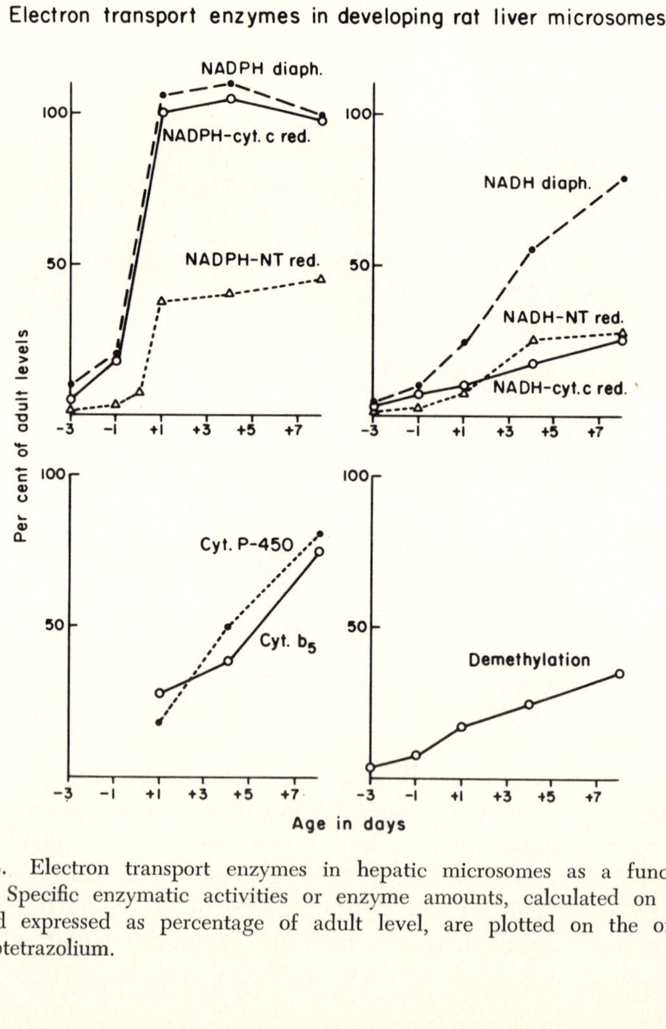

FIG. 3. Electron transport enzymes in hepatic microsomes as a function of age (4). Specific enzymatic activities or enzyme amounts, calculated on protein basis, and expressed as percentage of adult level, are plotted on the ordinate; NT = neotetrazolium.

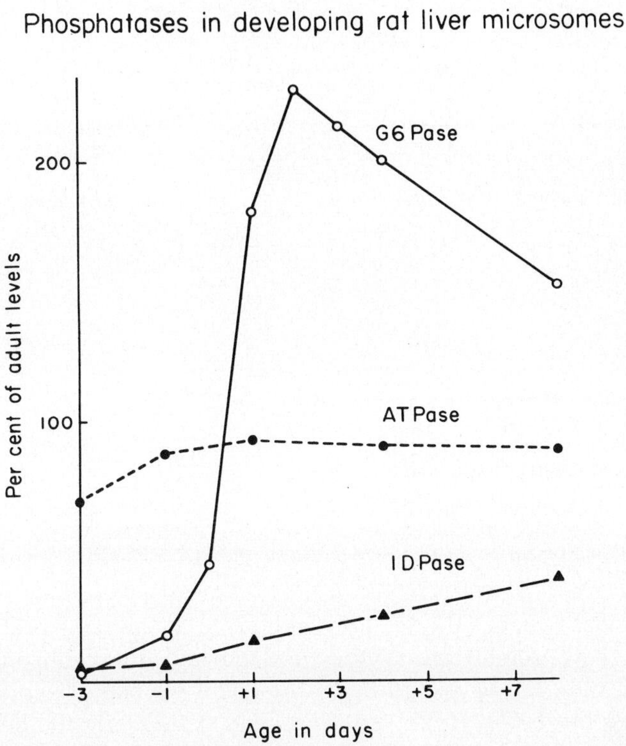

FIG. 4. Phosphatases in hepatic microsomes as a function of age (4). Ordinate value expressed as in Fig. 3. IDP = inosine diphosphate; G6P = glucose-6-phosphate.

in the fetal rat liver. All these enzymes are membrane constituents, some more firmly bound to the membrane than are others (6). We think that these enzymes are newly synthesized, for actinomycin D and puromycin, injected at birth, prevent the increases in enzymatic activities (4). Since EM examination shows no cytolysis of the liver,[5] we assume that these inhibitors act to prevent either transcription or translation of the genetic message. Thus we can make the initial assumption that messengers for these enzymes are synthesized and translated, and that, concomitantly, the synthesized proteins are laid down into membranes which are being assembled.

However, not all of the at least four components of the NADPH oxidase chain (Fig. 5) and not all of the at least four components of the

[5] Both the inhibitors do cause morphological changes in the cytoplasm, such as some vacuolization of the ER and a diminution in glycogen contents, but the polyribosomes still exist bound to membranes, as in the normal cell (4).

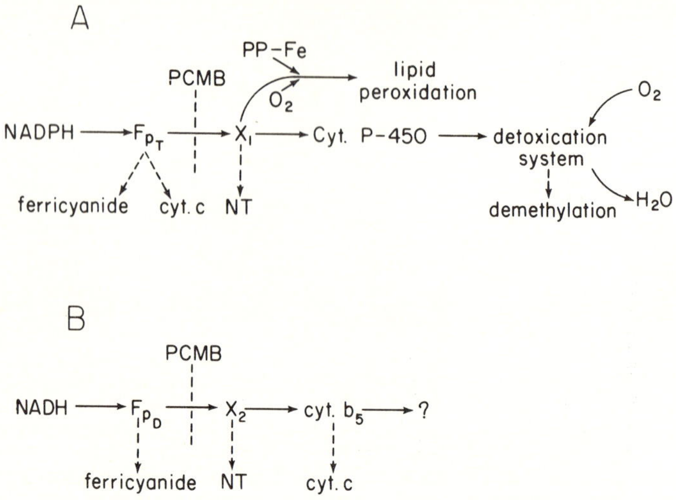

FIG. 5. Tentative scheme of the sequence of components in the two electron transport chains of rat liver microsomal membranes (4). PCMB = *p*-chloromercuribenzoate; PP–Fe = inorganic pyrophosphate–iron.

NADH oxidase chain (Fig. 5) are synthesized at the same rates. In both cases, the flavoproteins are synthesized and laid down into membranes at a rapid rate, but as we go down the chain we find a lagging behind these rates; the NADPH-NT reductase, and particularly the demethylating complex, are laggard in the first chain, while the NADH-NT reductase and the reduction of cytochrome *c* are laggard in the second. On the basis of these initial findings, we must modify the initial assumption made above and say that not all the proteins in these chains seem to be synthesized concomitantly: some of them appear to be synthesized and inserted into membrane structures at slower rates. In other words, it would appear that there is some membrane "ur-structure" into which newly synthesized enzymes are specifically inserted. Thus, membranes isolated at the various times of Figs. 3 and 4 will show varying ratios of the different enzymes to each other. Also, since these electron transport enzymes are thought to lie "in a line" next to each other along the membrane, for efficient electron flow, we would think a priori that they would be made off a polycistronic messenger (4), but at first glance this does not seem to be the case.

The G-6-Pase (Fig. 4) is interesting in that it comes up very rapidly into the membrane soon after birth; indeed, increases can be measured hour by hour during the first day of extrauterine life. Actually enzyme activity overshoots the adult value, as has already been noticed by others (*cf.* 4). This rapid increase in G-6-Pase activity (and in smooth ER amount) is accompanied by the long-known rapid diminution in glycogen content of the liver at the same time.

When we turn to the mutant algal cell, we come upon a situation which at first glance is quite different. From Dr. R. Sager's experiments it appears that we are dealing with a nuclear mutant in which one of the enzymes involved in chlorophyll production from protochlorophyll is missing or defective. Chlorophyll cannot be synthesized in the dark, but in the light, a photoactivation step occurs, as in higher plants, and the pigment is rapidly produced; thus light energy can be said to cause the cytoplasmic repair of the nuclear genetic defect. In this algal system, it is rather surprising that a defect in chlorophyll synthesis shows up as a defect in membrane formation. Essentially what happens is that, as soon as light is shut off, chlorophyll synthesis ceases, and the chlorophyll content of the cells, multiplying in the dark, is almost exactly halved at each generation; after six to eight generations, a minute amount is left, and the cells appear yellow. Concomitant with this change there is a similar reduction in the granar membranes (discs, thylakoids) of the chloroplast (Fig. 6*a*). Upon resumption of chlorophyll synthesis in the light (Fig. 7*a*), membrane synthesis resumes and grana are eventually formed (Fig. 2*b*). There is a distinct connection between chlorophyll formation and lamellae formation (*cf.* 9); it appears that chlorophyll acts as a "seed" around which the membrane structure is built.

As was mentioned earlier, there is a difference in the level at which membrane protein production is controlled in the two systems under discussion. While mRNA synthesis for all the constitutive ER enzymes mentioned in Figs. 3 and 4 (and probably others) seems to be a prerequisite for membrane formation, this is not the case in the mutant algae. For it can be seen in Fig. 6*a* and *b* that both chloroplast matrix enzymes (FDPase, G-3-P dehydrogenase, and RuDP carboxylase) and chloroplast lamellar enzymes (ferredoxin and cytochrome *f*) are capable of being synthesized in the absence of chlorophyll synthesis and in the absence of membrane formation. The proteins involved in electron transport in the chloroplast should be, in analogy with mitochondrial electron transport enzymes, localized in the membrane, and yet cytochrome *f*, ferredoxin, and ferredoxin reductase are being synthesized without being inserted into the finished membranes, for no lamellae are available. The simplest explanation is that, as soon as chlorophyll synthesis and membrane synthesis ensue, the enzymes already present, presumably in the chloroplast matrix but possibly in the small vesicular elements seen in Fig. 2*a*, begin to bind to the forming membranes. Some proof for this is given in Fig. 7*a*, *b*, and *c*, for the rate of the Hill reaction (Fig. 7*a*) (a measure of the electron transport reaction involving

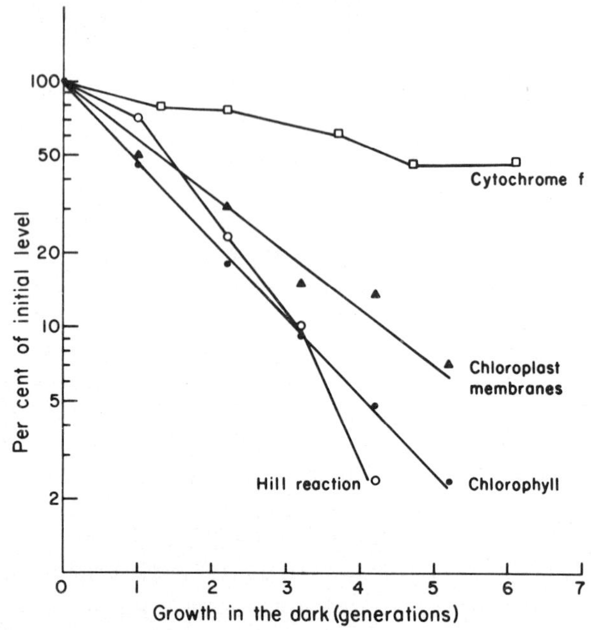

FIG. 6*a*. For legend see opposite page.

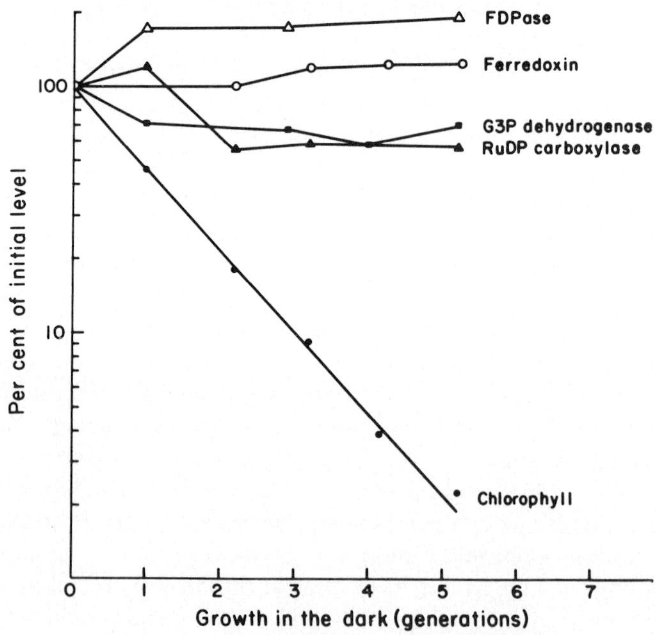

FIG. 6*b*. For legend see opposite page.

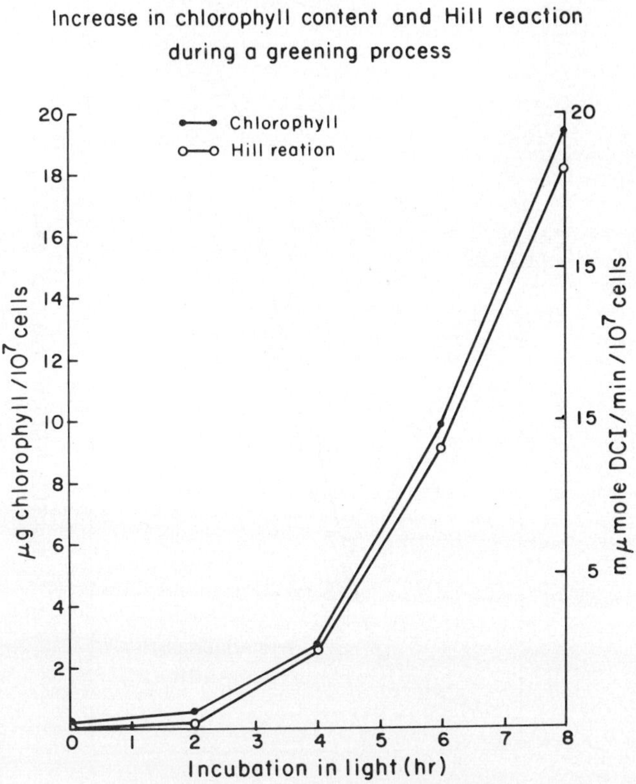

FIG. 7*a*. For legend see page 345.

system II), that of the ferredoxin-linked photoreduction of NADP (Fig. 7*b*) (a reaction involving system I), as well as that of the photo-oxidation of cytochrome *f* (Fig. 7*c*) (a reaction linking the two systems), are all linked stoichiometrically to the rate of chlorophyll synthesis. In rough terms, these enzymes are available for immediate insertion into chlorophyll-containing membranes, and become part of an electron-flow system, once these membranes are formed. We have never observed, for example, any lag in the Hill reaction once chlorophyll synthesis is resumed upon illumination. Here, then, we have a situation where, for some unknown reason, the formation of a membrane structure is dependent on chlorophyll formation; as these membranes are being assembled, enzymes already present in the nonlamellar parts of the chloroplast are attached

FIG. 6. Levels of chlorophyll (*a*, *b*), chloroplast membranes (*a*), and various chloroplast enzymes (*a*, *b*) during a degreening process (15).

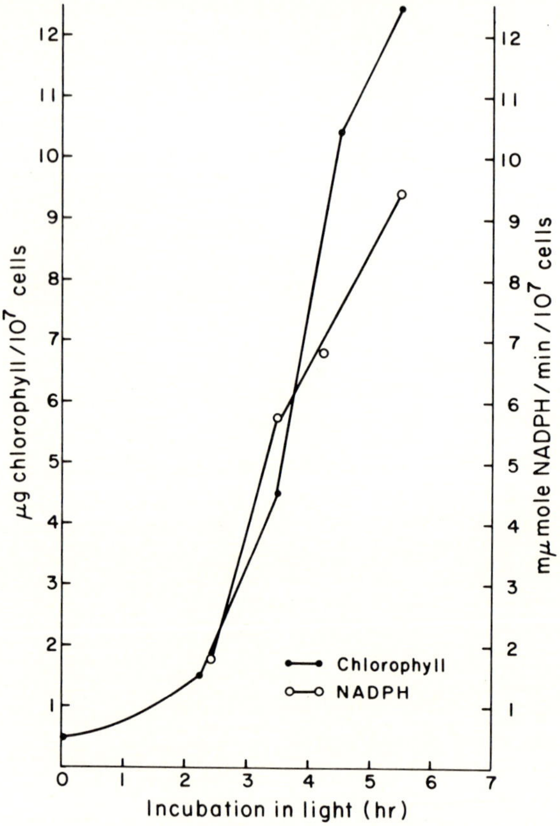

FIG. 7*b*. For legend see opposite page.

to specific sites on the membrane structure, to complete the electron transport chains.

The next point to be covered was the question as to the sites of synthesis of membrane proteins and their translocation onto membrane structures from the sites of synthesis. This problem was investigated with the ER system of the developing rat liver, for here we start from a cell containing predominantly ribosome-covered ER (rough ER), and we end up with one containing a mixture of smooth and rough ER; and, at variance with the algal system, we have a situation in which a net synthesis of most of the constitutive enzymes occurs. Figure 8 indicates that, after leucine injection into just-born rats, the proteins of the isolated membranes of the rough ER are more highly labeled at the early time intervals than are the proteins of the membranes of the smooth ER. We assume that the attached ribosomes synthesize the membrane proteins which, upon discharge from these ribosomes, become a part of the rough ER membranes; later on these proteins appear in the membranes of the smooth ER. Figures 9 and 10 give more supportive evidence to this point, in that, as new enzyme synthesis occurs, these new membrane enzymes appear first in the rough and later in the smooth ER. Since the distribution of these enzymes among the membrane types in the adult rat liver is roughly equal or even somewhat higher in the smooth ER (2, 14), we feel that these data are fairly good evidence that the first appearance of new membrane enzymes is in the membranes to which ribosomes are attached. It is also important that we have data (4) which imply that the newly synthesized enzymes are released from the ribosomes and immediately placed in their context in the membrane, without first becoming soluble constituents.

Because, as will be shown later, the membrane constituents of the ER in the adult rat liver are in rather rapid turnover, we also looked at the appearance of specific enzyme synthesis in the adult liver. Figure 11 shows that the incorporation of radioactive leucine is more rapid into NADPH–cytochrome *c* reductase and cytochrome b_5 isolated and purified from the rough membrane, than into the same enzymes similarly prepared from the smooth membrane. Although the curves in Figs. 8

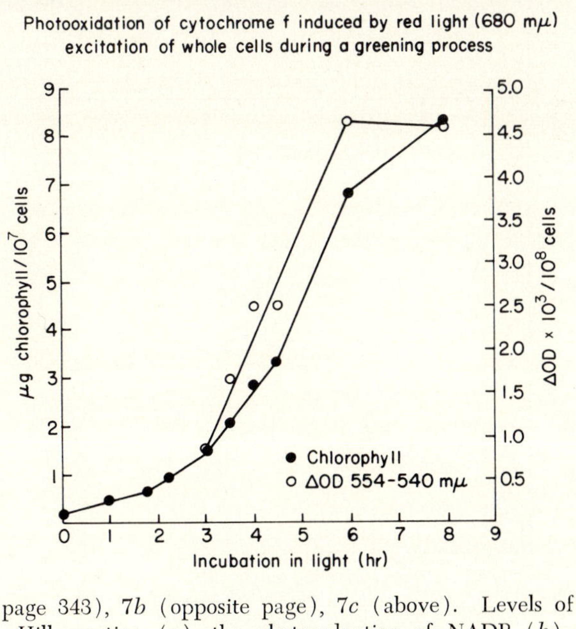

FIGS. 7*a* (page 343), 7*b* (opposite page), 7*c* (above). Levels of chlorophyll (*a*, *b*, *c*); the Hill reaction (*a*), the photoreduction of NADP (*b*), and photooxidation of cytochrome f (*c*) during a greening process (15).

through 11 are clearly not in the classical precursor-product relationship (*cf.* below), we feel that the data are indicative that in this particular case membrane proteins are synthesized by ribosomes attached to the ER membranes and incorporated into newly forming membrane at or near the site of synthesis. The data further suggest that the smooth ER membrane is derived, by means to be discussed below, from the rough ER membranes.

We next looked at the phospholipid involvement in membrane formation, specifically trying to answer the question as to whether specific

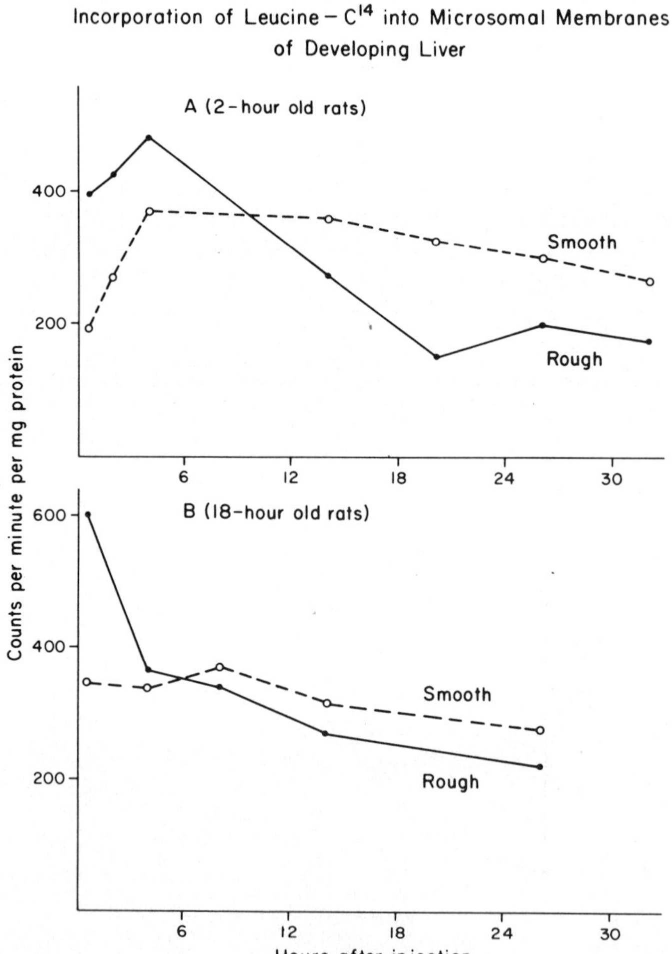

FIG. 8. Incorporation of leucine-1-^{14}C into protein of liver microsomal membranes of 2-hour-old (*A*) and 18-hour-old (*B*) rats (3).

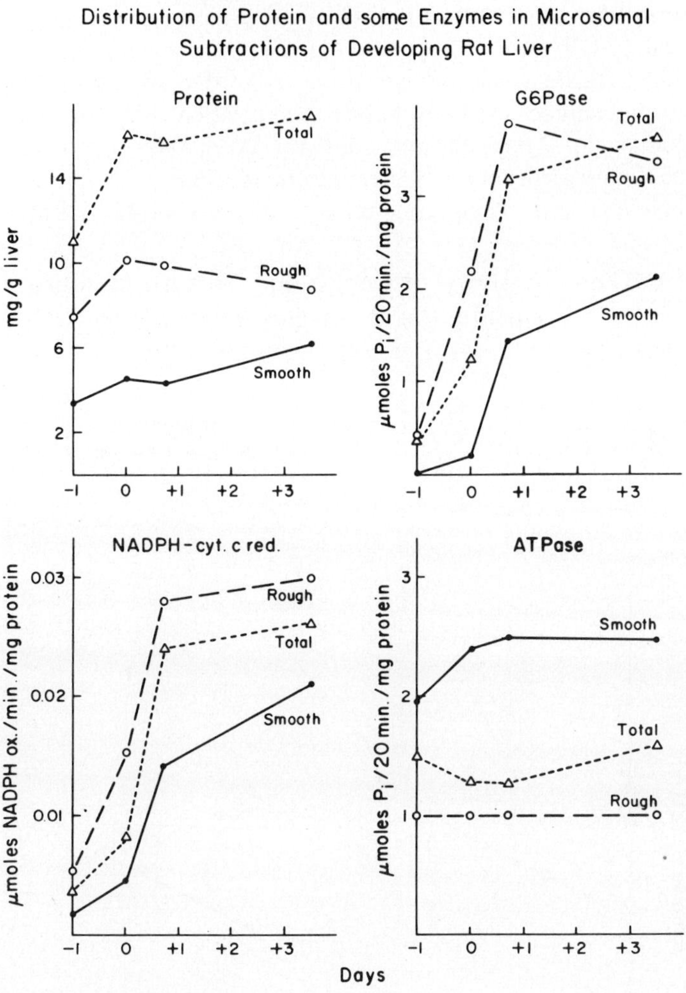

FIG. 9. Distribution of protein and some enzymes in rough and smooth microsomes of developing rat liver (4).

phospholipids are necessary for the assemblying of specific proteins into membrane structures (5, 7). As can be noticed from Figs. 3 and 4, liver microsomes isolated from animals of various ages have greatly varying proportions of microsomal membrane enzymes. Is this variation with time reflected in the phosphatide composition, since phospholipid is the major lipid class of these membranes (*cf.* 18)? Even though we are comparing microsomes containing mostly "old" membranes with microsomes containing both "old" and "new," we feel that the net increase of smooth membrane structures and of membrane enzymes is such that any gross changes in lipid composition in the new membranes should be discernible. Table I shows that there is hardly any difference in the relative percentage of phosphatide composition between microsomes from 2-hour-old, 2-day-old, and adult rat liver. Table II shows that the relative *in vivo* incorporation rates into the various phosphatides are also not much changed. There is a relative increase of PE labeling and a

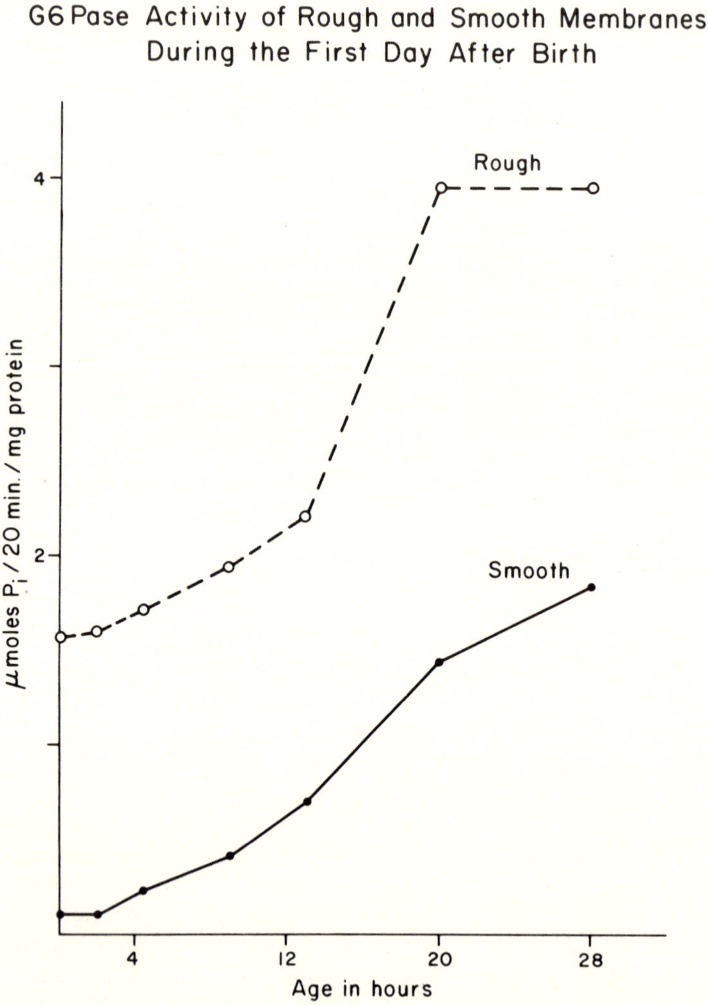

FIG. 10. Glucose-6-phosphatase of rough and smooth microsomes during the first day after birth (4).

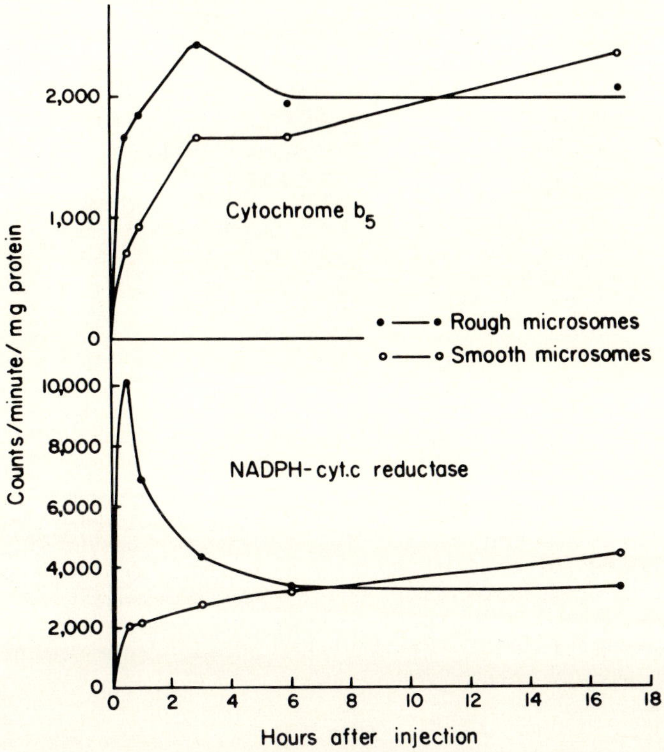

FIG. 11. Comparison of the specific radioactivities of cytochrome b_5 and of NADPH–cytochrome *c* reductase in rough microsomes with the specific radioactivities of the same enzymes in smooth microsomes. Leucine-1-^{14}C was injected into young adult rats. The two enzymes were isolated and purified from membranes prepared from rough and smooth liver microsomes.

relative decrease in SPH labeling in adult rat liver, but changes between 2-hour-old and 2-day-old rat liver—that is, when large changes in enzyme activity occur—are not marked. We did find (Table III) that the fatty acid composition of the microsomal phospholipid was quite different in the different age groups; this could be a matter of difference in the dietary regimes of the rats at these different periods. To try to exclude this dietary influence on enzyme levels of activity, we purposely put the pregnant females on a basic diet either alone or supplemented with corn oil or lard, and then determined the enzymic composition of the liver ER membranes from 5-day-old rats born from these females. As can be seen (Table IV), the fatty acid composition of the liver microsomal phospholipids of the 5-day-old rats was quite changed with respect to percentage composition of various fatty acids as a result of

TABLE I

PHOSPHATIDE COMPOSITION OF HEPATIC MICROSOMAL MEMBRANES AS A FUNCTION OF AGE IN RATS (3)[a]

Age of rats:	2 hours		2 days		Adult	
	μg PLP-P / gm liver	% of total	μg PLP-P / gm liver	% of total	μg PLP-P / gm liver	% of total
Total	97.5	100	115.0	100	196.0	100
PE	17.5	17.9	23.7	15.3	29.8	15.2
PS	8.7	8.9	19.7	12.7	15.3	7.8
PC	46.6	47.8	61.6	39.8	87.3	44.6
SPH	8.5	8.7	18.6	12.0	19.2	9.8
PI[b]	5.9	6.1	13.3	8.6	15.7	8.0
LPI	2.5	2.6	5.9	3.8	7.4	3.8
PA	1.8	1.8	1.9	1.2	3.1	1.6
Recovery	91.5	93.8	114.7	93.4	177.8	90.8

[a] Abbreviations are: PE, phosphatidyl ethanolamine; PS, phosphatidyl serine; PC, phosphatidyl choline; SPH, sphingomyelin; PI, phosphatidyl inositol; LPI, lysophosphatidyl inositol; PA, phosphatidic acid; PLP-P, phospholipid phosphorus. Representative experiment from a series of four. Variations from one experiment to another were within 6%.

[b] The content of the PI area of the chromatogram is heterogeneous: it comprises mono-, di-, and triphosphoinositides.

TABLE II

INCORPORATION OF ^{32}P INTO THE INDIVIDUAL PHOSPHATIDES OF DEVELOPING HEPATIC MICROSOMAL MEMBRANES (3)[a]

Age of rats[b]:	2 hours		2 days		Adult	
	cpm / μg PLP-P	% of total radio-activity	cpm / μg PLP-P	% of total radio-activity	cpm / μg PLP-P	% of total radio-activity
Total	5260	100	3790	100	1910	100
PE	7890	25.7	6440	23.5	5070	35.0
PS	3500	8.3	2790	8.4	1820	7.8
PC	6320	38.5	4470	42.5	1570	31.8
SPH	3380	8.0	2570	7.3	920	4.1
PI	2820	4.4	2370	4.8	1510	5.5
LPI	1620	1.2	1620	1.5	800	1.4
PA	1060	0.5	870	0.3	980	0.7
Recovery		86.6		88.3		86.3

[a] Abbreviations as in Table I. Representative experiment from a series of three. Livers were removed 3 hours after intraperitoneal injection of $^{32}PO_4$.

[b] At the time of injection.

TABLE III

FATTY ACID COMPOSITION OF MICROSOMAL PHOSPHOLIPIDS AS A FUNCTION OF AGE OF RAT (3)[a]

Type of fatty acid	Percent of total fatty acids at: 0 day	5 days	90 days
16:0	23.9	24.8	10.6
16:1	4.8	1.0	0.3
18:0	17.7	18.9	22.1
18:1	17.1	5.0	13.9
18:2	7.1	6.4	23.5
20:4	15.6	22.9	17.4
22:6	13.7	21.1	12.1

[a] The lipid extracts from 10 to 14 livers are pooled in each case, and fatty acid is determined by gas-liquid chromatography.

TABLE IV

EFFECT OF DIET ON COMPOSITION OF PHOSPHOLIPID FATTY ACIDS AND ON ENZYMIC ACTIVITIES OF HEPATIC MICROSOMES (4)

Diet:	Corn oil	Lard	Basic	
Age of rats:	5 days	5 days	5 days	Adult
Phospholipid/protein of microsomes	0.24	0.23	0.24	0.31
Fatty acid composition (% of total fatty acids)				
16:0	34.4	29.9	24.8	10.6
18:0	18.4	22.6	18.9	22.1
18:1	6.6	10.8	5.0	13.9
18:2	12.6	5.7	6.4	23.5
20:4	18.7	21.0	22.9	17.4
22:6	9.3	9.9	21.1	12.1
Specific enzyme activities				
NADPH-cytochrome *c* reductase[a]	0.023	0.017	0.019	0.023
NADH-cytochrome *c* reductase[a]	0.099	0.101	0.111	1.11
Demethylation[b]	0.84	0.89	0.98	4.37
IDPase[c]	2.34	2.02	2.04	12.1
ATPase[c]	0.91	1.04	1.17	1.17

[a] Micromoles NADPH or NADH oxidized/min/mg protein.

[b] Micromoles formaldehyde/min/mg protein.

[c] Micromoles P_i/20 min/mg protein.

these changes in the diet of their mothers. Nevertheless (Table IV), the rates of net enzyme synthesis and assembly into microsomal membranes were unaltered; enzymes which were normally rapidly synthesized, such as the NADPH–cytochrome c reductase, were still rapidly synthesized and put into membranes; enzymes that were normally slowly synthesized, such as the NADH–cytochrome c reductase, the IDPase, and the demethylation complex, were still slowly synthesized; while the activity of enzymes already present, such as the ATPase, was unaltered by the dietary changes. Finally, we performed an extraction (12) of the phospholipids of liver microsomes from both baby and adult rats, and, after appropriate controls were performed (*cf.* 4), we concluded, as Table V indicates, that the reactivation of the enzyme activity of the microsomal membrane enzymes can be accomplished by practically any combination of phospholipids, such as asolectin from soybean, or, better yet, by a specifically mitochondrial phospholipid such as cardolipin. All these experiments indicate that, within the limits tested, phospholipids, while necessary for the enzyme activity of the membrane proteins, are not

TABLE V
EFFECT OF VARIOUS PHOSPHOLIPIDS ON NADH–CYTOCHROME *c* REDUCTASE ACTIVITY OF LIPID-EXTRACTED MICROSOMES OF 8-DAY-OLD RATS (4)

Lipid	Mg lipid added		Micromoles NADH oxidized/ min/mg protein
	During preincubation	In assay	
None	—	—	0.042
Asolectin	100	—	0.092
Asolectin	100	8	0.092
Asolectin	100	16	0.093
Adult rat			
Microsomal	25	—	0.089
Microsomal	25	2	0.091
Microsomal	25	4	0.093
8-day-old rat			
Microsomal	18	—	0.079
Microsomal	18	1.4	0.083
Microsomal	18	2.8	0.083
Cardiolipin	24	—	0.062
Cardiolipin	24	1.5	0.126
Cardiolipin	24	3	0.117
Phosphatidyl choline	24	—	0.064
Phosphatidyl choline	24	1.5	0.066
Phosphatidyl choline	24	3	0.066
Mitochondrial	30	—	0.064
Mitochondrial	30	4	0.071
Mitochondrial	30	10	0.084

specific factors for the assembly of constitutive enzymes into membrane structures. These experiments do not exclude the possibility that small differences in lipid composition, as in types of fatty acids on the α or β positions of the glycerol, are important in this regard. But, at present the data suggest to us that the function of the lipid component of the membrane is to maintain hydrophobic conditions for catalytic processes, to bridge the gap between hydrophilic and hydrophobic functional groups (8), and possibly to confer different membrane configurations on specific membranes, such as cisternae in the ER, cristae in mitochondria, and lamellae in chloroplasts.

Also of interest are the findings which suggest that the synthesis of enzymes involved in the production of membrane lipids is not geared to the appearance of the membranes under study. For example, as far as we can tell from our data and from those of others (19) in the case of the developing rat liver, the necessary enzymes for phospholipid synthesis are already present when smooth membrane formation is called for just after birth. In addition, the labeled lipids are of higher specific radioactivity in the rough than in the smooth ER (Fig. 12), somewhat the same as with membrane proteins (Fig. 8). In the case of

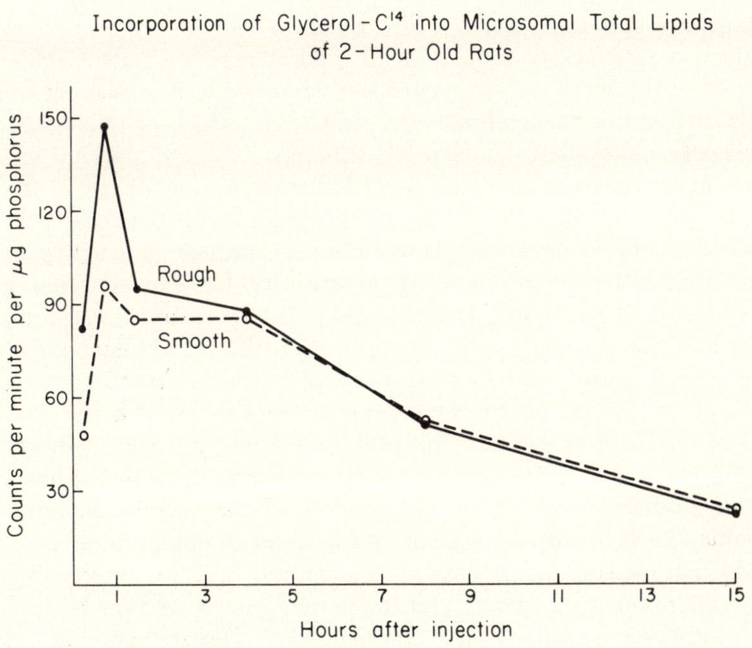

FIG. 12. Incorporation of glycerol-^{14}C into total lipid of rough and smooth microsomes of 2-hour-old rats (3).

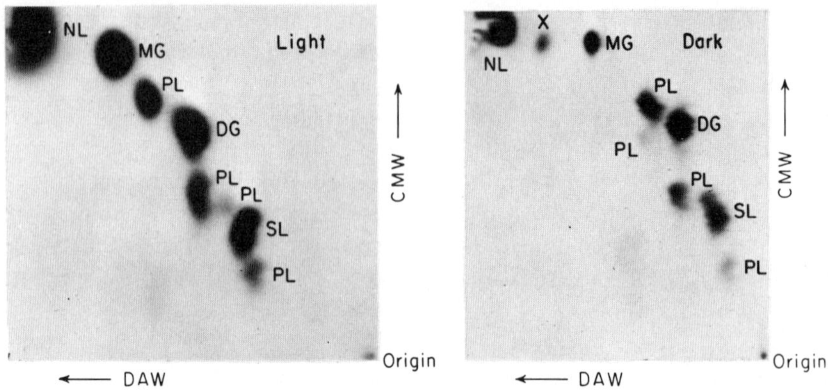

FIG. 13. Radioautograph of ^{14}C-labeled lipid extracts from dark-grown mutant *Chlamydomonas* cells incubated for 3.5 hours in light or dark with acetate-^{14}C (15). Separation by two-dimensional chromatography on silicic acid-impregnated paper. PL = phospholipid; SL = sulfolipid; MG = monogalactosyl glyceride; DG = digalactosyl glyceride; NL = neutral lipid and carotenoids; X = unknown.

the chloroplast lamellae in mutant *Chlamydomonas*, Fig. 13 indicates that there is no qualitative difference between the lipid composition of light-grown and dark-grown cells. Even though we have not as yet obtained a satisfactory separation of chloroplasts from algal cells, and have done the work on extracts of the whole cells, it is instructive that lipids such as the galactolipids and sulfolipids, which are thought to be specifically chloroplast lamellar constituents (1), are present in cells which have lost their lamellae; indeed their rate of synthesis in the full yellow cell is about 50% of that in the green cell. Thus the enzymes involved in the synthesis of these lipids are present and active in the algae, similar to the presumptive constitutive lamellar enzymes (Fig. 6*a*, *b*). Equally intriguing is the finding that these lipids are probably being laid down some place, perhaps within the chloroplast, other than their normal, and essentially final, destination, in the lamellar structures. We have no idea of the positioning mechanisms involved here.

The final results concern the problem of whether the membrane is turning over as a whole unit or whether a particular subunit has some sort of stability, while other components of the membrane are more dynamic. To this purpose we injected a series of young adult rats with radioactive leucine and acetate as protein and lipid precursors, isolated whole microsomes, or smooth and rough microsomes, and further isolated the membranes present in these fractions (16). Determination of specific activity at various times after injection gave us the regression lines of Figs. 14 through 16. It can be seen, first of all, that the half-lives for the

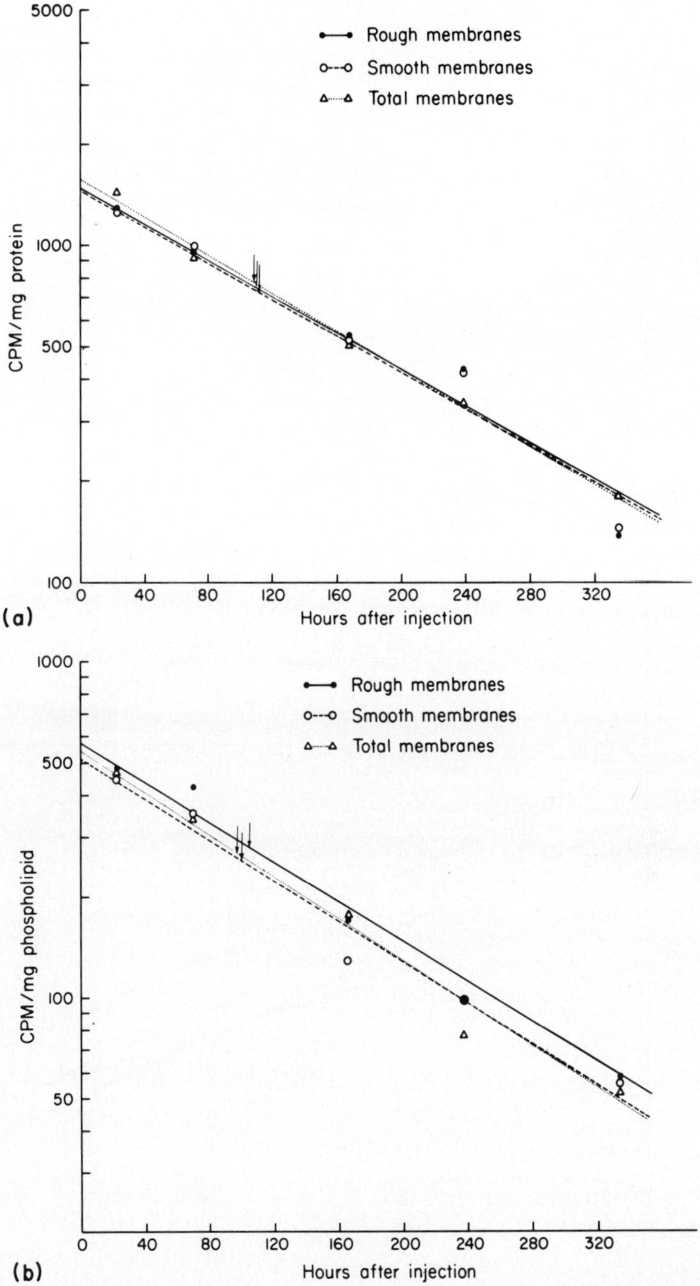

FIG. 14. Decrease with time of specific radioactivities of total proteins (*a*) and total phospholipids (*b*) of isolated total, rough, and smooth microsomal membranes (16). Leucine-1-^{14}C and acetate-1-^{14}C are the radioactive precursors. Each time point represents the pooled liver of 3 rats. Arrows denote half-life times, determined graphically.

total membrane constituents for the rough and smooth membranes were the same. They ranged from 108 to 113 hours for the total protein (Fig. 14*a*) and from 97 to 105 hours for the total lipid (Fig. 14*b*). At first glance this would indicate that the membrane is being replaced as a unit.

However, because the lipid half-life values, in these and other ex-

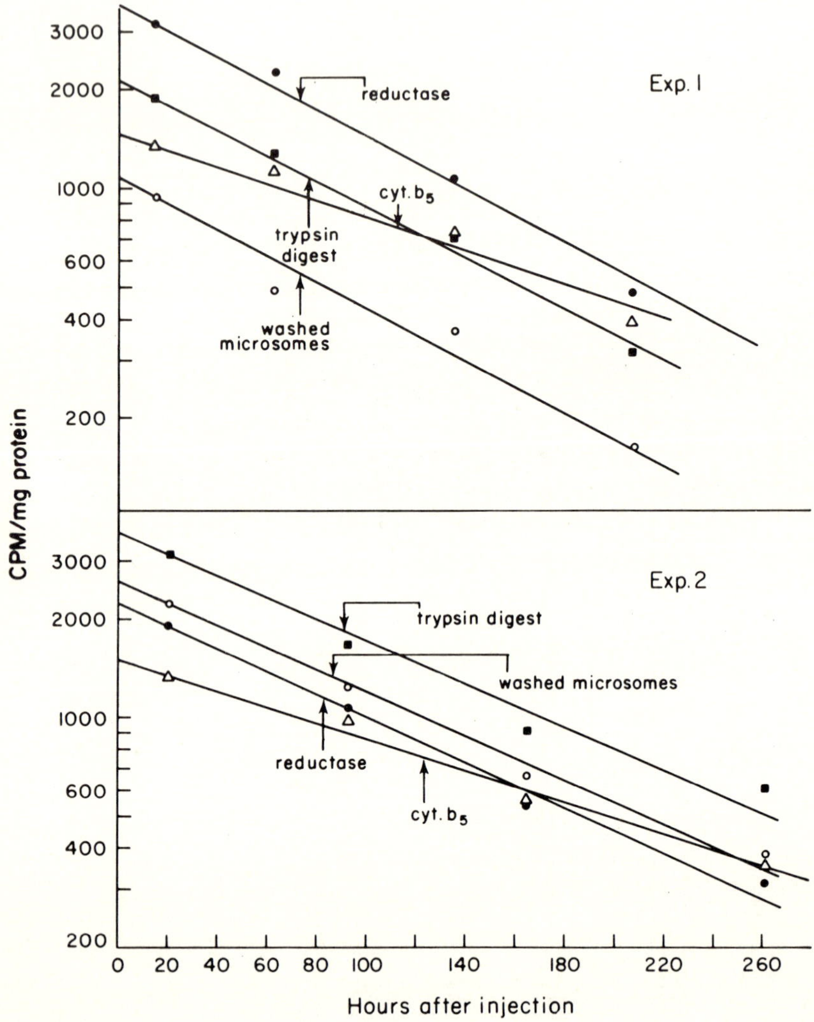

FIG. 15. Decrease with time of specific radioactivity of purified cytochrome b_5, NADPH–cytochrome *c* reductase, total protein of washed microsomes, and protein of microsomal residue (mostly membranes) after trypsin digestion (trypsin residue) (16). Leucine-1-^{14}C was the radioactive precursor. Each time point represents the pooled livers of 3 rats. Arrows denote half-life times, determined graphically.

periments, were always less than the protein half-life values, and because we wished to look more closely at the individual components of the membranes, we have found that we need to revise these conclusions, as a result of further experiments, exemplified by Figs. 15 and 16. Two loosely bound protein constituents of the ER membranes (6) were examined; these were isolated and purified as described (16). It is seen (Fig. 15) that in two separate experiments the half-life of the NADPH–cytochrome *c* reductase is equal to that of the membrane proteins as a whole (75 hours and 83 hours for the enzyme, and 75 hours and 87 hours for the membrane proteins). However, the half-life of cytochrome b_5 is significantly longer, being 117 and 123 hours in the two experiments. There are two possible explanations for this result: (a) that the bulk of the protein constituents of the membrane are turning over at identical rates, and a small number of individual proteins, exemplified by cytochrome b_5, are turning over differently; or (b) that the ER is made up of patches, or tesserae, as mentioned in the Introduction, that these tesserae turn over at different rates, and that cytochrome b_5 represents the proteins of one of the tesserae. At present, we have not enough data to be able to choose between these two alternatives, but we feel that the difference recorded between NADPH–cytochrome *c* reductase and cytochrome b_5 may be connected with the fact that the latter is a heme protein whose synthesis involves an additional step (heme complexing) which probably affects its rate of synthesis and may even also affect its rate of decay. We also have no conclusions regarding the turnover for a "structural protein," if such exists in the ER membranes, for the turnover of a single protein species may be quite different from that of the bulk of the membrane proteins and, unless isolated, cannot be deduced from the data of Figs. 14 and 15.

We then looked at subfractions of the lipid component of the membrane, and the relevant findings are shown in Fig. 16*a* and *b*. First, it is clear that the difference exemplified in Fig. 14*a* and *b* between turnover of total lipids and total proteins is real; the half-lives of total protein in Fig. 14*a* are 90 and 95 hours, and those of lipid 78 and 56 hours. Furthermore, the aqueous fraction, which after deacylation contains the glycerol backbone, is apparently turning over at a faster rate than the total lipids, the half-lives being 40 and 42 hours for the initial slopes in the two experiments. (The reason for the nonlinearity of the line will be mentioned below.) Because with acetate as precursor only a small percentage of the counts incorporated in total lipids appears in their aqueous fraction (only 15% at the initial time point), we repeated the experiments, replacing acetate by glycerol as precursor. The variance in components, shown in Fig. 16*b*, is exaggerated; the half-lives of total

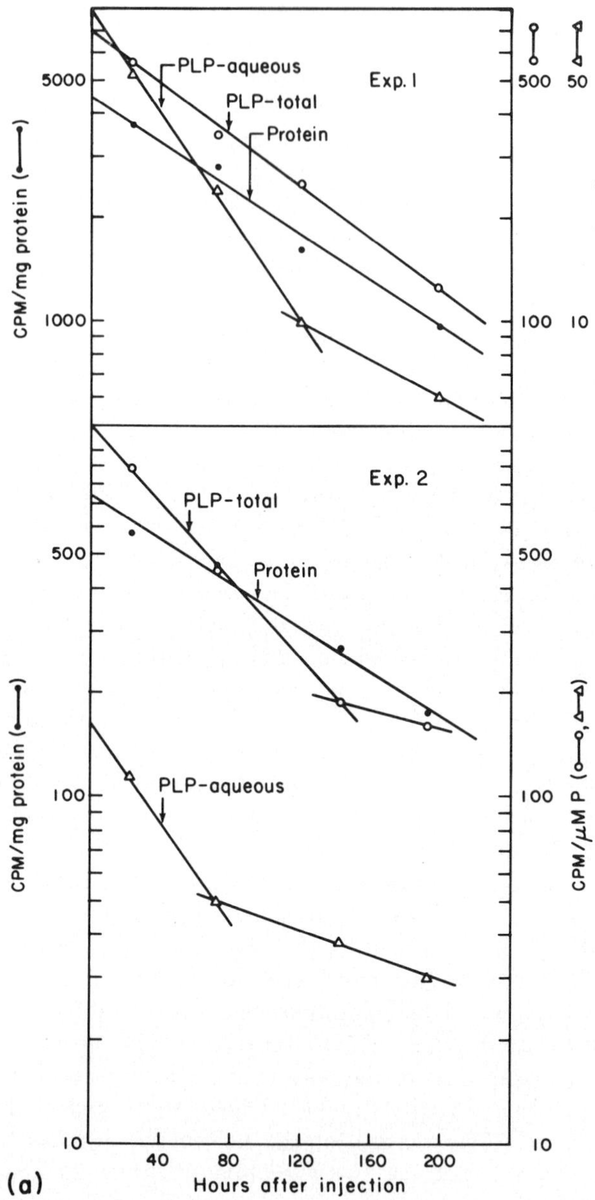

FIG. 16*a*. For legend see opposite page.

protein, total phospholipids, and aqueous extracts of the latter are 97, 44, and 29 hours, respectively. (Here, over 90% of the counts at the initial time points were in the aqueous fraction.) It is clear that the total lipid is turning over at rates quite higher than that of total protein, and that, within the lipid molecules, the glycerol backbone is turning over at a rate apparently higher than that of the fatty acids. The half-lives of the nonaqueous (mostly fatty acid) fractions can be approximately calculated from the radioactivity present there and, since recovery was the same at all time points, from the amount of phosphate in the original total lipid extract; these values were 81 and 72 hours for the two acetate experiments, and 140 hours for the glycerol experiments. The differences in half-lives between the glycerol and fatty acid components of the lipids can be explained by the presence in microsomes of enzymes which transacylate fatty acids from a phosphatide to a lysophosphatide, and elongate fatty acid chains beyond 16 carbon atoms (*cf.* 11). Thus it could be that the whole lipid molecule of the membrane is breaking

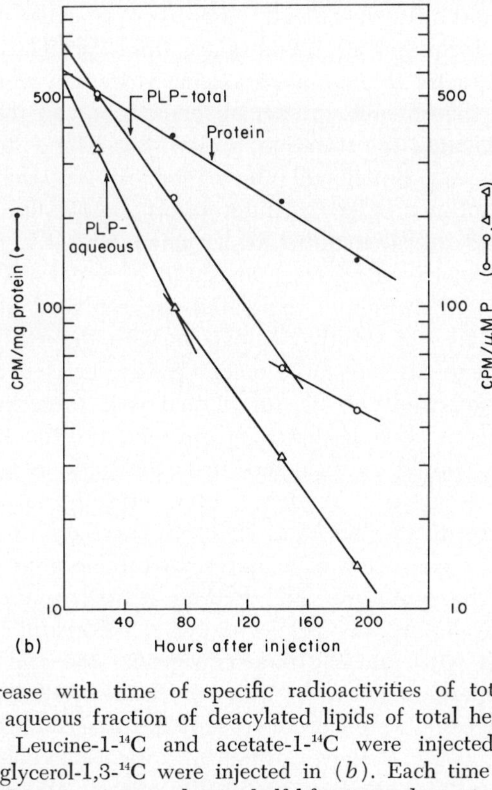

FIG. 16. Decrease with time of specific radioactivities of total protein, total phospholipid, and aqueous fraction of deacylated lipids of total hepatic microsomal membranes (16). Leucine-1-^{14}C and acetate-1-^{14}C were injected in (*a*), while leucine-1-^{14}C and glycerol-1,3-^{14}C were injected in (*b*). Each time point represents the pooled livers of 3 rats. Arrows denote half-life times, determined graphically.

down and being replaced as a unit, but, since its fatty acids are locally transacylated from one phosphatide to another and elongated without breaking down to 2-carbon units, their half-life in the membrane could be considerably longer than the half-life of the whole lipid molecules. The complexity of the situation is further indicated by the finding that the ratio of radioactivity in the nonaqueous fraction to that in the aqueous fraction of the total lipids is constantly changing with time after injection, with either acetate or glycerol as precursor (this ratio changes from 4.9 to 2.7 after acetate injection, and from 0.09 to 0.72 after glycerol injection). This finding may be related to the breaks in some of the decay curves (Fig. 16*a* and *b*) at late time points.

General Discussion

I think you will agree that one theme has been incessantly hammering in our ears, and this is that cellular membranes are dynamic structures. Their components are constantly being turned over and renewed, and yet the overall membrane structure is being preserved and, most importantly, specifically conserved. Specific proteins complex with lipids to form specific membranes; lipid specificity does not seem to be a critical factor in membrane assembly, though admittedly we have only looked at the obvious, grosser elements involved. To the extent that we have investigated the situation, and with reference to the absence of membrane tesserae, mentioned above, we can say that these proteins fall in and out of membranes at different rates. Also, lipids are assembled into membranes at rates quite at variance with the turnover of the protein constituents; moreover, the findings suggest that the various moieties of the phospholipid molecules are replaced at different rates. All these data are not strictly compatible with the idea of a membrane being made up of discrete units or complexes, hooked together to form a complete membrane (*cf.* 9), for in that case these complexes should be turning over as a whole, lipid and protein, and the average half-lives of the proteins should be expected to be the same as that of the lipids. However, this is negative evidence, and certainly not conclusive.

The problem of whether there is a "primary" or "ur-structure" of the membrane (17), consisting of a lipid–structural protein complex, is also not answered by our data. The finding regarding the appearance of enzymes in developing membranes of the newborn rat liver would tend to support this view, but the data concerning the appearance of functioning photosynthetic electron transfer chains in the reappearance of chloroplast lamellae would seem to indicate that such a primary structure is not necessary. However, these interpretations could be reconciled with each other by assuming a rate difference in the assembly of complete functioning membranes between the two cases.

I might say a word about morphological parameters of membrane formation, using the developing ER as an example. If we assume that the proteins of the assemblying membrane are made by ribosomes attached to existent complete membranes (*cf.* 18), and that the lipids, at least the phospholipids, are synthesized by enzymes localized in the same membranes (*cf.* 18), how is the new membrane formed? A priori it could be assembled and completed as an outgrowth of rough ER, or as a lengthening process between groups (polysomes) of membrane-bound ribosomes. From some morphological evidence (3), we tend to think that the latter may be the case, for in the liver of week-old rats the polysomes on membranes are spaced further apart than they are in the newborn rat liver, before new membrane synthesis has started, as if patches of membrane are being laid down between polysome arrays. In any case, we must also assume that ribosomes are bound to and fall off the membrane with comparative frequency, as if, after completion of the synthesis of membrane proteins, they are removed from that membrane site, to be translocated at another (smooth) membrane site; in other words, that the only difference between rough and smooth membranes is the temporary presence of ribosomes on the former. This rather rapid turnover of ribosomes on membranes is given credence by the half-life of ribosomal RNA, ~5 days (10, 13), and by the data in Fig. 11. These latter curves do not fit the classical precursor-to-product relationship; that is, they cannot be explained only in terms of enzyme synthesis in rough membranes followed by their transfer to smooth membranes. They could be explained, however, by further assuming that there is rapid exchange of membrane between the rough and the smooth ER, presumably by detachment–attachment of ribosomes, at a rate much higher than the turnover rate of the constitutive enzymes tested, so that in a short while (~17 hours in the experiments of Fig. 11) the two membrane types are equivalent with respect to the radioactivity of their newly synthesized enzymes.

All in all, our experiments have reached the end point of easy returns. Further work must somehow ascertain the mechanism involved in the coassembly of specific proteins and lipids to form a specific membrane, the "recognition" signals for this specificity, and the means the cell has at its disposal for the control of membrane biogenesis.

Acknowledgments

Figures 3, 4, 5, 8, 9, 10, and 12, and Tables I, II, III, IV, and V, are from refs. 3 and 4 and are reproduced by permission of the *Journal of Cell Biology* and The Rockefeller University Press. This work has been supported by Grants AM-01635, HD-01689, and GM-11325 from the National Institutes of Health.

REFERENCES

1. BENSON, A. A., *Ann. Rev. Plant Physiol.*, **15**, 1 (1964).
2. DALLNER, G., *Acta Patol. Microbiol. Scand. Suppl.*, p. 166 (1963).
3. DALLNER, G., SIEKEVITZ, P., AND PALADE, G. E., *J. Cell Biol.*, **30**, 73 (1966).
4. DALLNER, G., SIEKEVITZ, P., AND PALADE, G. E., *J. Cell Biol.*, **30**, 97 (1966).
5. DAS, M. L., AND CRANE, F. L., *Biochemistry*, **3**, 696 (1964).
6. ERNSTER, L., SIEKEVITZ, P., AND PALADE, G. E., *J. Cell Biol.*, **15**, 541 (1962).
7. FLEISCHER, S., BRIERLEY, C., KLOUWEN, M., AND SLAUTTERBACK, D. B., *J. Biol. Chem.*, **237**, 3264 (1962).
8. GREEN, D. E., AND FLEISCHER, S., *in* "Horizons in Biochemistry" (M. Kasha and B. Pullman, eds.), p. 381. Academic Press, New York, 1962.
9. GREEN, D. E., AND HECHTER, O., *Proc. Natl. Acad. Sci. U.S.*, **53**, 318 (1965); GREEN, D. E., AND PERDUE, J. F., *ibid.*, **55**, 1295 (1966).
10. HIATT, H. H., HENSHAW, E. C., HIRSCH, C. A., REVEL, M., AND FINKEL, M. D., *Israel J. Med. Sci.*, **1**, 1323 (1965).
11. LANDS, W. E. M., *Ann. Rev. Biochem.*, **34**, 313 (1965).
12. LESTER, R. L., AND FLEISCHER, S., *Biochim. Biophys. Acta*, **47**, 358 (1961).
13. LOEB, J., HOWELL, R., AND TOMKINS, G., *Science*, **149**, 1093 (1965).
14. MANGIELLO, V. C., AND PHILLIPS, A. H., *J. Biol. Chem.*, **240**, 3951 (1965).
15. OHAD, I., SIEKEVITZ, P., AND PALADE, G. E., *Federation Proc.*, **25**, 255 (1966); *J. Cell Biol.*, in press.
16. OMURA, T., SIEKEVITZ, P., AND PALADE, G. E., *J. Biol. Chem.*, **242**, 2389 (1967).
17. RICHARDSON, S. H., HULTIN, H. O., AND GREEN, D. E., *Proc. Natl. Acad. Sci. U.S.*, **50**, 821 (1963).
18. SIEKEVITZ, P., *Ann. Rev. Physiol.*, **25**, 15 (1963).
19. VILLE, C. A., AND HAGERMAN, D. D., *Am. J. Physiol.*, **194**, 457 (1958).

Invertase Biosynthesis and the Yeast Cell Membrane

J. Oliver Lampen, Norbert P. Neumann, Santiago Gascon,[1] and Bland S. Montenecourt

Institute of Microbiology, Rutgers, The State University, New Brunswick, New Jersey

As biologists and biochemists give increasing attention to the integrated functions of the living cell, they are frequently concerned with the problem of explaining how an organism produces an extracellular macromolecule which does not penetrate the cell membrane by free diffusion. One is interested in where the macromolecule (particularly an enzyme) is formed, how it reaches the outside of the cell, and how it is released. It is obvious that the membrane must play a role in this process, and hypotheses representing the extreme possibilities can be suggested for orientation. (*a*) The enzyme may be synthesized in the cytoplasm and exist there temporarily in a free state. It would subsequently be passed through the membrane by mechanisms as yet unknown. (*b*) The enzyme might be formed either within the membrane or on its inner surface and become external without a free stage in the cell. It is possible that the synthesis of carbohydrates, muropeptides, and lipids is truly external and occurs on the outer side of the membrane. These macromolecules contain a relatively small variety of repeating units, and the total polymer has no unique structure. Monomers or oligomers may pass through the membrane as lipid-bound or lipid-soluble intermediates and be polymerized on the outside (2). In contrast, synthesis of protein requires the specific ordering of twenty different amino acids, and only a very limited range of final structures would be active. The formation of exoenzymes is usually prevented by inhibitors of cytoplasmic protein synthesis (7) and the same complex machinery (ribosomes, messenger RNA, transfer RNA, etc.) is probably necessary for both processes. Thus it is extremely difficult to visualize how synthesis of exoenzymes could take place entirely on the exterior of the cell.

[1] Fellow of the Fundacion Juan March. Permanent address: Instituto de Biologia Celular, Velasquez 138, Madrid 6, Spain.

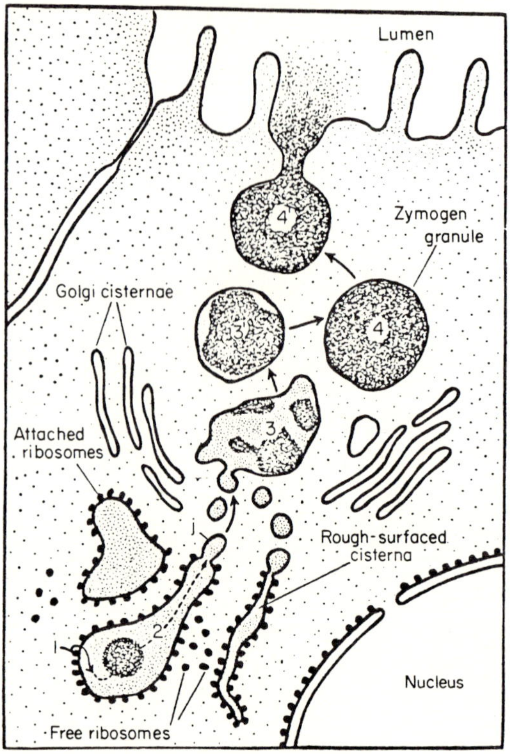

FIG. 1. Diagram of the stages in transport of the digestive enzymes inside the exocrine pancreatic cell. *1*, Synthetic phase and beginning of transport. *2*, Intracisternal transport to Golgi apparatus. *3*, Concentration of secretion products in smooth-surfaced vacuoles. *4*, Storage in zymogen granules. *4'*, Discharge into the acinar lumen. From Palade *et al.* (12).

Many types of cells produce external enzymes (bound or free), but microorganisms may present the phenomenon in its most simplified form. The stages in the secretion of an enzyme by mammalian cells are illustrated in Fig. 1 which represents the formation of digestive enzymes by pancreatic tissue (12). The enzyme is synthesized on the rough endoplasmic reticulum (step *1*) and then appears within the lumen of the reticulum. At this stage it can properly be considered external. Subsequently, the secreted enzyme is collected in the Golgi apparatus, stored in the zymogen granules, and ultimately released from the cell by fusion of the granules with the cell membrane (6); but those stages are secondary to the critical process and essential only because of the large size of the mammalian cell. A microorganism can be compared with an inverted sac of rough endoplasmic reticulum, and secretion of an enzyme by a microorganism with the first stage of the more elaborate process in the pancreatic cell.

INVERTASE–YEAST PROTOPLASTS

The system primarily studied in our laboratory is the formation and release of invertase by yeast protoplasts (for review, cf. ref. 7). The major phenomena and certain pertinent findings are illustrated schematically in Fig. 2. When a yeast culture is producing substantial amounts of invertase, the enzyme usually remains attached to the organisms, but at least 95% of it is external to the cell membrane and available to added substrate. Treatment of the organism with snail gut juice in the presence of an osmotic support will release the external invertase and convert the walled cell to a protoplast that with certain strains of yeast can be made essentially free of external enzyme (5). Under appropriate conditions the protoplast will form and release large quantities of invertase (and other proteins). Glucan and mannan are formed concurrently; N-acetylglucosamine is also present in the secreted macromolecules (3).

Gascon and Ottolenghi (4) observed that rupture of the yeast cell released about 95% of the invertase in a form that was "large" by the criterion of gel filtration on Sephadex G-200. Variable amounts of a "small" invertase were also present. Further investigation of this phenomenon with our strains of yeast has shown that the invertase released from cells during conversion to protoplasts (for example, by snail enzyme) is in the large form, as is the enzyme that is secreted by protoplasts. Essentially all the internal enzyme released by lysis of protoplasts is small. Present results are consistent with the generalization that

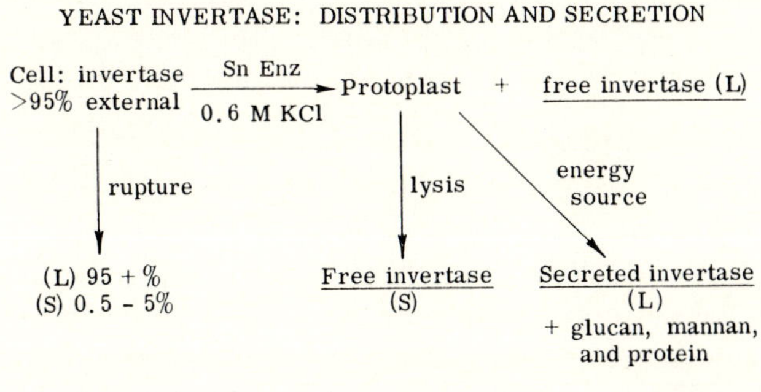

FIG. 2. Yeast invertase distribution and secretion. (L) and (S) refer to the large and small forms of invertase as distinguished by gel filtration on Sephadex G-200 (see Fig. 4).

invertase outside the cell membrane exists only as the large form, whereas the internal enzyme is exclusively small.

Nature of External Invertase

To investigate effectively the phenomenon of secretion, it was necessary to determine the characteristics of the external invertase of our strains of yeast. Previous preparations, though often of high specific activity, had been polydisperse perhaps because prolonged autolysis was generally used to liberate the enzyme. We attempted to avoid this by breaking the cells quickly in the cold and carrying out the purification as rapidly as possible. The fractionation is outlined in Table I. After

TABLE I
PURIFICATION OF YEAST INVERTASE FROM *Saccharomyces* STRAIN FH4C

Fraction	Volume (ml)	Protein (mg)	Invertase (units)	Specific activity (units/mg protein)
I Yeast suspension	318	10,400	254,000	24
II Crude juice	350	6,620	212,000	32
III Heat 50°C, pH 5.0	335	3,450	208,000	60
IV Ethyl alcohol 50%	40	1,030	157,000	152
V Ammonium sulfate	110	47	103,000	2190
VI SE and DEAE-Sephadex	125[a]	21	56,600	2700

[a] Lyophilization of the desalted material yielded 37 mg of enzyme.

some purification by heating at pH 5 and precipitation with alcohol, the preparation was made 0.8 saturated with $(NH_4)_2SO_4$. Under these circumstances, over 95% of the total protein precipitated, leaving the bulk of the invertase in the supernatant. Chromatography of this fraction on SE-Sephadex or DEAE-Sephadex yielded homogeneous material.

The purified enzyme is a glycoprotein containing approximately 50% mannan and 3% glucosamine (10, 11). By treating the invertase with proteolytic enzymes, mannan-peptides have been obtained. Analysis of these peptides implicates serine, threonine, and aspartic acid as possible points of linkage between the protein and the mannan. All fractions obtained thus far also contain glucosamine. The molecular weight of the mannan-peptides as determined by gel filtration is less than 10,000; since invertase is approximately half mannan and has a molecular weight of about 270,000, there must be a number of mannan moieties attached at different points to the protein portion of the enzyme. We cannot say if the mannan is required for activity, although the fact that some highly active preparations made by autolysis have been extremely low in carbohydrate (1) suggests that most of the mannan will not prove to be essential.

Mutant Strain Insensitive to Repression by Hexose

The formation of invertase by most strains of yeast is readily repressed by metabolites and especially by hexoses. This sensitivity limited severely the earlier studies on the secretion of invertase; even with sucrose as the carbon and energy source, there was an initial rapid phase of secretion followed by a slow repressed phase (5). A mutant strain (FH4C) was obtained (following ultraviolet irradiation) which forms invertase in the presence of high levels of hexoses or of succinate (13). Figure 3

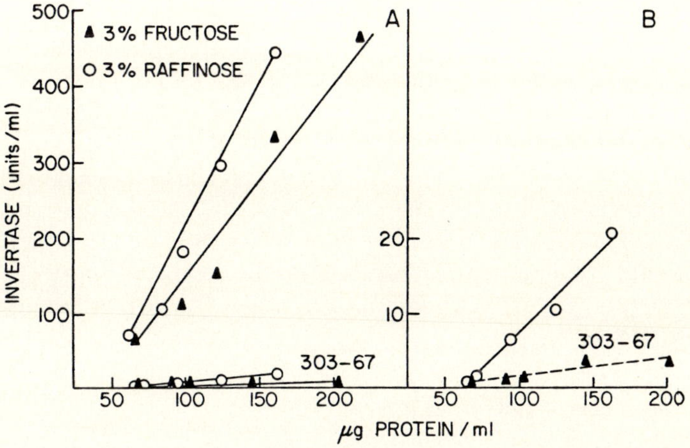

FIG. 3. Effect of hexose on invertase production by the hexose-resistant mutant FH4C and the parent strain 303-67. The organisms were grown overnight on yeast extract–peptone–2% raffinose medium and resuspended in Vogel's medium N with sugar as indicated for determination of the differential rate of enzyme formation. The data for strain 303-67 from graph *A* are replotted in *B* with different coordinates.

presents a differential plot of invertase formation in the presence of fructose or raffinose. The parent culture (strain 303-67) forms only small amounts of invertase with either sugar. Raffinose permits somewhat greater production of invertase by the mutant strain than does fructose; however, the mutant is clearly superior to the parent with either carbon source. Invertase represents 1 to 2% of the total protein of the mutant organisms. This hexose-resistant strain was the source of the purified invertase described earlier.

Nature of the Small Invertase

The behavior of the large and small forms of invertase during gel filtration using Sephadex G-200 is illustrated in Fig. 4. The fraction labeled L represents the invertase released from intact cells by the action of snail enzyme. A comparable pattern would be obtained with an extract of the mutant organisms or of the parent strain grown on low levels of glucose, although there would be some material present under the small invertase (S) peak. The small invertase fraction shown was obtained by lysis of protoplasts prepared from cells of the parent strain grown with high levels of glucose. It contains a small amount of large enzyme that probably was still attached to the outside of the membrane. A similar pattern would be obtained by gel filtration of an extract from intact cells. Small invertase has enzymic characteristics (pH optimum, K_m, and relative V_{max} for sucrose and raffinose) which are indistinguishable from those of the large enzyme. It differs in its gel filtration behavior and in its stability to pH. The internal invertase is stable at pH 7.5 and unstable at pH 5, precisely the reverse of the results obtained with the large enzyme.

One is, of course, anxious to know if the small enzyme contains mannan. The only indication that it may be low in mannan is that it is readily precipitated by 0.8 saturated $(NH_4)_2SO_4$ which does not precipitate the large mannan-containing form. We have no information concerning the biosynthetic relation between the two forms of invertase.

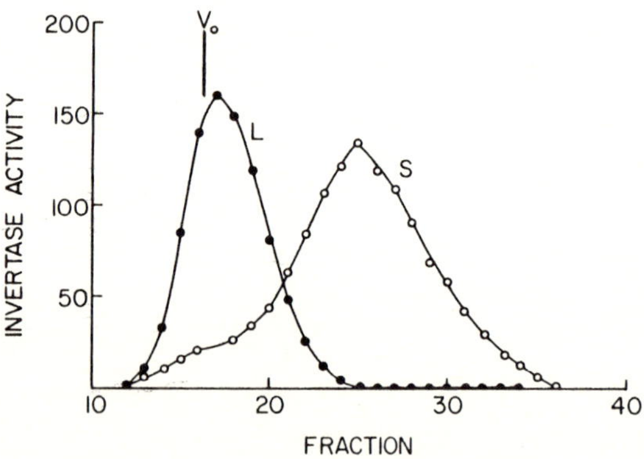

FIG. 4. Separation of large (L) and small (S) forms of invertase by gel filtration on Sephadex G-200. The column (2 × 50 cm) was equilibrated with 0.05 *M* Tris-HCl buffer, pH 7.5, and developed with the same buffer. V_0 = void volume of column. L = supernatant after protoplast formation from cells of parent strain 303-67 grown on a low level of glucose. S = lysate of protoplasts obtained from cells of strain 303-67 grown on 5% glucose.

Secretion by Protoplasts

As illustrated in Fig. 2, protoplasts secrete most of the major wall components—that is, polymers containing mannose or glucose and N-acetylhexosamine, protein, invertase, and acid phosphatase (7, 8). Two enzymes which appear to be exclusively intracellular, namely α-glucosidase and alkaline phosphatase, were not lost from the cell under these conditions; thus, there seems to be no gross leakage or damage to the cell membrane. One may conclude that release of the wall components (including invertase) by protoplasts occurs by a relatively specific mechanism and is not simply a result of cell membrane deterioration.

Secretion of invertase requires the presence of an energy source (a sugar is generally the best), but supplementation with amino acids has no effect, at least over a 2- or 3-hour period (7). The rate of formation and secretion of invertase is essentially independent of pH over the range of values from pH 4 to 7. By this criterion the system is "internal" and protected from changes in the external pH. Production of invertase is inhibited by cycloheximide (14) and thus presumably requires *de novo* protein synthesis. The polyene antifungal antibiotic, nystatin, will also inhibit invertase secretion. This antibiotic is effective at levels which cause sufficient disorganization of the cell membrane to permit leakage of small ions, but produce no gross disorganization of the cell membrane.

Discussion

One of the most intriguing questions arising from the study of invertase secretion is how an organism makes an enzyme that contains approximately equal parts of mannan and protein. One is interested in the degree to which the synthesis of mannan and that of protein are coupled and the possible reciprocal effect of inhibitors of the individual processes. For instance, will an inhibitor of mannan biosynthesis produce an invertase either low in or free of associated mannan? Characterization of the small internal invertase is under way. Knowledge concerning its composition may illuminate many features of the synthetic process and provide the basis for a sound experimental approach to the above questions.

It has been suggested (7) that formation of the exoenzyme penicillinase by *Bacillus licheniformis* may be directly coupled to membrane synthesis. For the invertase system, the only evidence bearing on this possibility is the observation that yeast protoplasts enlarge visibly during the period when invertase is being formed and thus that membrane biosynthesis is probably taking place. It should be emphasized that the synthesis and secretion of invertase by protoplasts seem to be part of a general unorganized synthesis of the components usually present in the cell wall. Metzenberg has reached the same conclusion from his studies with *Neurospora* (14). Polymers containing glucose and N-acetylhexosamine are laid down as a tangled fibrous mesh around the yeast protoplast. The other wall components (mannan, invertase, acid phosphatase, etc.) find no adequate receptor sites or matrix to hold them to the cell. They do not become attached and simply diffuse into the medium. Nevertheless, the availability of receptors is probably not a limiting factor in controlling invertase formation. The protoplast, which lacks effective receptor sites, makes about the same amount of enzyme as does the intact organism. Also in studies with cells whose walls had been partially removed by snail enzyme, the total production of invertase was relatively constant; the major variation was in the proportion retained by the cell (5).

Metabolite repression appears to be the major mechanism of control even with the protoplast system (5). The effect of hexose on invertase formation is illustrated in Fig. 5 for both the parent and the mutant strains. Secretion by the parent strain was effectively eliminated at 0.05 *M* glucose. The mutant was partially repressed at similar levels, but secretion was not further reduced by high concentrations of glucose. It seemed possible that the mutant strain produced two invertases, one repressible by hexose and the other not affected; however, it was shown that the enzyme made by the mutant in 0.125 *M* glucose and that made by the parent strain in 0.01 *M* glucose were indistinguishable. Both were the large invertase and had apparently identical enzymic characteristics.

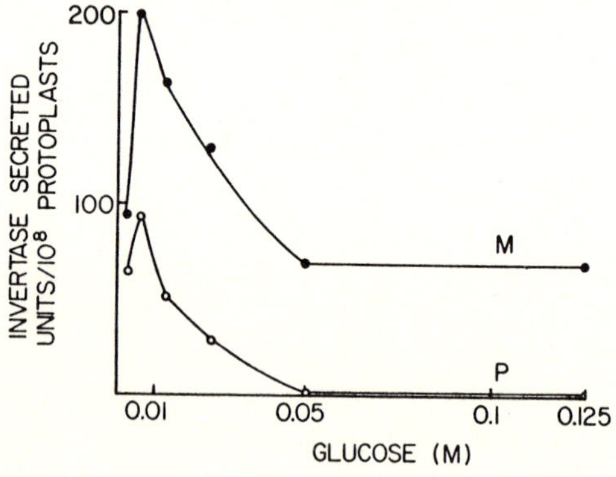

FIG. 5. Secretion of invertase by parent 303-67 (P) and mutant FH4C (M) protoplasts as a function of the concentration of glucose. Each incubation mixture contained 3.5×10^7 protoplasts/ml, 0.6 *M* KCl, 0.005 *M* $MgSO_4$, 0.02 *M* phosphate buffer, pH 6.0, and glucose as indicated. Incubated for 120 minutes at 30°C with shaking.

A final problem is the degree of dependence of invertase secretion on the fine structure of the membrane. Although Moor and Mühlethaler (9) have presented micrographs showing that large patches of the yeast cell membrane are relatively smooth with scattered lozenge-shaped depressions and hexagonal aggregates of particles, other investigators (see particularly the electron micrograph by Northcote shown in ref. 7) have observed considerable amounts of endoplasmic reticulum near the cell membrane and adjacent to the nucleus. One can only point out the limits set on our interpretations by our present information. The fine structure of the protoplast membrane has probably been altered by the rounding up after removal of the wall. Yet these forms still make invertase and other wall components at approximately normal rates. In contrast, the fact that nystatin inhibits the process shows that secretion can be eliminated by relatively specific damage to the cell membrane.

Conclusions

The major questions posed initially remain unanswered, but we have described here (*a*) purified forms of invertase and some information on their structure, (*b*) strains of yeast which enable us to eliminate or to manipulate metabolic repression, and (*c*) a convenient protoplast system forming and secreting large amounts of invertase. These improved tools should facilitate a more direct attack on the crucial problems in production of an external enzyme.

Acknowledgments

This work was supported in part by Public Health Service Grant AI-04572. One of us (Bland S. Montenecourt) has been a trainee under Public Health Service Training Grant 5 T1 GM 507.

References

1. Adams, M., and Hudson, C. S., *J. Am. Chem. Soc.*, **65**, 1359 (1943).
2. Anderson, J. S., Matsuhashi, M., Haskin, M. A., and Strominger, J. L., *Proc. Natl. Acad. Sci. U.S.*, **53**, 881 (1965).
3. Eddy, A. A., and Williamson, D. H., *Nature*, **183**, 1101 (1959).
4. Gascon, S., and Ottolenghi, P., 2nd International Symposium on Yeast, Bratislava, Czechoslovakia, July 1966, in press.
5. Islam, M. F., and Lampen, J. O., *Biochim. Biophys. Acta*, **58**, 294 (1962).
6. Jamieson, J. D., and Palade, G. E., *Proc. Natl. Acad. Sci. U.S.*, **55**, 424 (1966).
7. Lampen, J. O., *in* "Function and Structure in Micro-organisms" (M. R. Pollock and M. A. Richmond, eds.), p. 115, XV Symp. Soc. Gen. Microbiol., 1965.
8. McLellan, W. L., Jr., and Lampen, J. O., *Biochim. Biophys. Acta*, **67**, 324 (1963).
9. Moor, H., and Mühlethaler, K., *J. Cell. Biol.* **17**, 609 (1963).
10. Neumann, N. P., and Lampen, J. O., *Federation Proc.*, **25**, 588 (1966).

11. Neumann, N. P., and Lampen, J. O., *Biochemistry,* **6**, 468 (1967).
12. Palade, G. E., Siekevitz, P., and Caro, L. G., *Ciba Found. Symp. Exocrine Pancreas,* p. 23 (1962).
13. Symington, E. B., and Lampen, J. O., *9th Intern. Congr. Microbiol., Moscow,* p. 147 (1966).
14. Trevithick, J. R., and Metzenberg, R. L., *Biochem. Biophys. Res. Commun.,* **16**, 319 (1964).

Control of Membrane Differentiation and Quantum Conversion Efficiency in Chloroplasts

RODERIC B. PARK

Botany Department and Lawrence Radiation Laboratory,
University of California,
Berkeley, California

Chloroplasts in eucaryotic cells are particularly suitable material for studying membrane growth and differentiation. This is largely because many green cells can survive on a heterotrophic medium when the chloroplast is metabolically deficient, allowing chloroplast mutants to be carried in the host cell. In this sense, chloroplasts possess distinct advantages in the study of membrane growth and differentiation over their metabolic counterparts, the mitochondria, which are generally more essential to the survival of many cells than are chloroplasts.

The general features of mature chloroplast structure are shown in Fig. 1. The chloroplast is surrounded by an outer membrane, and an inner membrane which is often continuous with the internal membrane system (6, 21). The internal membranes are closed, sacklike structures, which Menke has called thylakoids (7). The outer membrane is very similar in ultrastructure to the cell membrane, while the inner membrane appears identical with the internal membranes of the chloroplast. Since chloroplast ribosomes (5) and DNA (3) are similar to that found in procaryotic cells, the outer chloroplast membrane may, in fact, serve as an isolation membrane which separates the host cell from the procaryotic invader. The inner membrane, which contains the photosynthetic pigments and electron transport apparatus of photosynthesis, may be homologous with a procaryotic cell membrane.

The proliferation and morphology of the internal membrane system in a developing chloroplast are controlled both by the interaction of internal genetic factors in the chloroplast and nucleus and by external factors such as metabolic inhibitors (15), application of plant hormones (18), mineral nutrition (14, 19), and exposure to various light regimes (20).

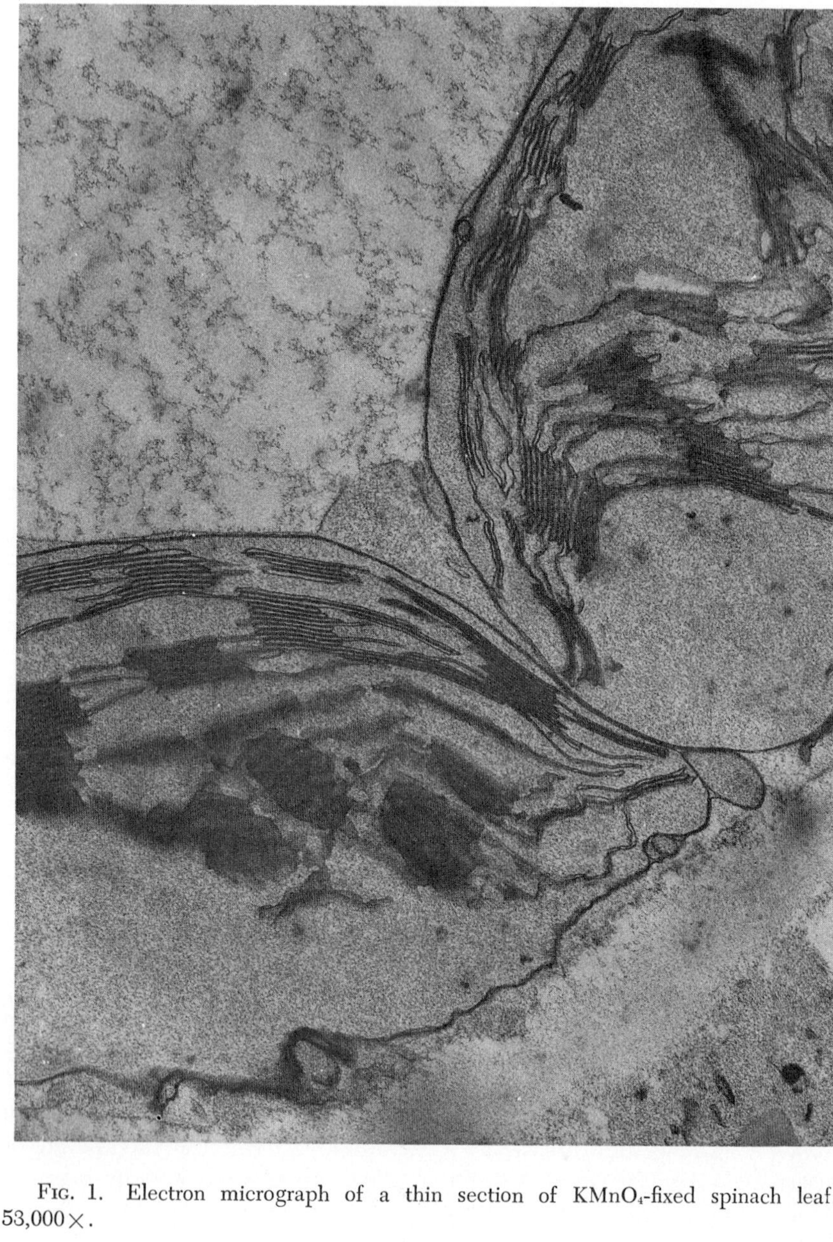

FIG. 1. Electron micrograph of a thin section of $KMnO_4$-fixed spinach leaf. 53,000×.

Of the plant hormones, kinins in particular appear to play an important role in thylakoid differentiation. Skoog and Miller (17) with tobacco tissue culture demonstrated that increasing kinetin concentrations produced buds and green shoots. Stetler and Laetsch showed further that the greening was accompanied by elaboration of the chloroplast internal membranes (18). In phosphate-deficient cells of tobacco Thomson and Weier (19) showed that the thylakoids of the chloroplasts did not unite to form grana structures, though the total amount of membrane material appeared to be even larger than in the controls. Manganese-deficient cells studied by Possingham *et al.* (14) yielded bizarre chloroplasts that sometimes appeared to combine with mitochondria to form an associated structure (though this is certainly to be construed as a deficiency condition).

The effect of light on development of chloroplast membranes has received much greater attention than the effect of mineral deficiency. While in some algae chlorophyll and chloroplasts are produced in total darkness, in higher plants chlorophyll is not formed unless the light-mediated conversion of protochlorophyll to chlorophyll occurs. In the absence of this conversion, the internal membrane material of the chloroplast forms a rectangular tubular array (the prolamellar body) which breaks down to form the thylakoids when chlorophyll is produced by illumination (4, 9, 20).

These developmental studies on chloroplasts have been concerned primarily with the extent and arrangement of thylakoid membranes. However, another type of differentiation, an intramembrane differentiation, also occurs during thylakoid growth. We discovered this phenomenon while studying quantasome arrangements in spinach thylakoids. We noted that, during the winter, spinach obtained in a local market possessed thylakoids not only micellar in character when viewed in shadowed preparations (11), but with a small percentage of the quantasomes arranged in paracrystalline arrays (Fig. 2). These arrays are also evident in freeze-etch preparations (Fig. 3). With the onset of summer these paracrystalline arrays disappeared. At the same time, we were aware that "winter spinach" gave generally higher photosynthetic phosphorylation rates and Hill reaction rates than "summer spinach." The above observations suggested that photoperiod (spinach is a long-day plant) might be an important factor in controlling both morphology and quantum conversion efficiency in spinach thylakoids. To test this hypothesis, we grew spinach under controlled photoperiod and temperature conditions and investigated the quantum requirement of the Hill reaction, chemical composition, and morphology of the thylakoids for several spinach varieties. Preliminary results from these experiments are pre-

FIG. 2. Electron micrograph of a metal-shadowed thylakoid showing a paracrystalline quantasome array (10). 110,000×.

FIG. 3. Freeze-etch preparation of spinach chloroplasts showing a paracrystalline quantasome array (2). 200,000×.

sented below. Some of these results have already been discussed by Park and Drury (12).

Methods

The plants used in these studies, *Spinacia oleracea* (variety Early Hybrid #7 or Viroflay), were grown from seed in soil–vermiculite mixtures for periods of 4 to 5 weeks, when they were harvested for study. The conditions of growth were either short days (10 hours) or long days (16 hours), with day temperatures 20°C and night temperatures 9°C. Light intensity was about 500 foot-candles at the level of the plants. Under these conditions long-day Early Hybrid plants began to bolt at 6 weeks, whereas the short-day plants had not flowered after 16 weeks, when they were discarded.

Chloroplasts were isolated according to Park and Pon (13), though in recent experiments we have substituted 0.05 *M* tricine, pH 7.5, for the 0.1 *M* phosphate used as grinding buffer in earlier experiments.

Chlorophyll was determined according to the method of Arnon (1). Nitrogen was measured by the micro-Kjeldahl method (13). Quantum requirements for the 2,6-dichlorophenol–indophenol (DCPIP) Hill reaction extrapolated to 0 light intensity were determined according to the methods of Sauer and Park (16). A typical reaction mixture for the quantum requirement calculation contained, in micromoles per milliliter, sucrose 440, potassium phosphate (pH 7.4) 90, methylamine 10.0, and DCPIP 0.03, in a reaction volume of 2 ml plus a concentration of chloroplasts to give $OD_{680\,m\mu}$ of 0.3 to 0.5. Freeze etching was carried out by following the methods described by Moor *et al.* (8).

Results

Quantum Requirements for DCPIP Reduction by Chloroplasts Isolated from Long-Day and Short-Day Plants

In Fig. 4, quanta absorbed per electron transferred from H_2O to DCPIP are plotted as a function of light intensity. These chloroplasts were isolated from 33-day-old plants of *Spinacia oleracea* L. (var. Early Hybrid #7). The true quantum requirement is obtained by extrapolating the data to 0 light intensity.

The decreased efficiency for quantum conversion by long-day Early Hybrid chloroplasts has now been observed in four separate experiments. The quantum requirements of long-day chloroplasts have always exceeded those of short-day chloroplasts by 20 to 100%. This difference does not appear to be due to preferential loss of activity of the long-day chloroplasts after isolation. We have not yet determined how quantum requirement varies with plant age. Perhaps the long-day chloroplasts are

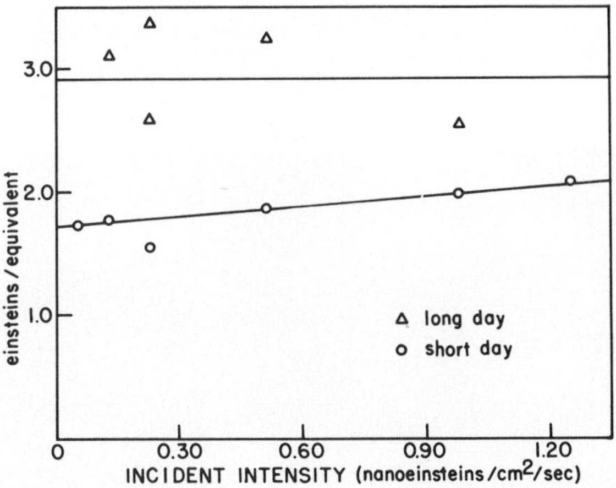

FIG. 4. Quantum requirements per electron for the DCPIP Hill reaction of chloroplasts isolated from short-day and long-day spinach plants. Actinic light 680 mμ (12).

never as efficient as those from short-day plants, or perhaps they are senescing more rapidly.

Not all spinach varieties respond to changes in photoperiod in the same way as Early Hybrid #7. The variety Viroflay is grown under summer conditions in California because it is less sensitive to long days than many other varieties and reaches edible size before flowering. Experiments by Susan Drury in our laboratory show that Viroflay, unlike Early Hybrid, retains its quantum conversion efficiency under long days for as long as six weeks.

We apparently have in spinach varieties not only a spectrum of photoperiodic response for flowering, but also a spectrum of thylakoid quantum conversion efficiencies related to photoperiod.

Chemical Composition of Thylakoids from Long-Day and Short-Day Spinach Plants

Chlorophyll/nitrogen ratios for chloroplasts and chloroplast functions are given in Table I. The data for chloroplasts indicate that the chloroplasts were broken and had lost most of their stroma protein. The chlorophyll/nitrogen ratios of the thylakoids, within experimental error, are identical. This relationship may not hold for the many other components we have not yet analyzed.

Morphology of Long-Day and Short-Day Spinach Chloroplasts

Figures 5*a* and 5*b* are a comparison of thin sections of long-day and short-day spinach mesophyll. Long-day chloroplasts contain more starch

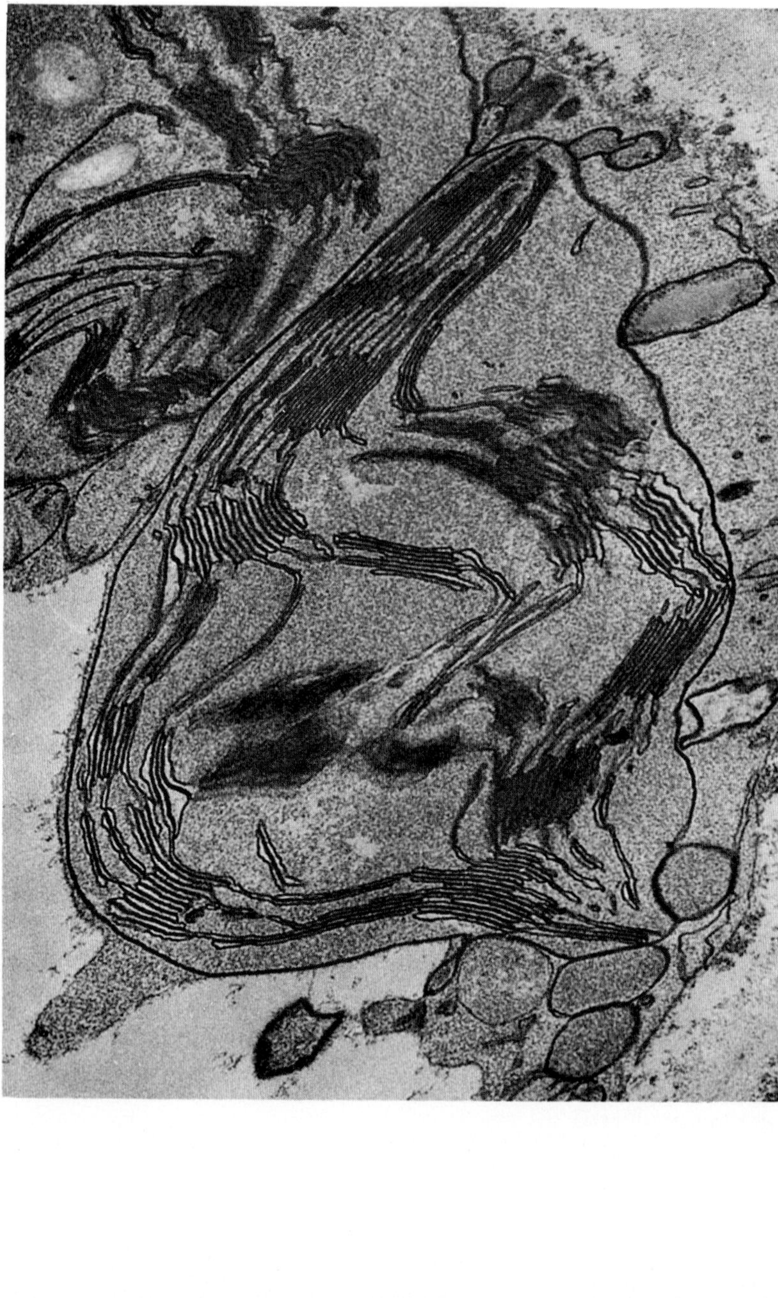

FIG. 5. (*a*) Electron micrograph of a thin section of Early Hybrid leaf grown under short days. 52,000×.

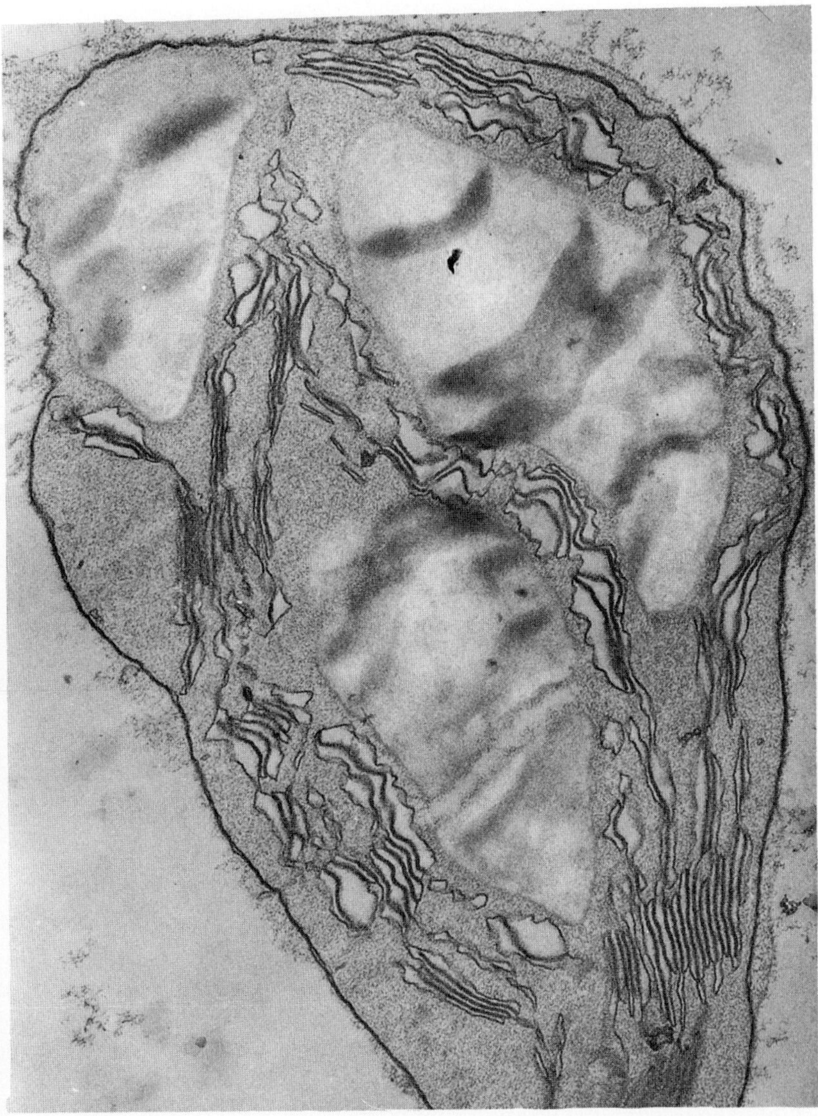

FIG. 5. (*b*) Electron micrograph of a thin section of Early Hybrid leaf grown under long days. 52,000×.

TABLE I
CHLOROPHYLL/NITROGEN WEIGHT RATIOS FOR CHLOROPLAST FRACTIONS (12)

Analyses	Short-day spinach			Long-day spinach		
	Chloro-plasts	Mem-branes	Stroma	Chloro-plasts	Mem-branes	Stroma
Total N, mg	3.14	2.58	0.45	1.14	1.00	0.16
Total N, mg/ml	0.314	0.258	—	0.114	0.100	—
Chlorophyll, mg/ml	0.318	0.318	—	0.123	0.123	—
Chlorophyll N, mg/ml	0.020	0.020	—	0.008	0.008	—
Nonchlorophyll N, mg/ml	0.294	0.238	—	0.106	0.092	—
Chlorophyll/nonchlorophyll N	1.08	1.34	—	1.16	1.34	—

and possess less well defined large thylakoids than short-day chloroplasts. Shadowed thylakoid preparations from short-day plants demonstrated occasional paracrystalline quantasome arrays, though the frequency of these arrays was never as great as in field-grown spinach. Long-day thylakoid preparations never demonstrated such arrays and generally showed less surface structure when observed by heavy metal shadowing.

DISCUSSION

In certain spinach varieties chloroplasts isolated from 4- to 5-week-old short-day plants are 20 to 100% more efficient than chloroplasts from long-day plants in quantum conversion at low light intensities. This difference in efficiency is associated with changes in the micellar character of the membrane, since paracrystalline quantasome arrays are associated only with short-day material both in the field and under controlled growth conditions. The abundance of the paracrystalline arrays is thus indicative of, but not responsible for, a change in the membrane structure which yields increased efficiency. Our initial measurements of membrane chemical composition indicate that the chemical composition of the two membranes with respect to chlorophyll and nitrogen content is similar. If this similarity holds true for other chemical components, the differences in membrane efficiency might be due totally to membrane organization.

Many important questions about this system remain unanswered. Is the less efficient long-day thylakoid distinguishable from a short-day senescing thylakoid? Are there variations in compounds other than chlorophyll and nitrogen in the membranes, and, finally, what factors are responsible for the relatively high abundance of paracrystalline quantasome arrays found in field-grown short-day spinach?

These differences in membrane structure which are related to quantum conversion efficiency probably occur during membrane growth.

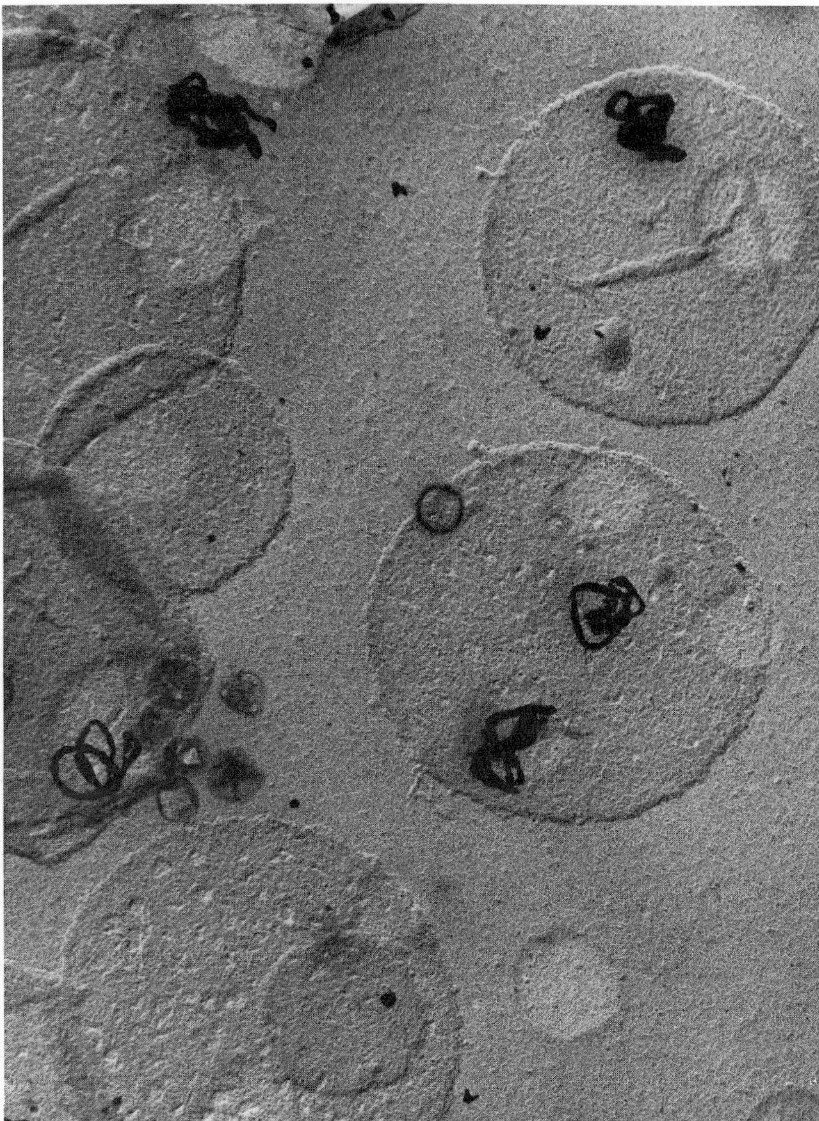

FIG. 6. An autoradiograph of chromium-shadowed spinach thylakoids isolated from a leaf which was fed 100 μc of H^3-acetate for a period of 5 hours in the light. 40,000×.

To understand how day length affects membrane structure, it will be necessary to understand in greater detail what the mechanism of growth is.

Membranes have no ends, they are continuous structures, and, in the case of thylakoids, their area is enormously increased during greening. Where is this new material added? It could be added uniformly over the surface, only where the inner and outer membranes are in contact, at the edges of grana stacks, or, perhaps, in mosaic fashion. We do know from the disposition of the paracrystalline arrays that material existing in the membrane can dictate the pattern according to which new material is added. But the site of this addition is unknown.

The most direct way of attacking the problem of membrane growth mechanisms is by autoradiography at the ultrastructural level. Admittedly this approach has many pitfalls: finding a specific precursor, differentiating between synthesis and exchange, and attaining sufficient resolution. In some initial experiments of this sort, we have made some progress in solving the last two problems, those of achieving sufficient resolution and specificity to tell that we are looking at membrane incorporation. In this method thylakoids are isolated and shadowed before autoradiography. One then places the emulsion on the side of the support film opposite the specimen and obtains a record of the distribution of radioactive material on the thylakoid surface (see Fig. 6). The thylakoids pictured here were subjected to swelling during preparation, so we cannot tell whether the silver grains all occur at the margins of the thylakoids or not. However, this is a promising method and should eventually tell us whether we have addition of new material over the entire surface or whether we have specific regions of growth within the thylakoid membrane.

How is the new material added? This poses an even more complex problem, since thylakoid proteins, like their mitochondrial counterparts, are highly insoluble in water. It would appear that these proteins must exist in association with solubilizing polar lipids in the cell before being added to the membrane proper. Research of the past ten years has given us an impressive array of thylakoid mutants and parameters which will affect thylakoid growth and differentiation. But we must still confess much ignorance on the site of membrane growth and almost total ignorance or its exact mechanism.

Acknowledgments

The above work was supported by National Science Foundation grant GB-4245 and U.S. Public Health Service grant GM-13943-01 issued by the National Institute of General Medical Science, and by the U.S. Atomic Energy Commission. The author wishes to thank Miss Susan Drury, Miss Ann Hughes, and Miss Victoria Koo for their aid in performing these experiments.

References

1. Arnon, D. I., *Plant Physiol.,* **24,** 1 (1949).
2. Branton, D., and Park, R. B., *J. Ultrastructure Res.,* **19,** 283 (1967).
3. Chun, Edward H. L., Vaughan, Maurice H., Jr., and Rich, Alexander, *J. Mol. Biol.,* **7,** 130 (1963).
4. Gunning, B. E. S., *Protoplasma,* **55,** 111 (1965).
5. Lyttleton, J. W., *Exptl. Cell Res.,* **26,** 312 (1962).
6. Manton, I., *J. Exptl. Botany,* **38,** 325 (1962).
7. Menke, W., *Ann. Rev. Plant Physiol.,* **13,** 27 (1962).
8. Moor, H., Mühlethaler, K., Waldner, H., and Frey-Wyssling, A., *J. Biophys. Biochem. Cytol.,* **10,** 1 (1961).
9. Mühlethaler, K., and Frey-Wyssling, A., *J. Biophys. Biochem. Cytol.,* **6,** 507 (1959).
10. Park, R. B., *J. Cell Biol.,* **27,** 151 (1965).
11. Park, R. B., and Biggins, J., *Science,* **144,** 1009 (1964).
12. Park, R. B., and Drury, S., presented at conference "La Croissance et La Viellesment des Chloroplastes," at Gorsem, Belgium, September 1965.
13. Park, R. B., and Pon, N. G., *J. Mol. Biol.,* **6,** 105 (1963).
14. Possingham, J. V., Vesk, M., and Mercer, F. V., *J. Ultrastruct. Res.,* **11,** 68 (1964).
15. Provasoli, L., Hutner, S. H., and Pintner, I. J., *Cold Spring Harbor Symp. Quant. Biol.,* **16,** 113 (1951).
16. Sauer, K., and Park, R. B., *Biochemistry,* **4,** 2791 (1965).
17. Skoog, F., and Miller, C. O., *Symp. Soc. Exptl. Biol.,* **11,** 118 (1957).
18. Stetler, D. A., and Laetsch, W. M., *Science,* **149,** 1387 (1965).
19. Thomson, W. W., and Weier, T. E., *Am. J. Botany,* **49,** 1047 (1962).
20. Von Wettstein, D., *Brookhaven Symp. Biol.,* **11,** 138 (1958).
21. Weier, T. E., *Am. J. Botany* **48,** 615 (1961).

Discussion of Part VII

F. LYNEN, E. RACKER, P. SIEKEVITZ, K. G. LARK, J. O. LAMPEN, A. W. GALSTON, R. B. PARK

CHAIRMAN LYNEN: The papers are open for discussion.

DR. RACKER: Dr. Siekevitz, you used the expression "constitutive enzymes" of the membrane. Are you planning to keep this expression in the published manuscript?

DR. SIEKEVITZ: Yes.

DR. RACKER: In that case, I have to continue.

DR. SIEKEVITZ: Any comments you might have before it is published are welcome.

DR. RACKER: Well, I think we ought to define what we mean by membranes before we go any further. If there can be a membrane without glucose-6-phosphatase, what is the meaning of constitutive enzymes?

DR. SIEKEVITZ: I think that is a good point, whether you can have a membrane without any of the enzymes being there, and this I think relates to whether one believes there is such a thing as a structural protein or not, a protein which has no enzyme activity, but which can complex various phospholipids to form a membrane structure which can then serve as a receptacle for various enzymes.

DR. RACKER: Your definition of the membrane is based on morphological or on enzymatic criteria?

DR. SIEKEVITZ: No, I believe it would be a morphological definition, one where you get a membrane out which has certain characteristics. It would have the characteristics of being a lipoprotein complex, I would say, which has a certain morphology which you can see in electron microscopes exemplified by a distance between the two outer layers of the membrane.

DR. RACKER: I really think it is a very basic question which ought to be settled. We are all talking about membranes. I think we all mean something different. You would exclude, for example, a membrane which doesn't have any lipoprotein even if you can see a structure which resembles a membrane?

DR. SIEKEVITZ: Which membrane wouldn't have a lipoprotein? I don't know.

DR. RACKER: I didn't say there are such membranes. I am asking for a definition and I want to know what you mean when you speak of a membrane.

Dr. Siekevitz: All right. I would say it would have to be a lipoprotein to perform some functions with regard to the hydrophilic and hydrophobic compartments of the cell. It doesn't have to have a sheet configuration. It can be a tubule.

Dr. Racker: Now, you retract the constitutive enzymes?

Dr. Siekevitz: No, no, I wouldn't. A constitutive membrane enzyme or a membrane protein would be one which exists in the cell as part of a membrane, tightly or loosely bound. It is one which, if the idea of structural protein is correct, does not have to be a part of the membrane for the membrane to exist as such. If the idea of a structural protein is wrong, then these constitutive membrane enzymes can also be considered as membrane structural proteins.

Dr. Lark: I was struck by the similarity, Dr. Lampen, between your results and the results of Melchers and Lennox who have been studying the secretion of macroglobulin from tumors. I believe they found that the carbohydrate was not coupled to the molecule until after the secretion process had occurred. So this could possibly be an exchange process that occurs as it goes through the membrane.

Dr. Lampen: We have no direct data on this point, but our preliminary findings with the internal enzyme suggest that the carbohydrate may be added outside of the membrane and not inside.

Dr. Siekevitz: Do you always find a correlation between invertase synthesis and invertase secretion, or can you disconnect, uncouple these two?

Dr. Lampen: Of course, in the intact cell invertase is secreted only in the sense that it becomes external to the membrane. It is loosely attached to the wall, possibly held into the mannan mesh by physical-chemical forces. With partial protoplasts, which don't have adequate receptor sites, some of the invertase will remain attached to the cell and the rest will be released into the medium. But we are not aware of any circumstances in which the organisms make only the small internal enzyme, and not the large external enzyme.

Dr. Galston: Dr. Park, your observation on the difference in the paracrystalline array of the quantasomes in the short-day and the long-day grown spinach brings to mind an observation of Withrow made some years ago that some phytochrome-controlled step is involved in the preparation of plastids for greening-up. I wonder whether there might be any relation between his observation on dark-grown plastids with a paracrystalline type of prolamellar body and photoperiodic contral of the kind of array that you see.

Dr. Park: My thought at this point is that the spatial relationship between the paracrystalline array and the prolamellar body is highly fortuitous. The materials are rather, well, somewhat different and the periodicities are certainly quite different that one sees. On the other hand, we of course think that phytochrome might have a role here, and we are going to be looking to find whether we can reverse the effect. We particularly like to work with a short-day plant rather than a long-day plant to see if we can do some photoperiodic control studies.

CHAIRMAN LYNEN: Any more questions? If not, I will close the session.

PART VIII

CHLOROPLAST SYNTHESIS

Chairman's Remarks

KONRAD BLOCH

Department of Chemistry,
Harvard University,
Cambridge, Massachusetts

One of the reasons why this meeting has been so successful is that the chairmen of the various sessions have shown remarkable restraint. They have attended to their business instead of giving lengthy speeches of their own. I am going to continue this tradition and discharge my duties by wrapping the microphone around the next speaker, Dr. Bogorad, who will speak on "Assembly of Chloroplasts."

Aspects of Chloroplast Assembly

LAWRENCE BOGORAD[1, 2]

Department of Botany,
University of Chicago,
Chicago, Illinois

Very young plastids in the growing point of a flowering plant appear as vesicles about 1 micron in diameter. Sometimes small evaginations from the inner portion of the limiting membrane protrude into the lumen of the vesicle. While still in darkness, the plastids enlarge, and an elaborate paracrystalline "prolamellar body" forms (Fig. 1), apparently by the pinching off, migration, and rearrangement of blebs from the inner element of the plastid membrane (22, 31). Some proplastids at this stage of development contain two prolamellar bodies and may also have a few lamellae. Ribosomes are abundant (13). Under other fixation conditions strands of DNA are visible in immature plastids (15).

Plastids in dark-grown *Euglena* resemble those in the growing point of a flowering plant. No further elaboration of *Euglena* plastid primordia occurs unless the organism is illuminated. Photosynthetic lamellae are formed in the light, but if irradiation is stopped development ceases (25). Obviously the steady-state plastid of a dark-grown (etiolated) leaf is much more complex morphologically and chemically than that of dark-grown *Euglena*. Investigations of plastid development start at different points in these two situations.

During a few seconds of illumination of an etiolated bean leaf the prolamellar body, which appears to be composed of partially fused tubes, dissociates into a group of loosely packed vesicles. This phenomenon has been termed "tube transformation" (29). The vesicles then disperse through the plastid, become aligned in rows of single vesicles, and the vesicles in each row fuse to form a primary lamella. Subsequently, at several places along each of these lamellae, additional flattened sack-like structures (thylakoids) form to produce grana which are characteristic of plastids of higher plants. Figure 2 is an electron micrograph of a plastid in a maize leaf which has been illuminated for 16 hours.

[1] Research Career Awardee of the National Institute of General Medical Sciences, U. S. Public Health Service.

[2] Present address: Harvard University, The Biological Laboratories, Cambridge, Massachusetts.

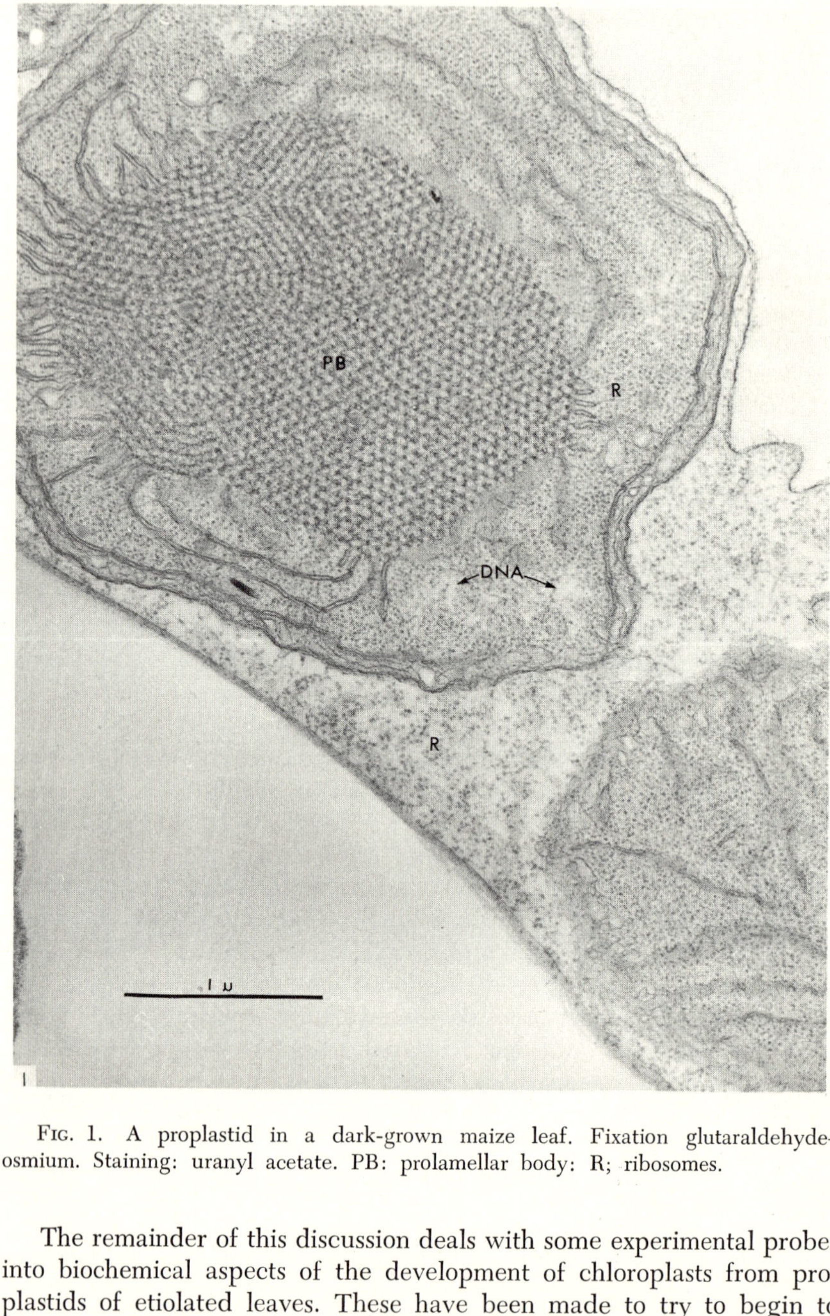

FIG. 1. A proplastid in a dark-grown maize leaf. Fixation glutaraldehyde–osmium. Staining: uranyl acetate. PB: prolamellar body: R; ribosomes.

The remainder of this discussion deals with some experimental probes into biochemical aspects of the development of chloroplasts from proplastids of etiolated leaves. These have been made to try to begin to understand the mechanisms of control of plastid maturation.

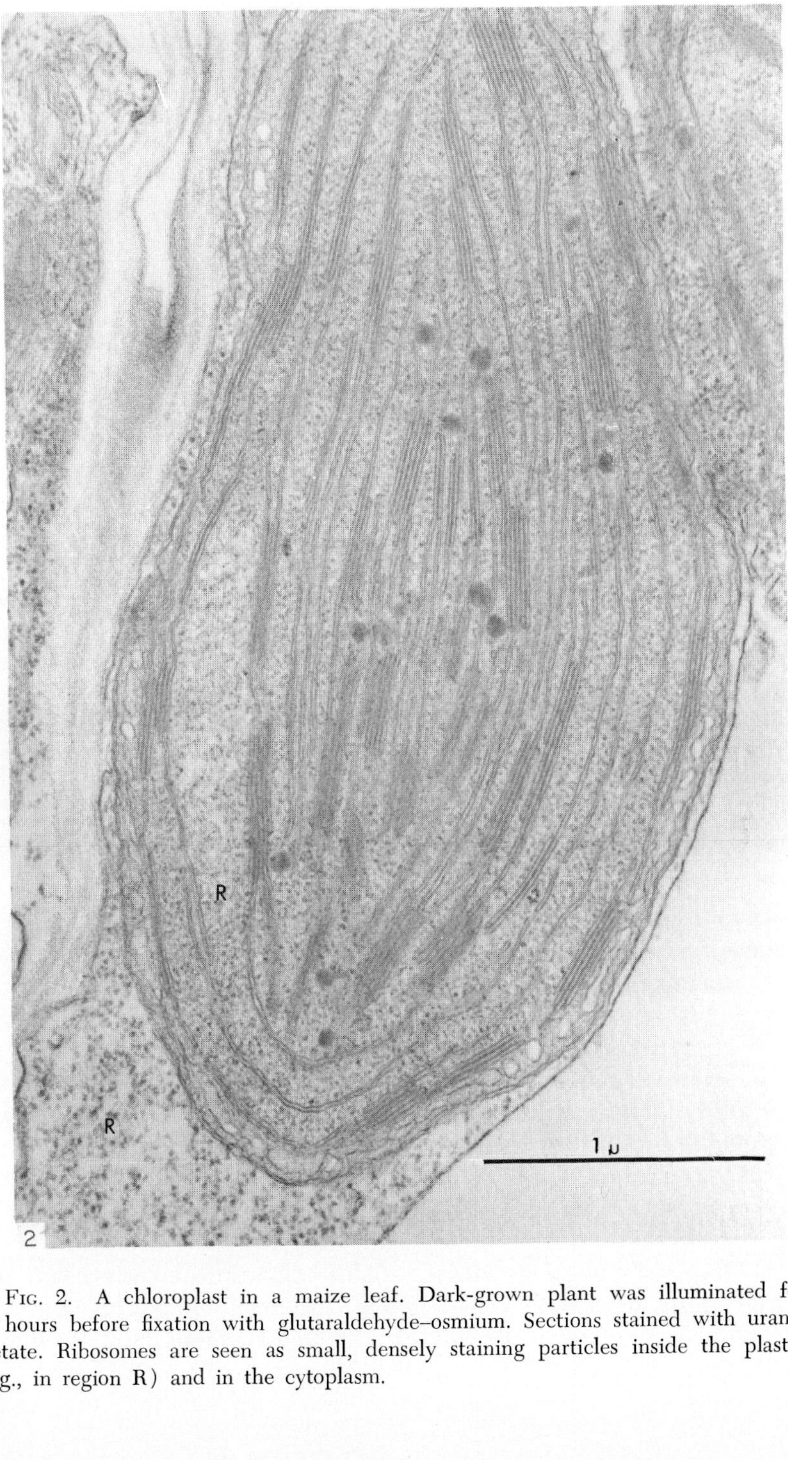

FIG. 2. A chloroplast in a maize leaf. Dark-grown plant was illuminated for 16 hours before fixation with glutaraldehyde–osmium. Sections stained with uranyl acetate. Ribosomes are seen as small, densely staining particles inside the plastid (e.g., in region R) and in the cytoplasm.

The Composition of Proplastids and Chloroplasts

Plastids of Black Valentine beans contain about 2.7 times as much protein and 3 times as much lipid, and are somewhat less than twice as heavy, as proplastids from the same source (21), but the relative concentrations of protein and lipid are not greatly different. Proplastids are about 50% protein and 20% lipid, while mature plastids contain 67% protein and 25% lipid. Differences in the concentration of specific compounds are more revealing.

The most conspicuous difference between a proplastid and a mature chloroplast is the amount of green porphyrin pigment: protochlorophyllide in the proplastid, and chlorophylls a and b in the mature plastid. The amount of protochlorophyllide in an etiolated proplastid is roughly $1/100$ to $1/300$ of the concentration of chlorophyll in a mature plastid of the same species. The protochlorophyllide appears to be attached to a protein of about 500,000 to 700,000 molecular weight (3) and is localized in the area of the prolamellar body of proplastid. The latter is judged from the observation of Boardman and Wildman (4) that proplastids contain one or two centers of red fluorescence rather than being diffusely fluorescent as would be the case if the protochlorophyllide were evenly distributed throughout the organelle. The protochlorophyllide–protein complex, protochlorophyllide holochrome, can be extracted from etiolated bean leaves and partially purified (3). When this complex is illuminated, either *in vitro* or *in vivo,* the pigment is reduced to chlorophyllide.

The availability of the porphyrin precursor δ-aminolevulinic acid (ALA) limits the production of protochlorophyllide in etiolated leaves. Granick (12) showed that etiolated barley leaves supplied with ALA accumulate about ten times the normal amount of protochlorophyllide while maintained in darkness; etiolated leaves of other species respond in the same manner. This demonstrates that the activity of some enzyme involved in ALA formation is limiting in proplastids but that all the other enzymes required for protochlorophyllide synthesis are present and active. The increased production of chlorophyll on illumination of etiolated leaves is thus dependent on the reinitiation of ALA synthesis.

Other enzymes which are known to be more active in green than in etiolated tissues are ribose-5-phosphate isomerase (present both in chloroplasts and outside them) and the chloroplast-limited enzymes (28) ribulose-5-phosphate kinase, ribulose diphosphate carboxylase, and NADP triose phosphate dehydrogenase. There is indirect evidence for greater activity of some enzymes of fatty acid synthesis (e.g., 9, 30, 33). Williams and Novelli (32) report that ribosome preparations from young dark-grown maize seedlings which have been illuminated for 1 hour and then maintained in darkness for 2 hours more are more active in protein synthesis than are those from unilluminated tissues; no significant effect was noticed in preparations made after the leaves had been in darkness for 1 hour after illumination, but full stimulation was achieved within 2 hours. The preparations were from unfractionated homogenates and probably contained both cytoplasmic and plastid ribosomes.

Increased activity of the enzymes mentioned above could be brought about simply by activation of enzyme molecules already present. On the other hand, light might somehow stimulate the synthesis of new enzyme molecules. Much of the ensuing discussion deals with attempts to separate these possibilities.

Chlorophyll Accumulation

As was pointed out, the first morphological consequence of illumination of an etiolated leaf is tube transformation—the dissociation of the elements of the prolamellar body in the proplastid. In bean plastids the action spectrum for this change corresponds to the absorption spectrum of protochlorophyllide holochrome *in vivo* (about 650 mμ), and, furthermore, the amount of light energy required for photoreduction of protochlorophyllide matches closely that which brings about tube transformation (29, 16). These data suggest that the dissociation of the elements of the prolamellar body may be a consequence of the photoreduction of the protochlorophyllide. This is an interesting problem in photochemistry and the physical chemistry of proteins, but there is little information available beyond that given here.

The pattern of chlorophyll accumulation when etiolated leaves are exposed to continuous light can be visualized as occurring as separable phenomena. First, as was pointed out above, the proplastids' protochlorophyllide is photoreduced to chlorophyllide. Then, after a few minutes, varying with the age and nature of the tissue, new synthesis of pigment commences at a relatively slow rate. Production of chlorophyll during the lag phase, in plants in which there is a lag, continues at a slow pace in some cases for a few hours. A period of rapid synthesis ensues until net chlorophyll production terminates 10 to 24 hours later; what controls the maximum chlorophyll concentration per plastid is not known.

If during the period of rapid chlorophyll synthesis bean plants are returned to darkness, the rate of accumulation of protochlorophyllide falls rapidly, and within 1 or 2 hours protochlorophyllide formation ceases—chlorophyllide, of course, fails to form because light is required for its synthesis by the reduction of protochlorophyllide (Fig. 3). These

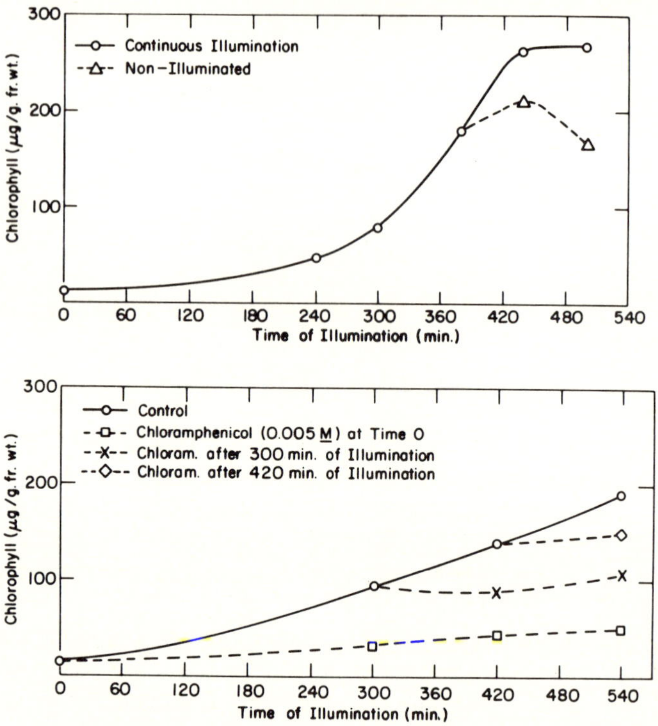

FIG. 3. *A*, effect of removing rapidly greening red kidney bean leaves from darkness on protochlorophyllide–chlorophyll synthesis. Leaves placed in darkness after 6 hours of illumination were illuminated for 5 minutes before harvest to convert accumulated protochlorophyllide to chlorophyllide. *B*, effect of administration of 5 m*M* chloramphenicol to isolated leaves of dark-grown bean plants at various times. Both control and chloramphenicol-treated tissues were under continuous illumination throughout the experiment. Leaves of 7- to 9 day-old dark-grown plants were used in experiments of these types. Isolated leaf halves were harvested under a green safelight and placd in petri dishes containing 0.2 *M* sucrose with or without inhibitors (11*a*).

data are obtained by extracting the pigment from leaf tissue into aqueous acetone and making spectrophotometric measurements. Is the arrest of protochlorophyllide formation a result of feedback inhibition regulating the production of ALA? Probably not.

The effect of the return of greening bean leaves to darkness can be mimicked in the light by the administration of puromycin or chloramphenicol (Fig. 3), suggesting that protein synthesis is required for continued chlorophyll production and that the maximum lifetime of whatever enzyme is limiting (presumably an enzyme involved in ALA formation) is of the order of 1 to 2 hours. Furthermore, since similar effects can be obtained by treatment of leaves with actinomycin D, it seems probable that some informational RNA as well as the enzyme itself is short-lived—with a maximum effective lifetime of less than 2 hours (10, 11*a*). The effects of darkness or chloramphenicol can be partially reversed by the administration of ALA (11*a*).

But what about the production of protochlorophyllide which is triggered by the initial illumination of etiolated leaves and continues at a relatively slow pace for a few hours?

The events at early stages of greening can be observed more effectively by a different kind of assay procedure than that used to obtain the data described above. As was shown originally by Shibata (27), the conversion of protochlorophyllide to chlorophyllide can be observed spectrophotometrically *in vivo*. Protochlorophyllide holochrome has an absorption maximum *in situ* at about 650 mμ. As an etiolated leaf is illuminated, the absorption in this region declines concomitantly with increased absorption at about 682 mμ; the latter is attributable to a chlorophyllide–protein complex. The absorption spectra of unilluminated etiolated maize leaves and of the same leaves after exposure to light are shown in Fig. 4. During a subsequent period in darkness any additional protochlorophyllide which may be formed can be detected by its absorption at about 650 mμ.

The set of curves in Figs. 5, 6, and 7 is from a series of experiments

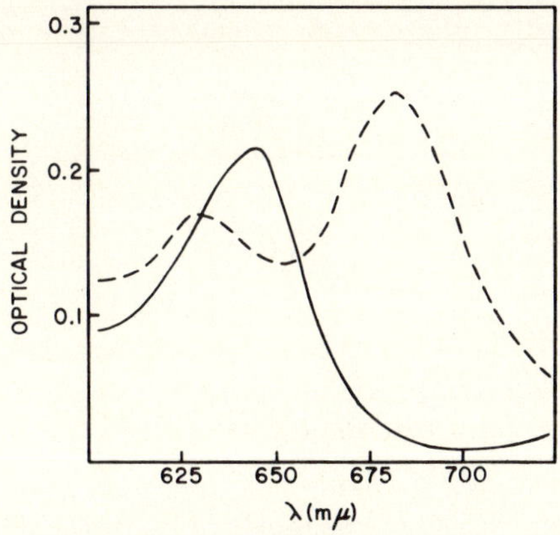

FIG. 4. Absorption spectrum of etiolated bean leaves (solid line) and of the same tissue after illumination for 30 seconds (broken line). The absorption maximum just below 650 mμ is that of the protochlorophyllide–protein complex (holochrome) plus some unconvertible pigment. On illumination the protochlorophyllide is reduced to chlorophyllide (absorption maximum at about 682 mμ)—the unconvertible pigment (maximum at about 632 mμ) remains.

with etiolated red kidney bean leaves (11*b*). These leaves were exposed to light for 1 minute—at an intensity high enough to convert all the protochlorophyllide to chlorophyllide. Then, 4 hours later, the absorption spectra of the leaves were determined (e.g., Fig. 5*a*); the chlorophyllide *a* formed during illumination 4 hours earlier now absorbs with a maximum near 675 mμ, and some new protochlorophyllide has formed; an absorption band with a maximum at about 650 mμ is apparent. Immediately after this observation the leaves were illuminated for another minute to determine whether any, or how much, of the newly accumulated protochlorophyllide could be converted to chlorophyllide (Fig. 6*a*). As can be seen by comparing Fig. 5*a* with Fig. 6*a* and by examining the calculated difference spectrum shown in Fig. 7*a*, the second minute of light leads to a drop in absorption at about 650 mμ and a marked increase at longer wavelengths, showing that the protochlorophyllide accumulated in darkness after the initial illumination is

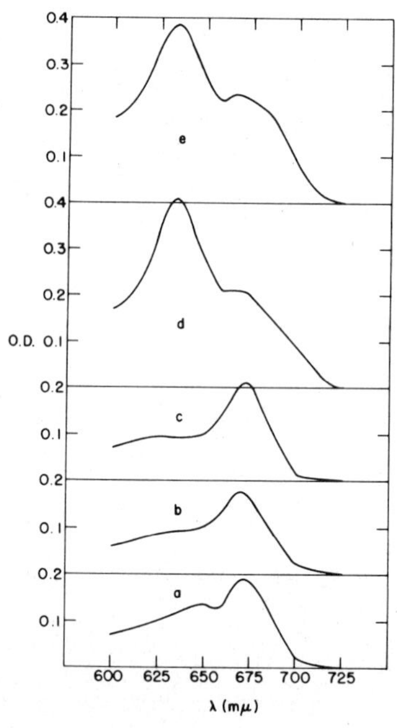

FIG. 5. Absorption spectra of 7-day-old dark-grown red kidney bean leaves taken 4 hours after illumination for one minute. *a,* Control. *b,* Incubated with 5×10^{-3} *M* chloramphenicol for 4 hours before illumination. *c,* Incubated with 10^{-3} *M* puromycin for 4 hours before illumination. *d,* Incubation with 10^{-2} *M* δ-aminolevulinic acid during the 4-hour period in darkness following illumination for 1 minute. *e,* As in *b,* except incubated with 10^{-2} *M* ALA in darkness after the 1-minute illumination (11*b*).

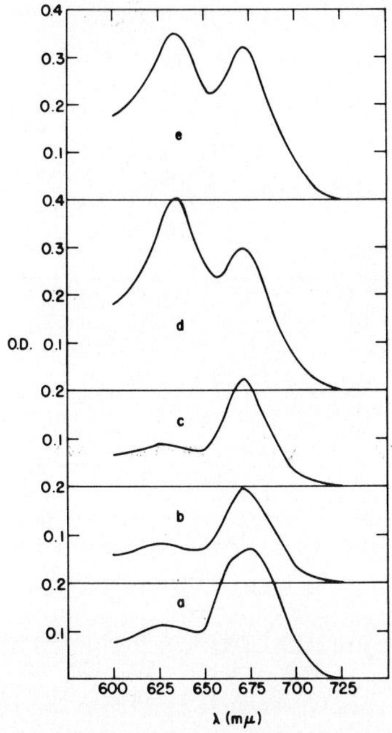

FIG. 6. As in Fig. 5, except absorption spectra redetermined after illumination for 1 minute. Total light and dark treatment: etiolated leaves, irradiated for 1 minute, maintained in darkness for 4 hours (absorption spectra shown in Fig. 5); irradiated for 1 minute (absorption spectra shown in this figure).

convertible to chlorophyllide a in the normal manner. (When ALA is administered to etiolated leaves in darkness (12), much of the newly formed protochlorophyllide absorbs *in vivo* at about 630 mμ; this is not converted to chlorophyllide when the leaves are illuminated.) Similar sets of spectra are shown for leaves preincubated with $5 \times 10^{-3}\,M$ chloramphenicol [a protein synthesis inhibitor which is known to interfere with massive greening (19)] (Figs. 5*b*, 6*b*, and 7*b*) or with $10^{-3}\,M$ puromycin (Figs. 5*c*, 6*c*, and 7*c*) for 4 hours prior to the initial illumination. Leaves treated with these inhibitors of protein synthesis produce little or no new protochlorophyllide after the initial illumination. Thus, the initial activity (after illumination) of the enzyme regulating the availability of ALA appears to result from the formation of new enzyme molecules (11*b*). Similar data have been obtained in experiments in which leaves have been treated with actinomycin D, suggesting that DNA-dependent RNA synthesis is also required (5, 11*b*).

This discussion brings us to the edge—but only to the edge—of an extremely interesting and important problem of organizational biosynthesis about which students of plastid development would like to know

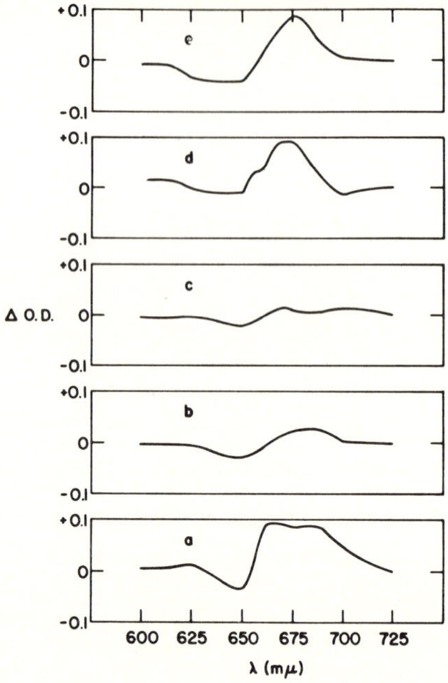

FIG. 7. Computed difference spectra. Figure 6 minus Fig. 5 for each.

more. What happens to the protein of the protochlorophyllide holochrome complex? Does it serve as an enzyme for protochlorophyllide photoconversion? That is, does a protochlorophyllide molecule become attached to the protein and then, after the pigment is reduced to chlorophyllide, move off to its permanent site in the plastid lamellae, freeing the protein to accept another protochlorophyllide molecule, etc.? Or is the holochrome protein a skeleton with which the newly formed chlorophyllide remains associated (although perhaps at a different site), to which additional protochlorophyllide molecules can become attached and reduced, and with which other components of the final lamellar structure can become associated? The latter suggestion implies that particles which form or become embedded in the lamellae might be built by accretion, with the protochlorophyllide holochrome protein serving as a nucleus. This question cannot be answered now, but there is some evidence which suggests at least that the holochrome protein can be reused.

Figures 5*d*, 6*d*, and 7*d* show the consequence on the production of convertible protochlorophyllide of supplying ALA to etiolated bean leaves which were illuminated for 1 minute, maintained in darkness for 4 hours (Fig. 5*d*), and then reilluminated (Fig. 6*d*). Comparable information about similar tissue pretreated with chloramphenicol (as in Figs. 5–7*b*) and supplied with ALA after illumination is provided in Figs. 5*e*, 6*e*, and 7*e*. A comparison of Figs. 5–7*b* and Figs. 5–7*e* reveals that convertible protochlorophyllide is formed in chloramphenicol-ALA-treated leaves. If the antibiotic blocks production of the holochrome protein, the protein is used more than once in protochlorophyllide reduction. (These spectra also show the accumulation of nonconvertible protochlorophyllide by ALA-fed etiolated leaves.)

To summarize observations on protochlorophyllide and chlorophyllide production in etiolated and greening leaves: (*a*) Etiolated leaves contain a small amount of protochlorophyllide but no chlorophylls. (*b*) Light absorbed by protochlorophyllide holochrome drives the reduction of the pigment to chlorophyllide. (*c*) Light also promotes the formation of additional protochlorophyllide. (See later discussion regarding the possible photoreceptive pigment for this process.) (*d*) The amount of protochlorophyllide present in etiolated leaves is limited by the availability of ALA; some enzyme involved in ALA production regulates the activity of the entire chain. (*e*) The enzymes required for forming protochlorophyllide from ALA are present and active in etiolated leaves, but experiments with chloramphenicol, puromycin, and actinomycin D demonstrate that additional ALA is produced only after some RNA and protein synthesis has occurred in response to illumination; the protein synthesized is presumably an enzyme required for ALA synthesis. (*f*) Production of ALA starts within minutes after illumination has begun. (*g*) The ALA-synthesizing system is labile; return of rapidly greening plants to darkness or administration of inhibitors of protein and RNA synthesis to continuously illuminated plants results in a rapid reduction in the rate of pigment formation and, finally, in its complete cessation within a few hours. (*h*) It seems unlikely that ALA production is controlled by feedback inhibition, but the possibility of repression by some product of the protochlorophyllide biosynthetic chain—for example, protochlorophyllide itself—exists.

Some Enzymes of the Photosynthetic Carbon Reduction Cycle

The enzyme ribulose diphosphate carboxylase responds to illumination about as rapidly as the δ-aminolevulinic acid-synthesizing system. The specific activity (on a protein basis) of ribulose diphosphate carboxylase is about 40% to 50% higher in etiolated maize leaves illuminated for 3 minutes at an intensity of 380 foot-candles than it is in unirradiated tissue. This initial rise is followed by a slower one, until after 3 hours of light the level of the enzyme is about 75% greater than before the tissues were exposed to light. The level falls somewhat during the next 9 hours, but then again, sometime between 12 and 18 hours after illumination has begun, a further increase to a level 50% above that in unilluminated tissues is observed. Treatment of leaf tissue with chloramphenicol or puromycin prior to the beginning of illumination blocks the increase in the enzyme level, suggesting that protein synthesis and not activation of the enzyme is occurring (7).

In contrast to the behavior of the ALA-synthesizing system and of ribulose diphosphate carboxylase, the enzyme ribulose-5-phosphate kinase shows no significant change in activity until some point between 6 and 12 hours after illumination has commenced. After 12 hours of illumination the enzyme activity is about 88% higher in illuminated than in unilluminated etiolated leaves. The activity of ribose-5-phosphate isomerase is not significantly altered until sometime between 12 and 18 hours of illumination; by the end of this time, its activity is 50% higher than before the start of irradiation (Table I). These increases in activity

TABLE I
INCREASES IN ACTIVITIES OF SOME ENZYMES OF THE PHOTOSYNTHETIC CARBON REDUCTION CYCLE DURING ILLUMINATION OF ETIOLATED MAIZE LEAVES

	Time of illumination and magnitude of increase[a]				
Enzyme	3 min	3 hr	6 hr	12 hr	18 hr
RuDP carboxylase	+40–50%				
RuP kinase				+88%	
R-5-P isomerase					+50%

[a] The entry for each enzyme indicates the time at which the first significant change in activity was observed. The magnitude of each change is referred to the activity in unilluminated leaves. The original data are based on activity per milligram of soluble protein in the leaf extract (7).

can be blocked by administration of chloramphenicol, indicating that they too are a consequence of formation of new enzymes. Added to this class of slow-responding enzymes is the NADP-linked triose phosphate dehydrogenase in beans. Marcus (18) found the first increase in the activity to occur sometime after 6 hours of illumination but before the leaves had been in light for 24 hours.

Ribulose diphosphate carboxylase activity and δ-aminolevulinic acid-producing capacity appear to respond more directly, or at least more promptly, to illumination than the other enzymes discussed whose increase in a sense is more indirect. These latter enzymes increase in activity only well after photosynthetic capacity has developed. (For example, Anderson and Boardman (1) observed that Hill reaction activity reached a maximum in etiolated bean leaves after they had been illuminated for 10 hours, and Butler's (6) data suggest that photosynthetic activity may begin at the end of the lag phase in chlorophyll production.)

The "late" increases in enzyme activity may depend on the accumulation of some products of photosynthesis. This is supported by the observation that treatment of etiolated leaves with $5 \times 10^{-5} M$ CMU (3-(4-chlorophenyl)-1-1-dimethylurea), a potent inhibitor of photosynthesis, has no effect on carboxylase accumulation, but the rise in the ribose-5-phosphate isomerase fails to occur (20). (A few experiments in which sucrose was supplied to etiolated leaves failed to reveal any changes in the level of the enzymes discussed—suggesting that it is not a simple case of starvation.)

To summarize to this point: (*a*) Ribose-5-phosphate isomerase (ribose-5-phosphate to ribulose-5-phosphate), ribulose-5-phosphate kinase (ribulose-5-phosphate to ribulose-1,5-diphosphate), and ribulose diphosphate carboxylase (ribulose-1,5-diphosphate plus CO_2 to 2 molecules of phosphoglyceric acid), three enzymes involved sequentially in photosynthetic carbon reduction, increase in activity when etiolated leaves are illuminated. (*b*) Experiments with inhibitors of protein synthesis indicate that the increase involves the synthesis of new protein. (*c*) The time course of increased activity is different for each of these three enzymes; the response to illumination is not that which would be expected if a single operon were being derepressed; the increase in ribose-5-phosphate isomerase seems to require photosynthetic activity.

There seem to be multiple kinds of stimulation systems in the development of the chloroplast. What starts off the first of these systems? What is the photoreceptor for the stimulating impulse, and what is the earliest biochemistry that one can observe?

The photoreceptor for the initial stimulation of enzyme production may be the bile pigment–protein complex, phytochrome. The following data are pertinent. Marcus (18) observed that the increase in activity of NADP triose phosphate dehydrogenase is promoted by red light and that this promotion can be reversed by exposure to far-red light. Mego and Jagendorf (21) in their study of the development of plastids in Black Valentine beans found that exposure to red light stimulated the increase of proplastid protein and lipid, but if red irradiation was followed by illumination of tissues with far-red light the rises in levels of these compounds failed to occur. The lag phase, the period of slow pigment synthesis preceding the rapid synthesis of chlorophyll by illuminated etiolated leaves, is affected by red and far-red light (23, 24). The general observation and the way in which experiments on the lag phase have been done are as follows: After a brief illumination of etiolated leaves, during which time the protochlorophyllide present is reduced to chlorophyllide, the leaves are returned to darkness for a few hours and then reilluminated. At this time chlorophyll production begins rapidly—as though the leaves had been in continuous light. Red light is more effective than blue in eliminating the lag phase of chlorophyll production; thus protochlorophyllide, which has a stronger absorption band in the blue than it does in the red region of the spectrum, appears not to be the photoreceptive pigment for the control of the lag phase. Furthermore, if an exposure of etiolated tissues to red light is followed with an appropriate exposure to far-red (730-mμ) light, chlorophyll production does not begin immediately upon reillumination of the leaves, but another lag period occurs, suggesting that the far-red light reverses the effect of red light in setting off some lag phase chemistry. Finally, the promotive effect of red light on the initial resynthesis of protochlorophyllide can also be reversed by exposure of leaves to far-red light (2). One disquieting complication is that far-red light can be as effective as red light in promoting lag phase elimination!

RNA Metabolism and Plastid Development

As was pointed out in an earlier part of this discussion, pigment production by illuminated etiolated leaves is blocked by actinomycin D (5, 11). This seems to implicate RNA synthesis as an early obligatory event in light-controlled plastid maturation. These observations have led to investigations of RNA metabolism in developing plastids of maize.

In the simplest experiments the leaves of 10- to 12-day-old etiolated maize plants were permitted to absorb P^{32}-phosphate in the dark. Then one set was exposed to light while another was maintained in darkness. Leaf RNA was extracted (8), centrifuged in a density gradient of sucrose, and the gradient was analyzed for the distribution of ultraviolet light-absorbing material and radioactivity. As is apparent from Fig. 8, the specific activity of each of the various kinds of RNA separable by this technique is much higher when the RNA is isolated from leaves illuminated for 2 hours than when it is isolated from similar ones kept in the dark. (The lag phase in pigment production lasts for about 2 to 3 hours in etiolated maize leaves of this age.)

As an extension of this sort of experiment—to determine whether the effect of illumination was on plastid RNA metabolism—RNA was isolated from plastid fractions obtained by differential centrifugation of homogenates of illuminated or unilluminated dark-grown leaves. Figure 9 shows the differences in specific activities of various RNA's after centrifugation on a sucrose density gradient. By contrast, the total RNA

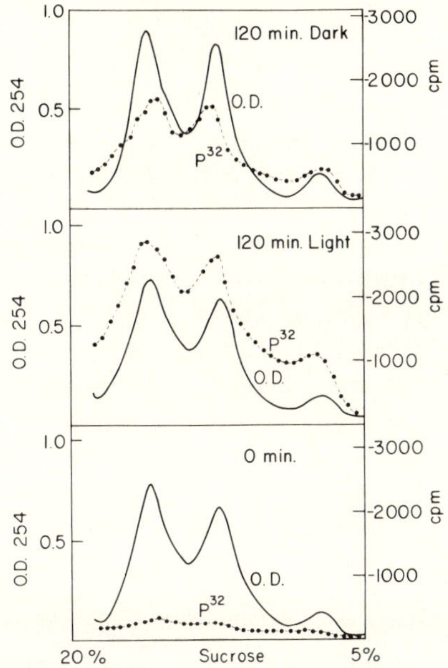

FIG. 8. Distribution of radioactivity (P^{32}) and 254-mμ absorption in a 5% to 20% sucrose density gradient. RNA preparations from whole dark-grown maize leaves. Cut leaves placed in P^{32}-phosphate (0.1 mCi) in darkness. Group "0 min." extracted after radiophosphate had been taken up; group "120 min. Light" illuminated for 2 hours after radiophosphate had been taken up in darkness; group "120 min. Dark" maintained in darkness for 2 hours after P^{32} had been taken up. Centrifugation for 19 hours at 23,000 rpm in SW 25.1 rotor of Spinco L-2 ultracentrifuge in a 5% to 20% sucrose gradient (containing 0.05 *M* Tris, pH 7.4; 0.1 *M* KCl; and 2 μg of polyvinyl sulfate per milliliter); 254 mμ absorption monitored continuously during collection of gradient. Radioactivity measured on aliquots of collected fractions.

obtained from a preparation of ribosomes from the supernatants of these chloroplast preparations (that is, cytoplasmic RNA contaminated with some from broken plastids) showed a maximum difference in specific activity of about one and one-half times when material from illuminated tissue was compared with that of unilluminated material. Thus there seems to be a specific effect of light on plastid RNA metabolism, as judged by the incorporation of P^{32}-phosphate into RNA. Furthermore, prior to the time of rapid synthesis of chlorophyll, the development of grana, and the production of most of the enzymes described above, a good deal of RNA, and presumably ribosome, synthesis has occurred.

The effect of light on plastid RNA production appears to be inductive in the sense that plastids of etiolated leaves supplied P^{32}-phosphate in the dark, illuminated for 30 minutes, and then maintained in darkness for 90 minutes contain the same sort of RNA's of about the same specific radioactivities as are found in plastids from leaves exposed to light continuously for 120 minutes. Thus, illumination directly or indirectly accelerates plastid RNA synthesis, and once accelerated it continues even in the darkness for at least 90 minutes; longer periods have not been studied.

Information about the time course of the development of photosynthetic capacity during illumination of etiolated bean leaves (e.g., 1, 6) makes it seem very unlikely that plastids have much, if any, capacity for photophosphorylation until some time after the lag phase in chlorophyll production has passed. This would appear to eliminate photo-

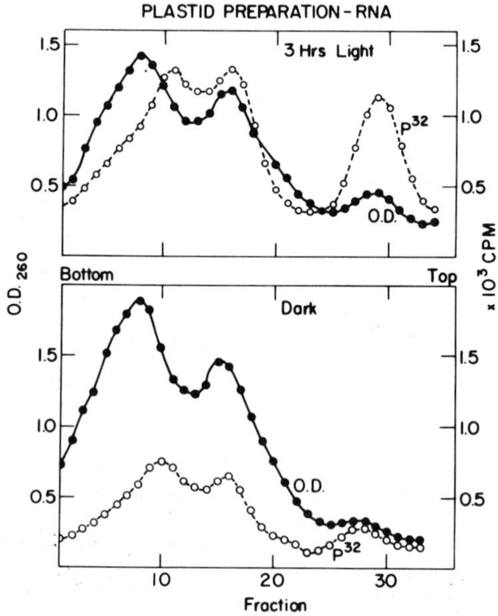

FIG. 9. Similar to Fig. 8, except showing RNA (260 mμ absorption and P^{32} incorporation) of plastid fraction isolated from leaves unilluminated or illuminated after administration of P^{32}-phosphate. Fractions collected manually, and optical densities at 260 mμ determined after dilution. The two heavier RNA's obtained from plastids are approximately 22 S and 17 S (uncorrected). These determinations were made with the Spinco analytical ultracentrifuge using absorption optics on RNA obtained from maize chloroplasts purified on a sucrose density gradient (see final entry in footnote *b* to Table IV for preparative conditions). The apparent base composition of the chloroplast 22 S and 17 S RNA's, recovered from a sucrose gradient, is approximately 25% adenine, 32% guanine, 20% uracil, and 23% cytosine; these values are based on the distribution of radioactivity, after paper electrophoresis, among AMP, GMP, UMP, and CMP obtained by alkaline hydrolysis of plastid RNA's prepared from sucrose gradient-purified chloroplasts of maize leaves supplied P^{32}-phosphate.

phosphorylation as the *cause* of the photostimulation of plastid RNA synthesis described. The inductive effect of light on plastid RNA production argues further against the phenomenon's being merely a reflection of photophosphorylation. What is the source of energy for the synthesis of proteins, lipids, nucleic acids, etc., in plastids not mature enough to do photosynthesis? Mature plastids cannot carry on oxidative phosphorylation; little is known about proplastids in this regard. This may be another example of interdependence of subcellular organelles—in this case involving mitochondria and proplastids.

The observation that all the kinds of RNA which could be seen on a sucrose density gradient increased in specific activity led to an examination of the relative activity of RNA polymerase in illuminated and unilluminated maize leaf tissue and chloroplasts.

TABLE II
RNA POLYMERASE IN MAIZE LEAVES[a]

Source of supernatant	Effect of addition of calf thymus DNA (cpm/mg protein)[b] −CT DNA	+CT DNA
Unilluminated	1808	3160
30 min light + 90 min dark	1432	2912

[a] Activity in supernatant fluid after centrifugation of leaf homogenate at 27,000 *g*.
[b] Incubation: 10 minutes, 30°C. Components of incubation mixture: 1.8 μmoles each CTP, GTP, UTP; 1.5 μCi ATP (0.14 μmole); 0.02 mmole Mg acetate; 0.04 mmole mercaptoethanol; 0.04 mmole Tris, pH 8.0; ±0.2 mg calf thymus DNA (CT DNA); and 0.1 mmole NH_4Cl; all in 0.4 ml *plus* 0.4 ml of leaf homogenate supernatant in 0.5 *M* sucrose; 0.5 *M* Tris, pH 8.0; 1 m*M* $MgCl_2$; and 10 m*M* mercaptoethanol. Leaves ground with 1 volume of sucrose–Tris–Mg–mercaptoethanol solution per weight of tissue.

Mans and Novelli (17) have shown that the supernatant fluid obtained after centrifugation of maize extracts at 27,000 g for 30 minutes contains a soluble RNA polymerase. The activity of this RNA polymerase is stimulated by the addition of calf thymus DNA. The presence of this enzyme and its stimulation by added DNA was confirmed (Table II), but no difference could be observed in preparations from comparable amounts of etiolated and briefly illuminated tissues.

On the other hand, the RNA polymerase activity of plastids from illuminated leaf tissue is considerably greater than that of etiolated proplastids (Table III). [Plastid RNA polymerases have been shown to occur in broad beans (14) and tobacco (26).] The enzyme was measured by the rate of incorporation of C^{14}-ATP into trichloroacetic acid-precipitable material. (GTP-C^{14} is incorporated about as well as

TABLE III
RNA POLYMERASE IN MAIZE PLASTIDS[a]

Source of plastids	Micromicromoles ATP-8-C^{14} incorporated by plastids from 1 gm leaf tissue[b]	
	−CT DNA	+CT DNA
Unilluminated	159	164
30 min light + 90 min dark	1036	1062

[a] Activity in washed 1000 *g* fraction.

[b] Incubation: 10 minutes, 30°C. Components of incubation mixture: 2 μmoles each CTP, UTP, GTP; 1.4 μCi ATP (0.14 μmole); 0.02 mmole Mg acetate; 0.04 μmole mercaptoethanol; 0.04 mmole Tris, pH 8.0; ±0.2 mg calf thymus DNA (CT DNA); and 0.1 mmole NH_4Cl; all in 0.4 ml *plus* 0.4 ml of plastid preparation (washed and resuspended 1000 *g* fraction) in 0.5 *M* sucrose; 0.05 *M* Tris, pH 8.0; 1 m*M* $MgCl_2$; and 10 m*M* mercaptoethanol. Leaves ground with 1 volume of 0.5 *M* sucrose; 0.5 *M* Tris, pH 8.0; 1 m*M* $MgCl_2$; and 10 m*M* mercaptoethanol per weight of tissue. 1000 *g* pellet suspended in sucrose; 0.05 *M* Tris, pH 8.0; $MgCl_2$; and mercaptoethanol; recentrifuged, resuspended, and used as noted.

ATP; the incorporation does not occur in the absence of all four ribonucleotide triphosphates; it is strongly inhibited by DNase, RNase, and actinomycin D.)

As is shown in Table III, the RNA polymerase activity of the 1000 g plastid fraction is not boosted by the addition of calf thymus DNA; this lack of response has varied from preparation to preparation, probably reflecting the level of contamination with DNA of broken nuclei. Stimulation of RNA polymerase activity can be demonstrated reproducibly with plastids purified by sucrose density gradient centrifugation (Table IV) or washed with sucrose-free buffer after sedimentation at 1000 g. Titration with calf thymus DNA of preparations of the latter type from plastids of illuminated and unilluminated etiolated maize leaves reveals that the difference in RNA-forming capacity (that is, incorporation of ATP-C^{14} or GTP-C^{14} into a trichloroacetic acid-precipitable product) of plastids from these two sources is attributable at least in part to a difference in RNA polymerase per se—in contrast to the possibility of the *entire* effect being due to an increase in the level of copiable DNA. It is not yet known whether any of the increase in the capacity of plastids to form RNA can be attributed to the availability of more endogenous template DNA.

With regard to rapidity of response to illumination, the stimulation of RNA polymerase activity roughly follows the pattern observed for changes in ALA-synthesizing capacity and the level of ribulose diphosphate carboxylase. For example, in a series of experiments in which

TABLE IV
RNA Polymerase in Maize Plastids[a]

Source of plastids	Experiment: 1	2	3	4	5
	Micromicromoles ATP-8-C^{14} incorporated by plastids from 1 gm leaf tissues[b]				
Unilluminated					
−CT DNA	27	22	151	55	106
+CT DNA	67	39	303	147	159
Illuminated					
−CT DNA	178	50	398	118	363
+CT DNA	247	83	466	317	490

[a] Activity in plastids isolated on sucrose density gradients.

[b] Incubation: Experiments 1, 2, 3: 10 minutes, 30°C. Experiments 4, 5: 15 minutes, 30°C. Components of reaction mixtures: as in Table III except 1.1 μCi ATP (0.14 μmole), and in experiment 4 ± 0.1 mg and experiment 5 ± 0.6 mg calf thymus DNA. Illumination: Experiments 1 and 2: 30 minutes light + 90 minutes darkness. Experiment 3: 120 minutes light. Experiment 4: 230 minutes light. Experiment 5: 230 minutes light. Preparation of plastids: As outlined in Table III, except plastids from 1000 *g* pellet were centrifuged for 45 minutes at 23,000 rpm (Spinco SW 25.1 rotor) in 17 to 68% sucrose; 0.05 *M* Tris, pH 8.0; 1 m*M* $MgCl_2$; and 10 m*M* mercaptoethanol.

leaves were illuminated for 5 minutes, the increase in RNA polymerase activity ranged from 30% to 100%. After illumination for 30 minutes, increases ranging from 60% to 300% have been observed. Here, as in changes in the specific activity of various types of RNA, the increase in plastid RNA polymerase activity is about the same if leaves are illuminated for 30 minutes and maintained in darkness for 90 minutes as it is if they are illuminated continuously for 120 minutes.

The usual rise in RNA polymerase activity is sharply reduced if leaves are treated with chloramphenicol prior to exposure to light. The data in Table V also show that when leaves are maintained in darkness and exposed to chloramphenicol the level of plastid RNA polymerase falls. This suggests that the enzyme may be turning over at a compara-

TABLE V
Effect of Administration of Chloramphenicol to Etiolated Maize Leaves on RNA Polymerase Activity of Isolated Plastids[a]

Dark	100%
Dark + chloramphenicol	60%
Light	650%
Light + chloramphenicol	176%

[a] Cut ends of leaves placed in chloramphenicol (4 mg/ml) for 16 hours in darkness. Leaves then either maintained in darkness or illuminated for 2 hours before plastids were isolated. Plastid isolation and assay of RNA polymerase as described in Table III.

tively rapid rate and that, although synthesis is occurring all the time in darkness, it is stimulated above the dark synthesis level by illumination. The activity of cytoplasmic RNA polymerase—that is, the enzyme in the 27,000 *g* supernatant of a maize leaf homogenate—is not affected by treating leaves with chloramphenicol.

Thus, one of the earliest detectable events in plastids following illumination of etiolated leaves is an increase in the activity, presumably by accelerated synthesis, of plastid RNA polymerase and the formation of at least ribosomal and transfer ribonucleic acids. But which comes first? Is the increase in RNA polymerase the result of some general stimulation of protein synthesis manifested as more ribulose diphosphate carboxylase, more RNA polymerase, and increased ALA synthesizing capacity? Or is the development of the RNA polymerase an earlier or near-primary event which leads to the other changes? One clue we do have is that administration of chloramphenicol not only affects the level of RNA polymerase but also depresses the normally observed illumination-stimulated increase in specific activity of plastid RNA's in etiolated leaves supplied P^{32}-phosphate—the new RNA polymerase is apparently working *in vivo*. But obviously this is a kind of merry-go-round—it is hard to tell *where* one gets on. Dr. Stent, in this Symposium, has reviewed many of the kinds of possible alternatives which could explain, for example, control of messenger RNA synthesis by ribosome synthesis, of protein synthesis by RNA synthesis, etc. There seems to be no point in repeating them even though they may be highly pertinent to interpretations of the effects described here.

Summary and Conclusions

Illumination of etiolated leaves results in the maturation of their proplastids. A few plastid enzymes and enzyme systems have been investigated with regard to the effect of illumination on their activity and persistence.

The proplastid of the etiolated leaf is an elaborate structure containing lipids, many kinds of proteins, ribosomes, polyribosomes, DNA, protochlorophyllide, etc. Some enzymes which have been investigated—for example, plastid RNA polymerase and some enzymes of the photosynthetic carbon reduction cycle—are present in proplastids but increase in amount at various times after illumination has commenced (Table VI). At least one system, that for the production of ALA, is not present, although it is clear from the presence of protochlorophyllide that it was operating at some time during the development of the proplastid. There is no complete catalog of constituents of etiolated proplastids and of green chloroplasts, but it may well be that all the components of the

TABLE VI
STRUCTURAL AND CHEMICAL CHANGES IN PLASTIDS DURING ILLUMINATION OF ETIOLATED LEAVES

Change	Time of initial change or increase		
	Seconds	Minutes	Hours
Structural changes			
	Tube transformation	Vesicle dispersal and fusion	
			Grana building
Chemical changes[a]			
	Protochlorophyllide → chlorophyllide		
		+RNA production	
		+RNA polymerase	
		+ALA production	
		+RuDP carboxylase	
			+RuP kinase
			+R-5-P isomerase

[a] Except for chlorophyllide, all compounds listed, including enzymes, are or have been present in proplastids before illumination but increase after leaves have been illuminated.

latter (or their immediate precursors—for example, protochlorophyllide) may be, or might have been, present in the proplastid. Perhaps the first effect of light on protein formation is *not* to bring about the synthesis of some types of proteins which had never existed in the proplastid before. But still—a mature chloroplast is qualitatively different from a proplastid in an etiolated leaf.

Disregarding for the moment the possible mechanism of stimulation (for example, production of new ribosomes, more informational or transfer RNA's, or an increase in some substrates or promotive factors), suppose the rate of synthesis of *all* constitutive (that is, always derepressed) plastid proteins were stimulated by light. How might such a general effect result in plastid maturation?

The amount of a particular enzyme present in an organelle, cell, or organism at any time obviously is affected by its rates of production and decay; its effective level may be altered further by endogenous inhibitors or stimulators, etc. If we assume that within the range of the normal cell environment the rate of decay is an inherent property of a particular kind of molecule and that proteins differ from one another in this regard, then, depending on the rate of protein synthesis, one enzyme might be absent, even if it is being synthesized, while another is being lost much more slowly than it is formed. In turn, once the enzyme catalyzes a reaction, the product might be used quickly as a substrate of another enzyme, might accumulate temporarily, or might persist and become a permanent part of the organelle or cell and thus be available, for example, for construction of permanent structures. Products of enzymes of the photosynthetic carbon reduction cycle are transitory, but it is possible that compounds produced by some enzymes can act as derepressors of other enzymes when the small molecules are being produced more rapidly than they can be consumed. In this way a single constitutive enzyme raised to a new high level of activity could later bring about increases in the production of other enzymes. On the other hand, the activity of the entire biosynthetic chain of chlorophyll seems to be limited by the availability of ALA, a compound produced by a labile system, but the end product of the chain is stable (evidence on the turnover of chlorophyll in green leaves is conflicting); its presence makes a permanent difference in the chloroplast. There is evidence that chlorophyll is not required for the formation of lamellae, and thus it is difficult to argue that the presence of this pigment alone facilitates lamella and grana building; but other similar control systems which lead to the formation of lamellar components by the increased activity of a labile enzyme early in a biosynthetic chain may exist and be involved in the formation of essential components of lamellae.

Some of the evidence and uncertainties regarding the possible participation of phytochrome in controlling plastid metabolism have been cited. This problem needs further exploration, but wherever it leads we will be left with the question of how absorption of visible light by some pigment is translated into the biochemical manifestations described.

It is not yet clear what events connect light absorption to RNA metabolism in plastids. Does any complete or partial derepression occur as an early event? If so, why of apparently unrelated enzymes such as RNA polymerase, ribulose diphosphate carboxylase, and some enzyme(s) involved in ALA synthesis? These three enzymes can be considered to have regulatory roles with regard to the possible production or action of other enzymes, and thus, if their synthesis were stimulated (at least the first two enzymes are present and working in etiolated proplastids, but their decay rate may be slower than that of the third), major changes in plastid metabolism might occur, as already discussed. Are the genes for these enzymes close to one another on the chloroplast chromosome? If there is any immediate derepressive effect, is it brought about by direct involvement of the light-modified photoreceptor or by some product(s) of its activity?

On the other hand, if the primary effect of the light-modified photoreceptor is stimulation of the synthesis of all plastid RNA's which are made in darkness (all constitutive RNA's) and qualitative control of plastid metabolism is secondary (through the action of enzymes, as discussed earlier), how does the photoreceptive pigment act? Is it itself a "cofactor" for RNA synthesis? Does it control the formation of a compound which is?

Another range of problems of organizational biosynthesis has been touched on only fleetingly in the preceding discussion. How are the lamellae formed and from what? Massive development of these structures in leaf plastids is dependent on illumination (compare Figs. 1 and 2). Little is known about the subject, and consequently any discussion would be too long to be handled here, but a few questions should bring the nature of the problem into better focus. What are the units of structure of the lamellae? What are the units of synthesis (this question was introduced briefly in the section on the protochlorophyllide holochrome)? Are any or most of the lamellar constituents present in etiolated proplastids—that is, are only a few new key compounds made in plastids when leaves are illuminated, or must most of the lamellar constituents be produced? Closely related to this question—what are the physical and chemical similarities and differences between the membranes of the prolamellar body, and the vesicles formed from it on brief illumination, and photosynthetically competent plastid lamellae?

Studies of plastid maturation seem to provide an almost overwhelming opportunity to examine problems of intracellular ecology and of integrated structural and chemical development. Unfortunately these studies still need to be carried out on these at best partially autonomous organelles growing in an undefined complex and probably constantly changing environment.

Acknowledgments

The research described here was made possible by research grants from the National Institute of General Medical Sciences, National Institutes of Health; the National Institute of Arthritis and Metabolic Diseases, National Institutes of Health; and the National Science Foundation; and by skilled technical assistance provided by Mrs. Louisa Ni and Mrs. Lumiko Shimada.

References

1. Anderson, J. M., and Boardman, N. K., *Australian J. Biol. Sci.*, **17**, 93 (1964).
2. Augustinussen, E., and Madsen, A., *Physiol. Plantarum*, **18**, 828 (1965).
3. Boardman, N. K., *in* "The Chlorophylls" (L. P. Vernon and G. R. Seeley, eds.), p. 437, Academic Press, New York, 1966.
4. Boardman, N. K., and Wildman, S. G., *Biochim. Biophys. Acta*, **59**, 222 (1962).

5. Bogorad, L., and Jacobson, A., *Biochem. Biophys. Res. Commun.*, **14**, 113 (1964).
6. Butler, W. L., *Arch. Biochem. Biophys.*, **93**, 413 (1961).
7. Chen, S., McMahon, D., and Bogorad, L., *Plant Physiol.*, **42**, 1 (1967).
8. DiGirolamo, A., Henshaw, E. C., and Hiatt, H. H., *J. Mol. Biol.*, **8**, 479 (1964).
9. Erwin, J., and Bloch, K., *Biochem. Z.*, **338**, 496 (1963).
10. Gassman, M., and Bogorad, L., *Plant Physiol.*, **40**, lii (1965).
11. (*a*) Gassman, M., and Bogorad, L., *Plant Physiol.*, **42**, 774 (1967); (*b*) Gassman, M., and Bogorad, L., *Plant Physiol.*, **42**, 781 (1967).
12. Granick, S., *Proc. 5th Intern. Biochem. Congr., Biochem. Moscow*, **5**, 176 (1961).
13. Jacobson, A. B., Swift, H., and Bogorad, L., *J. Cell Biol.*, **17**, 557 (1963).
14. Kirk, J. T. O., *in* "Biochemistry of Chloroplasts" (T. W. Goodwin, ed.), Vol. 1, pp. 319, Academic Press, London, 1966.
15. Kislev, N., Swift, H., and Bogorad, L., *J. Cell Biol.*, **25**, 327 (1965).
16. Klein, S., Bryan, G., and Bogorad, L., *J. Cell Biol.*, **22**, 433 (1964).
17. Mans, R. J., and Novelli, G. D., *Biochim. Biophys. Acta*, **91**, 186 (1964).
18. Marcus, A., *Plant Physiol.*, **35**, 126 (1960).
19. Margulies, M. M., *Plant Physiol.*, **37**, 473 (1962).
20. McMahon, D., and Bogorad, L., *Federation Proc.*, **26**, 807 (1967).
21. Mego, J. L., and Jagendorf, A. T., *Biochim. Biophys. Acta*, **53**, 237 (1961).
22. Menke, W., *Ann. Rev. Plant Physiol.*, **13**, 27 (1962).
23. Mitrakos, K., *Physiol. Plantarum*, **14**, 497 (1961).
24. Price, L., and Klein, W. H., *Plant Physiol.*, **36**, 733 (1961).
25. Schiff, J. A., and Epstein, H. T., *in* "Reproduction, Molecular, Subcellular and Cellular" (Michael Locke, ed.), p. 131, Academic Press, New York, 1965.
26. Semal, J., Spencer, D., Kim, Y. T., and Wildman, S. G., *Biochim. Biophys. Acta*, **91**, 205 (1964).
27. Shibata, K., *J. Biochem. (Tokyo)* **44**, 147 (1957).
28. Smillie, R. M., *Can. J. Botany*, **41**, 123 (1963).
29. Virgin, H. I., Kahn, A., and von Wettstein, D., *Photochem. Photobiol.*, **2**, 83 (1963).
30. Wallace, J. W., and Newman, D. W., *Phytochemistry*, **4**, 43 (1965).
31. Von Wettstein, D., *Brookhaven Symp. Biol.*, **11**, 138 (1958).
32. Williams, G. R., and Novelli, G. D., *Biochem. Biophys. Res. Commun.*, **17**, 23 (1964).
33. Wolf, F. T., Coniglo, J. G., and Bridges, R. B., *in* "Biochemistry of Chloroplasts" (T. W. Goodwin, ed.), Vol. I, p. 187, Academic Press, London, 1966.

The Nucleic Acids Associated with the Chloroplasts of *Euglena gracilis* and Their Role in Protein Synthesis[1]

George Brawerman and Jerome M. Eisenstadt

Departments of Biochemistry and Microbiology, Yale University School of Medicine, New Haven, Connecticut

Introduction

Chloroplasts exhibit physiological characteristics quite distinct from those of their intracellular environment. Aside from their specialized function, photosynthesis, they possess a genetic system relatively independent from the nucleus of the cell (14). Protein synthesis in chloroplasts has also been shown in many instances to be subject to regulatory agents which do not affect cellular protein synthesis. The occurrence of these autonomous processes indicates that chloroplasts may possess their own genetic material as well as their own protein-synthesizing system. It is the purpose of this paper to describe, in the case of *Euglena gracilis*, the nucleic acids associated with the chloroplasts, and to attempt a correlation between the characteristics of these substances and the physiological processes with which they are concerned.

Physiological Characteristics of Chloroplast Synthesis in *Euglena gracilis*

Protein Synthesis in Chloroplasts

Euglena gracilis possesses two remarkable properties which render it particularly suitable for the study of biochemical processes associated with chloroplasts. The cells can be grown in the dark in the presence of organic carbon sources, but they are unable to form chloroplasts under these conditions. Exposure of these colorless cells to light brings about the rapid appearance of chloroplasts. This process involves the synthesis of various chloroplast constituents. Since mature chloroplasts in *Euglena*

[1] Parts of this article are reproduced from G. Brawerman and J. M. Eisenstadt, *in* "Le Chloroplaste" (C. Sironval, ed.), Masson, Paris, 1966.

contain at least 40% of the total cellular protein, exposure of colorless cells to light results in the induced synthesis of a massive amount of specific proteins (6). The process of chloroplast formation can be affected by a variety of agents which do not alter materially cell growth and cytoplasmic protein synthesis. Removal of light results in the cessation of chloroplast formation. The synthesis of both chlorophyll and protein associated with the chloroplast structure is blocked. This is also true for soluble enzymes associated with the photosynthetic apparatus (5). The synthesis of chloroplasts has also been shown to be repressed by certain organic carbon sources (1). Chloramphenicol, an effective inhibitor of protein synthesis in bacteria, can inhibit preferentially chloroplast protein synthesis (19). Finally, growth of *Euglena* at 34°C, a temperature somewhat above optimal, results in a strong reduction in the rate of chloroplast synthesis without much effect on cell growth (3, 5). These examples of independent physiological behavior strongly suggest that the chloroplasts of *E. gracilis* possess a protein-synthesizing system with biochemical properties distinct from that of the cytoplasm. The discovery of ribosomes associated with the chloroplasts and containing RNA sharply different from the cytoplasmic RNA (7) provided evidence that a specific protein-synthesizing system associated with the chloroplasts does indeed exist in *E. gracilis*.

Loss of Ability to Form Chloroplasts

Euglena gracilis can also be subjected to treatments which result in the permanent loss of their ability to form chloroplasts. These treatments consist in exposure of the cells to a temperature of 34°C, to streptomycin, or to irradiation with ultraviolet light (20, 21). A self-replicating system associated with the chloroplasts appears to be inactivated, and is lost by dilution when the cells are allowed to multiply during the treatments (4). In the case of ultraviolet irradiation, the evidence suggests that the self-replicating component is a nucleoprotein (16). Thus the chloroplasts of *Euglena* appear to possess a genetic system far more sensitive than the nuclear chromosomal system to a variety of inactivating agents. The occurrence of such a system is supported by the presence in the chloroplasts of DNA with a nucleotide composition different from that of the nuclear DNA (8).

Nucleic Acids Associated with Chloroplasts of *Euglena gracilis*

Ribosomal RNA

Examination of the nucleotide composition of total cellular RNA of *E. gracilis* provided preliminary evidence for the occurrence of specific

TABLE I
NUCLEOTIDE COMPOSITION OF THE RNA OF GREEN AND COLORLESS *Euglena* CELLS[a]

Nucleotide	Green	Colorless
Adenylic acid	23.2 (0.2)	21.6 (0.3)
Guanylic acid	28.5 (0.5)	30.2 (0.7)
Cytidylic acid	26.3 (0.5)	27.5 (0.8)
Uridylic acid	22.0 (0.3)	20.7 (0.7)
A/C	0.88	0.79

[a] Values in parentheses represent standard deviations. Data reproduced from Brawerman and Chargaff (2).

RNA associated with the chloroplasts. Small but significant differences were observed in the RNA nucleotide composition of green cells as compared to that of cells lacking chloroplasts (Table I). These differences suggested that a species of RNA different from the cellular RNA of dark-grown cells was synthesized along with the chloroplasts. It was predicted that this RNA should be particularly rich in adenylic and uridylic acids. After extensive subcellular fractionations, an RNA species with these characteristics was localized in the chloroplasts, where it was found to be associated with ribosomal particles (7). The marked differences in the nucleotide composition of chloroplast and cytoplasmic ribosomes are illustrated in Table II.

Crude chloroplasts are heavily contaminated by ribosomes from the cytoplasm. Thus it was necessary to devise a purification procedure which would eliminate this cytoplasmic contamination. The nucleotide composition of the RNA in the chloroplast preparations provided a convenient criterion for purity during the purification steps. Excellent results were obtained when the crude chloroplasts were subjected to flotation in a sucrose solution of a density slightly above that of the chloroplasts

TABLE II
RNA NUCLEOTIDE COMPOSITION OF CHLOROPLAST AND CYTOPLASMIC RIBOSOMES[a]

Nucleotide	Ribosomes isolated from chloroplasts	Cytoplasmic ribosomes
Adenylic acid	30.7	22.7
Guanylic acid	27.1	29.5
Cytidylic acid	17.1	27.1
Uridylic acid	25.1	19.6
Pseudouridylic acid	None detected	1.1
A/C	1.80	0.84

[a] Data for chloroplast ribosomes, from Eisenstadt and Brawerman (12); and for cytoplasmic ribosomes, from Brawerman (7).

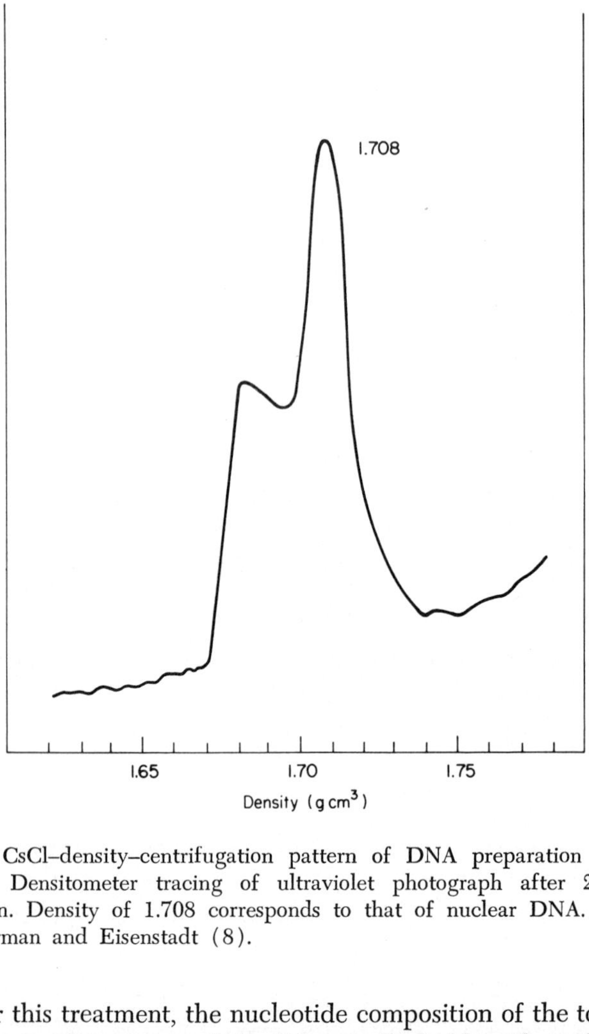

FIG. 1. CsCl–density–centrifugation pattern of DNA preparation from whole chloroplasts. Densitometer tracing of ultraviolet photograph after 26 hours of centrifugation. Density of 1.708 corresponds to that of nuclear DNA. Reproduced from Brawerman and Eisenstadt (8).

(12). After this treatment, the nucleotide composition of the total chloroplast RNA was the same as that of the purified chloroplast ribosomes.

The ribosomes can be obtained from purified chloroplasts by lysis with sodium deoxycholate followed by sedimentation at $100{,}000 \times g$. Most of the RNA in the chloroplasts sediments with the ribosomes. RNA extracted with phenol, either from the whole chloroplasts or from the purified ribosomes, shows two components with sedimentation values of approximately 19 S and 14 S. The two components appear to have similar nucleotide compositions (9). Cytoplasmic RNA, on the other hand, shows a single major component with a 19 S value and a nucleotide composition very similar to that of the total RNA of cells grown in the dark.

DNA

DNA extracted from purified chloroplasts contains a substantial proportion of a component with a buoyant density far lower than that of the bulk of the cellular DNA (Fig. 1). The latter, however, still represents the major component. Partial lysis of the chloroplast preparation, under conditions which leave the bulk of the chloroplast components in particulate form, resulted in the preferential solubilization of the nuclear DNA component (Fig. 2). It was observed that the efficiency of this separation procedure was dependent on the $MgCl_2$ concentration during deoxycholate treatment of the chloroplasts. With the appropriate concentration of $MgCl_2$, the nuclear component could be nearly completely removed (8). This result permits us to state that the nuclear component

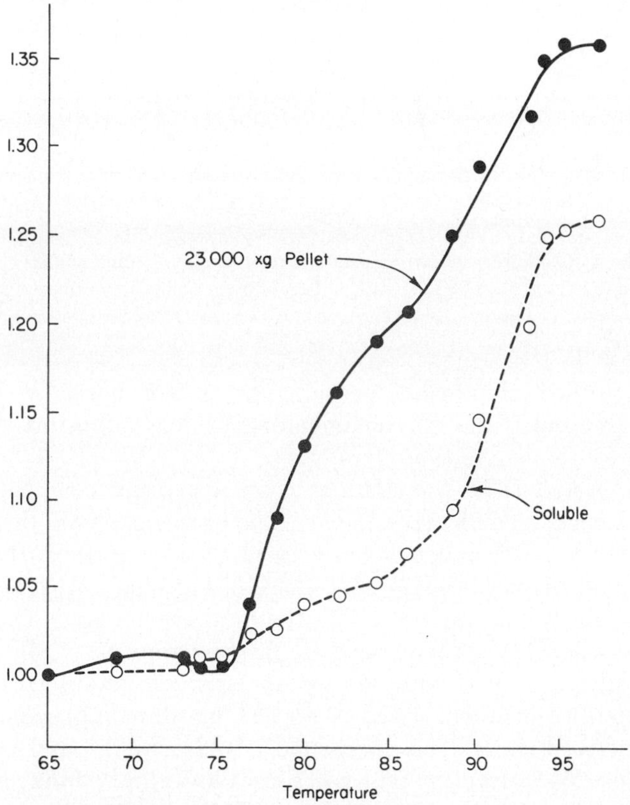

FIG. 2. Melting profiles of DNA preparations from centrifugal fractions of chloroplasts lysed in 0.01 *M* $MgCl_2$. Soluble refers to 100,000 × *g* supernatant fraction. Profile of DNA from soluble fraction is same as that of nuclear DNA. Reproduced from Brawerman and Eisenstadt (8).

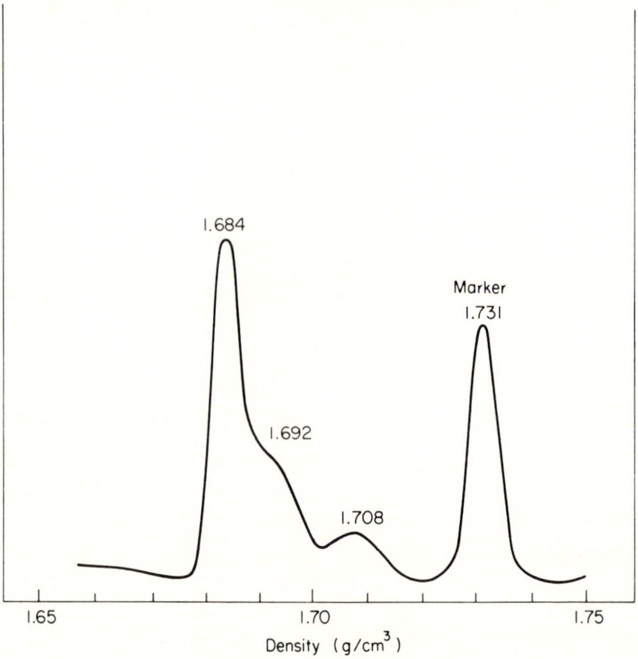

FIG. 3. CsCl–density–centrifugation pattern of DNA from 23,000 × *g* sediment of chloroplasts lysed in 0.005 *M* $MgCl_2$. Reproduced from Brawerman and Eisenstadt (8).

in the purified chloroplast preparations is not normally present in chloroplasts, but is due to contaminating nuclear fragments particularly sensitive to the deoxycholate treatment.

The purified DNA preparations contain a major component with a buoyant density of 1.684 (see Fig. 3). The component with the density of nuclear DNA, 1.708, is present in very small amount. It has been shown by Ray and Hanawalt (22) that the 1.684 component is absent from *Euglena* which has lost the ability to form chloroplasts. This confirms that the major DNA component of our preparations is physically and functionally associated with the chloroplasts. An additional component with a density of about 1.692 is present in substantial amount in our preparations (Fig. 3). This compound has also been found by Ray and Hanawalt (23) in *Euglena* which has lost the ability to form chloroplasts. It has been isolated recently from purified mitochondria and can be considered, therefore, as mitochondrial DNA (11, 15). The presence of the DNA contaminants in our preparations illustrates the difficulty in obtaining subcellular particles of absolute purity, and the caution which must be exercised in the interpretation of results obtained with "purified" preparations.

The chloroplast DNA has an unusually low content of guanylic and cytidylic acids (Table III). If allowance is made for the small amount of nuclear component still present in the preparations, the G-C content is about 25 mole %. 5-Methylcytosine, which is present in total *Euglena* DNA to the extent of 2.3 mole %, appears to be absent from the chloroplast DNA (8, 22).

TABLE III
NUCLEOTIDE COMPOSITION OF DEOXYRIBONUCLEIC ACID PREPARATIONS FROM *Euglena gracilis*[a]

Base	Cellular DNA	Chloroplast DNA preparations: 50% nuclear contamination	Chloroplast DNA preparations: 10% nuclear contamination
A	24.5	31.6	37.2
G	24.8	17.7	13.0
C	23.5	18.1	14.3
T	24.7	31.5	35.5
5-Methylcytosine	2.3	1.1	—

[a] Data reproduced from Brawerman and Eisenstadt (8).

THE PROTEIN-SYNTHESIZING SYSTEM ISOLATED FROM THE CHLOROPLASTS

Characteristics of Amino Acid Incorporation by Chloroplast Preparations

Chloroplasts isolated from cells grown with CO_2 as the sole carbon source are active in incorporating amino acids into acid-insoluble material (12). The characteristics of this reaction are very similar to those usually observed with ribosomal systems. The reaction proceeds rapidly for about 30 minutes, then slows down to a negligible rate. It is dependent on added transfer RNA and $100{,}000 \times g$ supernatant derived from *Euglena* (Table IV). The presence of an ATP-generating system is also required. The incorporation is highly sensitive to inhibitors of protein synthesis such as puromycin and ribonuclease. Polyuridylic acid specifically stimulates phenylalanine incorporation. The isolated chloroplasts probably have a damaged membrane, since they lose photosynthetic enzymes during the preparation. It is possible, therefore, that the soluble factors required for protein synthesis are normally present in the chloroplasts, but were also lost during the isolation.

TABLE IV
CHARACTERISTICS OF INCORPORATION OF AMINO ACIDS BY *Euglena* CHLOROPLASTS[a]

Conditions	Amino acid incorporated	
	Leucine	Phenylalanine
Complete system	430	192
Minus transfer RNA	52	127
Minus ATP system	86	
Plus puromycin (110 μg/ml)	65	
Plus RNase (50 μg/ml)	6	
Plus template RNA (100 μg/ml)	580	
Plus poly U (120 μg/ml)	403	397

[a] Values expressed as micromicromoles of amino acid incorporated per milligram of RNA in chloroplasts. Data reproduced from Eisenstadt and Brawerman (12).

Requirement of Chloroplast Ribosomes for Messenger RNA

Chloroplast ribosomes are about one-tenth as active as the chloroplasts from which they are derived (Table V). The amino acid incorporation activities can be compared because they are expressed in terms of the RNA of the preparations, rather than the protein. The cytoplasmic ribosomes, on the other hand, have an activity very close to that of the chloroplasts. The activity of the chloroplast ribosomes can be considerably enhanced by addition of RNA rich in template activity (Table V). This activity is defined as the ability to stimulate amino acid incorporation by the preincubated cell-free system from *Escherichia coli*, which was developed by Matthaei and Nirenberg (17). The stimulation of chloroplast ribosomes is a specific effect, since ribosomal RNA, which has little template activity, is largely ineffective. Template RNA has

TABLE V
INCORPORATION OF LEUCINE BY *Euglena* RIBOSOMES[a]

Conditions	Chloroplast ribosomes	Cytoplasmic ribosomes
Complete system	43	425
Minus supernatant, transfer RNA	2	8
Plus ribosomal RNA (215 μg/ml)	66	
Plus template RNA (100 μg/ml)	212	423
Plus puromycin (40 μg/ml)	58	
Plus actinomycin (56 μg/ml)	209	

[a] Values expressed as micromicromoles of leucine incorporated per milligram of RNA in ribosomes. Data reproduced from Eisenstadt and Brawerman (12).

little effect on the intact chloroplasts (Table IV). The requirement of chloroplast ribosomes for template RNA can be explained in the following manner. The chloroplasts contain messenger RNA in sufficient quantity for maximum activity of their protein-synthesizing system. When the ribosomes are sedimented from the chloroplast lysate, much of the messenger RNA remains in the supernatant. Thus the isolated ribosomes are dependent on exogenous messenger RNA for efficient amino acid incorporation into protein. This interpretation is supported by an examination of the RNA left in the 100,000 $\times$ *g* supernatant of the chloroplast lysate. The template activity of this RNA is about three times as high as that of the RNA derived from the ribosomes (9).

The cytoplasmic ribosomes, isolated by a procedure similar to that used to obtain the chloroplast particles, show no requirement for template RNA (Table V). This suggests that the cytoplasmic ribosomes do not lose their messenger RNA during the isolation procedure, and consequently that they bind messenger RNA more strongly than the chloroplast ribosomes.

Size of the Active Ribosomal Unit

The chloroplast ribosomes sediment more slowly than the cytoplasmic ribosomes (7). In order to determine the relative sizes of the units active in protein synthesis, the reaction mixtures after incorporation of amino acids were subjected to zone centrifugation, and the fractions collected through the bottom of the tubes were assayed for acid-insoluble radioactivity (12). With the cytoplasmic ribosomes, the major radioactivity peak coincided with the 70 S ribosomal peak (Fig. 4). In the case of the chloroplast ribosomes, the major radioactivity component sedimented well behind the 70 S position. After prolonged centrifugation, a 60 S component containing a substantial portion of the particle-bound radioactivity was resolved. This indicates that the active ribosomal unit in the chloroplasts has a sedimentation of approximately 60 S. The smaller size of the chloroplast ribosomes as compared to the cytoplasmic particles has been confirmed by electron microscopy of *Euglena* cells (Fig. 5).

Effect of Chloramphenicol

Chloramphenicol has no effect on the cytoplasmic cell-free system, even when used at a concentration of 360 μg/ml. The chloroplast system is strongly inhibited at comparable levels (Table VI). Amino acid incorporation by isolated chloroplast ribosomes is also sensitive to the antibiotic.

It has been shown by Pogo and Pogo (19) that chloramphenicol inhibits preferentially chloroplast protein synthesis in *Euglena*. Thus we

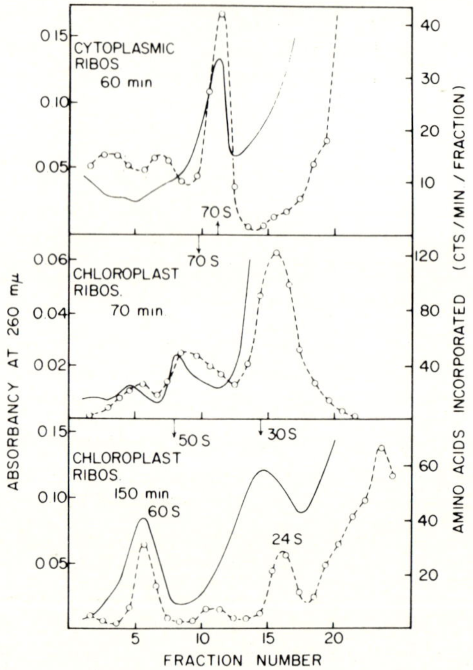

FIG. 4. Zone centrifugation of incubation mixtures with cytoplasmic and chloroplast ribosomes. Duration of centrifugation indicated in each figure. Solid lines represent absorbancy at 260 mμ, and dashed lines indicate radioactivity insoluble in hot acid. 70 S, 50 S, and 30 S positions obtained by running *E. coli* ribosomes in separate tubes. Data derived from Eisenstadt and Brawerman (12).

TABLE VI: EFFECT OF CHLORAMPHENICOL ON INCORPORATION OF LEUCINE BY THE CHLOROPLAST AND THE CYTOPLASMIC CELL-FREE SYSTEMS[a]

Chloramphenicol (μg/ml)	Chloroplasts	Ribosomes	
		From chloroplasts	From cytoplasm
60	99	99	99
120	84	74	103
210	55	61	94
360		59	94

[a] Values expressed as percentage of controls without chloramphenicol. Data reproduced from Eisenstadt and Brawerman (12).

FIG. 5. Portion of a thinly sectioned *Euglena* cell from a lag phase culture. Note the numerous small ribosomal granules in the chloroplasts and the larger ones in the cytoplasmic matrix. ×55,000. Preparations fixed in 1% glutaraldehyde in

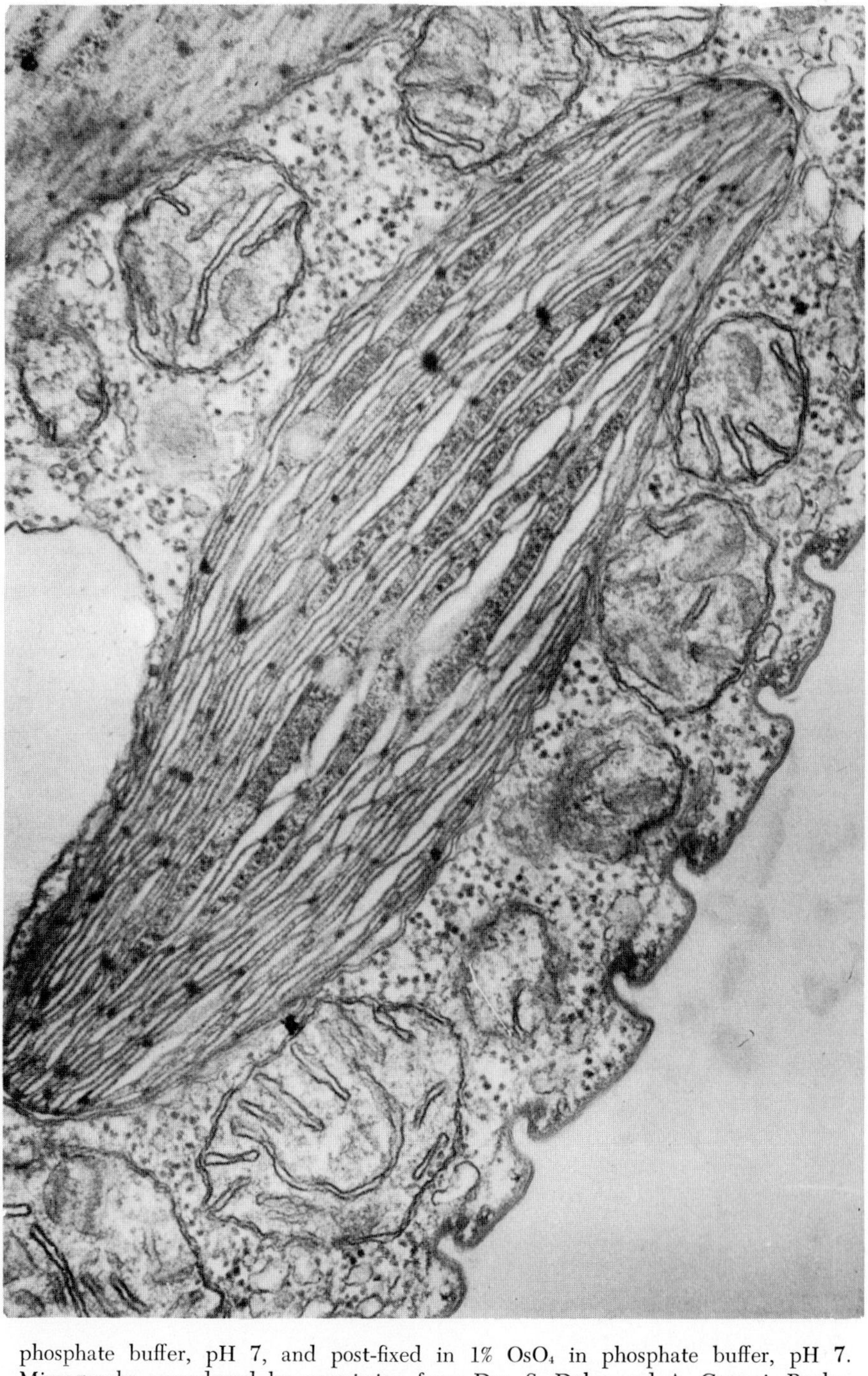

phosphate buffer, pH 7, and post-fixed in 1% OsO_4 in phosphate buffer, pH 7. Micrographs reproduced by permission from Drs. S. Dales and A. Cerami, Rockefeller Institute (unpublished observations).

have a good correlation between the *in vivo* behavior of chloroplast protein synthesis and the properties of the isolated protein-synthesizing system. This provides independent evidence that the particles obtained from the isolated chloroplasts are in fact concerned with the bulk of chloroplast protein synthesis in the intact cells.

Specificity of Chloroplast Ribosomes

The ability of the chloroplast ribosomes to respond to exogenous messenger RNA is rather unique among nonbacterial cell-free systems. While most ribosomes are stimulated by synthetic polynucleotides such as poly U, they remain essentially unaffected by natural messenger RNA. *Euglena* chloroplast ribosomes, like the *E. coli* particles, are strongly stimulated by viral RNA. This property has permitted a direct experimental examination of the possibility that ribosomes contribute to the information for specific protein synthesis. RNA from the bacterial virus f_2 has been shown to induce the synthesis of the viral coat protein by *E. coli* ribosomes (18). The polypeptide material produced by the chloroplast ribosomes in the presence of f_2 RNA was found to be similar to that formed by the *E. coli* cell-free system (24). Thus the ribosomes appear to have no specificity of their own with respect to determination of amino acid sequence in proteins.

It has been found recently that the stimulation of *Euglena* chloroplast ribosomes by f_2 RNA requires the presence of a factor distinct from the enzymes concerned with peptide bond formation (13). This factor appears to be involved in the initiation phase of protein synthesis. Such an "initiation" factor has also been demonstrated in the *E. coli* cell-free system (10, 13). What may be of particular interest is that the factor from *E. coli* does not function with the chloroplast ribosomes, and conversely the *Euglena* factor is inactive with *E. coli* ribosomes. The supernatant enzymes concerned with peptide bond formation, on the other hand, are active with the heterologous ribosomes. This is illustrated in Table VII, where the extent of polypeptide synthesis in the presence and in the absence of f_2 RNA is compared, using mixed systems consisting of supernatant fraction from one species and ribosomes from the other species. The occurrence of an initiation factor specific for a particular species of ribosomes provides an interesting possibility for the selective regulation of protein synthesis in chloroplasts.

DNA and Chloroplast Replication

Role of Chloroplast DNA

The occurrence of DNA in chloroplasts provides a biochemical basis for the genetic autonomy of these organelles. The precise role of the

TABLE VII
EFFECT OF F_2 RNA ON POLYPEPTIDE SYNTHESIS IN MIXED CELL-FREE SYSTEMS FROM *Euglena* AND *Escherichia coli*[a]

Source of ribosomes	Source of supernatant fraction	f_2 RNA None	200 μg/ml	400 μg/ml
E. coli	None	16		
	E. coli	110	720	1500
	Euglena	350	350	350
Euglena chloroplasts	None	23		
	Euglena	360	700	1300
	E. coli	540	490	560

[a] Values expressed as counts/min/ml reaction, of leucine-C^{14} incorporated into peptide linkage. Data are from Eisenstadt and Brawerman (13).

DNA, however, remains to be determined. The primary function of DNA is usually considered in terms of providing information for the amino acid sequences of all cellular proteins. The amount of DNA in *Euglena* chloroplasts is about the same as in *E. coli* (8). This is certainly sufficient to code for a large number of proteins. However, no chloroplast protein has yet been found to be under the control of a structural gene associated with the plastids. Genetic studies on various photosynthetic enzymes have shown that these proteins are under nuclear control. The possibility remains that some of the plastid proteins, such as the structural proteins, are coded by the chloroplast DNA.

Biochemical evidence for the role of the chloroplast DNA in *Euglena* is rather scant. The chloroplast ribosomal RNA tends to resemble the plastid DNA with respect to overall base composition (Table VIII). This suggests that this RNA may be transcribed from the chloroplast DNA. The presence in the plastids of RNA with template activity, presumably messenger RNA, suggests that these organelles are capable of transcription of information for protein synthesis. This messenger RNA, however, could be conceivably derived from the nucleus.

TABLE VIII
RELATIONS BETWEEN THE NUCLEOTIDE COMPOSITION OF DNA AND RIBOSOMAL RNA IN *Euglena gracilis*

	Ribosomal RNA		DNA	
	Chloroplasts	Cytoplasm	Chloroplasts	Nuclei
% (G + C)	44	57	25	51
A/C	1.9	0.84	2.6	0.95

DNA and Loss of Ability To Form Chloroplasts

Permanently bleached cells lack the chloroplast DNA component, and it is possible that the bleaching agents act directly at the level of the DNA. The data on the action spectrum ultraviolet bleaching tend to support this possibility. A particularly high sensitivity of the chloroplast DNA toward the various bleaching agents cannot at present be explained in terms of its biochemical properties. The chloroplast DNA is rich in thymine and is therefore more susceptible to ultraviolet-induced thymine dimer formation (25). Because of its low G-C content, it has a relatively low thermal denaturation temperature. It is doubtful, however, that these properties can account for the high sensitivity of the chloroplast-replicating system to the slightly elevated temperature of 34°C and to the ultraviolet treatment. The sensitivity of the chloroplast DNA could conceivably be based on its low degree of methylation as compared to the nuclear DNA. It is more likely, perhaps, that the sensitivity is determined either by the structural configuration of the DNA within the chloroplasts, or by a peculiar property of the DNA polymerase.

Physiological Significance of Specific Chloroplast Ribosomes

Ribosomes and Coding Specificity

In the formation of a subcellular structure, a mechanism must exist for the localization of the proteins associated with this structure. It is not inconceivable that proteins generated anywhere in the cytoplasm could have the capacity for migration and assembly into the right structures by virtue of their molecular properties. It may be fruitful, however, to search for a mechanism which ensures the synthesis of these proteins at the site of formation of the appropriate structure. In this case, specific ribosomes associated with a subcellular organelle could exert a selective influence on the nature of the polypeptides synthesized in this organelle. This would occur if the ribosomes were to participate in the determination of amino acid sequence in proteins, and if amino acid sequences peculiar to chloroplast proteins could be assembled only by chloroplast ribosomes. The experiments with f_2 RNA indicate, however, that the chloroplast ribosomes are not likely to contribute to the amino acid sequences of the chloroplast proteins.

Another possibility is that ribosomes are capable of selecting the appropriate messenger RNA molecules. This has been tested by comparing the response of *E. coli* and chloroplast ribosomes to RNA preparations from a variety of sources. No significant preference of the two types of ribosomes for any RNA, including the RNA isolated from each homologous source, could be detected. Thus the ribosomes do not appear to be capable of distinguishing between different types of messenger RNA. It is not impossible, however, that messenger RNA is transported to the ribosomes as a nucleoprotein; and that the protein moiety could be recognized by specific ribosomes.

A simpler mechanism could be postulated by assuming that the messenger RNA molecules for the structural proteins of the chloroplasts are generated *in situ* by the chloroplast DNA. The structural specificity of the proteins could then be ensured if the chloroplast membrane were impermeable to messenger RNA present in the cytoplasm. In this event the chloroplast proteins which are under nuclear control would have to be synthesized in the cytoplasm.

Ribosomes and Regulation of Protein Synthesis

The occurrence of a species of ribosomes involved in the synthesis of a particular group of proteins could provide for a mechanism for the selective inhibition or stimulation of the synthesis of these proteins. This could be readily achieved if these ribosomes were sensitive to specific regulatory agents. The sensitivity of chloroplast ribosomes to chloramphenicol provides a good model for such a selective mechanism. The repression of chloroplast protein synthesis by removal of light and by some organic carbon sources could be explained by such a selective effect on chloroplast ribosomes. In this context it may be of interest to note some similarities between bacterial and chloroplast ribosomes. Both types are sensitive to chloramphenicol and appear to be easily depleted of messenger RNA. These properties may perhaps be related to some physiological characteristics of protein synthesis common to bacteria and to *Euglena* chloroplasts.

ACKNOWLEDGMENTS

The authors wish to thank Mrs. Audrey Eisenstadt for excellent technical assistance.

This work was supported by USPHS Research Career Program Award GM-K3-3295 (GB) from National Institute of Medical Science and by research grants from USPHS, AM-07189 and GM-11527.

REFERENCES

1. APP, A. A., AND JAGENDORF, A. T., *J. Protozool.*, **10**, 340 (1963).
2. BRAWERMAN, G., AND CHARGAFF, E., *Biochim. Biophys. Acta*, **31**, 172 (1959).
3. BRAWERMAN, G., AND CHARGAFF, E., *Biochim. Biophys. Acta*, **31**, 178 (1959).
4. BRAWERMAN, G., AND CHARGAFF, E., *Biochim. Biophys. Acta*, **37**, 221 (1960).
5. BRAWERMAN, G., AND KONIGSBERG, N., *Biochim. Biophys. Acta*, **43**, 374 (1960).

6. Brawerman, G., Pogo, A. O., and Chargaff, E., *Biochim. Biophys. Acta,* **55,** 326 (1962).
7. Brawerman, G., *Biochim. Biophys. Acta,* **72,** 317 (1963).
8. Brawerman, G., and Eisenstadt, J. M., *Biochim. Biophys. Acta,* **91,** 477 (1964).
9. Brawerman, G., and Eisenstadt, J. M., *J. Mol. Biol.,* **10,** 403 (1964).
10. Brawerman, G., and Eisenstadt, J. M., *Biochemistry,* **5,** 2784 (1966).
11. Edelman, M., Epstein, H. T., and Schiff, J. A., *J. Mol. Biol.,* **17,** 463 (1966).
12. Eisenstadt, J. M., and Brawerman, G., *J. Mol. Biol.,* **10,** 392 (1964).
13. Eisenstadt, J. M., and Brawerman, G., *Biochemistry,* **5,** 2777 (1966).
14. Gibor, A., and Granick, S., *Science,* **145,** 890 (1964).
15. Krawiec, S., and Eisenstadt, J. M., unpublished observations.
16. Lyman, H., Epstein, H. T., and Schiff, J. A., *Biochim. Biophys. Acta,* **50,** 301 (1961).
17. Matthaei, J. H., and Nirenberg, M. W., *Proc. Natl. Acad. Sci. U.S.,* **47,** 1580 (1961).
18. Nathans, D., Notani, G., Schwartz, J. H., and Zinder, N. D., *Proc. Natl. Acad. Sci. U.S.,* **48,** 1424 (1962).
19. Pogo, B. G. T., and Pogo, A. O., *J. Protozool.,* **12,** 96 (1965).
20. Pringsheim, E. G., and Pringsheim, O., *New Phytologist,* **51,** 65 (1952).
21. Provasoli, L., Hutner, S., and Shatz, A., *Proc. Soc. Exptl. Biol. Med.,* **69,** 279 (1948).
22. Ray, D. S., and Hanawalt, P. C., *J. Mol. Biol.,* **9,** 812 (1964).
23. Ray, D. S., and Hanawalt, P. C., *J. Mol. Biol.,* **11,** 760 (1965).
24. Schwartz, J. H., Eisenstadt, J. M., Brawerman, G., and Zinder, N. D., *Proc. Natl. Acad. Sci. U.S.,* **53,** 195 (1965).
25. Setlow, J. K., *Photochem. Photobiol.,* **3,** 393 (1963).

Discussion of Part VIII

K. Bloch, R. B. Park, L. Bogorad, A. W. Galston, J. R. Warner, O. E. Landman, G. Brawerman, F. F. Davis, R. O. Burns

Chairman Bloch: Are there any questions?

Dr. Park: I wasn't quite sure from your presentation, Dr. Bogorad, to what extent you regard a prolamellar body as an obligate part of the greening process. Would you elaborate on this?

Dr. Bogorad: It obviously isn't, because if one doesn't have darkness, prolamellar-body-containing proplastids don't accumulate. All I am saying is that this situation is useful for these experiments—we can start with all the plastids in this one state and then move on from there.

Dr. Galston: I don't understand the significance which you wish to attribute to the lack of calf thymus DNA effect on the chloroplast ribosomes as opposed to the cytoplasmic ribosomes.

Dr. Bogorad: When we make not-so-clean preparations, that is, 1,000 *g* pellets, which obviously contain lots of other things besides plastids, we may not get a calf thymus DNA effect. There are probably many broken nuclei in that fraction. However, if we test plastids cleaned by repeated washing or by centrifugation through a sucrose density gradient, we get a much more consistent response to added DNA. But the effect of light is clear in both kinds of preparations. On the other hand, as Mans and Novelli have shown, the soluble RNA polymerase obtained after centrifugation of maize homogenates responds strongly to added calf thymus DNA.

Dr. Warner: One of the very interesting problems is whether these ribosomes are made like the ribosomes of higher organisms or maybe like the ribosomes of bacteria or whatever. I wondered if you looked for 45S type of ribosome precursor RNA.

Dr. Bogorad: I showed one slide of P^{32} incorporation during a 15 minute illumination and you saw the curve. Sometimes when we do short term irradiations, the radioactivity patterns show a peak quite clearly in front of the heavier ribosomal RNA. Base analyses of this material done by looking at the P^{32} distribution among the four nucleotides obtained after alkaline digestion are like those for the ribosomal RNA's. However, we have almost always gotten incorporation into mature ribosomal RNA's after 15 minutes and I don't know why the other pattern has been observed only occasionally.

Dr. Landman: I wanted to ask you whether you know whether any of these changes can be triggered in proplastids outside the cell.

DR. BOGORAD: Well you know, everytime an astronaut is sent up, we figure we can waste a day too so we try such an *in vitro* experiment, and so far they have been more successful than we have.

DR. F. F. DAVIS: Dr. Brawerman, do you have any evidence for chloroplast transfer RNA?

DR. BRAWERMAN: No, unfortunately we don't. *Euglena* chloroplasts, as they are isolated, have lost many of their soluble components. That is why we have to add the supernatant fraction. One would have to find a way to isolate the chloroplast without damaging the membrane.

DR. WARNER: I just wondered if your chloroplast ribosomes were similar to others in having two parts analogous to 50 and 30S and two kinds of RNA.

DR. BRAWERMAN: We have not been very successful in dissociating the ribosomes. However, the ribosomal RNA has two peaks, one at 14S and one at 19S, which is about what you would expect from the size of the ribosomes.

DR. BURNS: Are the sedimentation coefficients of the ribosomes obtained by extrapolating to zero RNA concentration?

DR. BRAWERMAN: No, these are very crude values which we obtain by comparing sedimentation in the sucrose gradient to that of *E. coli* ribosomes, so I wouldn't consider them as absolute values. However, relatively speaking, the cytoplasmic particles sediment to about the same location as *E. coli* ribosomes.

DR. BURNS: One further question. I believe nitrosoguanidine is a bleaching agent for *Euglena.* Does it act preferentially on the chloroplasts rather than the nuclear DNA?

DR. BRAWERMAN: I am not sure. I have no experience with it.

DR. BURNS: The point is, can you offer any explanation why the chloroplasts when bleached would lose the DNA or have I got the whole idea reversed here, that they lose the DNA and therefore they are bleached.

DR. BRAWERMAN: Well, we really don't know which comes first. The assumption has been that the bleaching agent affects the DNA directly; however, the only piece of evidence that we have for that concerns bleaching by UV irradiation. In Schiff's laboratory, they found that the action spectrum of UV bleaching resembles that of a nucleoprotein. In other words, it has a peak around 260 mμ and another one around 280 mμ, which would suggest that a nucleic acid is involved at any rate. This is all we really know about the mechanism of bleaching. There is no direct evidence of what exactly the sensitive factor is.

DR. SIEKEVITZ: I wonder if you would want to comment on another possibility of a DNA regulation. That is, if the chloroplasts are like the mitochondria in that they divide, perhaps in conjunction with cellular division, there might be some regulatory role here in that the impulse for the division of the nuclear DNA is the same impulse for the division or the duplication of the chloroplast DNA. The result is that you have a synchrony of new cells having the same number of chloroplasts.

DR. BRAWERMAN: Well, I guess under normal conditions you would have to have some sort of synchrony. However, one can very easily destroy the synchrony when one bleaches the cells. The bleaching is only effective or takes place only during cell multiplication, so it probably involves a dilution of the chloroplast replicating units, and this can only take place if one has dissociated nuclear replication from chloroplast replication. This must be the key to the bleaching effect.

CHAIRMAN BLOCH: If there are no further comments or questions, the session is adjourned. Thanks to all the speakers and discussants.

PART IX

RIBOSOME SYNTHESIS

Chairman's Remarks

RICHARD B. ROBERTS

Department of Terrestrial Magnetism,
Carnegie Institution of Washington,
Washington, D. C.

The astronauts have been scrubbed[1] but our count-down is right on time, I believe, and it is 9:15 and time to call this meeting to order. It is customary to take a few minutes here while the latecomers find their seats. So I would like to remind you of the name ribosome. We have been using it all day and it is almost nine years old, that name. It was suggested by Dr. Howard Dintzis at a symposium held by the Biophysics Society in early 1958, and it replaced what we had to say before, which was "the ribonucleoprotein particles of the microsomal fraction." So it is easy to see why a nice word like ribosome rapidly became accepted. Eight years is quite a long time, and eight years ago there was evidence of rapidly-labeled RNA which accumulated protein and became ribosomes (2), but other concepts were completely lacking at that time. The messenger idea came along quite a bit later and in fact the messenger diverted some attention from ribosome biosynthesis because everybody was looking for rapidly-labeled RNA and thinking that it was entirely messenger and ignoring the ribosome precursor part of the rapidly-labeled material. Even four years ago when this symposium series began, the concept of polysome was entirely lacking. It was badly needed. There were too many ribosomes for the number of messenger molecules available and the messenger was much too long to fit on any one ribosome in any reasonable sort of way. It was suggested then (1) that possibly more than one ribosome was engaged in the unit of protein synthesis, but it wasn't considered very seriously. It was only when the polysomes were actually seen and separated that they became acceptable. So now we continue with the study of ribosome biosynthesis and the way to learn more about these things is usually to break them down into the units, so today we are going to hear something about the subunits that go into making up ribosomes.

REFERENCES

1. ROBERTS, R. B., *in* "Informational Macromolecules" (H. J. Vogel, V. Bryson, and J. O. Lampen, eds.), p. 373, Academic Press, New York, 1963.
2. ROBERTS, R. B., BRITTEN, R. J., AND BOLTON, E. T., *in* "Microsomal Particles and Protein Synthesis" (R. B. Roberts, ed.) p. 84, Pergamon Press, New York, 1958.

[1] A Gemini flight was cancelled 30 minutes before the session opened.

Taking the Ribosome Apart and Putting It Back Together Again

THEOPHIL STAEHELIN, HESCHEL RASKAS, AND MATTHEW MESELSON

The Biological Laboratories,
Harvard University,
Cambridge, Massachusetts

In the course of its adventures as the organelle of protein synthesis, the ribosome must faithfully recognize and guide the multiple interactions of a large array of molecules, including aminoacyl-sRNA, peptidyl-sRNA, various transfer factors, GTP, and messenger RNA. Furthermore, the protein and RNA subunits of the ribosome must almost certainly be able to guide their own assembly during ribosome biosynthesis. This complexity and precision of ribosome activity must depend upon a corresponding complexity and precision of ribosome structure. These considerations incline us to visualize the ribosome, as Max Delbrück did once in conversation, as a tiny machine with many precisely matched parts which selects and presides over the interactions of the components of the translation process. (To me, TS, being of Helvetic origin, the nearest example is, of course, the Swiss precision watch.) Our approach to studying the relation between ribosome structure and function has accordingly been more mechanical than chemical. We have attempted to disassemble the ribosome and put it together again, testing the subunits and the reassembled particles for the ability to perform various ribosomal functions. Our approach starts from the discovery of Meselson *et al.* (7) that ribosomes subjected to equilibrium density gradient centrifugation in CsCl solution at appropriate concentrations of magnesium ions dissociate into a discrete protein component and a protein-deficient ribonucleoprotein "core." At equilibrium, the dissociated proteins float at the meniscus of the density gradient, and the cores form a band in the gradient.

We can summarize the properties of ribosomes in CsCl as follows: If native 70 S ribosomes are centrifuged in CsCl solution containing magnesium ions, one finds all the ribonucleoprotein in two distinct bands. The denser of these bands is called A, and the less dense B. At high magnesium ion concentration, the B band predominates, but at lower magnesium ion concentrations progressively more A is found until, at about 0.001 *M*, all the ribonucleoprotein is in the A band. (At even lower magnesium ion concentrations, the ribosomal RNA is released and sediments to the bottom of the gradient.) Particles recovered from the B band sediment in a sucrose gradient containing a low concentration of magnesium ions as normal 50 S and 30 S ribosomes. We presume that the B band contains 70 S ribosomes. In contrast, the material in the A band consists of protein-deficient core particles. The 50 S ribosome gives rise to a 42 S core, and the 30 S ribosome to a 23 S core. We refer to the cores according to their approximate S values: 42 and 23. The dissociated proteins found at the meniscus are designated according to the particle from which they are derived: M50 and M30.

Reconstitution of Ribosomes from Cores and Meniscus Fractions

In these experiments (10), purified 50 S and 30 S ribosomes were separately centrifuged in CsCl solution to obtain the four ribosomal fractions 42, 23, M50, and M30. Reconstruction of functional ribosomes was attempted by mixing either core with either meniscus fraction and dialyzing against 0.01 *M* Tris, pH 7.4, and 0.01 *M* Mg acetate (T-M buffer) to remove CsCl. These reconstitution mixtures, as well as cores and meniscus proteins dialyzed alone, were compared with native 50 S and 30 S ribosomes for activity in a cell-free protein-synthesizing system using polyuridylic acid messenger and ^{14}C-phenylalanine. The results of some typical experiments are summarized in Table I.

We conclude from these experiments that:

1. Synthetic activity is reconstituted from synthetically inactive cores and meniscus fractions when they are appropriately combined.

2. The only combinations which possess the activity of 50 S ribosomes are those in which the 42 S core is dialyzed with M50; the activity of the 30 S ribosome is found only when the 23 S core is combined with M30. Thus, not only do both cores and meniscus fractions comprise or contain structures indispensible for synthetic activity, but also, the mere presence of all these components is not sufficient to give synthetic activity. Rather, as is seen from the two starred assays of Table I, there occurs during the removal of CsCl by dialysis a specific and essential interaction between cores and their homologous meniscus fractions.

The interactions between cores and meniscus fractions which occur during dialysis have been investigated by sedimentation analysis. Homologous and heterologous [core + M] fractions were prepared and sedimented through sucrose density gradients with either native 50 S or 30 S ribosomes taken as references. The results are summarized in Table II.

It is seen that the reconstitution process gives particles with sedimentation values close to the native ones in the case of homologous

TABLE I
POLY-U-DIRECTED POLYPHENYLALANINE SYNTHESIS BY RECONSTITUTED RIBOSOMES[a]

Ribosomal input	Percent activity relative to native
Native 50 + native 30	(100)
Native 50 alone	1–11
Native 50 + [23]	1–7
Native 50 + [23 + M30]	39–64
Native 50 + [23 + M50]	4–11
Native 50 + [M30]	4
Native 30 alone	2–7
Native 30 + [42]	2–8
Native 30 + [42 + M50]	40–88
Native 30 + [42 + M30]	6–14
Native 30 + [M50]	3
*[23 + M30] + [42 + M50]	15–41
*[23 + M50] + [42 + M30]	2–4
[23] + [42]	0–3
[23] + [42 + M50]	2–4
[23 + M30] + [42]	<1–4
[M30] + [M50]	~1

[a] The assay is described by Staehelin and Meselson (10). Fractions in brackets are mixtures dialyzed against T-M buffer to remove CsCl. One hundred percent activity corresponds to four to twelve amino acids polymerized per ribosome equivalent.

reconstitution mixtures and somewhat less for heterologous mixtures. The behavior of homologously and heterologously reconstituted particles in the polyphenylalanine-synthesizing system suggests that the two types of particle do not readily interconvert under the conditions of assay (Table I, starred assays).

TABLE II
SEDIMENTATION OF CORES AND RECONSTITUTED PARTICLES[a]

Particle	S value
Native 50	50
[42]	42–43
[42 + M50]	~49
[42 + M30]	~47
Native 30	30
[23]	23–24
[23 + M30]	29–30
[23 + M50]	~27

[a] Sedimentation values of ribosomes, cores, and reconstituted particles were measured in sucrose density gradients with native 30 S ribosomes as reference in all gradients with 50 S derivatives, and *vice versa*.

mRNA and sRNA Binding by Ribosome Derivatives and Reconstituted Particles

In another series of experiments (8) cores, meniscus fractions, and reconstituted particles were tested for the ability to bind mRNA and sRNA by using conditions in which it is known that the 30 S ribosome binds messenger (12) and the 50 S ribosome binds sRNA (2).

Messenger binding to 30 S ribosomes and their derivatives was studied with both poly-U and poly-C. The particles were incubated in the presence of the radioactive polynucleotide and then subjected to zone centrifugation in sucrose density gradients to separate bound from unbound messenger.

The messenger-independent binding of sRNA to 50 S ribosomes and their derivatives was investigated with ^{14}C-aminoacyl-sRNA. Particles and ^{14}C-aminoacyl-sRNA were incubated at low temperature and then centrifuged to pellet ribosomes and ribosome-bound sRNA.

The results of these binding studies are summarized in Table III. They parallel those found for *in vitro* protein synthesis.

Homologously reconstituted particles have the same binding ability as native ribosomes, but heterologously reconstituted particles and cores are inactive. In a sense, this result is disappointing, since it would have

TABLE III
mRNA and sRNA Binding by 30 S and 50 S Ribosomes and Their Cores and Reconstituted Derivatives[a]

Ribosomal input	mRNA (^{14}C-poly-U) binding
Native 30	+
[23]	−
[23 + M30]	+
[23 + M50]	−
[42 + M30]	−
Native 50	−
	sRNA binding
Native 50	+
[42]	−
[42 + M50]	+
[42 + M30]	−
[23 + M50]	−
Native 30	−

[a] The binding of ^{14}C-poly-U and of sRNA was studied as described in the text. About 130 uracil residues were bound per ribosome in positive assays (+), and no measurable amount in the negative assays (−). sRNA binding ranged from 0.44 to 0.96 molecule per ribosome in positive assays (+) and from 0.00 to 0.11 molecule per ribosome for negative assays (−).

facilitated the study of the binding reactions themselves if the binding sites could have been localized to a distinct ribosome fraction. Apparently, the binding sites are composed of structures from both the meniscus and the core fraction, or else some interaction between core and meniscus structures is required to stabilize the necessary configuration for binding.

Determination of Streptomycin Sensitivity by a Subunit of the 30 S Ribosome

The sensitivity of bacteria to streptomycin appears to be determined by some aspect of ribosome structure (9). Moreover, the determinant of sensitivity seems to reside in the 30 S ribosome—for cell-free protein-synthesizing systems are inhibited by streptomycin if they are prepared with 30 S ribosomes from sensitive bacteria but not if they are prepared with 30 S from resistant strains, regardless of the source of 50 S ribosomes employed.

In sensitive systems, the addition of streptomycin not only inhibits total synthesis, but also causes severe misreading of the mRNA code (3). Thus, a streptomycin-sensitive poly-U-directed system containing phenylalanine and isoleucine incorporates phenylalanine almost exclusively in the absence of streptomycin, but produces polypeptides containing a substantial proportion of isoleucine if streptomycin is added.

In order to learn if the determination of streptomycin sensitivity can be ascribed to the core or the meniscus fraction of the 30 S particle, sensitive and resistant 30 S ribosomes were isolated from sensitive and resistant bacterial strains and dissociated into cores and meniscus fractions. These were then reconstituted either homologously or heterologously. The various reconstitution mixtures were tested for protein synthesis in the presence and in the absence of streptomycin in a poly-U-directed system (containing native 50 S ribosomes) with phenylalanine and isoleucine. The results (11) are given in Table IV. It is clear that streptomycin sensitivity and resistance is determined by the 23 S core and not by the meniscus proteins.

The following points should be noted:

First, phenylalanine incorporation in the absence of streptomycin is almost the same in all assays with reconstituted ribosomes and on the average is only about 30% less than incorporation with native 30 S ribosomes.

Second, inhibition of polyphenylalanine synthesis in the presence of streptomycin is observed in all cases, but the inhibition is considerably greater in assays with sensitive 30 S ribosomes or reconstituted derivatives containing cores from sensitive rather than resistant bacteria.

TABLE IV
PHENYLALANINE AND ISOLEUCINE INCORPORATION IN THE PRESENCE AND IN THE ABSENCE OF STREPTOMYCIN[a]

Ribosomal input	Phenylalanine: Incorporation (amino acids/50 S ribosome) No Sm[b]	Phenylalanine: Incorporation (amino acids/50 S ribosome) Plus Sm	Phenylalanine: Inhibition (%)	Isoleucine: Incorporation (amino acids/50 S ribosome) No Sm	Isoleucine: Incorporation (amino acids/50 S ribosome) Plus Sm	Isoleucine: Stimulation factor
50R + 30S	8.7	2.7	69	0.09	1.58	18
50R + 30R	6.6	5.2	21	0.02	0.10	5
50R + [23S + M30S]	4.9	2.4	51	0.05	0.99	20
50R + [23S + M30R]	5.5	2.4	56	0.06	1.08	18
50R + [23R + M23S]	4.6	3.4	26	0.01	0.05	5
50R + [23R + M30R]	5.6	4.9	22	0.02	0.06	3

[a] Ribosomes, cores, and meniscus fractions from streptomycin-sensitive and streptomycin-resistant bacteria are designated with the letters S and R, respectively. Reactions with streptomycin contained 20 μg/ml of the antibiotic added simultaneously with polyuridylic acid. Phenylalanine incorporation was conducted in the presence of 0.025 *M* magnesium acetate, 10 mμmoles/ml of each of 17 cold amino acids, and 13 mμmoles of L-^{14}C-phenylalanine. Isoleucine incorporation was studied in the presence of 0.03 *M* magnesium acetate, 10 mμmoles/ml of L-^{12}C-phenylalanine, and 5 mμmoles/ml of L-^{14}C-isoleucine. Duplicate samples run with 20% less magnesium acetate gave results essentially the same as those in the table. Reactions with 50 S ribosomes in combination with cores or meniscus fractions alone gave negligible incorporation.

[b] Sm = streptomycin.

This indicates that the determinant of sensitivity resides in the 23 S core, not in the meniscus fraction. This suggestion is confirmed by the results for isoleucine incorporation.

Even in the absence of streptomycin, when isoleucine incorporation is low in all assays, reaction mixtures containing sensitive 30 S ribosomes or their cores incorporated about three times as much isoleucine as did those with their counterparts from resistant bacteria, again indicating that the core rather than the meniscus fraction distinguishes sensitive from resistant ribosomes. (This result is consistent with the finding that mutations at the streptomycin locus can alter the pattern of genetic suppression even in the absence of streptomycin itself (1).

The most pronounced difference between streptomycin sensitivity and resistance is seen in the amounts of isoleucine incorporated in the presence of streptomycin. In all assays streptomycin stimulates the incorporation of isoleucine, but the stimulation is much greater in mixtures containing native sensitive 30 S ribosomes or reconstituted 30 S ribosomes with cores from sensitive rather than resistant bacteria. These results demonstrate that streptomycin sensitivity is determined by the 23 S ribonucleoprotein core.

Electrophoretic Characterization and Separation of Ribosomal Core and Meniscus Proteins

The experiments described so far show that the meniscus fraction derived from 50 S ribosomes differs in important respects from that derived from the 30 S. However, a much more detailed analysis of the core and meniscus proteins of 50 S and 30 S ribosomes has been achieved by disc electrophoresis in acrylamide gels, as first applied to ribosomal proteins by Leboy *et al.* (6). They reported the resolution of 34 protein bands, 13 from the 30 S and 21 from the 50 S ribosome, with apparently only one protein from each subunit having identical mobility. Figure 1 shows the complex band patterns of proteins extracted from 70 S, 50 S, and 30 S ribosomes. The different bands, stained with amido black, show characteristic colors varying from turquois and other shades of blue to gray and black. It is obvious that the patterns from 50 S and 30 S ribosomes differ from each other and that the 70 S pattern represents, at least very nearly, the sum of the two.

The abundance of electrophoretic components does not necessarily mean that ribosomes contain an equally large number of chemically different polypeptides. Aggregation of a small number of different subunits could produce a complex series of composite molecules with differing electrophoretic mobilities. Nonspecific and reversible aggregation has, however, been ruled out. Waller (16) demonstrated that individual bands could be removed from starch gels and rerun without the appearance of new components. Traut (15) showed that groups of radioactive protein bands, eluted from acrylamide gels, rebanded quantitatively at their original positions even in the presence of a large excess of added unfractionated ribosomal protein.

Figures 2 and 3 show acrylamide gel electrophoresis patterns from 30 S ribosomes, 23 S cores, and M30 proteins as well as from 50 ribosomes and their corresponding fractions. It is clear that all four protein fractions—M30, M50, 42, and 23—give different patterns, although there may be a few bands common to either core and its corresponding meniscus fraction. It should be noted that just as the gel pattern obtained from 70 S is the sum of the 30 S and 50 S gel patterns, so is the 50 S pattern the sum of the M50 and 42 S core patterns, and likewise for the 30 S pattern. This nice additivity reassures us that the bands we observe correspond to species present in intact ribosomes, rather than to artifacts formed during extraction or electrophoresis. This view is in accord with the results of Waller and of Traut, cited earlier.

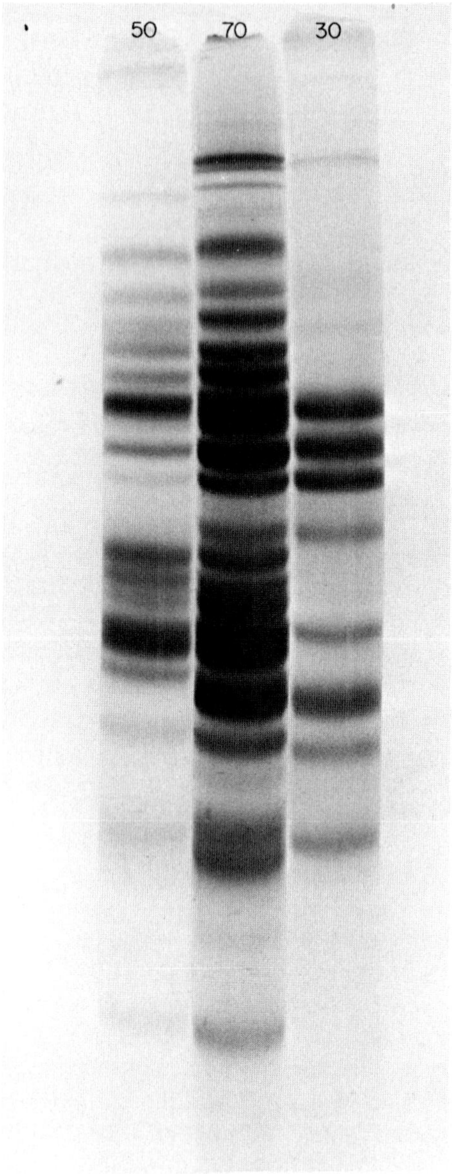

Fig. 1. Acrylamide gel electrophoresis band patterns of ribosomal proteins from 70 S ribosomes, 50 S ribosomes, and 30 S ribosomes. For details of methods see Gesteland and Staehelin (4). The arrow indicates a double band in M50, whose components are resolved in sucrose gradient electrophoresis into an acidic protein [most prominent in fraction (a)] and a basic one [most prominent in fraction (d)].

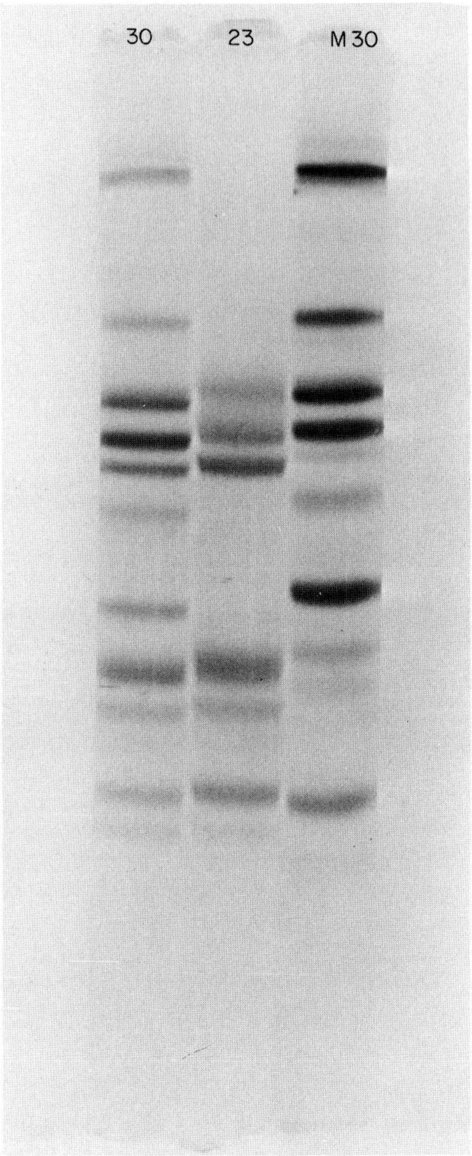

FIG. 2. Acrylamide gel electrophoresis of proteins from 30 S ribosomes, 23 S cores, and M30.

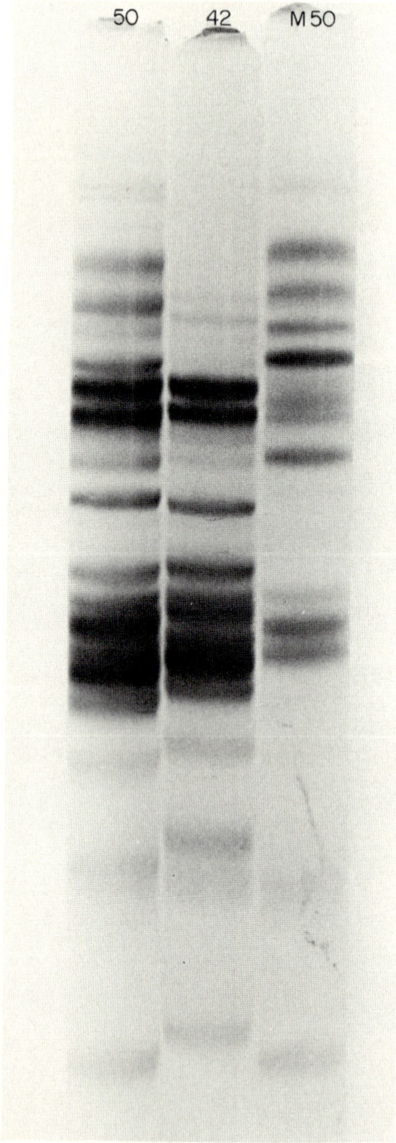

FIG. 3. Acrylamide gel electrophoresis of proteins from 50 S ribosomes, 42 S cores, and M50.

Whatever all the components observed in disc gel electrophoresis may be, either different single proteins or various highly specific and stable aggregates formed *in vivo* from only a few different subunits, we should like to be able to isolate them preparatively in order to begin to determine their possible functions in the ribosome, along lines similar to those we have followed in studying cores and meniscus fractions. The method we chose for separation of ribosomal proteins into subgroups is preparative sucrose gradient electrophoresis. Proteins are separated ac-

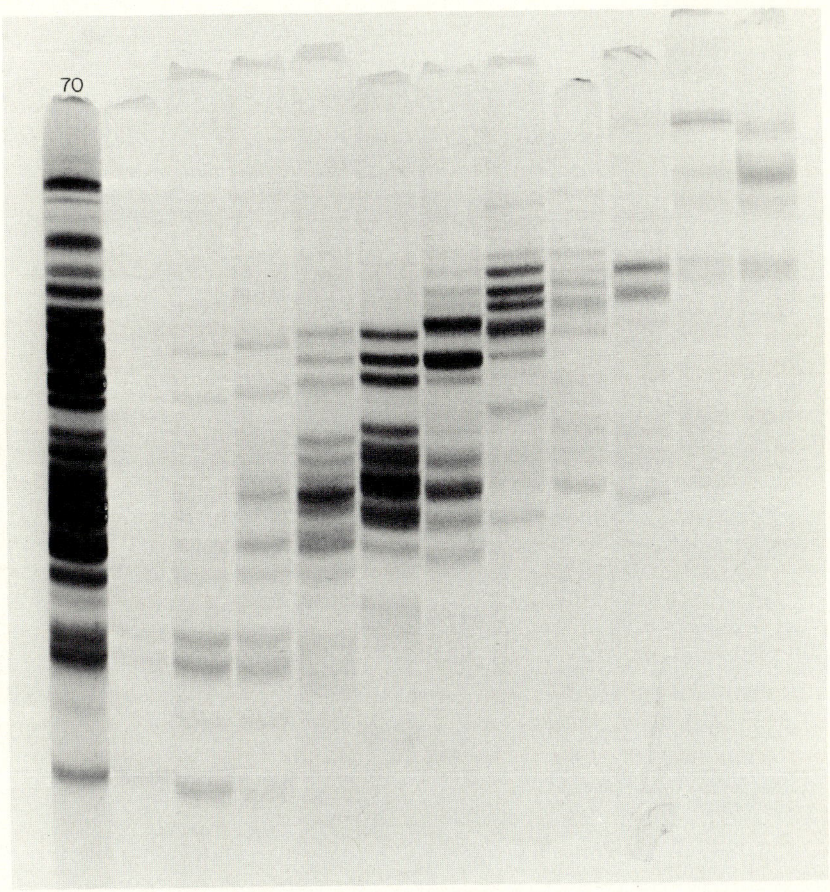

FIG. 4. Acrylamide gel electrophoresis of proteins from 70 S ribosomes (left), and from fractions of a preparative sucrose gradient electrophoresis of total 70 S ribosomal proteins. The more acidic fractions are to the right.

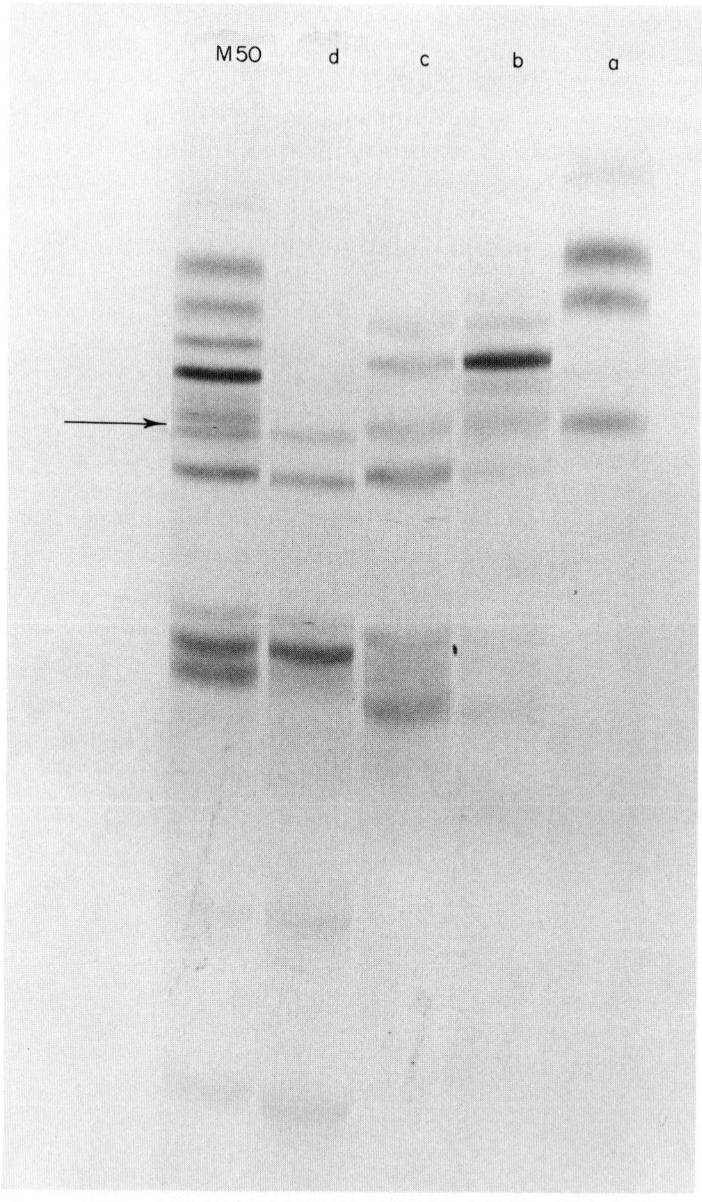

FIG. 5. Acrylamide gel electrophoresis of total M50 proteins and subgroups of M50 proteins obtained after separation by sucrose gradient electrophoresis (*a–d*) (*a*) Most acidic fraction. (*d*) Most basic fraction.

cording to electrophoretic mobility in a column of buffered 6 *M* urea solution stabilized by a sucrose density gradient. After electrophoresis, the gradient is collected in fractions, and an aliquot of each is analyzed by disc gel electrophoresis. Such an analysis is shown in Fig. 4, where proteins from whole 70 S ribosomes (gel shown on the extreme left) have been spread in a sucrose gradient by electrophoresis. The other gels shown in Fig. 4 are from successive fractions from a preparative sucrose gradient electrophoresis. By selecting appropriate fractions, one should be able to obtain groups of proteins which have no common components. Using this technique we have separated the meniscus proteins from CsCl density gradients into several groups, in order to attempt the reconstitution of ribosomal particles. Such a fractionation of the meniscus proteins from 50 S ribosomes is shown in Fig. 5. Fraction (*a*) contains the most acidic proteins without detectable contamination by other components. Fraction (*b*) consists of one major band and traces of proteins which are prominent in the neighboring fractions. The two rather weak bands marked with an arrow in M50 move nearly together in acrylamide gels, but move quite differently in free solution electrophoresis. One of them is more prominent in the acidic fraction (*a*), while the other is prominent in the basic fractions (*c*) and (*d*). Studies on selective and stepwise reconstitution of different cores lacking different amounts of protein with fractionated meniscus proteins are now in progress.

Future Prospects

We are now at a point where we have disassembled our little Swiss watch into many parts and have successfully reassembled functional ribosomes from a few major subassemblies. We have associated certain functions with certain subunits or groups of proteins. However, we are still far from understanding ribosome structure and its relation to biological function. Since we are at this time so far from our goal, I decided to accept some unscientific help from my wife. With remarkable intuition, she reassembled all the parts we so poorly separated into different test tubes by centrifugation and electrophoresis. The picture she came up with (Fig. 6*a*) most strikingly confirms our initial model and working hypothesis. Only one minor and functionally not very significant detail of our assumptions turned out to be wrong when she pictured the rear view of our ribosome (Fig. 6*b*). This came as a bit of a surprise to us, but undoubtedly was a well-known fact to the next speaker, Dr. Nomura, who with his colleagues has obtained many results which parallel and extend our own (5, 13, 14).

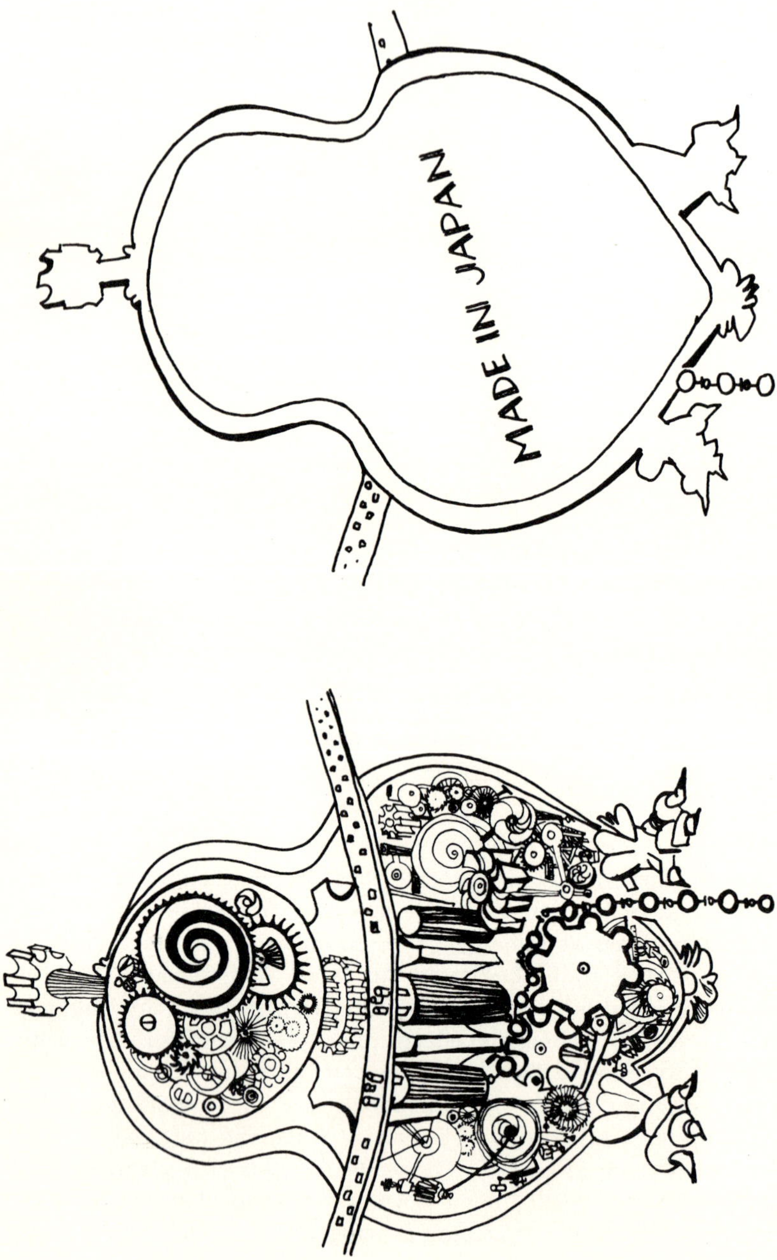

FIG. 6. (*a*) The ribosome, reconstituted and portrayed by Mrs. Gabrielle Staehelin. (*b*) Rear view of Fig. 6*a*.

Acknowledgments

Part of the disc gel electrophoresis studies were performed in collaboration with Dr. Raymond Gesteland (4). Our experiments have been supported by grants from the National Science Foundation and the National Institutes of Health.

References

1. Anderson, W. F., Gorini, L., and Breckenridge, L., *Proc. Natl. Acad. Sci. U.S.*, **54**, 1076 (1965).
2. Cannon, M., Krug, R., and Gilbert, W., *J. Mol. Biol.*, **7**, 360 (1963).
3. Davies, J., Gilbert, W., and Gorini, L., *Proc. Natl. Acad. Sci. U.S.*, **51**, 883 (1964).
4. Gesteland, R. F., and Staehelin, T., *J. Mol. Biol.*, **24**, 149 (1967).
5. Hosokawa, K., Fujimura, R., and Nomura, M., *Proc. Natl. Acad. Sci. U.S.*, **55**, 198 (1966).
6. Leboy, P. S., Cox, E. C., and Flaks, J. G., *Proc. Natl. Acad. Sci. U.S.*, **52**, 1367 (1964).
7. Meselson, M., Nomura, M., Brenner, S., Davern, D., and Schlessinger, D., *J. Mol. Biol.*, **9**, 696 (1964).
8. Raskas, H. J., and Staehelin, T., *J. Mol. Biol.*, **23**, 89 (1967).
9. Spotts, C. R., and Stanier, R. Y., *Nature*, **192**, 633 (1961).
10. Staehelin, T., and Meselson, M., *J. Mol. Biol.*, **16**, 245 (1966).
11. Staehelin, T., and Meselson, M., *J. Mol. Biol.*, **19**, 207 (1966).
12. Takanami, M., and Okamoto, T., *J. Mol. Biol.*, **7**, 323 (1963).
13. Traub, P., Hosokawa, K., and Nomura, M., *J. Mol. Biol.*, **19**, 211 (1966).
14. Traub, P., Nomura, M., and Tu, L., *J. Mol. Biol.*, **19**, 215 (1966).
15. Traut, R. R., *J. Mol. Biol.*, **21**, 571 (1966).
16. Waller, J. P., *J. Mol. Biol.*, **10**, 319 (1964).

Structure and Function of Ribosomes and Subribosomal Particles[1]

M. NOMURA AND P. TRAUB

Laboratory of Genetics,
University of Wisconsin,
Madison, Wisconsin

INTRODUCTION

Cell organelles are organized structures which are morphologically distinct and which perform certain specific cellular functions—for example, protein synthesis in the case of ribosomes, or energy supply in the case of mitochondria. Most of the cell organelles are aggregates of heterogeneous macromolecular components and possess a supramolecular structure. Studies of such cell organelles are confronted with two major problems: first, to obtain a detailed correlation between the structures and their function, and second, to elucidate the mechanism through which these complex and specific structures are elaborated.

One such organelle is ribosomes. Ribosomes are the site of protein synthesis. They can be isolated in a reasonably pure form, and their function in protein synthesis can be demonstrated in *in vitro* experiments. Numerous studies have been performed to elucidate the mechanism of protein synthesis, and yet we know surprisingly little about the function of the ribosomes. Previously, the role of ribosomes was thought to be the provision of nonspecific sites for the binding of messenger RNA and the aminoacyl transfer RNA (see reviews by Watson, 38, 39). However, the recent study of the mode of action of streptomycin (Davies *et al.*, 5; Gorini and Kataja, 10, 11) suggests that the structure of ribosomes may be profoundly involved in the translation of the genetic code. The structural complexity of ribosomes (Waller, 36; Waller and Harris, 37; Leboy *et al.*, 15) also hints of greater complexity in their function than has yet been determined.

Recently success has been reported in the splitting of part of the

[1] Paper No. 1104 from the Laboratory of Genetics, University of Wisconsin, Madison, Wisconsin.

protein components from ribosomes, leaving inactive ribonucleoprotein cores, and in the reconstitution of functionally active ribosomes from these inactive components (Hosokawa *et al.*, 12; Staehelin and Meselson, 30; see also Spirin and Belitsina, 28). These studies have opened a new way to study the relationship between the structure and the function of ribosomes.

In this article, we shall review some of our more recent work along these lines on *Escherichia coli* ribosomes. The proteins split from ribosomes were further fractionated (see Fig. 1), and various artificial sub-

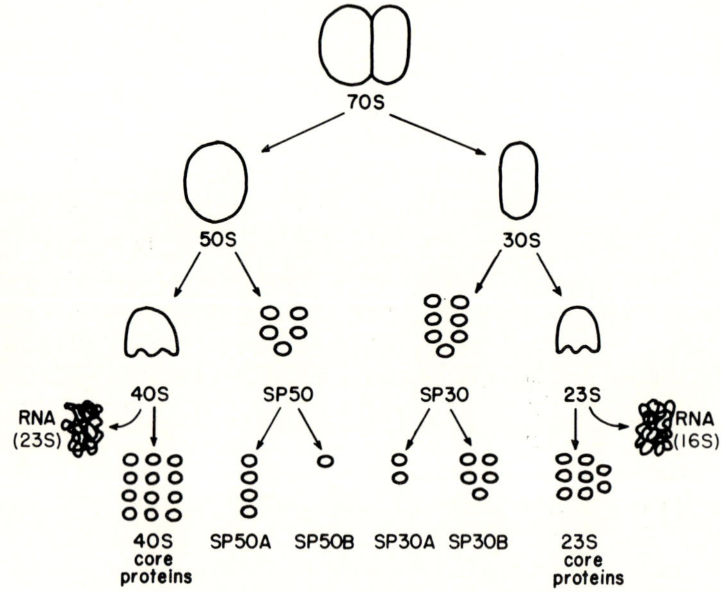

FIG. 1. A schematic representation of ribosomal components.

ribosomal particles were prepared from the core particles and one protein fraction with the omission of another protein fraction. These subribosomal particles were then analyzed for the several known ribosomal functions (Traub and Nomura, 35). We shall then discuss the reconstitution process itself in relation to the problem of ribosome assembly. Finally, we shall review our studies on biologically altered ribosomes, namely Sm^r ribosomes and colicin E3-inactivated ribosomes.

FRACTIONATION OF RIBOSOMAL PROTEINS AND PREPARATION OF VARIOUS SUBRIBOSOMAL PARTICLES

When 50 S or 30 S ribosomes are centrifuged to equilibrium in 5.2 *M* CsCl in the presence of 0.04 *M* Mg^{++}, they dissociate into smaller ribonucleoprotein particles and free proteins, called "split" proteins (Meselson *et al.*, 17). The 50 S ribosomal subunits yield 40 S "core" particles, while the 30 S ribosomal subunits yield 23 S "core" particles, both of which are inactive in protein synthesis. Mixing the split proteins with these core particles results in reconstitution of ribosomes which are active in *in vitro* polypeptide synthesis (Staehelin and Meselson, 30; Hosokawa *et al.*, 12). Functionally active 30 S particles can be reconstituted by the combination of 23 S core particles and the split proteins from the 30 S ribosomes, but not by the combination of 23 S core particles and the split proteins from the 50 S ribosomes. Similarly, active 50 S particles can be reconstituted by the combination of 40 S core particles and the split protein from the 50 S ribosomes, but not by the combination of 40 S core particles and the split protein from the 30 S ribosomes. These results suggest that at least some of the 30 S ribosomal proteins are distinct from the 50 S ribosomal proteins, and that the restoration of synthetic activity to core particles requires the addition of some specific proteins.

Split proteins from the 50 S ribosomes (called SP50) have been further fractionated by adsorption and elution from DEAE in the presence of 6 *M* urea. Two fractions were obtained: the basic fraction (SP50B), which was not adsorbed by the DEAE column at pH 8, and the acidic fraction (SP50A), which was adsorbed by the DEAE column at pH 8 and then eluted with 2 *M* LiCl at pH 4.5. Both SP50A and

TABLE I

FRACTIONATION OF RIBOSOMAL COMPONENTS AND THEIR REASSOCIATION

		(a) 50 S	
50 S	$\xrightarrow[\text{centrifugation}]{\text{in 5.2 } M \text{ CsCl}}$	40 S (core particles)	+ SP50 (split proteins)
SP50	$\xrightarrow[\text{in 6 } M \text{ urea}]{\text{DEAE}}$	SP50B	+ SP50A
40 S + SP50B	→	[40, SP50B]	or [40, B]
40 S + SP50A	→	[40, SP50A]	or [40, A]
40 S + SP50B + SP50A	→	[40, SP50A, SP50B]	or [40, A, B]
40 S + SP50	→	[40, SP50]	
		(b) 30 S	
30 S	$\xrightarrow[\text{centrifugation}]{\text{in 5.2 } M \text{ CsCl}}$	23 S (core particles)	+ SP30 (split proteins)
SP30	$\xrightarrow[\text{in 6 } M \text{ urea}]{\text{DEAE}}$	SP30B	+ SP30A
23 S + SP30B	→	[23, SP30B]	or [23, B]
23 S + SP30A	→	[23, SP30A]	or [23, A]
23 S + SP30B + SP30A	→	[23, SP30A, SP30B]	or [23, A, B]
23 S + SP30	→	[23, SP30]	

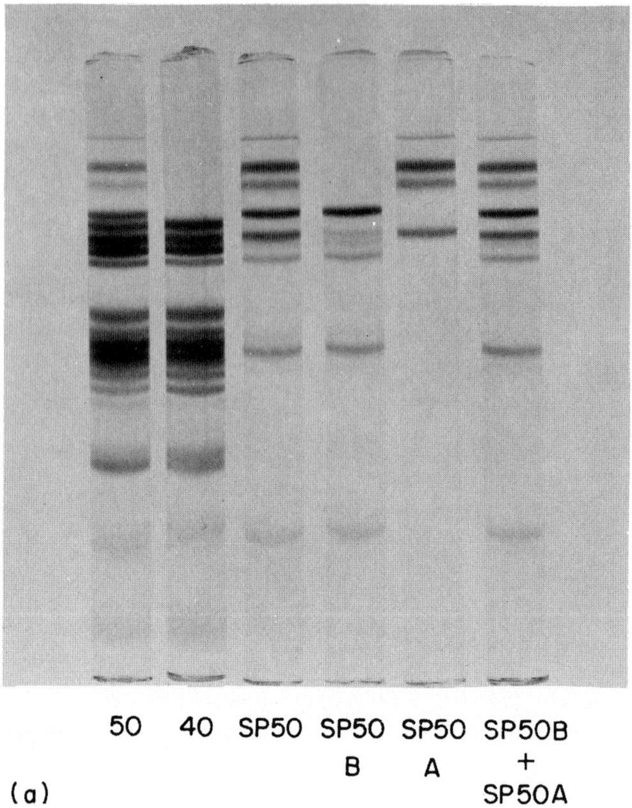

FIG. 2. Polyacrylamide gel electrophoresis of ribosomal proteins at pH 4.5 (*a*) Proteins from (left to right) 50 S, 40 S, SP50, SP50B, SP50A, and a mixture of SP50B and SP50A. (*b*) Proteins from 50 S, 40 S, [40, B], [40, A], and [40, A, B] particles. Proteins were prepared from the particles or the protein fractions and analyzed as described by Traub *et al.* (34). The top in the figure is the anode side, and the proteins migrated in the direction from the top to the bottom (the cathode side).

SP50B are soluble in buffers at neutral pH and low salt concentrations. Subribosomal particles [40, B] were prepared by mixing the 40 S core with SP50B in the presence of CsCl, dialyzing the mixture to remove CsCl, and then centrifuging the resultant aggregate to remove free proteins. Similarly, [40, A] subribosomal particles were prepared from 40 S core plus SP50A, and the control reconstituted particles were prepared either from 40 S core particles plus both SP50A and SP50B ([40, A, B]) or from 40 S core plus SP50 ([40, SP50]). Table I*a* summarizes these methods of making the protein fractions and the subribosomal particles.

Analyses of proteins in these protein fractions and in the subribosomal particles were done with polyacrylamide gel electrophoresis (Fig.

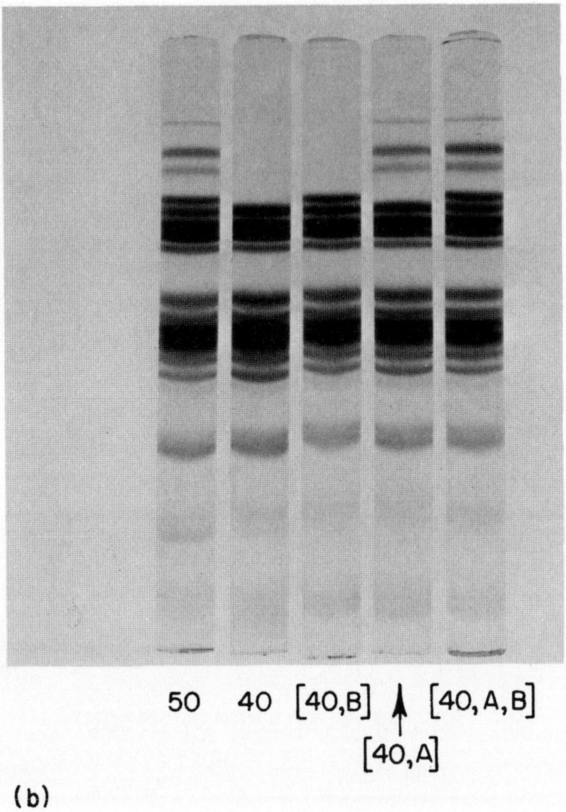

FIG. 2*b*. For legend see opposite page.

2*a* and *b*); the results from several experiments are summarized in Fig. 4*a*. It can be seen that the 40 S core particles lack some specific protein components which appear in the SP50 fraction, and that clear separation of SP50 proteins into two fractions, SP50A and SP50B, has been achieved by the DEAE technique.

The results (Fig. 2*b*) also show that the particles [40, A] lack the protein which appears as the major component of the SP50B fraction, but contain all the other ribosomal protein components. The particles [40, B] lack the proteins in the SP50A fraction, namely 50 S proteins I, II, III, and perhaps VI. The control samples of reconstituted 50 S particles, [40, A, B], showed a band pattern almost indistinguishable from that of native 50 S ribosomes. Thus, three different subribosomal particles—40 S, [40, A], and [40, B]—can be obtained from 50 S ribosomes, each deficient in some specific protein component(s) of the original ribosomes.

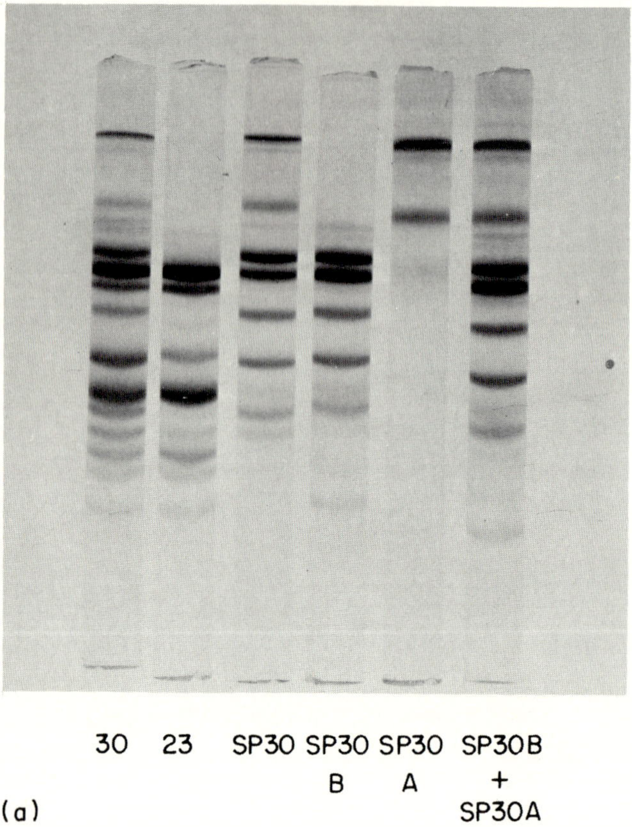

FIG. 3. Polyacrylamide gel electrophoresis of ribosomal proteins. (*a*) Proteins from 30 S, 23 S, SP30, SP30B, SP30A, and a mixture of SP30B and SP30A. (*b*) Proteins from 30 S, 23 S, [23, B], [23, A], and [23, A, B].

Essentially the same technique was applied to the 30 S ribosomes (Table I*b*). The 30 S ribosomes were dissociated into 23 S core and split proteins, SP30. The split proteins were then fractionated into two fractions, SP30B and SP30A. The basic fraction, SP30B, represents proteins which are not adsorbed on DEAE, and the acidic fraction, SP30A, represents proteins which are adsorbed on DEAE. A new artificial subribosomal particle, [23, A], was prepared from the mixture of 23 S core particles plus SP30A, and another new subribosomal particle, [23, B], was prepared from the mixture of 23 S core particles plus SP30B. The control reconstituted 30 S particles were prepared either from 23 S core particles plus both SP30A and SP30B ([23, A, B]) or from 23 S core particles plus SP30 ([23, SP30]). Analyses of proteins in the various protein fractions and subribosomal particles with gel electrophoresis are

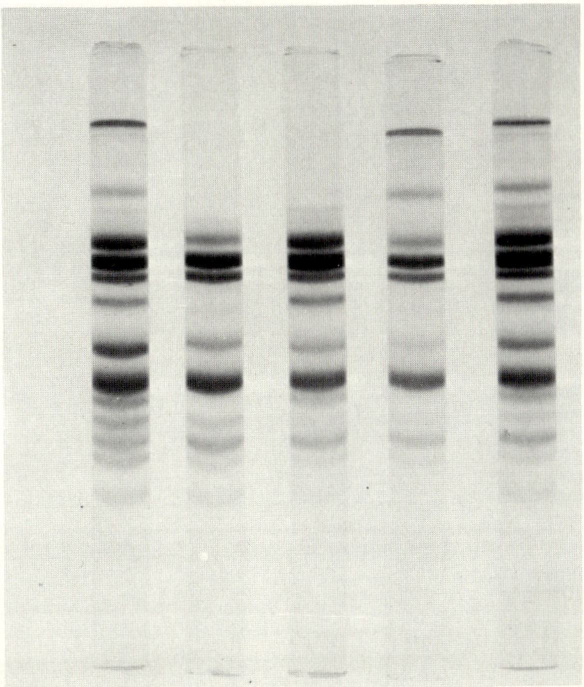

FIG. 3*b*. For legend see opposite page.

shown in Fig. 3*a* and *b*, and the results are summarized in Fig. 4*b*. Although some of the 30 S protein bands are shared by both the 23 S and the SP30 fractions, it can be seen that the 23 S core particles lack some specific protein components which appear in the SP30 fraction. Clear separation of SP30 proteins into two fractions was also achieved. The results also show that the particles [23, A] lack (or have reduced amounts of) the proteins contained in the SP30B fraction—namely, proteins III, VI, and IX, and possibly IV and VII—and that the particles [23, B] lack the proteins contained in the SP30A fraction—namely, proteins I and II. The 23 S core particles lack (or have reduced amounts of) both the SP30A and the SP30B proteins. Control reconstituted 30 S particles, [23, A, B], showed a band pattern very similar to that of native 30 S ribosomes. Thus, three different subribosomal particles—23 S, [23, A], and [23, B]—can be obtained from 30 S ribosomes.

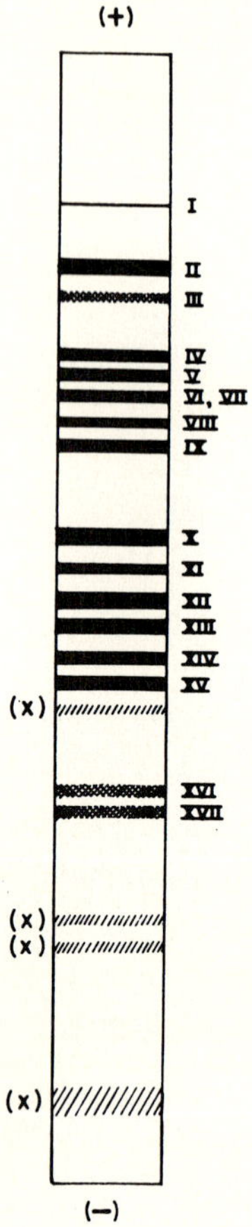

No.	SP50	SP50 A	SP50 B	40	[40, A]	[40, B]	50 or [40, A, B]
I	+	+			+		+
II	+	+			+		+
III	+	+			+		+
IV	+		+			+	+
V				+	+	+	+
VI	+	+			+		+
VII				+	+	+	+
VIII				+	+	+	+
IX	(+)		(+)	+	+	+	+
X				+	+	+	+
XI				+	+	+	+
XII				+	+	+	+
XIII	(+)		(+)	+	+	+	+
XIV				+	+	+	+
XV				+	+	+	+
XVI				+	+	+	+
XVII				+	+	+	+

(a)

FIG. 4. Summary of the gel electrophoretic analysis of protein components of the ribosomes, subribosomal particles, and protein fractions. (*a*) Particles and fractions derived from 50 S. (*b*) Those derived from 30 S.

Although a detailed physical and chemical analysis of these subribosomal particles has not been completed, examination of 40 S core particles and 23 S core particles with the electron microscope by Dr. H. Ris, University of Wisconsin, has shown that these core particles retain a compact structure. These core particles are consequently quite different from the unfolded ribosomes obtained by removal of Mg^{++} (Spirin, 27; Gesteland, 8). The latter particles have sedimentation coefficients of 24 S and 14 S and retain all the ribosomal proteins. The core particles, 40 S and 23 S, are also different from incomplete ribosomes (CM particles) (Nomura and Watson, 23), isolated from chloramphenicol-inhibited *E. coli*. CM particles, which are also deficient in some of

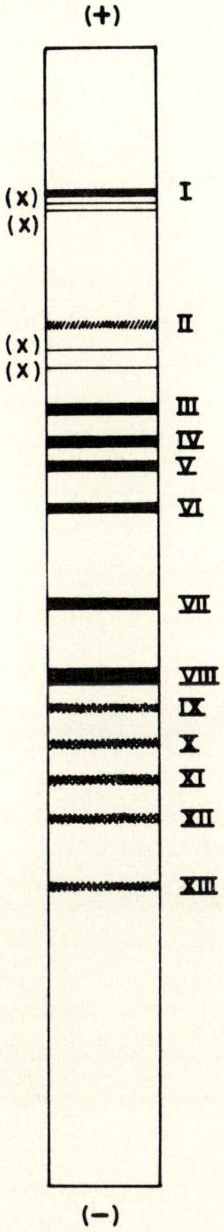

No.	SP30	SP30A	SP30B	23	[23, A]	[23, B]	30 or [23, A, B]
I	+	+			+		+
II	+	+			+		+
III	+		+			+	+
IV	+		+	+	+	+	+
V				+	+	+	+
VI	+		+			+	+
VII	+		+	+	+	+	+
VIII				+	+	+	+
IX	+		+			+	+
X	(+)		(+)	+	+	+	+
XI				+	+	+	+
XII				+	+	+	+
XIII				+	+	+	+

(b)

FIG. 4*b*. For legend see opposite page.

the ribosome protein components, have been shown to have average sedimentation coefficients of 25 S and 18 S, and to have a loosely coiled structure different from the compact structure of the ribosomes (Kurland *et al.*, 14).

FUNCTIONAL CAPACITY OF VARIOUS SUBRIBOSOMAL PARTICLES

The various subribosomal particles described in the previous section have been examined for their ability to perform known ribosomal functions (Traub and Nomura, 35). In the case of derivatives of 50 S ribosomes, the following functions were assayed: (1) The activity in poly U-dependent polyphenylalanine synthesis (Nirenberg and Matthaei, 20). This was assayed in the presence of native 30 S ribosomes. (2) The ability to bind a mixture of radioactive tRNA in the absence of messenger RNA. The binding of tRNA by 50 S in the absence of mRNA was first described by Cannon *et al.* (2). In the case of derivatives of 30 S ribosomes, the following functions were assayed: (1) The activity in poly U-dependent polyphenylalanine synthesis assayed in the presence of native 50 S ribosomes. (2) The ability to bind a specific tRNA in the *presence* of mRNA (Matthaei *et al.*, 16; Suzuka *et al.*, 32). (3) The ability to bind a synthetic mRNA, poly U (Okamoto and Takanami, 24). The results of these studies are summarized in Table II.

TABLE II
FUNCTIONAL CAPACITY OF VARIOUS SUBRIBOSOMAL PARTICLES[a]

(*a*) *Particles derived from 50 S*

Particles	Amino acid incorporation	tRNA binding
50 S	100	100
[40, A, B]	50	32
40 S	3	5
[40, A]	27	5
[40, B]	9	26

(*b*) *Particles derived from 30 S*

Particles	Amino acid incorporation	Specific tRNA binding[b]	Poly U binding
30 S	100	100	100
[23, A, B]	74	64	86
23 S	1	0	5
[23, A]	2	0	20
[23, B]	28	34	56

[a] Figures show percent relative activity obtained in a typical experiment (Traub and Nomura, 35). The activity of [40, SP50] and [23, SP30] is similar to that of [40, A, B] and [23, A, B], respectively.

[b] Binding of radioactive Phe-tRNA was assayed in the presence of poly U (Nirenberg and Leder, 19).

The subribosomal particles [40, B] as well as the 40 S core particles were found to be essentially inactive in polyphenylalanine synthesis, whereas the subribosomal particles [40, A] were found to retain considerable activity, usually about one half of the control reconstituted 50 S particles ([40, A, B] or [40, SP50]). The difference between [40, A] and [40, A, B] particles is due to the lack of the protein component of SP50B in the [40, A] particle, since the addition of SP50B proteins stimulated the activity of [40, A] particles. The inactivity of [40, B] particles is not due to an inactivation during the preparation or storage of the particles, because the addition of SP50A proteins restored the activity to the same extent as that of [40, A, B]. This stimulatory action of SP50A proteins is specific. The SP50B fraction does not stimulate the [40, B] particles. Similarly, the inactivity of the 40 S core particles is not due to inactivation, but is due to the lack of the split proteins (SP50A plus SP50B). The inactive 40 S core can be stimulated partially by SP50A protein(s) but not by SP50B protein(s). They can be nearly completely restored by the addition of a mixture of SP50A and SP50B proteins.

It is concluded from these experiments that some protein(s) in the SP50A fraction are indispensable for the synthesis of polypeptide, and that the proteins in the SP50B fraction may be dispensable. However, the SP50B fraction does have a stimulatory effect which remains to be accounted for.

Gilbert (9) has demonstrated that, after poly U-directed phenylalanine incorporation, the growing nascent polypeptide chains are bound to the 50 S ribosomes through the binding of their terminal transfer RNA to the 50 S ribosomes. Cannon *et al.* (2) observed that, in the presence of 10^{-2} M Mg^{++}, tRNA reversibly binds to a site on 50 S without added messenger RNA. It was suggested that this site is the site that holds the growing peptidyl-tRNA. Our experiments have shown that both 40 S and [40, A] particles are essentially devoid of the tRNA-binding activity, whereas [40, B] particles have a binding ability comparable to that of the [40, SP50] or [40, A, B] particles, which is about 30% of that of the native 50 S ribosomes. Inability of 40 S core particles to bind tRNA has been independently shown by Raskas and Staehelin (personal communication, 25). Thus, the presence of the SP50B protein is essential for this tRNA binding. However, this protein is *not* essential for the 50 S particles to function in polypeptide synthesis. The subribosomal particle [40, A] is essentially devoid of this protein and lacks most of the tRNA-binding function, and yet it is active in the synthesis of polyphenylalanine (Table II). Since this SP50B protein(s) is not contained in the *E. coli*-soluble protein preparation used in our incorporation assay, these experimental results suggest that full activity of the tRNA-binding function of 50 S is not obligatory for polypeptide synthesis.

It has been found that there is a close parallelism between the specific tRNA-binding ability and the phenylalanine incorporation activity of the various subribosomal particles derived from 30 S ribosomes. Both 23 S core and [23, A] particles are essentially inactive in these two functions. Subribosomal particles [23, B] are active both in specific tRNA binding and in polypeptide synthesis, but the activity is only about one half of that of the control reconstituted particles [23, A, B]. SP30A proteins stimulate this partially active particle [23, B], but SP30B proteins do not. Similarly, the inactive [23, A] particles can be stimulated by SP30B, but not by SP30A proteins. The inactive 23 S core particles can be stimulated to some extent by SP30B proteins, but not by SP30A proteins. Better stimulation can be obtained by the mixture of SP30B and SP30A proteins.

From these experiments, it is concluded that some proteins in the SP30B are indispensable for both the specific tRNA-binding function and the incorporation activity, while the SP30A proteins may be dispensable but have a stimulatory action in some indirect way.

The ability of the various subribosomal particles to bind poly U was studied by using the Milipore filter technique as described by Moore (18). The 23 S core particles are nearly completely inactive in this reaction. This has also been shown by Raskas and Staehelin (personal communication, 25), independently. The subribosomal particles [23, B] have a binding ability comparable to that of the control reconstituted 30 S ([23, SP30] or [23, A, B]) particles, whereas the [23, A] subribosomal particles showed only a weak activity. It is concluded that SP30B proteins are indispensable, and SP30A proteins may be dispensable for the mRNA binding.

The presence of dispensable components (SP50B and SP30A) in ribosomes poses questions as to their possible functions. These components might have some important roles *in vivo* which are not detectable in our *in vitro* assay system. However, they have a definite stimulatory action in the *in vitro* assay, and the nature of this stimulation must be explained. We have examined the nature of the difference between the functional ability of [23, B] particles and of [23, A, B] particles. The binding of ^{14}C-phenylalanyl tRNA by [23, B] particles has been studied in more detail and compared with that of the [23, A, B] particles. So far, we have studied the effect of Mg^{++} concentration, inhibition by LiCl at different NH_4Cl concentrations, and the effect of temperature on the binding reaction. Although several small quantitative differences have been observed, we have found no remarkable qualitative differences. We have also looked for a possible increased frequency of error in the translation of the genetic code in the system when native 30 S is replaced by the partially defective subribosomal particle [23, B]. No remarkable increase in the error frequency with poly U messenger was detected in the system consisting of [23, B] particles and native 50 S ribosomes. These negative experiments suggest, but do not prove, that the stimulation by the SP30A component is due to a quantitative increase in the functional capacity of [23, B] particles without any specific qualitative change. It is possible that the [23, B] particle can assume several different structures, some active and others inactive, and that, in the presence of the SP30A components, only the active form of the [23, B] particle is structurally permitted in [23, A, B] particles.

Kinetics of the Reconstitution

As was already mentioned, the fractionated proteins (SP50A, SP50B, SP30A, and SP30B) are soluble in buffers at moderate salt concentrations. With these soluble split protein preparations available, we are now able to study the reconstitution in a more quantitative way in a system which does not utilize high concentration of CsCl.

We have studied the rate of reconstitution to see whether there is any detectable time lag before the appearance of the functional structure in our reconstitution system. We have found that the reconstitution of 50 S ribosomes takes place very fast. For example, 40 S core particles were mixed with SP50A and SP50B proteins at 0°C, and the incorporation assays were immediately (in less than 1 minute) started. No difference was observed in the initial rate of the incorporation between the experimental reconstitution mixture and an equivalent amount of the control [40, A, B] particles. The reconstitution process is essentially complete within a minute or so in this system.

In the formation of functional 30 S structure from inactive 23 S core particles, SP30A proteins, and SP30B proteins, or from inactive [23, A] particles and SP30B proteins, a definite time lag of about 2 minutes was observed in the tRNA binding by the experimental reconstitution mixture in our experimental condition, whereas no such time lag was observed in the tRNA binding by the native 30 S or by the reconstituted 30 S ([23, A, B]) particles. The formation of functional 30 S structure from [23, A] particles and SP30B proteins was studied in more detail. The time lag can be abolished by preincubating the [23, A] particles with the SP30B proteins for 5 minutes at 28°C before the start of the binding assay.

In the course of these experiments, it was observed that a strong turbidity appears upon the addition of SP30B proteins to [23, A] particles, but that the turbidity disappears quickly. This suggests that some intermediate structure was formed, and they converted to a final functionally active form. The nature of the observed time lag and of the possible intermediate structure is currently being studied.

It should be noted that the reconstitution of 30 S particles was followed by measuring the specific tRNA binding of the reconstituted 30 S particles. The assay system does not contain any soluble proteins or ribosomal particles other than the components necessary for the reconstitution. Thus, our experiments have clearly established that the reconstitution takes place reasonably fast in a simple environment without any other cellular component or preformed complete ribosomes.

Studies on Ribosomes Altered *in Vivo*

So far we have discussed our work on the several different subribosomal particles constructed *in vitro*. Another useful approach to the study of ribosomes would be to examine ribosomes altered *in vivo* either by mutation or by other causes, and to characterize their structural alteration in relation to the altered function.

Several examples of mutational alterations of ribosomes have been reported. One is the ribosomes from streptomycin-resistant (or streptomycin-dependent) mutants of *E. coli*. *In vitro* experiments showed that streptomycin (Sm) inhibits poly U-directed polyphenylalanine synthesis by ribosomes isolated from Sm-sensitive *E. coli* strains, but does not inhibit polypeptide synthesis by ribosomes isolated from Sm-resistant mutant strains (Flaks *et al.*, 6; Speyer *et al.*, 26). Furthermore, the sensitivity has been shown to be associated with the 30 S ribosomes, but not with the 50 S ribosomes (Davies, 4; Cox *et al.*, 3). Another example of mutational alteration in ribosomes was described by Apirion (1), who showed an increased temperature sensitivity of the ribosomes from a suppressor strain in their capacity to function in the *in vitro* polypeptide synthesis. The alteration is in the 50 S part of the ribosomes. In both cases, proteins from altered ribosomes were examined by gel electrophoresis and compared with proteins from the normal ribosomes. No difference was found by these workers. Flaks *et al.* (7) have recently reported a mutational alteration in the gel electrophoresis pattern of 50 S ribosomal proteins from a temperature-sensitive strain of *E. coli*.

With the availability of the ribosomal reconstitution technique, it has now become possible to locate these mutational alterations within the ribosomal structure. The alteration in the 30 S ribosomes from the Sm-resistant mutant was studied by our group (33) as well as by Staehelin and Meselson (31).

The 30 S ribosomes from a Sm-sensitive strain and from a Sm-resistant strain were dissociated into the 23 S core particles and the split protein fraction. The 23 S core particles were then mixed with the split protein fraction in four different combinations (Table III), and sensitivity of the resultant reconstituted 30 S ribosomes ([23, SP30]) to Sm was then tested by following poly U-directed phenylalanine incorporation in the system containing native 50 S ribosomes from the sensitive strain. Streptomycin inhibited the incorporation only when the reconstituted 30 S contained the 23 S component derived from sensitive cells, and the degree of inhibition was about the same as that found with the sensitive native 30 S ribosomes. The system containing the 23 S component derived from resistant cells was resistant to Sm, as was the system with

TABLE III
SENSITIVITY TO STREPTOMYCIN (SM) OF RECONSTITUTED RIBOSOMES[a]

Reconstituted 30 S			
23 S	SP30	Native 30 S	Inhibition by Sm (%)
—	—	s	46
—	—	r	0 (−4)
s	s	—	46
s	r	—	41
r	s	—	0 (−2)
r	r	—	0 (−9)

[a] The reconstituted 30 S or control native 30 S were mixed with native 50 S ribosomes obtained from a Sm-sensitive strain, and the poly U-directed incorporation of phenylalanine was assayed in the presence and in the absence of Sm (5×10^{-6} *M*). Data taken from Traub *et al.* (33).

the resistant native 30 S ribosomes (Table III). It has been concluded that the sensitivity to Sm is determined by the 23 S core particles, not by the split protein fraction.

The second system we have studied is the ribosomes from *E. coli* treated with a specific colicin, colicin E3. Colicins are bactericidal proteins, which are synthesized by certain strains of Enterobacteriaceae harboring a cytoplasmic genetic element called the colicinogenic factor. One of the colicins, colicin E3, inhibits protein synthesis in sensitive cells without inhibiting the synthesis of other macromolecules (Nomura, 21, 22). Recent experiments done in our laboratory (Konisky and Nomura, 13) have shown that ribosomes obtained from E3-inhibited cells are physically intact, but are inactive when assayed with the poly U-directed phenylalanine incorporation system. The alteration is in the 30 S part of the ribosome; the 50 S part is unaltered. The altered 30 S retains most of the original ability to bind mRNA, but fails to show the specific tRNA-binding function. Using the present reconstitution system, we have examined whether the defect resides in the 23 S core or in the split protein fraction. 30 S particles were reconstituted from 23 S core particles and SP30 proteins in four different combinations, as shown in Table IV, and their activities were tested. It is clear from the results that the defect is in the 23 S core particles; the split proteins from this defective 30 S are essentially intact. Somehow, colicin E3 causes a specific alteration in the 23 S core part of the pre-existent 30 S ribosomes and abolishes the specific tRNA-binding function without causing a gross effect on the ribosome structure.

Thus, in the two cases that we have studied (streptomycin-resistant ribosomes and colicin E3-inactivated ribosomes) we have found the alteration in the 23 S core. To gain more information, it would obviously be very useful to be able to dissociate 23 S core further, hopefully to free RNA and the mixture of proteins, and then to reconstitute functionally active ribosomes from these components. On the other hand, a systematic search of altered ribosomes may enable us to find a system in which the alteration is in the split protein fraction; for this fraction we already have the reconstitution assay available which would allow us to study more deeply the problem of the structure and function of the protein components of the ribosomes.

TABLE IV
POLY U-DIRECTED PHE-tRNA-BINDING ACTIVITY OF RECONSTITUTED 30 S PARTICLES[a,b]

Reconstituted 30 S		Activity (% of the control reconstituted 30 S)
23 S	SP30	
Control	Control	100
Control	E3	116
E3	Control	16
E3	E3	20

[a] Data taken from unpublished experiments by Konisky and Nomura. Native 30 S ribosomes from E3-treated cells showed an activity about 30% of the control when assayed with the poly U-directed Phe-tRNA binding and an activity about 10% of the control when assayed with the poly U-directed polyphenylalanine synthesis.

[b] Control: the component from control cells. E3: the component from cells treated with colicin E3.

SUMMARY AND CONCLUSIONS

Our studies on the reconstitution of ribosomes have established first that the ribosomal proteins are heterogeneous. We have previously shown that 23 S core proteins are physically and functionally different from the split proteins and cannot replace the split proteins in the reconstitution of active 30 S particles (Traub *et al.*, 34). Now we have shown that the split proteins SP30 are separated into two functionally and physically distinct fractions, SP30A and SP30B. Similarly, 50 S proteins consist of at least three different protein fractions, which can be distinguished from each other by physical and functional criteria. We have also obtained some information on the function of several ribosomal components separated. Thus, SP50A is indispensable for the polypeptide synthesis but not for the nonspecific tRNA-binding activity of the 50 S ribosomes. SP50B is dispensable for the polypeptide synthesis but has a stimulatory activity in some way. On the other hand, SP50B is indispensable for the observed nonspecific tRNA-binding activity of 50 S ribosomes. As to the functions of the 30 S ribosomes, we have shown that SP30B is indispensable for the amino acid incorporation and the specific tRNA binding as well as for the poly U binding. SP30A is dispensable in these functions but has a stimulatory activity in both the amino acid incorporation and the specific tRNA binding. Thus, our experimental results strongly suggest that there are at least six different proteins in the ribosomes. In the two systems that we have studied, Sm-resistant ribosomes and colicin E3-inactivated ribosomes, we have found the alteration in the 23 S core.

Finally, our experiments have shown that the reconstitution of ribosomes from the core particles and proteins is spontaneous and takes place reasonably rapidly in a physiological environment without any other cellular component or preformed complete ribosomes. It is pleasing to note that the results of our experiments have so far been consistent with the belief that the apparently complex structure of various cell organelles can be generated by simple physicochemical interactions of the component macromolecules in a given physiological environment without other pre-existing cell structures.

Acknowledgments

The authors wish to thank Dr. K. Hosokawa and Mr. J. Konisky, who participated in some aspects of the present project. The authors would also like to express their thanks to Drs. W. F. Dove, C. G. Kurland, O. Smithies, and H. Temin for their critical reading of the manuscript. This work was aided by research grants from the National Science Foundation (GB-3947) and from the National Institutes of Health, U.S. Public Health Service (AI-05641).

References

1. Apirion, D., *J. Mol. Biol.*, **16**, 285 (1966).
2. Cannon, K., Krug, R., and Gilbert, W., *J. Mol. Biol.*, **7**, 360 (1963).
3. Cox, E. C., White, J. R., and Flaks, J. G., *Proc. Natl. Acad. Sci. U.S.*, **51**, 703 (1964).
4. Davies, J. E., *Proc. Natl. Acad. Sci. U.S.*, **51**, 659 (1964).
5. Davies, J. E., Gilbert, W., and Gorini, L., *Proc. Natl. Acad. Sci. U.S.*, **51**, 883 (1964).
6. Flaks, J. G., Cox, E. C., Witting, M. L., and White, J. R., *Biochem. Biophys. Res. Commun.*, **7**, 390 (1962).
7. Flaks, J. G., Leboy, P. S., Burgi, E., and Kurland, C. G., *Cold Spring Harbor Symp. Quant. Biol.*, **31**, 623 (1966).
8. Gesteland, R. F., *J. Mol. Biol.*, **16**, 67 (1966).
9. Gilbert, W., *J. Mol. Biol.*, **6**, 389 (1963).
10. Gorini, L., and Kataja, E., *Proc. Natl. Acad. Sci. U.S.*, **51**, 487 (1964).
11. Gorini, L., and Kataja, E., *Proc. Natl. Acad. Sci. U.S.*, **51**, 1955 (1964).
12. Hosokawa, K., Fujimura, R. K., and Nomura, M., *Proc. Natl. Acad. Sci. U.S.*, **55**, 198 (1966).
13. Konisky, J., and Nomura, M., *J. Mol. Biol.*, **26**, 181 (1967).

14. Kurland, C. G., Nomura, M., and Watson, J. D., *J. Mol. Biol.*, **4**, 388 (1959).
15. Leboy, P. S., Cox, E. C., and Flaks, J. G., *Proc. Natl. Acad. Sci. U.S.*, **52**, 1367 (1964).
16. Matthaei, H., Amelunxen, F., Eckert, K., and Heller, G., *Ber. Bunsenges.*, **68**, 735 (1964).
17. Meselson, M., Nomura, M., Brenner, S., Davern, C., and Schlessinger, D., *J. Mol. Biol.*, **9**, 696 (1964).
18. Moore, P. B., *J. Mol. Biol.*, **18**, 8 (1966).
19. Nirenberg, M., and Leder, P., *Science,* **145**, 1399 (1964).
20. Nirenberg, M., and Matthaei, J. H., *Proc. Natl. Acad. Sci. U.S.*, **47**, 1589 (1961).
21. Nomura, M., *Cold Spring Harbor Symp. Quant. Biol.*, **28**, 315 (1963).
22. Nomura, M., *Proc. Natl. Acad. Sci. U.S.*, **52**, 1514 (1964).
23. Nomura, M., and Watson, J. D., *J. Mol. Biol.*, **1**, 204 (1959).
24. Okamoto, T., and Takanami, M., *Biochim. Biophys. Acta,* **68**, 325 (1963).
25. Raskas, H. J., and Staehelin, T., *J. Mol. Biol.*, **23**, 89 (1967).
26. Speyer, J. F., Lengyel, P., and Basilio, C., *Proc. Natl. Acad. Sci. U.S.*, **48**, 684 (1962).
27. Spirin, A. S., *in* "Macromolecular Structure of Ribonucleic Acid," p. 170. Reinhold Publishing Corp., New York, 1964.
28. Spirin, A. S., and Belitsina, N. V., *J. Mol. Biol.*, **15**, 282 (1966).
29. Spitnik-Elson, P., *Biochim. Biophys. Acta,* **55**, 741 (1962).
30. Staehelin, T., and Meselson, M., *J. Mol. Biol.*, **16**, 245 (1966).
31. Staehelin, T., and Meselson, M., *J. Mol. Biol.*, **19**, 207 (1966).
32. Suzuka, I., Kaji, H., and Kaji, A., *Biochem. Biophys. Res. Commun.*, **21**, 187 (1965).
33. Traub, P., Hosokawa, K., and Nomura, M., *J. Mol. Biol.* **19**, 211 (1966).
34. Traub, P., Nomura, M., and Tu, L., *J. Mol. Biol.*, **19**, 215 (1966).
35. Traub, P., and Nomura, M., manuscript in preparation.
36. Waller, J. P., *J. Mol. Biol.*, **10**, 319 (1964).
37. Waller, J. P., and Harris, J. J., *Proc. Natl. Acad. Sci. U.S.*, **47**, 18 (1961).
38. Watson, J. D., *Science,* **140**, 17 (1963).
39. Watson, J. D., *Bull. Soc. Chim. Biol.*, **46**, 1399 (1964).

The Preparation of Protein-Depleted Particles from Rat Liver Ribosomes

ANNE-MARIE REBOUD AND MARY L. PETERMANN

Sloan-Kettering Division, Institute for Cancer Research, New York, New York

When *Escherichia coli* ribosomes are centrifuged in a gradient of concentrated cesium chloride, half their proteins dissociate and float to the top of the tube, leaving subribosomal nucleoprotein particles. These particles retain their structure and can combine with the dissociated proteins to give active ribosomes (6, 2). This procedure cannot be used on animal ribosomes, since their proteins are completely dissociated in moderately concentrated salt solutions. Partially stripped particles have, however, been prepared by Sephadex fractionation.

Rat liver ribosomes were isolated in the presence of bentonite to adsorb RNase. They were freed of extraneous proteins by deoxycholate treatment, a wash at pH 8, and magnesium precipitation, and stored at −20°C (4). To strip off the easily dissociable structural proteins, 30 mg of ribosomes were dissolved in 15 ml of 0.5 *M* LiCl, 0.00186 *M* $MgCl_2$, 0.0046 *M* glycine, and 0.0023 *M* KOH, pH 9.3. After standing for 30 minutes at 5°C the solution was passed through a 95 by 2-cm column of Sephadex G-100, in the same solvent, at 5°C. The stripped nucleoprotein particles emerged at the void volume, 130 ml. The leading fractions, whose absorbancies at 260 and 235 mμ had a ratio of about 1.73, were pooled; the yield was about 10 mg. The trailing portion of the peak, which had a lower ratio, and presumably contained more protein, was discarded. The rest of the proteins were eluted gradually from the column. The stripped particles were dialyzed for 2 days against 0.02 *M* $KHCO_3$, pH 8.5. For comparative studies, ribosomal subunits were freed of magnesium (5) and dialyzed overnight against 0.02 *M* $KHCO_3$. Ribosomal RNA was isolated and freed of magnesium as previously described (4, 5), and dialyzed against 0.02 *M* $KHCO_3$ for 2 hours. Protein and RNA were determined by the Lowry and orcinol methods (3). The ribosomal subunits and the stripped particles were fixed in formaldehyde and banded in CsCl gradients in a preparative ultracentrifuge (1).

Comparison with the ribosomal subunits (Table I) shows that about half the ribosomal proteins had been removed from the stripped particles. Their density was correspondingly increased to that of whole *E. coli* ribosomes, which are also two-thirds RNA (1). The particles tended to aggregate, particularly in buffers that contained magnesium, but were fairly soluble in 0.02 *M* $KHCO_3$ at pH 8.5; analytical ultracentrifugation showed 36% of 17 S, 47% of 29 S, and 17% of aggregated material. The two sedimentation coefficients, at infinite dilution, fell between those of the subunits and the RNA's (Table I). All these values were low; all three types of macromolecules had probably assumed more extended configurations because of the scarcity of counter-ions in this solvent.

TABLE I
PROPERTIES OF RIBOSOMAL SUBUNITS, STRIPPED PARTICLES, AND RNA'S

	Protein/RNA	Density[a] in CsCl (gm/ml)	$s^0_{20,w}$ [b,c] (S)	Mobility,[b,d] $\times 10^5$ (cm^2/sec/volt)
Subunits	1.0	1.59	21.0 and 35.5	−15.7
Particles	0.5	1.64	17.0 and 29.2	−15.7
RNA's			14.7 and 24.5	−17.4

[a] After fixation in formaldehyde.
[b] In 0.02 *M* $KHCO_3$.
[c] Measured at 25°C.
[d] Measured at 0°C.

The electrophoretic mobilities were measured by the moving-boundary method, on solutions containing 1.5 mg/ml. The subunits and stripped particles both gave single but asymmetrical boundaries, and the RNA gave single symmetrical peaks. The descending mobilities of the subunits and particles were the same, while that of the RNA was higher. The particles may have had a greater negative charge than the subunits, but were retarded more by frictional drag, or the stripped proteins may not have contributed to the net charge of the subunits at pH 8.5.

The properties of these stripped particles are those of well-defined structures, not random aggregates of RNA and ribosomal proteins. Whether they resemble ribosomal precursors is not known, since incomplete ribosomes have not been found in animal cells. When protein synthesis is inhibited and RNA synthesis continues, as in HeLa cells treated with cycloheximide, only complete ribosomes are seen; animal cells apparently contain large pools of ribosomal proteins (7). Further characterization of these stripped particles, now in progress, may add to our knowledge of ribosomal structure.

Acknowledgments

Anne-Marie Reboud holds a fellowship from the Centre National de la Recherche Scientifique, Paris, France. This investigation was supported by Grant 08748 from the National Cancer Institute and U.S. Atomic Energy Commission Contract AT(30-1)-910.

References

1. Lerman, M. I., Spirin, A. S., Gavrilova, L. P., and Golov, V. F., *J. Mol. Biol.*, **15**, 368 (1966).
2. Nomura, M., and Traub, P., this volume.
3. Petermann, M. L., "The Physical and Chemical Properties of Ribosomes," Elsevier, Amsterdam, 1964.
4. Petermann, M. L., and Pavlovec, A., *J. Biol. Chem.*, **238**, 3717 (1963).
5. Petermann, M. L., and Pavlovec, A., *Biochim. Biophys. Acta*, **114**, 264, (1966).
6. Staehelin, T., Raskas, H., and Meselson, M., this volume.
7. Warner, J., Girard, M., Latham, H., and Darnell, J., *J. Mol. Biol.*, **19**, 373 (1966).

Discussion of Part IX

R. B. ROBERTS, P. SIEKEVITZ, M. NOMURA, T. STAEHELIN, J. MARMUR, E. RACKER, M. MESELSON, R. P. NOVICK

CHAIRMAN ROBERTS: The meeting is now open for discussion.

DR. SIEKEVITZ: I would like to ask two questions. The first one is to both Dr. Staehelin and Dr. Nomura and it has to do with something that Dr. Petermann said. Do they think that any of the split proteins are the polypeptide chains which have been newly synthesized? These proteins might be on the inside and they might be nascent chains which are still on the isolated ribosomes. The second question is to Dr. Nomura and has to do with the observation, which I think is not so good, by Dr. Staehelin, and I wonder if you have done this experiment. He has shown some specificity in reconstitution of certain proteins with certain particles in the incorporation experiments. But when he did the experiments on reconstitution to show which proteins go with which particles in the density gradient, he finds that there is no specificity. That is, the split proteins from the 50 S went on, apparently, to the core particles of the 30 S. And I wonder if you had done this kind of experiments also.

DR. NOMURA: We did similar experiments when we started reconstitution and found the same results as Dr. Staehelin showed. We made radioactive split proteins from radioactive 30 S and then mixed these with nonradioactive 40 S core and we asked how much radioactive protein sticks to the 40 S core. We found that about 50% or 60% stuck. Similar nonspecific binding of radioactive SP50 to 23 S cores was also reported by us (Hosokawa *et al.*, 1966). But then we found that the radioactive proteins in such nonspecific aggregates can be removed by washing with a high concentration of salt. So now my feeling is that this kind of nonspecific binding can be eliminated by a rigorous purification method. The same was true when we added homologous proteins. We asked how much homologous protein sticks and, in fact, we could show that extra protein sticks to the reconstituted core. But as I mentioned, activity didn't increase by adding extra protein; the activity reaches a certain plateau value. Yet, we can add more and more radioactive protein. In this case, again by rigorous purification, we may be able to remove such nonspecifically bound proteins.

DR. SIEKEVITZ: Is the presence of nascent polypeptide chain excluded rigorously?

DR. NOMURA: I cannot rigorously exclude it, but our ribosomes are purified by extensive ammonium sulphate precipitation to start with. That doesn't exclude it, although I think it rather unlikely.

Dr. Staehelin: You could maybe say the following to this question. If nascent protein, presumably peptidyl-tRNA, is bound to the ribosomes, you would expect this to be a large variety of proteins, hundreds maybe, of different proteins. Therefore, you would not expect to see it as any single band showing up in gel electrophoresis. Actually, I am quite convinced that the proteins we are using for reconstitution and see in gel electrophoresis are indeed ribosomal proteins, because the nascent polypeptide chains would be scattered all over. If there is one nascent polypeptide chain per ribosome but maybe only every hundredth or every thousandth ribosome has the same nascent polypeptide chain, we would not expect to see it.

Dr. Nomura: But it doesn't exclude the presence of a small amount of nascent polypeptide in the split protein fractions.

Dr. Staehelin: Yes, the only thing is that the analysis in acrylamide gel would certainly not show these proteins unless you have one particular protein which is synthesized in great excess so that every fourth or every third ribosome has an identical nascent protein. And then again you would expect that these nascent polypeptide chains are in different stages of assembly and would move completely differently in electrophoresis even if they are translated from identical messengers. So certainly we cannot exclude the presence of these peptide chains, but I think we can definitely exclude that any of the visible bands in acrylamide gel electrophoresis represent nascent polypeptide chains.

Dr. Marmur: What happens to the 5 S RNA when the 50 S ribosomes are centrifuged in CsCl?

Dr. Staehelin: We both didn't pay any attention to it. You can only look at so many things at one time. If it is split, then you would expect it at the bottom of the CsCl gradients and we would never recover it, which would mean that at least for poly U directed polyphenylalanine synthesis 5 S RNA would not be essential. If it is remaining in the core, then of course it is there. [*Note added in proof:* We have now analyzed the 42 S cores for the presence of 5 S RNA. It is definitely present, approximately to the same extent as in native 50 S ribosomes (Brownlee and Staehelin, 1967, in preparation).]

Chairman Roberts: Does that answer that?

Dr. Nomura: With radioactive 4 S we showed that it goes to the bottom of the centrifuge tube. So contamination by such 4 S RNA is small in our core fraction. Therefore, if 5 S RNA dissociates in 5 *M* CsCl, we would be essentially eliminating 5 S RNA from both the core fraction and the split protein fraction.

Dr. Racker: First of all I would like to say that I am very impressed by these very exciting developments in the resolution of the ribosomes. But I would like to make three very brief comments. First of all, I agree with Dr. Petermann. I also don't like the word "core particles." A noncommittal name like "subribosomal particles" would be more appropriate since we don't know what is core and what is not core. The second point is that I am a little worried about the statement that one component is essential and another is not essential. With the mitochondria we have found that depending on the method of resolution you may get completely different results. In one system, addition of a component may be necessary, in another, addition of another component may be essential. I venture to say that all thirty proteins you find in ribosomes may be essential. The third point is a suggestion with regard to trouble you are likely to get into, in your studies on resolution. Once you get solubilization of a protein component, you might expect that its properties will change during the course of purification; you may find differences in solubility and particularly differences in stability. It may be worth while for you to familiarize yourself with difficulties and troubles we encountered in resolving the mitochondria.

Chairman Roberts: Thank you, Dr. Racker. Dr. Meselson has a question. I think we have time for about three more and I hope one can be my own.

Dr. Meselson: I would like to ask Dr. Nomura if the soluble A and B fractions form a precipitate when mixed.

Dr. Nomura: In our experience, yes. But we never carefully studied it in terms of concentration of proteins and other variables, so I can't give a definite answer.

Dr. Novick: I would like to ask Dr. Nomura whether one of the soluble fractions, I think the B fraction from the 50 S ribosomes, has any affinity for transfer RNA or polyuridylic acid in solution by itself.

Dr. Nomura: We have tried experiments concerning binding, especially with respect to poly U binding. We found that split proteins from 30 S in fact bind poly U. So at first we got excited. Then we found that split proteins from 50 S also bind poly U. We also found that basic proteins from core particles bind poly U. These latter two bindings are certainly nonspecific in nature. To test the specificity of such binding further, we examined whether these aggregates, that is, protein-poly U complexes, pick up phenylalanyl-sRNA. No positive result was observed. Proteins from 30 S bind poly U, but they do not pick up phenylalanyl-transfer RNA.

Dr. Staehelin: May I just add a single experience. Actually we tested the binding of poly U in sucrose gradients, as you remember.

What you really get if you test meniscus or split protein from the 30 S with poly U is a precipitate which goes right to the bottom of the gradient tube. The poly U almost completely disappears from the gradient. One can even show quite nicely that preferentially the high-molecular-weight poly U disappears, presumably precipitating with the M30 protein, leaving some poly U of very low S value near the top of the gradient. So we too disregarded actually these results as an unspecific and probably not very significant interaction between poly U and M30 proteins.

CHAIRMAN ROBERTS: I am afraid the time has come when other questions, including my own, should be reserved to catch the speakers later. So I wish to thank the speakers for excellent presentations and call this session adjourned.

PART X

MITOCHONDRION SYNTHESIS

Chairman's Remarks

EFRAIM RACKER

Section of Biochemistry and Molecular Biology,
Cornell University,
Ithaca, New York

Since we shall hear a great deal about a structural protein from *Neurospora* this morning, I would like to make a few comments with regard to structural proteins in general. Unfortunately the name, structural protein, is ill-defined. For example, it was used about twenty years ago by W. T. Astbury, when he was working with keratin and other fibrous proteins; it has been applied to proteins involved in muscular contraction as well as to proteins which contribute to the morphology of virus particles. Usually, the name structural protein is being used when a morphological rather than a catalytic function is implied. Frequently it is inferred that a structural protein is and should be rather insoluble. In most instances, however, there is little direct evidence for either of these two criteria: (*a*) lack of catalytic activity and (*b*) insolubility.

In the case of mitochondria, a structural protein has been obtained after extraction with sodium dodecylsulfate. Unfortunately this reagent denatures many soluble enzymes and renders them insoluble in water. Other detergents and solvents that have been used to purify "structural protein" from mitochondria (2, 6) have similar effects. It is not surprising, therefore, that several investigators including our next speaker, Dr. Woodward, have found evidence for considerable heterogeneity when "structural protein" from mitochondria was analyzed by acrylamide gel electrophoresis. Moreover, Dr. Woodward finds now the same "structural protein" in the soluble fraction of *Neurospora.* We agree with him that another name should perhaps be used.

Four years ago we reported on the isolation from mitochondria of a protein which we called coupling factor 4 (F_4). It stimulated phosphorylation in submitochondrial particles that were obtained by sonic oscillation of mitochondria either in the presence of phospholipid or at an alkaline pH (1). F_4 was found to be similar to structural protein in several respects, e.g., in its capacity to bind phospholipid and cytochromes (8). However, in contrast to "structural protein," F_4 was isolated from mitochondria without the aid of detergents and was soluble at pH 7.5. Although we have not claimed that F_4 is pure, it appears to contain one major component when analyzed by acrylamide gel electrophoresis (3). In view of these observations we would like to suggest that preparations of structural proteins isolated from mitochondria with the aid of detergents and solvents contain an insoluble form of coupling factor 4. This conclusion is supported by the observations (*a*) that structural protein prepared according to Richardson *et al.* (6) in spite of its insolubility, contained some coupling factor activity (8) and (*b*) that F_4 precipitated with acetone became insoluble and yet retained coupling factor activity (1).

When soluble F_4 was mixed with phospholipid, a precipitate was formed which was shown to consist of characteristic membranous structures which were vesicular (5). Moreover, F_4 was shown to be an essential constituent of the purified membrane preparation that conferred oligomycin sensitivity to added mitochondrial ATPase (4). In view of these and other observations we propose (but have not proved) that F_4 plays a role in the spatial organization of the members of the electron transport chain and the enzymes concerned with phosphorylation. In fact our current research efforts on the reconstitution of the electron transport chain is based on this working hypothesis.

It is apparent from our findings that an organizational protein such as F_4 does not necessarily by itself have to be insoluble in water. Insolubility, therefore, appears to be a poor criterion for a "structural protein." We can, furthermore, challenge the second criterion for a structural protein, namely, the lack of a catalytic activity. We have shown recently that another coupling factor of mitochondria, F_1, has both a catalytic and structural role (7). Although we have not claimed that F_4 is a catalyst as well as organizational protein, we have not ruled out this possibility and are indeed considering it.

If we take stock of what we have just heard about structural protein we come to the conclusion that it is premature to apply this term indiscriminately to proteins that are insoluble for one reason or another. I think you will find support for this reservation in the data on *Neurospora* which will be presented by Dr. Woodward.

References

1. Conover, T. E., Prairie, R. L., and Racker, E., *J. Biol. Chem.*, **238**, 3831 (1963).
2. Criddle, R. S., Bock, R. M., Green, D. E., and Tisdale, H., *Biochemistry*, **1**, 827 (1962).
3. Haldar, D., Freeman, K., and Work, T. S., *Nature*, **211**, 9 (1966).
4. Kagawa, Y., and Racker, E., *J. Biol. Chem.*, **241**, 2461 (1966).
5. Kagawa, Y., and Racker, E., *J. Biol. Chem.*, **241**, 2475 (1966).
6. Richardson, S. H., Hultin, H. O., and Fleischer, S., *Arch. Biochem. Biophys.*, **105**, 254 (1964).
7. Schatz, G., Penefsky, H. S., and Racker, E., *J. Biol. Chem.*, **242**, 2552 (1967).
8. Zalkin, H., and Racker, E., *J. Biol. Chem.*, **240**, 4017 (1965).

Genetic Control, Function, and Assembly of a Structural Protein in *Neurospora*

DOW O. WOODWARD AND K. D. MUNKRES[1]

Department of Biological Sciences,
Stanford University,
Stanford, California

Experimental results collected at several laboratories during the past two or three years have led to a number of generalizations which relate to mitochondrial structure and function. Among those results and generalizations, the following should be emphasized as a preface to the approach that we have taken: (*a*) Numerous proteins and phospholipids, as well as significant amounts of DNA (17, 21, 24, 26) and RNA (20, 27), have been isolated and identified as normal components of mitochondria. (*b*) Most of the essential components of a protein-synthesizing system have been specifically identified in mitochondria (3, 5, 21, 24, 27); moreover, isolated mitochondria incorporate labeled amino acids into membrane protein (3, 5, 30, 35) and into oligomyosin-sensitive ATPase (10). (*c*) Localization of proteins in different parts of the cell, as well as at specific sites within a mitochondrion, is becoming increasingly evident (4, 23). Perhaps there are no freely soluble, unbound enzymes within a cell at all. The fact that polyribosomes are membrane-bound may indicate that enzymes are synthesized at their site of localization within the cell. Exceptions might involve the transport of enzymes from the site of synthesis to the site of localization, and/or the amount of unbound protein resulting from an equilibrium established between bound and unbound protein, even though the amount of unbound protein may be negligible *in vivo*. The difficulty is in finding conditions suitable for the isolation of an enzyme *in situ* without destruction of the native complex.

Recent investigations in our laboratory have revealed that mutations of a cytoplasmic gene result in single amino acid replacements in the structural protein of the mitochondrial membrane (36). The results of

[1] Present address: Laboratory of Molecular Biology, University of Wisconsin, Madison, Wisconsin.

comparative studies of this protein from two *Neurospora* mutants (*mi-1* and *mi-3*) illustrate the importance of the structural integrity of this protein in determining the function and conformation of a nuclear-determined mitochondrial enzyme, malate dehydrogenase (MDH) (23). Mutant forms of malate dehydrogenase were found to differ from wild-type MDH in their affinities for malate when complexed with wild-type mitochondrial structural protein, mitochondrial membranes, or mitochondria. These mutant forms of MDH have wild-type malate affinities when free in solution (Table I). This phenomenon is interpreted to

TABLE I[a]
MICHAELIS CONSTANTS[b] OF WILD-TYPE AND MUTANT MALATE DEHYDROGENASES COMPLEXED WITH WILD-TYPE AND MUTANT MITOCHONDRIAL STRUCTURAL PROTEINS

Complexing agent	Wild-type MDH	Mutant MDH strains 24	10	20	7
None	0.72	1.1	2.1	2.9	1.4
Intact mitochondria (wild type)	2.5	<u>100</u>	<u>140</u>	<u>250</u>	<u>660</u>
Mitochondrial membranes (wild type)	2.5	<u>25</u>	<u>50</u>	<u>100</u>	<u>250</u>
Wild-type MSP	0.72	<u>3.3</u>	<u>7.0</u>	<u>10</u>	<u>33</u>
mi-1 MSP	<u>500</u>	<u>6.0</u>	3.0	3.0	1.4
mi-3 MSP	<u>40</u>	<u>4.0</u>	1.7	1.2	<u>4.5</u>

[a] Data taken from Munkres and Woodward (23).

[b] Michaelis constants are for the substrate malate (K_m, mM). Underlined values indicate significant differences from the value measured in the absence of a complexing agent.

indicate that the mutant enzymes are altered in locational specificity rather than at the active center (23). In a reciprocal manner, the affinity for malate of different forms of MDH (and, in this case, particularly the wild-type enzyme) is markedly altered when in complex with mutant structural protein. These data suggest re-evaluation of the functional variation that may exist in a given enzyme *in vitro* versus *in vivo*. Binding to a site of localization could conceivably alter not only enzyme kinetics but also enzyme specificity and function (16). This may be especially relevant in the case of an enzyme such as DNA polymerase in which the purified enzyme does not seem to duplicate the *in vivo* function (29).

Other mitochondrial enzymes have also been shown to complex with mitochondrial structural protein (4). Assuming that the activity of these and other membrane-bound proteins can be modified to the same extent as MDH (by altering the primary structure of the structural protein), then it should not be surprising to find a number of differences between wild type and mutant with an alteration in the mitochondrial structural protein. The differences expected might involve the function of any of the proteins which are bound to mitochondrial structural protein *in vivo*. The data in Table II include some of the differences that have been reported between wild type and *mi-1* since this mutant was first described over 10 years ago (22). This interaction between MDH, which is specified by nuclear genes (23), and mitochondrial structural protein, which is specified by a cytoplasmic gene, we refer to as a nucleocytoplasmic interaction at the molecular level.

TABLE II
SUMMARY OF DIFFERENCES BETWEEN WILD TYPE AND *mi-1*[a]

	Number of days of growth	Wild type	Strain *mi-1*	Reference
Cytochrome c	2	100	1500	(33)
Cytochrome b		100	—	(34)
Cytochromes $a+a_3$		100	—	(12, 34)
Cytochrome oxidase	3	100	13	(33)
Succinic acid oxidase	3	100	10	(33)
Fatty acid		100	1500	(11)
Succinic acid dehydrogenase	3	100	111	(33)
Catalase	2	100	119	(34)

[a] Levels of various mitochondrial components normalized in arbitrary units with wild type set at 100.

At least one function of mitochondrial DNA can be implied—namely, to code for the primary structure of a mitochondrial structural protein. The function and/or regulation of numerous proteins whose primary structure is specified by nuclear genes, but which complex with structural protein, could be indirectly controlled by mitochondrial DNA. Nucleocytoplasmic interactions of this type may prove to be a significant clue to the problem of regulation of protein synthesis and protein function (9) in *Neurospora* mitochondria.

A number of criteria have been established by which the mitochondrial structural protein can be identified. Some of these are pertinent to the data presented; hence, brief mention will be made of some characteristics and physical properties of mitochondrial structural protein.

The structural protein isolated from mitochondria has a molecular weight of about 23,000 (36). This was determined by sedimentation equilibrium in the Model E analytical centrifuge and by comparing amino acid composition and tryptic peptides. Moreover, the minimum combining weight of structural protein with both NADH and ATP is about 23,000. From thirty-six separate analyses, the average amino acid composition of hydrolyzed structural protein was calculated. Based on molecular weights ranging from 20,000 to 30,000, the summation of deviations from even integers for all the individual amino acids revealed that the closest approximation to even integers was at a molecular weight of 23,000. The $s_{20,w}$ value calculated from sedimentation velocity studies of mitochondrial structural protein was 2.1 to 2.2.

There is nothing particularly striking about the amino acid composition of structural protein, but its solubility properties are unique. Solubility varies considerably, depending on whether the structural protein is free or complexed. For example, it is only very slightly soluble between pH 5 and pH 6, but it increases in solubility in the presence of myoglobin with which it forms a complex. It is most soluble at pH 10.5 and higher or at pH 3. Between pH 7 and pH 9, structural protein preparations differ from each other with respect to solubility depending in part on the degree of binding with other components. Certain phosphate compounds which have a high affinity for structural protein (15) tend to reduce its solubility. At a neutral pH, solubility is increased by the addition of urea; structural protein is soluble in 8 *M* urea.

Structural protein preparations have been observed with Schlieren optics in the Model E ultracentrifuge. Some preparations sediment as a single symmetrical boundary, while others, purified by the same procedure and having the same amino acid composition, show considerable heterogeneity by this criterion. A similar type of variability has been observed on polyacrylamide gels after electrophoresis (37). Regardless of whether a preparation of structural protein shows a single band or a broad diffuse smear of protein in the acrylamide gel, the amino acid composition and peptide maps remain the same. This strongly suggests that the heterogeneity is caused by different states of aggregation or interaction with nonprotein components. Since the standard criteria of purity of structural protein are not very useful because of its unique properties, additional criteria have been routinely employed. These include peptide maps and amino acid analyses, neither one of which is capable of detecting the presence of small amounts of foreign protein.

With this background information, the question of mutation induction and expression of cytoplasmic genes naturally arises. We shall not discuss this aspect of the general problem of genetic control of mitochondria except to say that preliminary experiments in our laboratory suggest that mutation expression of the cytoplasmic gene mutations in *mi-1* and *mi-3* requires segregation of a cytoplasmic factor (presumably mitochondria).

The fact that *mi-1* and *mi-3* are morphologically distinguishable from wild type (Fig. 1) when grown on sorbose agar plates suggests an interesting question concerning the relationship between mitochondrial membranes and other cell membranes. In addition to the mitochondrial membrane, are other membranes of *mi-1* and *mi-3* also altered, resulting in differences in colony morphology? (Note that nuclear gene modifiers affect colony morphology of both *mi-1* and *mi-3*. In other words, do cellular membranes contain a common structural protein? Analyses of electron micrographs have been interpreted to indicate structural continuity between cellular membranes (6).

In investigating the relationship between the structural proteins of the various cell membranes, we have isolated structural protein from five different sources. They include (in addition to mitochondria) nuclei, microsomes, a 100,000 $\times$ g supernatant, and an acetone powder prepared from intact mycelia. These sources were separated from each other by sucrose shelves and density gradients. From each one, protein was isolated by the same procedure as that used to purify the mitochondrial structural protein. From the mitochondrial protein, 28% was recovered as structural protein, while 10% of the nuclear protein, 15% of the microsomal protein, and 40% of the nonsedimentable protein were recovered as structural protein. About 16% of the extractable protein from acetone powder was isolated as structural protein.

Structural protein isolated from each of the above fractions was tested on Öuchterlony plates (25) to determine precipitation response, using antibody prepared against mitochondrial structural protein. A positive precipitation reaction was observed with all the proteins. Furthermore, the same pattern of a diffuse precipitin band was observed in all cases.

Peptide maps from these structural protein preparations are illustrated in Fig. 2. Tryptic peptide maps of structural proteins from the same four cell fractions were made by using two-dimensional separations of the tryptic digests (13). Twenty-nine ninhydrin-positive spots were identified as common to peptide maps from digests of structural protein from mitochondria, nuclei, microsomes, and the microsomal supernatant. About 26 tryptic peptides are expected, based on the amino acid composition and molecular weight of mitochondrial structural protein. Some of the ninhydrin-positive spots are probably due to free amino acids and/or secondary peptides due to incomplete trypsin digestion. In addition to the 29 ninhydrin-positive spots common to all four preparations, 7 ninhydrin-positive spots were observed in only three of the four maps (nuclear structural protein excepted). A few other (mostly weakly staining) ninhydrin-positive spots were observed in one or two of the

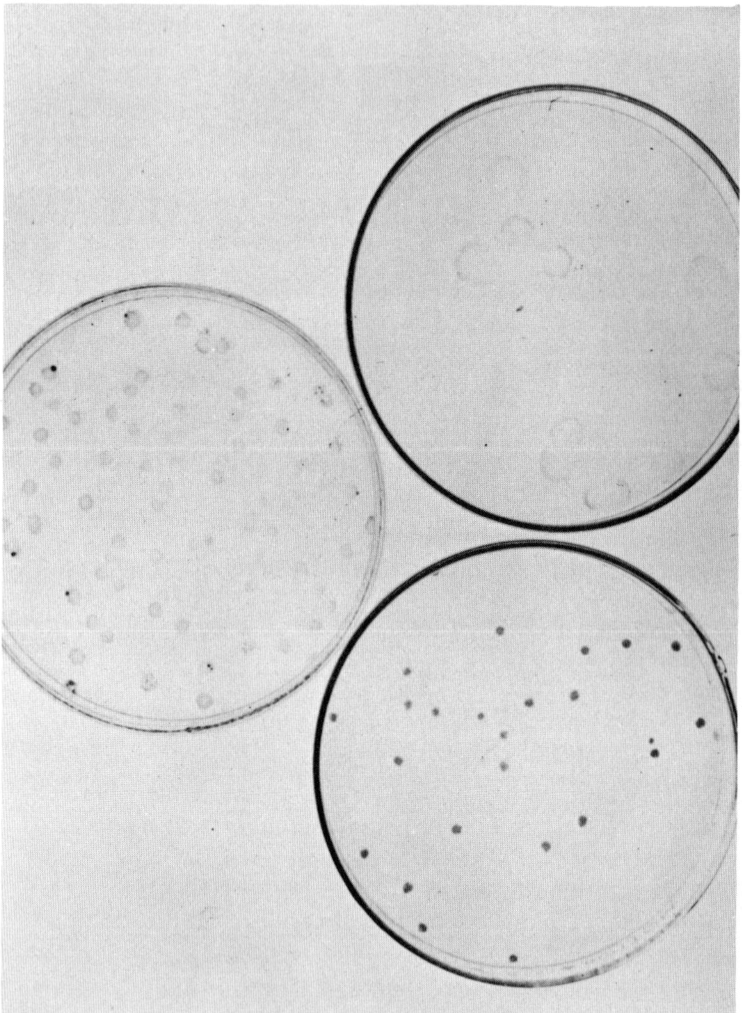

FIG. 1. Comparison of colony morphology of ST-A4 wild type, *mi-1*, and *mi-3*. Colonies were overplated with Magdala Red dye. These colonies were grown on Vogel's minimal medium with sorbose. The strains are: (top) ST-A4 wild type; (bottom left) *mi-3*; (bottom right) *mi-1* (Poky). The colony morphology of both *mi-1* and *mi-3* shown above is characteristic of a specific genetic background.

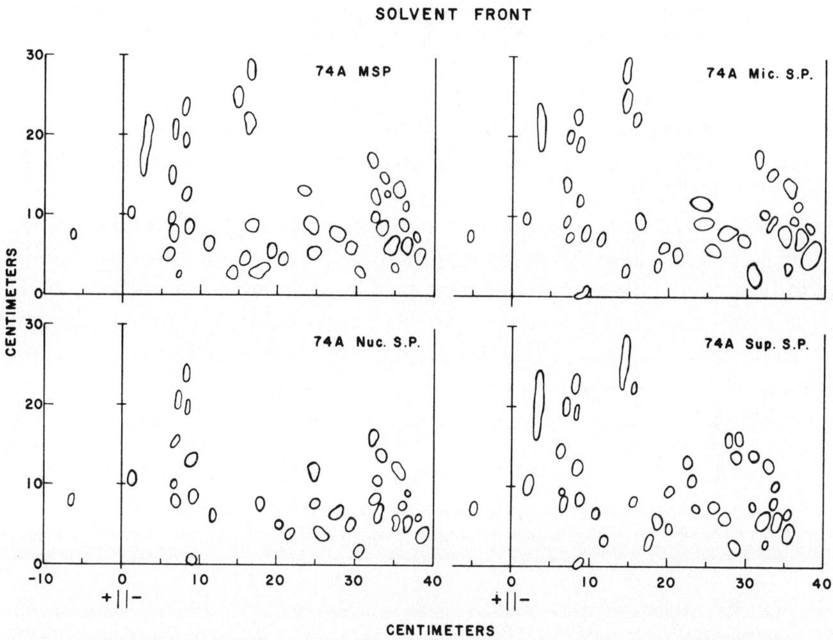

FIG. 2. Peptide maps of structural protein trypsin digests. These peptide maps were made after 12-hour trypsin digestion of N-(4-dimethylamino-3,5-dinitrophenyl)-maleimide (DDPM)-treated structural protein from the following sources: (upper left) mitochondrial fraction; (upper right) microsomal fraction; (lower left) nuclear fraction; (lower right) microsomal supernatant fraction (100,000 $\times$ *g* supernatant).

preparations. The absence of complete correlation of all ninhydrin-positive spots can be accounted for in at least three ways: The ninhydrin spots that differ may be due to different amounts of contaminating proteins and/or secondary peptides due to incomplete digestion of structural protein; the structural protein may be composed of two polypeptide chains (similar to insulin), only one of which is common to all the structural protein fractions; or microheterogeneity may be present in the structure of the structural protein in a manner analogous to the heterogeneity found in globulins of higher organisms.

We are unable to distinguish between these possibilities at present, but our preliminary data (37) suggest that structural protein may be composed of two polypeptide chains. This is based on two lines of evidence: (1) Sedimentation equilibrium analyses reveal the presence of a component within a preparation of structural protein which is smaller than the 23,000 molecular weight component; and (2) a fraction has been isolated from a structural protein preparation which yields only 9 tryptic peptides. However, these smaller polypeptide components could equally well be explained by cleavage of a peptide bond resulting from treatment at certain stages of purification with very basic solutions (36).

Many structural protein preparations were analyzed by polyacrylamide disc electrophoresis (32). The only information to be obtained from numerous acrylamide gels is that structural proteins from the different cell fractions show a similar diffuse pattern under the conditions utilized. Some mitochondrial structural protein preparations show a diffuse pattern, and other preparations show a single discrete band in addition to a diffuse smear of protein. Controls included purified *Neurospora* fumarase (14), which showed a single band, and crude ribosomal protein, which shows many bands that are well resolved. The variation among structural protein fractions from the sources mentioned above was not any greater than the variation among different preparations of mito-

TABLE III
AMINO ACID COMPOSITION OF STRUCTURAL PROTEIN OF SUBCELLULAR FRACTIONS OF *Neurospora*

	Residues amino acid per mole protein from source indicated[a]					
Amino acid	M	N	P	S	No. 1	No. 2
lys	14	12	14	14	15	13
his	4	4	4	4	4	5
arg	12	11	12	11	12	12
asp	19	18	19	19	20	19
thr	12	12	11	11	11	13
ser	12	13	12	12	11	13
glu	22	20	23	23	23	22
pro	10	10	10	10	10	10
gly	18	19	17	17	17	19
ala	20	18	19	20	20	20
val	18	17	16	16	16	15
met	3	4	3	3	3	3
ile	12	11	11	11	11	11
leu	18	19	18	18	18	17
tyr	6	6	6	6	6	7
phe	8	8	8	8	8	9
try	3	3	3	3	3	3
½ cys	4	4	4	4	4	4
Number of analyses	36	4	3	2	2	2

[a] M, mitochondria; N, nuclei; P, microsome; S, soluble; No. 1, buffer extract of mycelial acetone powder; No. 2, extract of residue from No. 1 with mixture of sodium dodecyl sulfate, cholate, and deoxycholate. Number of residues is rounded to nearest even integer. Methodology is the same as described in an earlier publication (36).

TABLE IV

TRYPTOPHAN AND CYSTEINE COMPOSITION OF STRUCTURAL PROTEIN[a] FROM SUBCELLULAR FRACTIONS ISOLATED FROM WILD TYPE AND *mi-1*

Source of structural protein	Tryptophan[b]	Cysteine[c]
Mitochondria		
1. Wild type	3.0	4.3
2. *mi-1*	1.7	5.0
Microsomes		
1. Wild type	3.3	3.6
2. *mi-1*	2.0	5.1
Nuclei		
1. Wild type	3.3	3.7
2. *mi-1*	2.0	5.1
100,000 × *g* supernatant		
1. Wild type	3.5	3.9
2. *mi-1*	1.9	4.8

[a] Based on 23,000 molecular weight.

[b] Spectrophotometric determinations (8).

[c] HMB (2) and N-(4-dimethylamino-3,5-dinitrophenyl)-maleimide (DDPM) (7) titrations.

chondrial structural protein. The acrylamide gel comparisons, however, were not useful criteria of homogeneity, presumably owing to many aggregation states of structural protein.

A comparison of the amino acid compositions of these structural protein fractions is given in Table III. Structural protein from these fractions did not show any significant difference in composition from that of the mitochondrial structural protein.

Perhaps more significant than the above data, in reference to the relationship between structural proteins isolated from different membranes, is the comparison of different structural proteins from wild type and *mi-1*. Since the mitochondrial structural proteins of *mi-1* and wild type differ with respect to cysteine and tryptophan content, analyses of these amino acids were carried out on structural proteins isolated from different membrane fractions of wild type and *mi-1* (Table IV). Because these structural protein fractions all carry the same amino acid replacement (tryp → cys), these data strongly suggest that a component of the mitochondrial structural protein is common to other cellular structures (presumably membranes).

The above data raise an additional question; that is, if all the membranes within a cell contain a structural protein which serves a similar complexing function as that observed in the case of the mitochondrial structural protein, how does a given protein distinguish between one membrane and another? An example of this problem is illustrated by the two membranes of the mitochondrion, both of which appear to contain the same structural protein (18, 19). Completely different groups of proteins have been found associated with the inner (respiratory enzymes) and outer (Krebs cycle enzymes) membranes of the mitochondrion (1, 18, 19, 31). Cytochromes from the inner membrane and malate dehydrogenase from the outer membrane all complex with mitochondrial structural protein (4, 23).

The answer to the question of locational specificity is not clear, but a possible explanation is suggested on the basis of observations of structural protein isolated from the soluble protein fraction. The view expressed here in answer to the above question is based on the assumption that most, if not all, of the so-called "soluble proteins" within a cell are in fact localized *in vivo*. The supernatant from the microsomal pellet ($100{,}000 \times g$ supernatant) was treated in the same way as all the membrane fractions in order to isolate a structural protein. Surprisingly, about 40% of the total protein in this fraction was isolated and characterized as structural protein. Amino acid composition (Table III), peptide map (Fig. 2), and electrophoretic properties were indistinguishable between this structural protein and mitochondrial structural protein. Moreover, tryptophan and cysteine analyses on structural protein obtained from the soluble fraction of both wild type and *mi-1* revealed the same replacement of tryptophan by cysteine as in membrane structural proteins (Table IV).

The possibility that the structural protein found in the soluble fraction is a degradation product from mechanical disruption during extraction has been considered. However, the amount of structural protein obtainable from mitochondria does not differ under conditions in which malate dehydrogenase is released from the mitochondria. When mitochondria are isolated in sucrose solutions ranging from 0.1 *M* sucrose to 1.3 *M* sucrose, the amount of MDH that remains associated with the mitochondria ranges from about 10% at the low to about 95% at the high sucrose concentration. Under these same conditions there is no detectable difference in the amount of structural protein that remains associated with the mitochondria.

Assuming that most of the so-called "soluble enzymes" within a cell are, in fact, localized *in vivo*, the large amount of structural protein present in a soluble form may provide an answer to the question of locational specificity. Each newly formed enzyme might bind a structural protein monomer after being released from the ribosome. The specificity necessary to determine into which membrane the complex would be incorporated would depend on the particular enzyme involved in the complex. A complex involving structural protein and cytochrome c would be incorporated into the inner membrane of the mitochondrion, while a structural protein–invertase complex would only be incorporated into the plasma membrane. Such complexes might also provide the specificity determining the quantity and quality of phospholipid bound to the complex either before or after being incorporated into the appropriate membrane. This type of incorporation of protein complexes into growing membranes is compatible with Luck's (20) data on mitochondrial biogenesis. Alternatively, one might regard the structural protein found in the soluble fraction as representing structural protein from free ribosomes which would be found in this supernatant fraction and/or nascent structural protein still attached to ribosomes. This explanation, however, does not account for the vast amount of structural protein found in the soluble fraction.

Speculation about the selective advantage of a gene which codes for a membrane structural protein being located in the cytoplasm is perhaps premature. However, one suggestion could be made on the basis of some of the observations described in this paper. In considering the large amount of structural protein synthesized compared with other proteins, a different system of regulation of mRNA and protein synthesis might be expected on the basis of what is known about regulation of nuclear gene activity. Most of the cellular protein is required in catalytic amounts, whereas regulation of a protein required in such quantities as structural protein would perhaps be more easily accomplished outside the nuclear environment under a completely different set of conditions. Perhaps more significant is the relationship between gene dosage and the amount of protein synthesized. Assuming 50 to 100 mitochondria per nucleus, each having a copy of a gene coding for structural protein, one might predict the synthesis of 50 to 100 times as much structural protein relative to a protein encoded by one copy of a nuclear gene (assuming equivalent rates of transcription and translation). In *Neurospora* about 16% of the total extractable protein was recovered as structural protein. A constitutive enzyme such as MDH (23) with heterologous subunits encoded by two nuclear genes comprises about 0.4% of the total extractable protein. The ratio of the amounts of these two proteins is directly proportional to gene dosage. Moreover, since the number of mitochondria may vary to a much greater degree than the number of nuclei, the rate of synthesis of structural protein may be varied to a larger degree than if it were encoded by a nuclear gene.

In summary, the role of structural protein is probably different than its name implies. Perhaps it should have been called an organizer protein, which would better agree with what we postulate as a major function of this so-called structural protein. A major structural role appears even less likely, based on the observation that membrane-like material forms in the absence of structural protein (18). Care must be taken, however, in attempting to demonstrate the absence of structural protein; it is equally difficult to equate membranes formed *in vitro* with naturally occurring membranes. Membrane-like material has even been observed in phospholipid preparations completely free of protein (28).

One interpretation of the data presented here which emphasizes the organizer role of structural protein can be visualized in the following manner. A gene which controls the primary structure of structural protein is located in mitochondrial DNA. Since there are many more copies of this gene than there are of nuclear genes, many copies of structural protein mRNA are produced; hence, much more structural protein is formed than is formed in the case of proteins encoded by nuclear genes.

The structural protein monomer complexes with phospholipid or some other component which renders it soluble. Soluble forms of structural protein have been observed under a variety of conditions (37). Being produced in large quantities, the structural protein saturates the cytoplasm but cannot be incorporated into membrane until it complexes with an enzyme. The formation of the structural protein–enzyme complex confers specificity to the complex. The additional specificity determines the extent to which phospholipid or other nonprotein components can bind the complex. But more crucial to the hypothesis is that the specificity of the structural protein–enzyme complex permits it to be incorporated into a growing membrane, but only into membranes composed of similar subunits. This hypothesis implies nothing concerning the function of the structural protein in maintaining structural integrity of membranes. Yet, it is difficult to conceive of such an abundant protein's not being involved in some way to help maintain structural and functional integrity of the membrane.

Experiments in our laboratory are under way to determine the extent to which extramitochondrial enzymes complex with mitochondrial structural protein, and also to test the affinity of mitochondrial enzymes for structural proteins from other cell membranes. If such a multifunctional role can be ascribed to structural protein as is postulated here, it is not surprising that little variation is found in structural proteins of distantly related organisms (36).

Acknowledgments

The valuable technical assistance of Geraldine Holland, Barbara Andrews, and Phyllis Monroe are gratefully acknowledged. This work was supported by National Science Foundation (GB-2373) and U.S. Public Health Service (GM-10067) Grants.

REFERENCES

1. ALLMANN, D. W., BACHMANN, E., AND GREEN, D. E., *Arch. Biochem. Biophys.*, **115**, 153, 165, 172 (1966).
2. BOYER, P. D., *J. Am. Chem. Soc.*, **76**, 4331 (1954).
3. CHATTERJEE, S. K., DAS, H. K., AND ROY, S. C., *Biochim. Biophys. Acta*, **114**, 349 (1966).
4. CRIDDLE, R. S., BOCH, R. M., GREEN, D. E., AND TISDALE, H. D., *Biochemistry*, **1**, 827 (1962).
5. DAS, H. K., CHATTERJEE, S. K., AND ROY, S. C., *J. Biol. Chem.*, **239**, 1126 (1964).
6. FREY-WYSSLING, A., AND MUHLETHALER, K., "Ultrastructural Plant Cytology," Elsevier, Amsterdam, 1965.
7. GOLD, A. H., AND SEGAL, H. L., *Biochemistry*, **4**, 1506 (1965).
8. GOODWIN, T. W., AND MORTON, R. A., *Biochem. J.*, **40**, 628 (1946).
9. GREEN, D. E., AND WHARTON, D. C., *Biochem. Z.*, **338**, 335 (1963).
10. HALDAR, D., FREEMAN, K., AND WORK, T. S., *Nature*, **211**, 9 (1966).
11. HARDESTY, B. A., AND MITCHELL, H. K., *Arch. Biochem. Biophys.*, **100**, 330 (1963).
12. HASKINS, F. A., TISSIÈRES, A., MITCHELL, H. K., AND MITCHELL, M. B., *J. Biol. Chem.*, **205**, 423 (1953).
13. HELINSKI, D. R., AND YANOFSKY, C., *Biochim. Biophys. Acta*, **63**, 10 (1962).
14. HILL, J. H., unpublished data.
15. HULTIN, H. O., AND RICHARDSON, S. H., *Arch. Biochem. Biophys.*, **105**, 288 (1964).
16. JURTSHUK, P., JR., SEKUZU, J., AND GREEN, D. E., *J. Biol. Chem.*, **238**, 3595 (1963).
17. KALF, G. F., *Biochemistry*, **3**, 1702 (1964).
18. KOPACZYK, K., PERDUE, J., AND GREEN, D. E., *Arch. Biochem. Biophys.*, **115**, 215 (1966).
19. LAUWERS, A., AND ALLMAN, D., unpublished observations.
20. LUCK, D. J., *Am. Naturalist*, **99**, 241 (1965).
21. LUCK, D. J., AND REICH, E., *Proc. Natl. Acad. Sci. U.S.*, **38**, 442 (1952).
22. MITCHELL, M. B., AND MITCHELL, H. K., *Proc. Natl. Acad. Sci. U.S.*, **38**, 442 (1952).
23. MUNKRES, K. D., AND WOODWARD, D. O., *Proc. Natl. Acad. Sci. U.S.*, **55**, 1217 (1966).
24. NASS, S., NASS, M. M. K., AND HENNIX, U., *Biochim. Biophys. Acta*, **95**, 425 (1965).
25. ÖUCHTERLONY, O., *Acta Pathol. Microbiol. Scand.*, **26**, 507 (1949).
26. PARSONS, J. A., *J. Cell Biol.*, **25**, 641 (1965).
27. RENDI, R., AND WARNER, R. C., *Ann. N.Y. Acad. Sci.*, **88**, 741 (1960).
28. REVEL, J. P., ITO, S., AND FAWCETT, D. W., *J. Biophys. Biochem. Cytol.*, **4**, 495 (1958).
29. RICHARDSON, C. C., SCHILDKRAUT, C. L., AND KORNBERG, A., *Cold Spring Harbor Symp. Quant. Biol.*, **28**, 19 (1963).
30. ROODYN, D. B., SUTLIE, J. W., AND WORK, T. W., *Biochem. J.*, **85**, 177 (1962).
31. SOTTOCASA, G., KUYLENSTIERNA, B., AND ERNSTER, L., personal communication.
32. TAKAYAMA, K., MACLENNAN, D. H., TZAGOLOFF, A., AND STONER, C. D., *Arch. Biochem. Biophys.*, **114**, 223 (1966).

33. Tissières, A., and Mitchell, H. K., *J. Biol. Chem.*, **208**, 241 (1954).
34. Tissières, A., Mitchell, H. K., and Haskins, F. A., *J. Biol. Chem.*, **205**, 423 (1953).
35. Wintersberger, E., *Biochem. Z.*, **341**, 409 (1965).
36. Woodward, D. O., and Munkres, K. D., *Proc. Natl. Acad. Sci. U.S.*, **55**, 872 (1966).
37. Woodward, D. O., and Munkres, K. D., unpublished data.

Circularity and Other Properties of Mitochondrial DNA of Animal Cells

MARGIT M. K. NASS

Department of Therapeutic Research,
University of Pennsylvania School of Medicine,
Philadelphia, Pennsylvania

INTRODUCTION

The analysis and interpretation of problems pertaining to mitochondrial DNA is aided by ideas that originated early in mitochondrial history and by modern developments. Some of these concepts will be outlined briefly.

The problem of the mechanism of mitochondrial duplication has been the subject of investigation and argumentation since the turn of the century when mitochondria were first described by Altmann (1) as particles resembling bacteria in size and shape. Hypotheses that mitochondria carry genetic information, divide, or even represent symbiotic bacteria (cf. 6, 43, 51, 58, 74, 81) flourished and were then abandoned pending the development of suitable experimental techniques. Although the discovery that oxidative respiration is associated with mitochondria led to the concentration of most research activities on problems of electron transport and oxidative phosphorylation, some early studies had shown the non-Mendelian transmission of certain mutations related to mitochondrial cytochrome components (16, 37). Diverse mechanisms of formation of new mitochondria have been postulated, including the synthesis *de novo,* formation from other cell structures (nuclear envelope, plasma membrane, endoplasmic reticulum, Golgi vesicles, microbodies), and formation by division of pre-existing mitochondria (cf. 51, 58, 81). Strong evidence for the latter type of mechanism was first presented by Luck (33) in *Neurospora,* based on quantitative autoradiographic analyses.

Following the recognition of DNA as hereditary material, the development of new techniques has led within the past few years to an exponentially growing body of literature dealing with the detection, isolation, and properties of mitochondrial DNA (5, 11, 13, 15, 24, 27, 29, 34, 38–46, 48, 52, 55, 56, 60, 63–65, 68–70, 72, 77), RNA or ribosomes (36, 45,

46, 53, 54, 71, 76), DNA polymerase (62, 84), RNA polymerase (27, 34, 47, 49, 50, 85), mitochondrial protein synthesis (2, 3, 21, 27, 28, 31, 57, 75, 83), continuity (33, 52, 56, 68), and genetic function (38, 39, 56). Not all available references can be cited here. Analogous developments occurred several years earlier in the field of chloroplast biology.

Essentially three general areas of study must be integrated in order to gain some understanding of the process and molecular basis of mitochondrial synthesis: (1) Analysis of mitochondrial multiplication in relation to cell division (33, 52, 68) and turnover of mitochondrial components (19, 82); (2) detection and characterization of mitochondrial nucleic acids, their mechanism of replication, and their capacity to code for specific mitochondrial proteins; and (3) the demonstration that mitochondrial DNA does indeed carry genetic information essential for the physical continuity of mitochondria. Some evidence for the genetic function of mitochondrial DNA was recently obtained for mutants of *Neurospora* (56) and yeast (38, 39). In cells of most higher organisms, however, where altered mitochondrial characteristics are usually lethal, the function and significance of mitochondrial DNA must be studied by more indirect means—for example, the demonstration in isolated mitochondria of DNA replication and DNA-dependent RNA and protein synthesis. Such studies suggest the possibility that at least some types of mitochondria can be cultivated *in vitro*. Some attempts of cultivation have been reported (79). Since by present knowledge mitochondria more than superficially resemble prokaryotic microorganisms in various aspects of structure and function (cf. 35, 53, 66), comparisons of these systems will undoubtedly lead to increased insight into the mechanism of mitochondrial replication. Such analogies are further strengthened by the recent finding of circularity in mitochondrial DNA (40, 65, 77), a property which it shares with the DNA of some viruses and bacteria but apparently not with that of nuclei of eukaryotes.

Some Properties of Isolated Mitochondrial and Nuclear DNA

DNA has been localized by electron cytochemical techniques in mitochondria of all major animal phyla (43), and it has been isolated in purified form from mitochondria of several types of plants (69), yeast (11, 38, 39, 60, 72), *Neurospora* (34, 56), amphibians (13), birds (5, 55, 77), and various mammals (11, 24, 29, 40, 63, 65, 77). In most cases the buoyant density of mitochondrial DNA differs from that of nuclear DNA, which indicates a difference in guanine plus cytosine content (61). These differences tend to be greater in lower than in higher organisms and are small (63) or undetectable (11) in a few mammalian cell types. A typical sedimentation profile of the mitochondrial DNA of mouse fibroblast cells

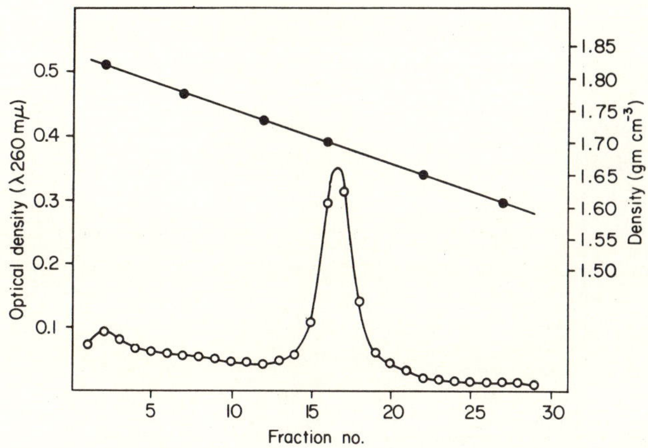

FIG. 1. Cesium chloride gradient of phenol-extracted mitochondrial DNA of L cells after 66 hours of centrifugation at 35,000 rpm (rotor SW 50L) in the Spinco L2 ultracentrifuge. Mitochondrial DNA appears homogeneous and bands at a density of 1.698 g cm^{-3}.

(L cells) during preparative centrifugation in cesium chloride is shown in Fig. 1. The DNA was extracted from highly purified mitochondria with phenol and then treated with pancreatic and T_1 ribonucleases (40). Mitochondrial DNA bands at a density of 1.698 g cm^{-3} as compared with 1.703 g cm^{-3} for nuclear DNA. The calculated guanine plus cytosine contents (61) of 39% and 44%, respectively, is in good agreement with the results obtained by direct base analysis by quantitative thin-layer chromatography, as shown in Table I. Thermal denaturation curves for DNA also indicated a difference of about 3% GC for these two types of DNA (41).

When purified DNA of L cells was prepared for electron microscopy (cf. 40), mitochondrial DNA showed the striking feature of circularity (Fig. 2). Most molecules consisted of lightly twisted forms with several

TABLE I
BASE COMPOSITION OF MITOCHONDRIAL AND NUCLEAR DNA OF L CELLS

DNA	% Molar Proportions of: Adenine	Guanine	Cytosine	Thymine	Guanine plus cytosine
Mitochondrial DNA[a]	28.9 ± 0.3	20.8 ± 0.3	20.1 ± 0.4	30.2 ± 0.5	40.9
Nuclear DNA[a]	28.0 ± 0.9	22.2 ± 1.1	21.7 ± 0.9	28.1 ± 0.9	43.9

[a] Mean of six determinations by quantitative thin-layer chromatography.

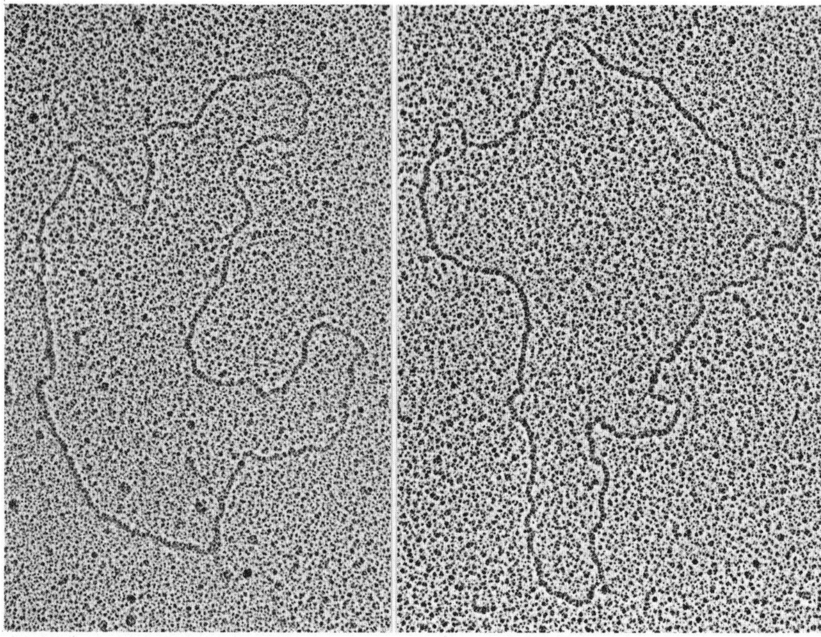

FIG. 2. Electron micrographs demonstrating circularity of mitochondrial DNA of L cells. DNA was purified in a cesium chloride gradient, as shown in Fig. 1, spread on a monolayer of cytochrome c, and rotary-shadowed with platinum-iridium. The population of molecules consisted of twisted and untwisted forms. $\times 48{,}000$.

points of crossing; a smaller proportion of cyclic molecules was either completely untangled as illustrated in Fig. 2 or highly twisted with occasional rosette patterns (Fig. 3). A few linear fragments either of the same length or shorter than the circles were also encountered and are assumed to represent broken pieces of mitochondrial DNA and/or fragments of nuclear DNA that were not completely removed by deoxyribonuclease routinely used in the purification procedure of mitochondria.

Nuclear DNA isolated and prepared by the same methods, however, contained only linear molecules which were considerably longer than mitochondrial DNA (Fig. 4).

Mitochondrial DNA had a uniform length of 4.74 ± 0.02 microns, as summarized in Fig. 5. This length corresponds to a molecular weight of 9.1×10^6 (40). Mitochondrial DNA from some other cell types examined thus far are circular and of similar length—for example, mitochondrial DNA of mouse Sarcoma I tumor cells with 4.76 microns (Fig. 6), beef heart and chicken liver with 5.45 microns (77), and mouse liver with

4.96 ± 0.10 microns (65). Although the narrow distribution of size of these circles in each cell type suggests a uniform species of molecules, the possibility remains open that molecules within a given population are not all identical in their nucleotide sequences.

Clear evidence has been lacking thus far that circular DNA exists in other parts of animal cells (cf. 23, 40). The bacterial chromosome of *Escherichia coli* appears circular (4, 8), and the DNA of a number of viruses may assume a circular structure during some part of the viral replication cycle—for example, the DNA's of ΦX 174 (18, 20), polyoma (14, 78), and papilloma (12, 30) viruses. The DNA of bacteriophages T_2 (73) and lambda (22, 26, 32) exist in linear form, but can be induced artificially to assume ring forms.

The ease of renaturation after denaturation by heat or alkali has been described as a characteristic feature of circular polyoma virus DNA (14, 80). The two components which are obtained by sedimentation analyses of this DNA were shown to contain twisted circular structures in the faster-moving fraction and open circular forms in the slow-moving band (78). The former structure could be converted to the latter by introducing single-chain scissions.

An analogous situation appears to exist for mitochondrial DNA,

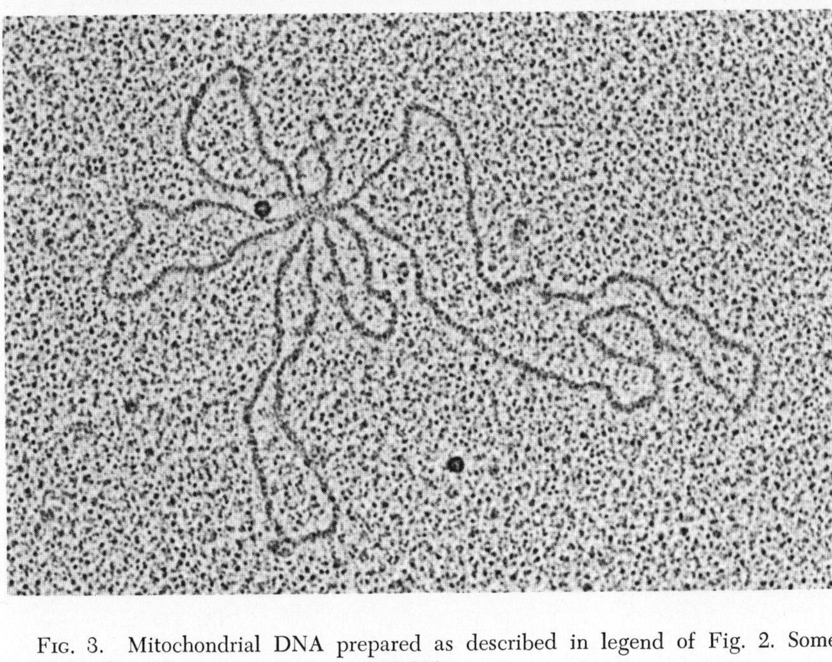

FIG. 3. Mitochondrial DNA prepared as described in legend of Fig. 2. Some molecules formed rosette patterns. ×100,000.

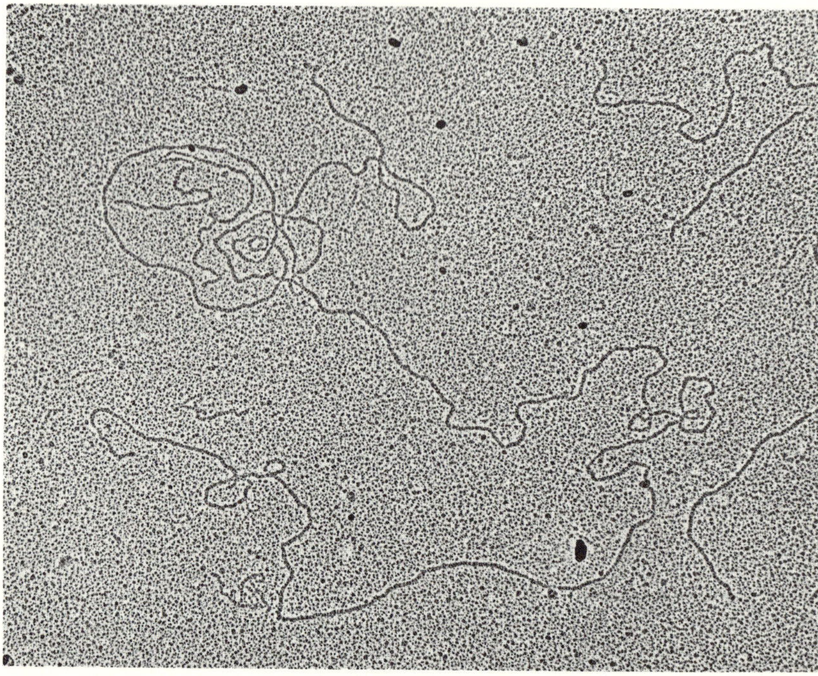

FIG. 4. Electron micrograph of nuclear DNA of L cells prepared by the same methods used for mitochondrial DNA (see legend of Fig. 2). Nuclear DNA consisted of linear molecules only. ×30,000.

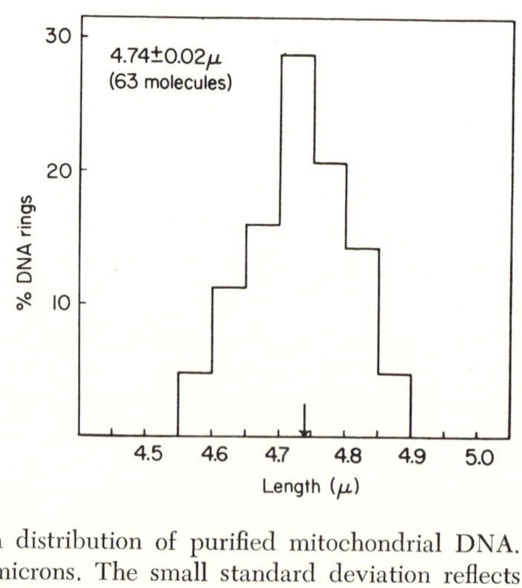

FIG. 5. Length distribution of purified mitochondrial DNA. The mean molecular length is 4.74 microns. The small standard deviation reflects the close similarity in size of these molecules. From Nass (40).

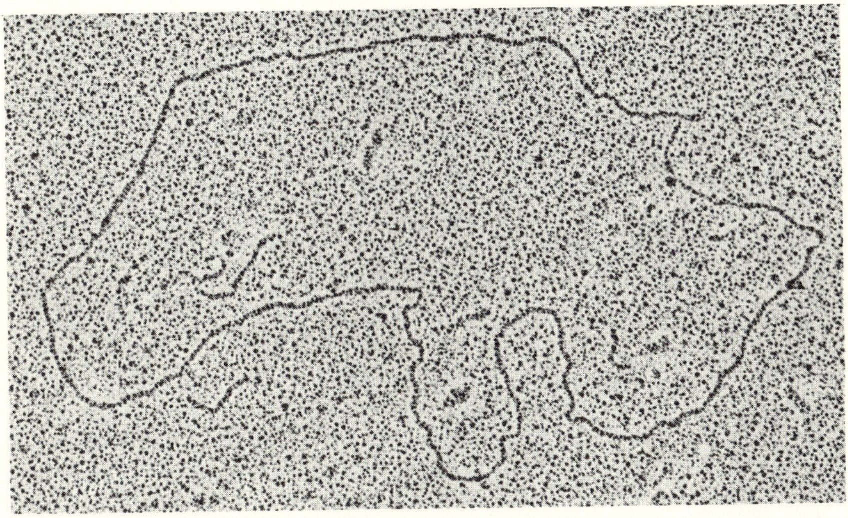

FIG. 6. Electron micrograph of the DNA isolated from mitochondria of mouse Sarcoma I tumor cells, as described in the legend to Fig. 2. These molecules are also circular and have the same size as mitochondrial DNA of L cells. ×65,000.

although this type of study is still in its infancy. Readiness to renature after denaturation by heat or alkali was noted for a number of mitochondrial DNA's from mammalian species (5, 11, 41) and yeast (11), in contrast to nuclear DNA, which remains denatured. Two components sedimenting with an S_{20} of 39 to 42 and an S_{20} of 27 to 29 were reported for the mitochondrial DNA of chicken liver (5). It remains to be determined whether all mitochondrial DNA's have typical renaturation properties and contain circular forms and whether the differences in coiling are artifactual characteristics introduced during the preparative procedure. Extensive treatment of mitochondria by deoxyribonuclease during their purification, freezing, and thawing, and exposure of the DNA to mild shear, to name but a few adverse conditions, may affect the structural integrity of the circular DNA. Experiments relating to the conformation of mitochondrial DNA *in vivo* are discussed in the next section.

DNA of Intact and Osmotically Disrupted Mitochondria

DNA can be identified electron microscopically in ultrathin sections of mitochondria (42–45) by utilizing the methods of Ryter and Kellenberger (59) developed for the preservation of the bacterial nucleoplasm. Most types of mitochondria display one or several electron-lucid areas in their matrices which contain DNA, as shown in Fig. 7, resembling multinucleated bacteria. It was suggestive that the DNA located in various compartments of a mitochondrion represented individual molecules. That this is indeed the case is shown in the following experiments (cf. 40).

Freshly isolated mitochondria were disrupted osmotically on a monolayer of cytochrome c, according to the methods of Kleinschmidt *et al.* (30) for the osmotic release of viral DNA. The DNA of lysed mitochondria should adsorb immediately to the protein film with little chance for gross molecular alterations. The structural appearance of the DNA would, therefore, give information on three specific problems: (1) the tertiary structure of the DNA immediately after release from the organelle; (2) the number and size of the DNA molecules associated with one ruptured mitochondrion; and (3) whether DNA exists in circular form inside the mitochondrial organelle or whether the circular conformation of isolated purified DNA was possibly an artifact of preparation; for example, the linear form of isolated lambda bacteriophage DNA, which has cohesive ends (22), was found to join intramolecularly on storage in 2 *M* ammonium acetate at 5°C (26).

The osmotic shock treatment released the DNA fibers from approximately 10 to 20% of the mitochondria. It is assumed that the remaining mitochondria did not respond to changes in osmotic pressure due to some change of their structural integrity during the extensive purification procedure. Typical results are illustrated in Figs. 8 and 9. The first indication that mitochondrial DNA may be circular *in vivo* was the

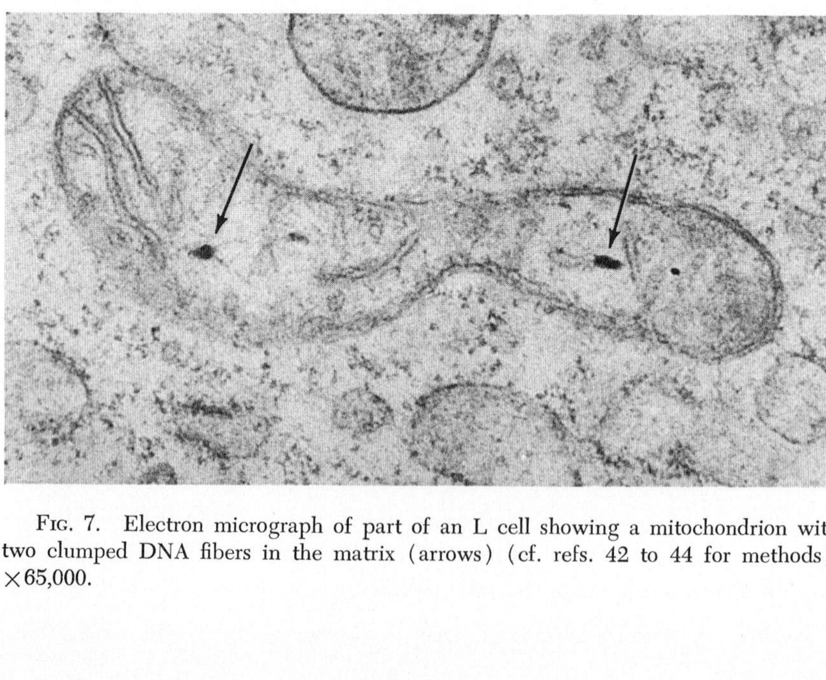

FIG. 7. Electron micrograph of part of an L cell showing a mitochondrion with two clumped DNA fibers in the matrix (arrows) (cf. refs. 42 to 44 for methods). ×65,000.

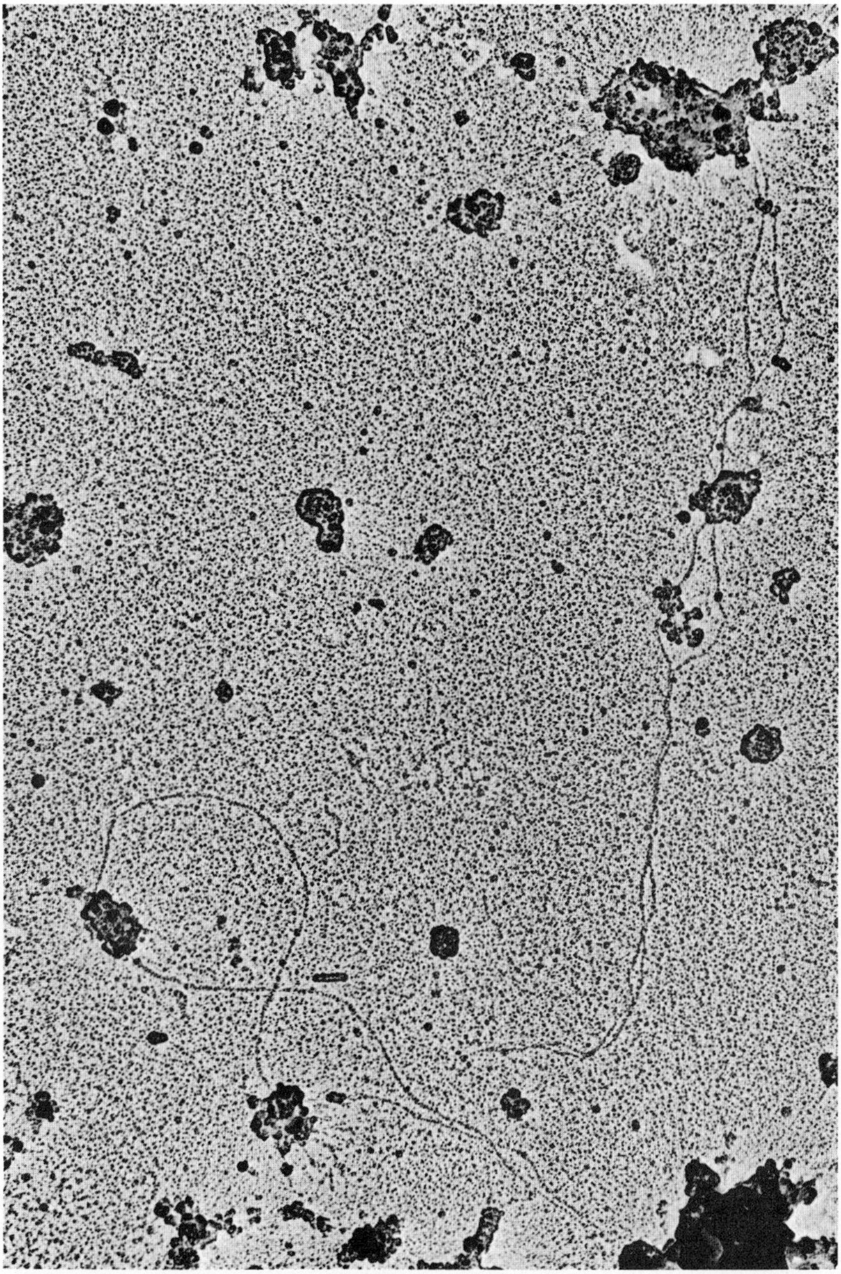

FIG. 8. Electron micrograph of DNA extruded from two osmotically disrupted mitochondria (located in bottom and upper right corner). One coiled DNA loop is seen connected with each fragment. Two additional molecules were located outside the range of this micrograph. ×60,000.

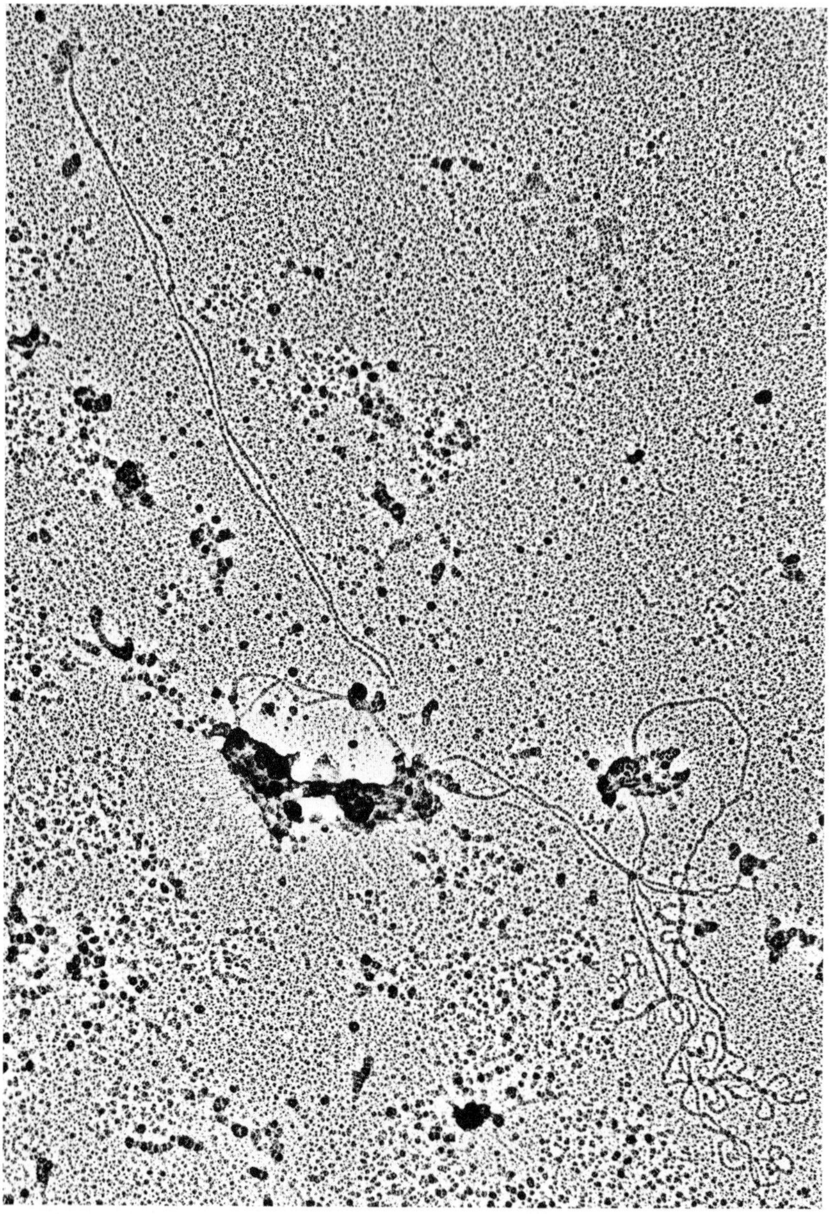

FIG. 9. Electron micrograph of four DNA molecules extruded from a single mitochondrion (located in center). There are no free ends visible. The molecule to the left is a flattened uncoiled circle, about 5 microns in circumference. The coiled molecule in the center remained attached to the mitochondrial membrane at several points and crosses the third and fourth molecules to the right, which are both tangled. ×38,000.

consistent observation that DNA fibers emerged from ruptured organelles in pairs, with no free ends visible, forming more or less intensely coiled loops of similar length (Figs. 8 and 9). These fibers were sensitive to brief treatments by deoxyribonuclease (Fig. 10) but not ribonuclease. Small fragments of membrane were frequently observed in association with the DNA fibers at various points along their length (Figs. 8 and 9), and examination of a large number of micrographs has given reasons to assume that these points of contact are not merely coincidental. Also, in ultrathin sections of mitochondria DNA fibers were frequently seen attached to mitochondrial cristae (43). Whether these regions of contact reflect the sites of a functional involvement of mitochondrial DNA in RNA and protein synthesis is an attractive interpretation, but at this time is merely speculative, based on similar concepts in the function of bacterial DNA.

The fact that the DNA loops emerging from mitochondria frequently remain attached to the main mass of the ruptured organelle raises the questions of (1) whether these "basic" regions of attachment represent

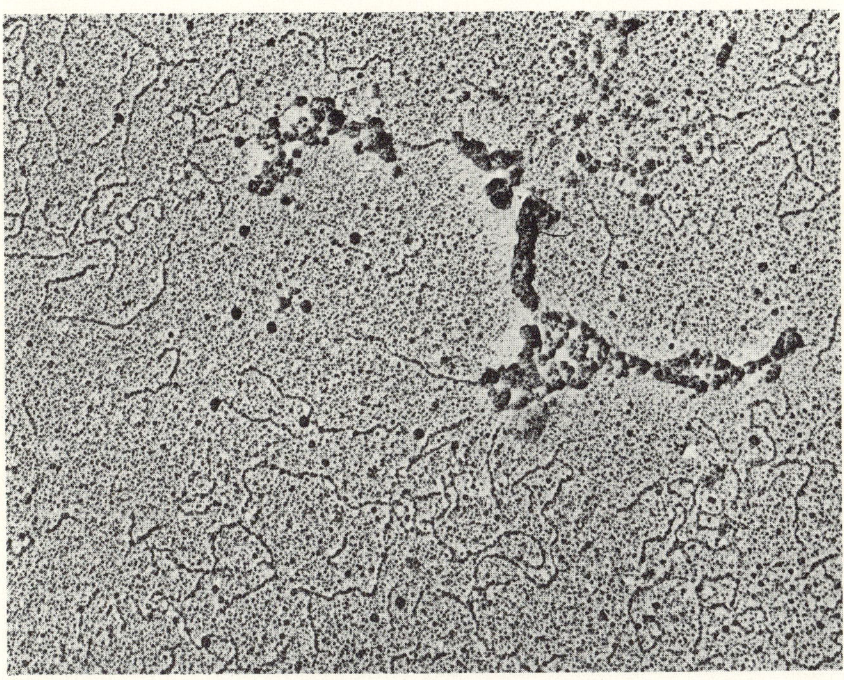

FIG. 10. Similar preparation to that shown in Figs. 8 and 9, except that DNA is partially degraded by exposing the fibers during spreading to deoxyribonuclease. ×50,000.

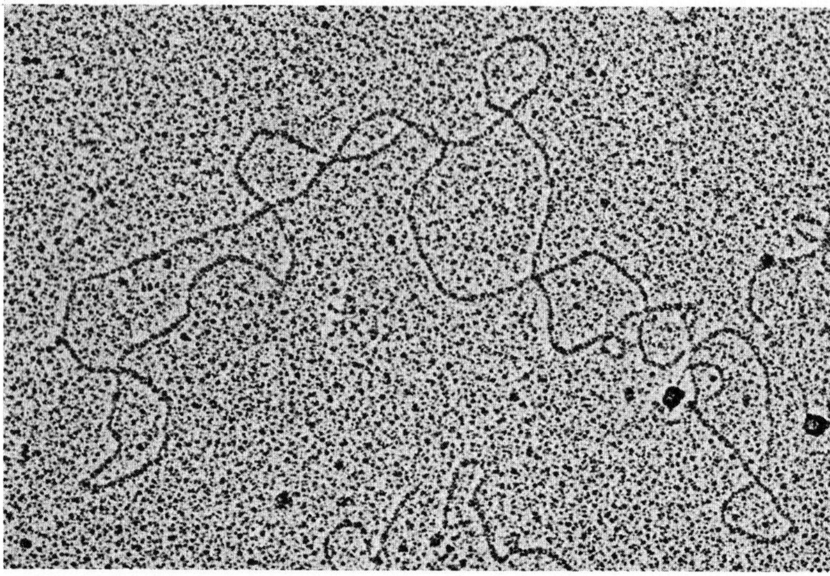

FIG. 11. Electron micrograph of a single circular DNA molecule, about 5 microns in length, extruded from an osmotically disrupted mitochondrion. ×100,000.

sites where replication is initiated and (2) whether the DNA fibers are actually continuous at these points. Obviously only the latter question can be answered at this time. Some molecules were found completely removed from fragments of mitochondria—for example, the molecule shown in the upper left of Fig. 9 and that represented in Fig. 11. Especially the latter illustrates the circularity of this DNA. Since almost negligible amounts of free ends were encountered in many different preparations, it is assumed that the predominant form of mitochondrial DNA is circular.

Nuclear DNA extruded from nuclei by the same methods as were used for mitochondria was found as long strands with some free ends visible (Fig. 12). Linearity of nuclear DNA seen in autoradiograms of lysed Chinese Hamster cells was reported by Huberman and Riggs (23). However, since very long strands of DNA are particularly prone to mechanical breakage, it is difficult to judge the conformation of nuclear DNA *in vivo*.

It appears that the DNA circles of mitochondria are packed inside the organelle in a coiled form, since most extended DNA does appear twisted, and even the few relaxed circles seen consist of both halves lying parallel as if they had just unwound (Fig. 9). On well-shadowed preparations the direction of twist of the emerging DNA fibers could be identified; it was right-handed for most molecules, although a few left-handed turns were seen.

The contour lengths of the mitochondrial DNA of many preparations were measured and are summarized in Table II. The basic length of 5.24 ± 0.24 microns and molecular weight of 10×10^6 for individual molecules are consistent with the findings for purified DNA. The

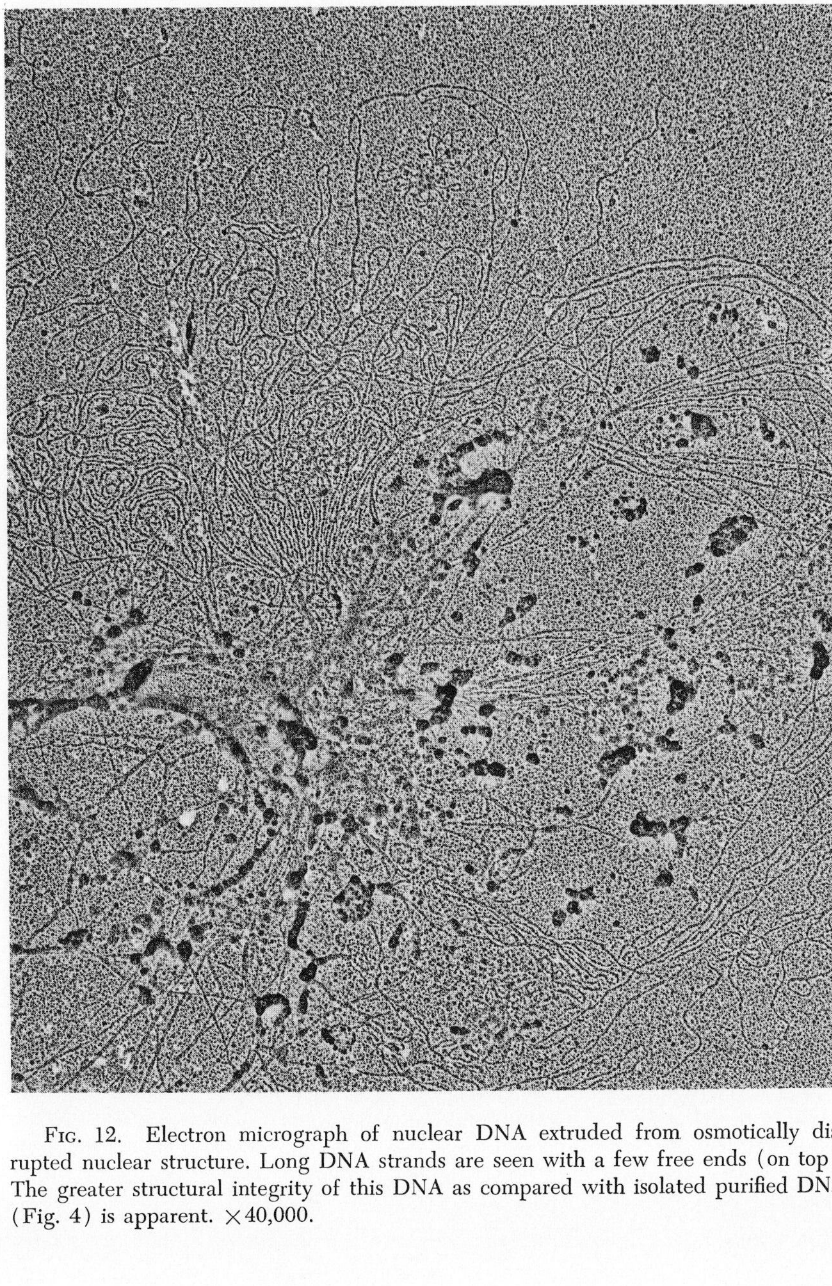

FIG. 12. Electron micrograph of nuclear DNA extruded from osmotically disrupted nuclear structure. Long DNA strands are seen with a few free ends (on top). The greater structural integrity of this DNA as compared with isolated purified DNA (Fig. 4) is apparent. $\times 40{,}000$.

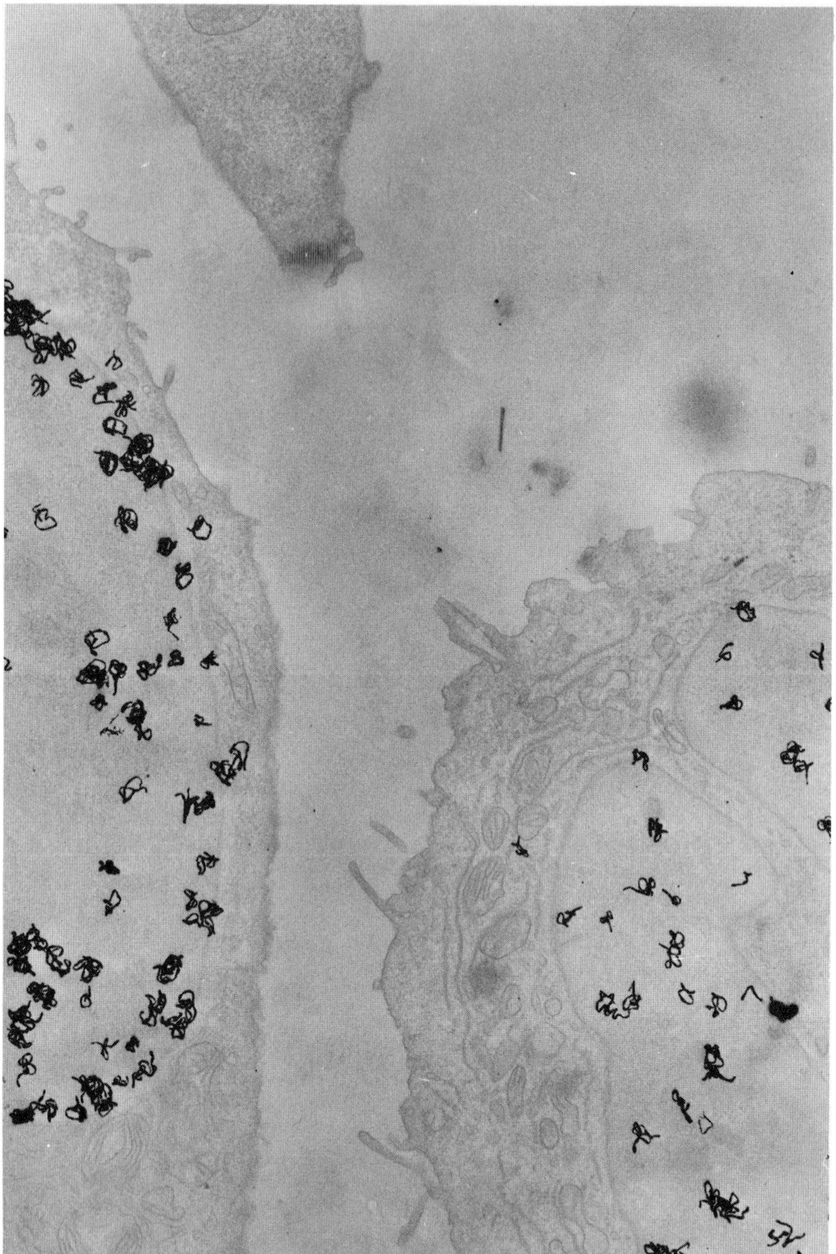

FIG. 13. High-resolution autoradiograph of L cells growing in monolayers after exposure to 10 μCi/ml of H^3-thymidine for 4 hours. The nuclei of many cells are specifically labeled; the cytoplasm is unlabeled; there are very few nonspecific background grains. (Method of L. G. Caro, *in* "Methods in Cell Physiology," Vol. 1, Academic Press, New York, 1964). ×11,000.

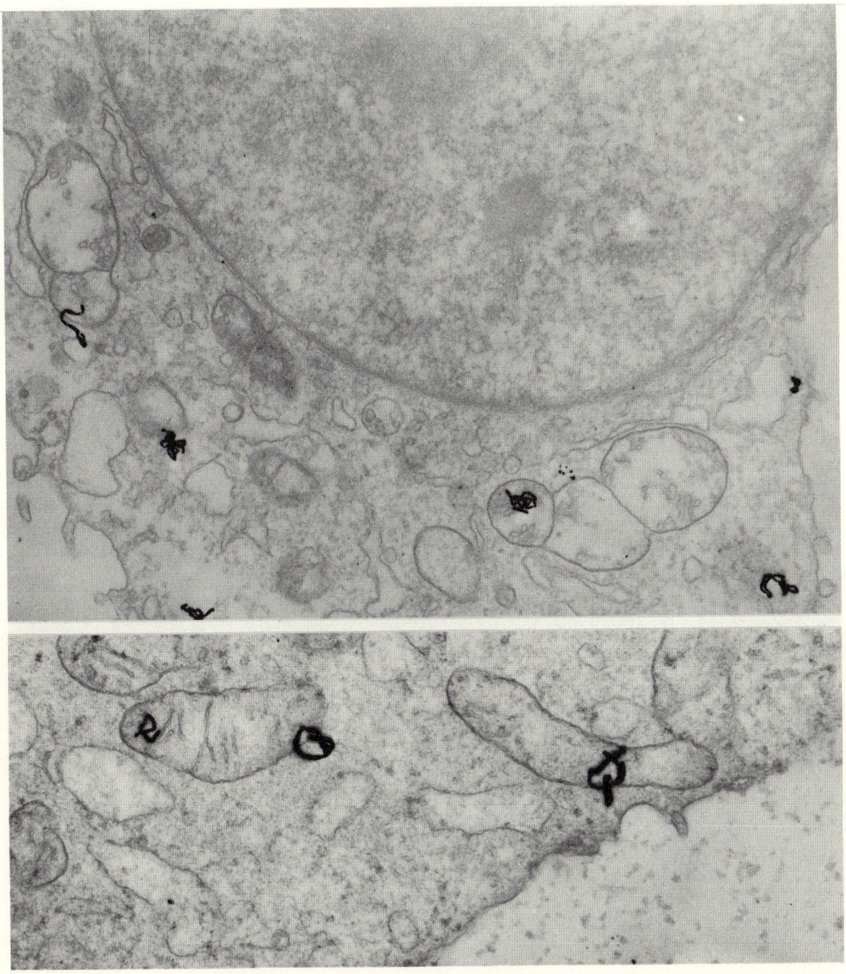

FIG. 14. High-resolution autoradiographs of the same cell culture shown in Fig. 13. A small proportion of cells had completely unlabeled nuclei but showed silver grains in the cytoplasm, on or near mitochondria. $\times$20,000.

molecular weights of the circular DNA of various viruses listed above range from 1.6 to 130×10^6 (cf. 25). The length of the circular DNA of *E. coli* is of the order of 700 to 900 microns (8).

Although some aggregates of fibers were difficult to measure, the results in Table II also indicate that multiple amounts of the basic length of 5.24 microns were frequently associated with remnants of a single mitochondrion. That these multiple amounts actually represent several

TABLE II
MEAN LENGTH OF MITOCHONDRIAL DNA PREPARED BY OSMOTIC SHOCK[a]

DNA	Number of molecules or aggregates of molecules	Length (microns)	$\frac{\text{Length}}{5.24}$
Individual twisted ring molecules	30	5.24 ± 0.24	1.0
Complex tangled structures, no free ends visible	9	10.7 ± 1.2	2.0
	3	14.2 ± 0.3	2.7
	3	21.6 ± 0.8	4.1
	1	31.5	6.0

[a] From Nass (40).

individual molecules is best illustrated in Fig. 9. Two molecules can be distinguished clearly, and two more molecules are closely entangled. Generally two to six circular molecules are present in one mitochondrion. In agreement with these findings, chemical measurements of the DNA per mitochondrion, based on actual counts of mitochondria, gave values of 53×10^6 daltons of DNA per mitochondrion or the equivalent of five to six molecules of molecular weight 9×10^6 (cf. 40). It may be calculated that one circular DNA molecule represents approximately 16,000 nucleotide pairs. Assuming a triplet code, this amount of DNA conceivably could code for a maximum of 5300 amino acids or 32 polypeptide chains of molecular weight 20,000. Other available estimates are based on total DNA per mitochondrion and are given as 150 polypeptide chains of molecular weight 20,000 for rat liver mitochondria (46) and 60 polypeptides of molecular weight 15,000 for yeast mitochondria (72). Among the many important questions that remain to be answered are: (1) Is mitochondrial DNA, if it does code for the synthesis of mitochondrial protein, sufficient in quantity to be responsible for the formation of all mitochondrial protein, or does nuclear DNA assume at least partial control? (2) Do all DNA molecules contained in one mitochondrion possess the same nucleotide sequences, and, furthermore, are the DNA molecules of all mitochondria in a given cell genetically identical?

Little is known thus far about the synthesis of mitochondrial DNA. There is suggestive evidence, based on the incorporation of tritiated thymidine into cytoplasmic particles and nuclear structure of *Tetrahymena pyriformis* (9, 52), *Allium cepa* (7), *Physarium polycephalum* (17), *Euglena* (10), and marine algae (67), that the uptake of label into the cytoplasm (mitochondria or chloroplasts) is not necessarily synchronized with nuclear DNA synthesis. A similar observation was made with L cells, as shown in Figs. 13 and 14. The majority of labeled cells showed radioactivity specifically in the nucleus and not in the cytoplasm, whereas a smaller percentage of cells was labeled in the cytoplasm only, mostly on or near mitochondria. Further studies are in progress, especially bearing on the nature and exact localization of the label in relation to mitochondrial and nuclear structure and division. The apparent periodicity in nuclear and mitochondrial DNA synthesis may provide a key to a future understanding of the controlling mechanisms regulating the interaction of cellular organelles.

Conclusion

The mitochondrion studied with modern techniques and understanding resembles prokaryotic microorganisms not only in size, shape, and mode of duplication, as intuitively maintained by scientists of the previous generation, but also in aspects of structure and function at the molecular level. The present knowledge of mitochondria provides a more substantial basis for the possibility of a prokaryotic origin of mitochondria during the evolution of the cell. Many homologies in the localization of respiratory enzymes in bacterial cells and mitochondria have been pointed out (35, 66). The similarities are further emphasized by the presence of circular DNA molecules in mitochondria, a property which these organelles share with bacteria and some viruses. The location of more than one DNA molecule in a mitochondrion resembles the structure of multinucleated bacteria, although the DNA molecules of the latter are considerably longer. The evidence for division, DNA-dependent RNA synthesis, and protein synthesis in mitochondria increases the probability of partial autonomy of these organelles. To obtain a fundamental understanding of the mechanism of mitochondrial duplication it will be useful to learn more about the replication mechanisms of bacteria and of some viruses which have circular DNA during or at one stage of their life cycle. In addition, the relation of nuclear activities to mitochondrial function and division will need extensive investigation.

Acknowledgments

This work has been supported by grant PO1-AI07005, U.S. Public Health Service. Excellent technical assistance was contributed by Miss Anneke Theunissen and Mr. John R. W. Hobbs. Mouse Sarcoma I tumor cells were kindly provided by Dr. Elizabeth Miller, University of Pennsylvania.

References

1. Altmann, R., "Die Elementarorganismen und ihre Beziehungen zu den Zellen," Veit Co., Leipzig, 1890.
2. Bachelard, H. S., *Biochem. J.*, **100**, 131 (1966).
3. Beattie, D. S., Basford, R. E., and Koritz, B., *Biochemistry*, **5**, 926 (1966).

4. Bleecken, S., Strohbach, G., and Sarfert, E., *Z. allgem. Mikrobiol.*, **6**, 121 (1966).
5. Borst, P., and Ruttenberg, G. J. C. M., *Biochim. Biophys. Acta*, **114**, 645 (1966).
6. Buchner, P., "Endosymbiose der Tiere mit pflanzlichen Mikroorganismen," Verlag Birkhauser, Basel/Stuttgart, 1953.
7. Budd, G. C., and Mills, G. M., *Nature*, **205**, 524 (1965).
8. Cairns, J., *J. Mol. Biol.*, **6**, 208 (1963).
9. Cameron, I. L., *Nature*, **209**, 630 (1966).
10. Cook, J. R., *J. Cell Biol.*, **29**, 369 (1966).
11. Corneo, G., Moore, C., Sanadi, D. R., Grossman, L. I., and Marmur, J., *Science*, **151**, 687 (1966).
12. Crawford, L. V., *J. Mol. Biol.*, **8**, 489 (1964).
13. Dawid, I. B., *Proc. Natl. Acad. Sci. U.S.*, **56**, 269 (1966).
14. Dulbecco, R., and Vogt, M., *Proc. Natl. Acad. Sci. U.S.*, **50**, 236 (1963).
15. Edelman, M., Epstein, H. T., and Schiff, J. A., *J. Mol. Biol.*, **17**, 463 (1966).
16. Ephrussi, B., "Nucleo-Cytoplasmic Relations in Microorganisms," Clarenden Press, Oxford, 1953.
17. Evans, T. E., *Biochem. Biophys. Res. Commun.*, **22**, 678 (1966).
18. Fiers, W., and Sinsheimer, R. L., *J. Mol. Biol.*, **5**, 408 (1962).
19. Fletcher, M. J., and Sanadi, D. R., *Biochim. Biophys. Acta*, **51**, 356 (1961).
20. Freifelder, D., Kleinschmidt, A. K., and Sinsheimer, R. L., *Science*, **146**, 254 (1964).
21. Graffi, A., Butschak, G., and Schneider, E. J., *Biochem. Biophys. Res. Commun.*, **5**, 418 (1965).
22. Hershey, A. D., Burgi, E., and Ingraham, L., *Proc. Natl. Acad. Sci. U.S.*, **49**, 748 (1963).
23. Huberman, J. A., and Riggs, A. D., *Proc. Natl. Acad. Sci. U.S.*, **55**, 599 (1966).
24. Humm, D. G., and Humm, J. H., *Proc. Natl. Acad. Sci. U.S.*, **55**, 114 (1966).
25. Josse, J., and Eigner, J., *Ann. Rev. Biochem.*, **35**, 789 (1966).
26. Kaiser, A. D., and Inman, R. B., *J. Mol. Biol.*, **13**, 78 (1965).
27. Kalf, G. F., *Biochemistry*, **3**, 1702 (1964).
28. Kalf, G. F., and Gréce, M. A., *Biochem. Biophys. Res. Commun.*, **6**, 674 (1964).
29. Kalf, G. F., and Gréce, M. A., *J. Biol. Chem.*, **241**, 1019 (1966).
30. Kleinschmidt, A. K., Kass, S. J., Williams, R. C., and Knight, C. A., *J. Mol. Biol.*, **13**, 749 (1965).
31. Kroon, A. M., *Biochim. Biophys. Acta*, **108**, 275 (1965).
32. Lipton, A., and Weissbach, A., *J. Mol. Biol.*, **21**, 517 (1966).
33. Luck, D. J. L., *J. Cell Biol.*, **16**, 483 (1963).
34. Luck, D. J. L., and Reich, E., *Proc. Natl. Acad. Sci. U.S.*, **52**, 931 (1964).
35. Marr, A. G., *Ann. Rev. Microbiol.*, **14**, 241 (1960).
36. Mil'man, L. S., and Kuzyaeva, V. A., *Tsitologiya*, **4**, 42 (1962).
37. Mitchell, H. K., and Mitchell, M. B., *Proc. Natl. Acad. Sci. U.S.*, **38**, 442 (1952).
38. Mounolov, J. C., Jakob, H., and Slonimski, P. P., *Biochim. Biophys. Res. Commun.*, **24**, 218 (1966).
39. Moustacchi, E., and Williamson, D. H., *Biochem. Biophys. Res. Commun.*, **23**, 56 (1966).
40. Nass, M. M. K., *Proc. Natl. Acad. Sci. U.S.*, **56**, 1215 (1966).

41. Nass, M. M. K., unpublished observations.
42. Nass, M. M. K., and Nass, S., *J. Cell Biol.*, **19**, 593 (1963).
43. Nass, M. M. K., Nass, S., and Afzelius, B. A., *Exptl. Cell Res.*, **37**, 516 (1965).
44. Nass, S., and Nass, M. M. K., *J. Cell Biol.*, **19**, 613 (1963).
45. Nass, S., and Nass, M. M. K., *J. Natl. Cancer Inst.*, **33**, 777 (1964).
46. Nass, S., Nass, M. M. K., and Hennix, U., *Biochim. Biophys. Acta*, **95**, 426 (1965).
47. Neubert, D., and Helge, H., *Biochem. Biophys. Res. Commun.*, **18**, 600 (1965).
48. Neubert, D., Helge, H., and Bass, R., *Arch. Exptl. Pathol. Pharmakol.*, **252**, 258 (1965).
49. Neubert, D., Helge, H., and Merker, H. J., *Biochem. Z.*, **343**, 44 (1965).
50. Neubert, D., Helge, H., and Teske, S., *Arch. Pharmakol. Exptl. Pathol.*, **252**, 452 (1966).
51. Novikoff, A. B., *in* "The Cell" (J. Brachet and A. E. Mirsky, eds.), Vol. 2, p. 299, Academic Press, New York, 1961.
52. Parsons, J. A., *J. Cell Biol.*, **25**, 641 (1965).
53. Pollard, C. J., Stemler, A., and Blaydes, D. F., *Plant Physiol.*, **41**, 1323 (1966).
54. Rabinowitz, M., Desalle, L., Sinclair, J., Stirewalt, R., and Swift, H., *Federation Proc.*, **25**, 581 (1966).
55. Rabinowitz, M., Sinclair, J., Desalle, L., Haselkorn, R., and Swift, H., *Proc. Natl. Acad. Sci. U.S.*, **53**, 1126 (1965).
56. Reich, E., and Luck, D. J. L., *Proc. Natl. Acad. Sci. U.S.*, **55**, 1600 (1966).
57. Roodyn, D. B., Reis, P. J., and Work, T. S., *Biochem. J.*, **80**, 9 (1961).
58. Rouiller, C. H., *Intern. Rev. Cytol.*, **9**, 227 (1960).
59. Ryter, A., and Kellenberger, E., *Z. Naturforsch.*, **13b**, 597 (1958).
60. Schatz, G., Haslbrunner, E., and Tuppy, H., *Biochem. Biophys. Res. Commun.*, **15**, 127 (1964).
61. Schildkraut, C. L., Marmur, J., and Doty, P., *J. Mol. Biol.*, **4**, 430 (1962).
62. Schmieder, M., and Neubert, D., *Arch. Pharmakol. Exptl. Pathol.*, **255**, 68 (1966).
63. Schneider, W. C., and Kuff, E. L., *Proc. Natl. Acad. Sci. U.S.*, **54**, 1650 (1965).
64. Schuster, F. L., *Exptl. Cell Res.*, **39**, 329 (1965).
65. Sinclair, J. H., and Stevens, B. J., *Proc. Natl. Acad. Sci. U.S.*, **56**, 508 (1966).
66. Stanier, R. Y., and Van Niel, C. B., *Arch. Mikrobiol.*, **42**, 17 (1962).
67. Steffensen, D. M., and Sheridan, W. F., *J. Cell Biol.*, **25**, 619 (1965).
68. Stone, G. E., and Miller, O. L., Jr., *J. Exptl. Zool.*, **159**, 33 (1965).
69. Suyama, Y., and Bonner, W. D., Jr., *Plant Physiol.*, **41**, 383 (1966).
70. Suyama, Y., and Preer, J. R., Jr., *Genetics*, **52**, 1051 (1965).
71. Swift, H., Kislev, N., and Bogorad, L., *J. Cell Biol.*, **23**, 91 A (1964).
72. Tewari, K. K., Votsch, W., and Mahler, H. R., *J. Mol. Biol.*, **20**, 453 (1966).
73. Thomas, C. A., Jr., and MacHattie, L. A., *Proc. Natl. Acad. Sci. U.S.*, **52**, 1297 (1964).
74. Trager, W., *in* "The Cell" (J. Brachet and A. E. Mirsky, eds.), Vol. 4, p. 151, Academic Press, New York, 1960.
75. Truman, D. E. S., *Biochem. J.*, **91**, 59 (1964).
76. Truman, D. E. S., and Korner, A., *Biochem. J.*, **84**, 40 P (1962).

77. Van Bruggen, E. F. J., Borst, P., Ruttenberg, G. J. C. M., Gruber, M., and Kroon, A. M., *Biochim. Biophys. Acta,* **119,** 437 (1966).
78. Vinograd, J., Lebowitz, J., Radloff, R., Watson, R., and Laipis, P., *Proc. Natl. Acad. Sci. U.S.,* **53,** 1104 (1965).
79. Vogel, F. S., and Kemper, L., *Lab. Invest.,* **14,** 1868 (1965).
80. Weil, R., *Proc. Natl. Acad. Sci. U.S.,* **49,** 480 (1963).
81. Wilson, E. B., "The Cell in Development and Heredity," 3rd ed., The Macmillan Co., New York, 1928.
82. Wilson, J. E., and Dove, J. L., *J. Elisha Mitchell Sci. Soc.,* **81,** Suppl. 1, 21 (1965).
83. Wintersberger, E., *Biochem. Z.,* **341,** 409 (1965).
84. Wintersberger, E., *Biochem. Biophys. Res. Commun.,* **25,** 1 (1966).
85. Wintersberger, E., and Tuppy, H., *Biochem. Z.,* **341,** 399 (1965).

Discussion of Part X

E. Racker, M. V. Simpson, M. Meselson, M. M. K. Nass, P. Siekevitz, D. O. Woodward, N. Neumann, S. Granick

Chairman Racker: I will open the discussion for comments and questions.

Dr. Simpson: The discovery some years ago of a protein biosynthesis system in mitochondria raised the question then of the extent of the autonomy possessed by these organelles, particularly since it was already known at that time that mitochondria of certain species were under the control of cytoplasmic as well as Mendelian genetic factors. The discovery of DNA in mitochondria and the recent work on their biogenesis added even more force to the question of just how much independence these organelles have. One approach toward an answer to this question is to consider the problem of the site of replication of mitochondrial DNA.

Does the mitochondrion make its own DNA or is the DNA made for it by some other apparatus or system, for example the nucleus, and then inserted into the mitochondrion later on? Dr. Peter Parsons and I have attacked this problem by working with isolated mitochondria to see whether we could develop a system which would incorporate deoxynucleotides into DNA, and have found in the case of rat liver mitochondria that such a reaction occurs. These mitochondria will incorporate TTP and dATP into mitochondrial DNA. In 2 hours, the equivalent of about 1% of the DNA present is "synthesized." There is a partial dependence, either singly or together, on the other three deoxynucleotides. Actinomycin D inhibits strongly at concentrations that inhibit DNA polymerase. The intact mitochondrial system is resistant to DNase.

I will take one more moment now to summarize the evidence that incorporation is being carried out by mitochondria and that the product is DNA. In the first place, we have been able to detect no intact nuclei by light microscopy nor anything that can be identified as nuclear fragments by electron microscopy. The mitochondrial fraction, prepared by differential ultracentrifugation and sucrose density gradient centrifugation, appears very clean. With respect to bacterial contamination, I think we have laid that spectre to rest both in the protein synthesis system and in this system by using aseptic techniques and germ-free animals. The bacterial contamination has been reduced from about 100,000 bacteria per ml to less than 200 per ml with no decrease either in amino acid incorporation or in deoxynucleotide incorporation.

With respect to whether the product is DNA, it bands as DNA in CsCl, is sensitive to DNase, resistant to mild alkaline hydrolysis, and resistant to ribonuclease. To answer the question of whether the DNA is truly mitochondrial, we have made use of the appreciable difference in the buoyant densities of nuclear and mitochondrial DNA species, after they have been subjected to renaturation procedures. The buoyant densities of native nuclear and mitochondrial DNA from rat liver are the same, but we have been able to distinguish between the two species by making use of some observations from Marmur's and Borst's laboratories. Both nuclear and mitochondrial DNA can be denatured and this process is accompanied by equal increases in buoyant density. When both species are subjected to renaturation procedures, nuclear DNA shows almost no change in density, but mitochondrial DNA renatures and returns to virtually its original density. In such experiments with our *in vitro*-labeled DNA, we have been able to show that the label follows the mitochondrial DNA peak, rather than the nuclear. We also know that the counts are in the interior of the molecule; we see very little terminal labeling at the 3′-OH end of the chains. Very preliminary results on nearest-neighbor frequency analysis suggest that the DNA is not spuriously labeled in the sense that the labeling does not result from long stretches of, say, labeled thymidylate attached to DNA. We have not gone far enough with this technique to be able to say that a faithful copy of mitochondrial DNA is being made. Moreover, whether we are dealing with a repair mechanism or the *de novo* synthesis of new strands, is also still a mystery and we intend to attack this problem in the near future.

DR. MESELSON: Dr. Nass, do the apparently circular DNA molecules possess the high sedimentation coefficient expected for supercoils?

DR. NASS: That we don't know yet. We can just see by looking in the electron microscope that many circular molecules are highly twisted structures and some are partly twisted and some open circles. We have not yet done sedimentation velocity runs.

DR. SIEKEVITZ: Dr. Woodward, one other possibility from the data is that it very much resembles a cell fractionation scheme in which you have a soluble enzyme in which the enzyme is also coming down in the nuclear, mitochondrial, and microsome fractions as a contaminant. And thus you have some of this protein there in the soluble fraction which is not really in the mitochondrial or microsome fraction.

DR. WOODWARD: Perhaps I should have explained something concerning the nature of the purification procedure because we eliminate the soluble protein before we start the purification procedure for mitochondrial, nuclear, or microsomal structural protein. This, of course, includes washing the particulate fractions thoroughly.

DR. SIEKEVITZ: But this comes from the soluble fraction—the supernatant. The supernatant would also contain about one half of the proteins of the cell, and one half of these proteins you characterize as structural protein, and you only have a small percentage of that protein in the nuclei and the microsomes. I would think that perhaps these latter are the result of a contamination from some of the proteins which are in the supernatant.

DR. NEUMANN: The question that came to my mind is sort of related to what Phil just said. What really is your assay for structural protein? How do you know this is structural protein?

DR. WOODWARD: Well, actually, I haven't taken the time to explain all of the criteria that have been used, but these include the solubility properties of the enzyme, polyacrylamide disc electrophoresis, molecular weight, amino acid composition, peptide maps, and the binding characteristics of the protein as well, which include ATP, NADH, and myoglobin binding in addition to MDH. In different preparations, several of these criteria will have been used.

CHAIRMAN RACKER: Do all these other fractions also bind malic dehydrogenase?

DR. WOODWARD: We haven't checked the other fractions for the binding of malic dehydrogenase.

DR. SIMPSON: Just a general comment. In the light of Dr. Nass' results, I just wanted to suggest that perhaps we should drop the term "nonchromosomal inheritance" and use something like "non-Mendelian" or "nonsegregational," but I think perhaps now that "nonchromosomal" could be very misleading. Mitochondrial DNA could very well be a legitimate chromosome.

CHAIRMAN RACKER: One final question.

DR. GRANICK: You have mutants which show a difference in amino acid composition, Would you then say that the mitochondria produce this particular protein and it gets everywhere, including the nucleus? Would that be correct?

DR. WOODWARD: There are two possibilities for explaining this. Either it is produced at the site of the mitochondrion, or messenger RNA is produced at the site of the mitochondrion or in some other place and transported to all parts of the cell. An interesting facet of this question is that there could have been selection for this particular gene being located in the cytoplasm. Because of the fact that it must produce a protein in large amounts compared with those genes that are responsible for catalytic proteins, controlled by nuclear genes. This simply suggests that since mitochondria outnumber nuclei, gene dosage could be responsible for the large quantities of structural protein synthesized.

CHAIRMAN RACKER: I think as the last speaker it is up to me to express our very sincere thanks to the organizers, Dr. Vogel, Dr. Lampen, and Dr. Bryson, for this very stimulating symposium which I think we all agree was fully satisfying in respect to both spiritual and physical nourishment.

Author Index

Numbers in parentheses are reference numbers and indicate that an author's work is referred to although his name is not cited in the text. Numbers in italics show the page on which the complete reference is listed.

C

H

Y

Z

Subject Index

D

E